NATO ASI Series

Advanced Science Institutes Series

A series presenting the results of activities sponsored by the NATO Science Committee, which aims at the dissemination of advanced scientific and technological knowledge, with a view to strengthening links between scientific communities.

The Series is published by an international board of publishers in conjunction with the NATO Scientific Affairs Division

A	Life Sciences	Plenum Publishing Corporation
B	Physics	London and New York
C	Mathematical and Physical Sciences	Kluwer Academic Publishers Dordrecht, Boston and London
D	Behavioural and Social Sciences	
E	Applied Sciences	
F	Computer and Systems Sciences	Springer-Verlag Berlin Heidelberg New York
G	Ecological Sciences	London Paris Tokyo Hong Kong
H	Cell Biology	Barcelona Budapest
I	Global Environmental Change	

NATO-PCO DATABASE

The electronic index to the NATO ASI Series provides full bibliographical references (with keywords and/or abstracts) to more than 30000 contributions from international scientists published in all sections of the NATO ASI Series. Access to the NATO-PCO DATABASE compiled by the NATO Publication Coordination Office is possible in two ways:

- via onlIne FILE 128 (NATO-PCO DATABASE) hosted by ESRIN, Via Galileo Galilei, I-00044 Frascati, Italy.

- via CD-ROM „NATO Science & Technology Disk" with user-friendly retrieval software in English, French and German (© WTV GmbH and DATAWARE Technologies Inc. 1992).

The CD-ROM can be ordered through any member of the Board of Publishers or through NATO-PCO, Overijse, Belgium.

Series I: Global Environmental Change, Vol. 13

The ASI Series Books Published as a Result of
Activities of the Special Programme on
Global Environmental Change

This book contains the proceedings of a NATO Advanced Research Workshop held within the activities of the NATO Special Programme on Global Environmental Change, which started in 1991 under the auspices of the NATO Science Committee.

The volumes published as a result of the activities of the Special Programme are:

Vol. 1: **Global Environmental Change.**
Edited by R. W. Corell and P. A. Anderson. 1991.
Vol. 2: **The Last Deglaciation: Absolute and Radiocarbon Chronologies.**
Edited by E. Bard and W. S. Broecker. 1992.
Vol. 3: **Start of a Glacial.** Edited by G. J. Kukla and E. Went. 1992.
Vol. 4: **Interactions of C, N, P and S Biogeochemical Cycles and Global Change.** Edited by R. Wollast, F. T. Mackenzie and L. Chou. 1993.
Vol. 5: **Energy and Water Cycles in the Climate System.**
Edited by E. Raschke and D. Jacob. 1993.
Vol. 6: **Prediction of Interannual Climate Variations.**
Edited by J. Shukla. 1993.
Vol. 7: **The Tropospheric Chemistry of Ozone in the Polar Regions.**
Edited by H. Niki and K. H. Becker. 1993.
Vol. 8: **The Role of the Stratosphere in Global Change.**
Edited by M.-L. Chanin. 1993.
Vol. 9: **High Spectral Resolution Infrared Remote Sensing for Earth's Weather and Climate Studies.**
Edited by A. Chedin, M.T. Chahine and N.A. Scott. 1993.
Vol. 10: **Towards a Model of Ocean Biogeochemical Processes.**
Edited by G. T. Evans and M.J.R. Fasham. 1993.
Vol. 11: **Modelling Oceanic Climate Interactions.**
Edited by J. Willebrand and D.L.T. Anderson. 1993.
Vol. 12: **Ice in the Climate System.** Edited by W. Richard Peltier. 1993.
Vol. 13: **Atmospheric Methane: Sources, Sinks, and Role in Global Change.** Edited by M. A. K. Khalil. 1993.
Vol. 14: **The Role of Regional Organizations in the Context of Climate Change.** Edited by M. H. Glantz. 1993.
Vol. 15: **The Global Carbon Cycle.**
Edited by M. Heimann. 1993.
Vol. 16: **Interacting Stresses on Plants in a Changing Climate.**
Edited by M. B. Jackson and C. R. Black. 1993.

Atmospheric Methane: Sources, Sinks, and Role in Global Change

Edited by

M. A. K. Khalil

Oregon Graduate Institute
P. O. Box 91000
Portland, Oregon 97291
USA

Springer-Verlag
Berlin Heidelberg New York London Paris Tokyo
Hong Kong Barcelona Budapest
Published in cooperation with NATO Scientific Affairs Division

Proceedings of the NATO Advanced Research Workshop on the Atmospheric
Methane Cycle: Sources, Sinks, Distributions, and Role in Global Change, held
at Mt. Hood near Portland, OR, USA, October 7-11, 1991

ISBN-13:978-3-642-84607-6   e-ISBN-13:978-3-642-84605-2
DOI: 10.1007/978-3-642-84605-2

Library of Congress Cataloging-in-Publication Data
Atmospheric methane : sources, sinks, and role in global change / edited by M.A.K. Khalil. p. cm. –
(NATO ASI series. Series I, Global environmental change ; vol. 13)
"Published in cooperation with NATO Scientific Affairs Division."
"Proceedings of the NATO Advanced Research Workshop on the Atmospheric Methane Cycle:
Sources, Sinks, Distributions, and Role in Global Change, held at Mt. Hood near Portland, OR, USA,
October 7-11, 1991" – T.p. verso.
ISBN-13:978-3-642-84607-6
1. Atmospheric methane–Congresses. 2. Sinks (Atmospheric chemistry)–Congresses. 3. Methane–
Environmental aspects–Congresses. I. Khalil, M. A. (Muhammad Ahsan Khan), 1950- . II. North
American Treaty Association. Scientific Affairs Division. III. NATO Advanced Research Workshop on
the Atmospheric Methane Cycle: Sources, Sinks, Distributions, and Role in Global Change (1991 :
Portland, Or.) IV. Series. QC879.85.A78 1993 551.5'112–dc20 93-31890

This work is subject to copyright. All rights are reserved, whether the whole or part of the material is
concerned, specifically the rights of translation, reprinting, reuse of illustrations, recitation, broadcasting, reproduction on microfilm or in any other way, and storage in data banks. Duplication of this
publication or parts thereof is permitted only under the provisions of the German Copyright Law of
September 9, 1965, in its current version, and permission for use must always be obtained from
Springer-Verlag. Violations are liable for prosecution under the German Copyright Law.

© Springer-Verlag Berlin Heidelberg 1993
Softcover reprint of the hardcover 1st edition 1993

Typesetting: Camera ready by authors
31/3145 - 5 4 3 2 1 0 - Printed on acid-free paper

Preface

Atmospheric methane is thought to be the most important trace gas involved in man-made climate change. It may be second only to carbon dioxide in causing global warming. Methane affects also the oxidizing capacity of the atmosphere by controlling tropospheric OH radicals and creating O_3, and it affects the ozone layer in the stratosphere by contributing water vapor and removing chlorine atoms. In the long term, methane is a natural product of life on earth, reaching high concentrations during warm and biologically productive epochs. Yet the scientific understanding of atmospheric methane has evolved mostly during the past decade after it was shown that concentrations were rapidly rising.

Because of the environmental importance of methane, North Atlantic Treaty Organization's Scientific and Environmental Affairs Division commissioned an Advanced Research Workshop. This book is the result of such a conference held during the week of 6 October 1991 at Timberline Lodge on Mount Hood near Portland, Oregon. About 100 scientists were invited, or accepted as participants, in recognition of their significant scientific achievements or potential for leading new developments.

Review lectures on the main topics were presented to the entire assembly by those who had been invited to write the chapters in this book. Specialist subgroups met during the next three days, coordinated by session chairs who documented the discussions in the working group reports included in this volume. The meeting was extremely active with discussions that continued late into the nights.

This book began in 1990 when the framework was conceived and writers were invited. Drafts of most chapters were available at the meeting. After the meeting, session chairs and other participants met to decide whether there were gaps in the existing material. Several new chapters were identified and commissioned. My own writings for the book are in the areas for which I could not find writers or where there were gaps. During most of 1992 we reviewed, edited, and refined each chapter. We abandoned the pursuit of perfection early on, but we sought completeness.

My approach was to define the areas that constitute the current understanding of methane with a chapter devoted to each subject. Above all, the book was not to be a collection of specialized research articles but of definitive and critical reviews that formed a coordinated whole. Research results, I believe, belong in peer-reviewed journals. Accordingly, it was announced before the meeting that papers dealing with original research would be considered for publication in the journal as a companion volume to the book. This allowed formal presentation and debate on current research results. Much of the research presented in the journal special issue was discussed in the specialty working groups, and some papers that were more complete are referenced by the session chairs in the working group reports. In the end, 55 technical papers were published in *Chemosphere, 26* #s 1-4 (1993) with 814 pages written by 125 authors from all over the world (see Appendix 2 for a Table of Contents of the journal special issue).

In putting together this book, I was fortunate to have the support of my family, of many friends and colleagues, and of generous sponsors. I especially want to thank Martha J. Shearer and Francis Moraes who, as members of the technical review team, contributed significantly to editing, reviewing, and correcting each chapter. I thank Edie Taylor who copy-edited the entire book and produced the final photo-ready version. I received much valuable advice, encouragement, and major contributions from the organizing committee, which consisted of Paul Crutzen, Robert Harriss, Rei Rasmussen, Dominique Raynaud, and Wolfgang Seiler. Major financial support from NATO's Scientific and Environmental Affairs Division was the foundation for this work and is gratefully acknowledged. The critical involvement of scientists from many non-NATO countries and the excess of U.S. scientists was supported by supplementary grants from the National Science Foundation (ATM-9120070) and the United States Environmental Protection Agency (Order No.1-W-1085-NASA). Both these grants were given to Andarz Company which provided additional support.

M.A.K. Khalil, Professor and Director, Global Change Research Center
Oregon Graduate Institute of Science and Technology, Portland, Oregon 97291-1000 U.S.A.

ATMOSPHERIC METHANE:

Sources, Sinks, and Role in Global Change

CONTENTS

Preface		v
Chapter 1.	Introduction	1

Record of Atmospheric Methane

Chapter 2.	Measurement and Research Techniques: Working Group Report R. Conrad and R.A. Rasmussen	7
Chapter 3.	The Record of Atmospheric Methane D. Raynaud and J. Chappellaz	38
Chapter 4.	Isotopic Abundances in the Atmosphere and Sources C.M. Stevens	62
Chapter 5.	Atmospheric Methane Concentrations: Working Group Report B. Stauffer, M. Wahlen, and F. Moraes	89

Formation and Consumption of Methane

Chapter 6.	Biological Formation and Consumption of Methane D.R. Boone	102
Chapter 7.	Formation and Consumption of Methane: Working Group Report N.T. Roulet and W.S. Reeburgh	128

Sources and Sinks

Chapter 8.	Stable Isotopes and Global Budgets	
	M.J. Whiticar	138
Chapter 9.	Methane Sinks and Distributions	
	M.A.K. Khalil, M.J. Shearer, and R.A. Rasmussen	168
Chapter 10.	Sources of Methane: An Overview	
	M.A.K. Khalil and M.J. Shearer	180

Methane Emissions from Individual Sources

Chapter 11.	Ruminants and Other Animals	
	D.E. Johnson, T.M. Hill, G.M. Ward, K.A. Johnson, M.E. Branine, B.R. Carmean, and D.W. Lodman	199
Chapter 12.	Rice Agriculture: Emissions	
	M.J. Shearer and M.A.K. Khalil	230
Chapter 13.	Rice Agriculture: Factors Controlling Emissions	
	H.-U. Neue and P.A. Roger	254
Chapter 14.	Biomass Burning	
	J.S. Levine, W.R. Cofer, III, and J.P. Pinto	299
Chapter 15.	Wetlands	
	E. Matthews	314
Chapter 16.	Waste Management	
	S.A. Thorneloe, M. A. Barlaz, R. Peer, L.C. Huff, L. Davis, and J. Mangino	362
Chapter 17.	Industrial Sources	
	L.L. Beck, S.D. Piccot, and D.A. Kirchgessner	399
Chapter 18.	Minor Sources of Methane	
	A.G. Judd, R.H. Charlier, A. Lacroix, G. Lambert, and C. Rouland	432

Chapter 19. Sources and Sinks of Methane: Working Group Report
                D. Bachelet and H.-U. Neue   457

The Environmental Role of Methane & Current Issues

Chapter 20. The Role of Methane in the Global Environment
                D.J. Wuebbles and J.S. Tamaresis   469

Chapter 21. The Current and Future Environmental Role of Atmospheric Methane: Model Studies and Uncertainties: Working Group Report
                J.P. Pinto, C.H. Brühl, and A.M. Thompson   514

Appendices

Appendix 1. List of Participants and Photographs of Working Groups   533

Appendix 2. Contents of the *Chemosphere* Special Issue of Contributed Papers   545

Index   553

Chapter 1

An Introduction to Atmospheric Methane

M.A.K. KHALIL

Global Change Research Center, Department of Environmental Science and Engineering Oregon Graduate Institute, P.O. Box 91,000, Portland, Oregon 97291-1000 USA

I have two purposes in writing this chapter. The first is to give a brief eyewitness report on the evolution of the current knowledge of methane trends, and the second is to provide a guide to the logic behind this book.

The increasing trend, now so well accepted, is the single most important reason for the current interest in methane. The trend itself is not complicated nor is it very profound, but it is a gauge of deeper changes in the methane cycle. Its existence is the foundation of much of the research during the past decade, particularly on the global budget, which is tied directly to explaining why methane is increasing.

The existence of methane in the Earth's atmosphere has been known since 1948 when Migeotte published the first measurements showing concentrations of about 2 ppmv. It was listed under "non-variable components of atmospheric air" by Glueckauf in 1951 and similarly in many textbooks since. Glueckauf also pointed out in his 1951 paper that, from the isotopic analysis by F.W. Libby, atmospheric methane appeared to be of mostly biogenic origin, and there was evidence that the concentration was likely to be around 1200 ppbv rather than the more commonly accepted 2 ppmv; nonetheless, he reported the higher concentration in his table, which is used to this day (CRC, 1991). From the 1950s to the early part of the 1980s, there was a prevalent belief that methane was a stable gas in the Earth's atmosphere. By 1975, well before systematic measurements could establish globally increasing trends, there were three important published results that already told us that methane must be increasing and that it had more than doubled over the last century, both effects due to human activities, and pointed to the important role of methane emissions in man-made global change, including tropospheric and stratospheric chemistry. The three results, largely ignored at the time, were the following.

1. During the 1960s and the 1970s, sporadic measurements of methane had been taken by many scientists, but there was no systematic measurements program. These data were not analyzed to look for trends, or at least no results of such investigations were published. We showed early on that the published data supported rapidly increasing trends during the 1960s and 1970s (Rasmussen and Khalil, 1981). Later, an expanded and refined version of these calculations was published resolving issues of calibration that had emerged in the interim (Khalil et al., 1989).

2. Ehhalt (1974) published a global budget of methane that was accepted for a long time. This budget attributed 100-220 Tg/yr to animals (mostly cattle) and 280 Tg/yr to rice agriculture out of total emissions of 590-1060 Tg/yr. Accordingly, about half of the global source of methane was controlled by human activities (although in the papers these emissions were reported as biogenic and not included under anthropogenic sources). With so much of the source being controlled by human activities and with rice fields and cattle populations still increasing at 2%-3%/yr, we were convinced that atmospheric methane must be increasing. Our calculations suggested atmospheric trends of 1%-2%/yr. This was the support we needed to publish our direct observations of methane increase as recorded at Cape Meares, Oregon, even though we had data only for one and a half years (Rasmussen and Khalil, 1981).

Remarkably, in a paper I found only recently, Fred Singer (1971) had come to the same conclusions a decade earlier. He estimated emissions of 45 Tg/yr from domestic ruminants and 140 Tg/yr from rice fields and increasing at 2%-3% per year. In his paper, Singer argued that the rising component of atmospheric methane from human activities, "cultural" sources as he called them, would have a significant effect on water vapor and ozone in the stratosphere.

3. In 1973, Robbins et al. showed that methane concentrations were only 560 ppbv in bubbles or air extracted from polar ice. The authors attributed it to chemical processes in ice that transformed methane ultimately to CO. When we tried to test the hypothesis, our results suggested a different interpretation of the Robbins data: that their observations actually represented atmospheric concentrations (Rasmussen et al., 1982). Soon afterwards we published our first results on methane in the ancient atmosphere and its implications (Khalil and Rasmussen, 1982). This work also contained our reservation about future increases and suggested that the trends would eventually slow down and stop

because of limitations to the growth of population. It seems that we had the right idea, but the trends started falling sooner and much more rapidly than expected (Khalil and Rasmussen, 1990 and elsewhere; Steele et al., 1992).

Early papers on increasing methane met with considerable skepticism and encountered opposition from prominent quarters. Eventually the atmospheric record became long enough that the increasing trend became undeniable. Controversies still persist as to the rates of increase and how much they have changed during the past 15 years, but these are perhaps of more esoteric interest.

During the intervening years, there has been much argument as to whether the increase of methane is caused by increasing emissions or decreasing sinks (particularly declining OH). For those who want to control methane concentrations, this is a most important issue. There has been an enormous bias towards the sink explanation, perhaps because early papers emphasized this mechanism for methane trends, with CO leading to a decrease of OH and a consequent increase of CH_4 (Sze, 1977; Chameides et al., 1977). Any evaluation of the trends in methane sources clearly shows that they have increased significantly during the past 100-300 years. When these changes are taken into account, it leaves a small role, if any, for declining sinks. Consequently, we have argued that the increase of methane between now and 100 years ago is dominated by increases in emissions (Rasmussen and Khalil, 1981; Khalil and Rasmussen, 1982, 1985, and elsewhere). But what about the latest decade? Is the measured trend dominated by increasing sources or declining sinks? This is a more difficult question, but current research suggests that, in recent years, OH is constant or even increasing slightly (Pinto and Khalil, 1991; Madronich and Granier, 1992; Prinn et al., 1992).

We now stand at yet another crossroad. Methane concentrations are still increasing but the rate is rapidly declining. What does this mean? Will methane stabilize without any effort on our part or any need for inter-governmental agreements? Or are we in a cusp when one class of sources is reaching stability because of limits of growth while another class of sources may be on the rise? For instance, sources related to food production such as rice and cattle may be reaching their limits, while sources related to energy and waste disposal may be rising. If so, methane trends may revert to faster increases in the future and rise to levels we cannot predict. So we come to this book at a time when we are beginning to understand past changes, but the past is getting uncoupled from the future.

This book is designed to cover all aspects of the global methane cycle as we now understand it. Its chapters form a coordinated whole rather than a dissociated collection of research articles. The logic of the organization is to answer four questions that follow each other. How do we measure methane in the atmosphere? What do the measurements tell us about atmospheric distributions and trends? How do we explain the observed concentrations in terms of the specific sources and sinks? And finally, what are the implications of these observations for the environment? We go from the most experimental to the most theoretical.

These questions relate fundamentally to the mass balance of methane, which, for any infinitesimal or finite or even large-sized region or box, can be expressed as the balance between processes that increase methane concentration in the box (direct emissions from sources inside and transport of methane from outside) and processes that reduce concentrations in the box (transport of methane outward and removal by chemical reactions or deposition). Trends arise when these competing processes are out of balance. The elements of the mass balance are sources, sinks, transport, and concentrations. Although all these elements are discussed in the book, we have put special emphasis on estimates of methane emissions from its various sources and the factors that control these emissions; this subject is more than half of the book. Although this is a complex and difficult aspect of the global methane cycle, it is most amenable to direct experimental observations and is valued because it provides the opportunity to control global warming.

On the global scale, the present understanding of sources, sinks, atmospheric burden and the trends are quantitatively balanced with only small remaining uncertainties. The fraction of anthropogenic emissions is constrained by the existing atmospheric and ice core data and likely to be quite accurate. The contribution of individual sources to the present atmospheric concentrations is substantially more uncertain and is an active area of research. On yet smaller scales of time and space, such as emissions from countries, small regions or over seasons, there are practically no constraints leading to mass balances that are not unique and completely unreliable in many cases. At these small scales current technology for the assessment of methane creates irreducible uncertainties that will forever prevent accurate estimates of emissions.

This book is supported by 55 original research articles divided along the same lines as the contents of the book and published in *Chemosphere* (Contents in Appendix 2). Together, the journal and the book provide a view of where we stand on the methane cycle and point to where we must go.

References

Chameides, W.L., S.C. Liu, R.J. Cicerone. 1977. Possible variations in atmospheric methane. *J. Geophys. Res., 82*:1,795-1,798.
CRC. 1991. *Handbook of Chemistry and Physics*, 72nd Edition (CRC Press, Boca Raton, Florida).
Ehhalt, D.H. 1974. The atmospheric cycle of methane. *Tellus, 26*:58-70.
Glueckauf, E. 1951. The composition of atmospheric air. In: *Compendium of Meteorology* (Thomas F. Malone, ed.), American Meteorological Society, Boston, U.S.A., pp 3-9.
Khalil, M.A.K., R.A. Rasmussen. 1982. Secular trend of atmospheric methane. *Chemosphere, 11*:877-883.
Khalil, M.A.K., R.A. Rasmussen. 1985. Causes of increasing methane: Depletion of hydroxyl radicals and the rise of emissions. *Atmos. Environ., 19*:397-407.
Khalil, M.A.K., R.A. Rasmussen. 1990. Atmospheric methane: Recent global trends. *Environ. Sci. Technol., 24*:549-553.
Khalil, M.A.K., R.A. Rasmussen, M.J. Shearer. 1989. Trends of atmospheric methane during the 1960s and 1970s. *J. Geophys. Res., 94*:18,279-18,288.
Madronich, S., C. Granier. 1992. Impact of recent total ozone changes on tropospheric ozone photodissociation, hydroxyl radicals and methane trends. *Geophys. Res. Lett., 19*:465-467.
Migeotte, M.V. 1948a. Spectroscopic evidence of methane in the Earth's atmosphere. *Phys. Rev., 73*:519-520.
Migeotte, M.V. 1948b. Methane in the Earth's atmosphere. *J. Astrophys., 107*:400-403.
Pinto, J.P., M.A.K. Khalil. 1991. The stability of tropospheric OH during ice ages, inter-glacial epochs and modern times. *Tellus, 43B*:347-352.
Prinn, R.G., et al. 1992. Global average concentration and trend of hydroxyl radicals deduced from ALE/GAGE trichloroethane (methyl chloroform) data for 1978-1990. *J. Geophys. Res., 97*:2,445-2,461.
Rasmussen, R.A., M.A.K. Khalil. 1981. Atmospheric methane: Trends and seasonal cycles. *J. Geophys. Res., 86*:9,826-9,832.
Rasmussen, R.A., M.A.K. Khalil, S.D. Hoyt. 1982. Methane and carbon monoxide in snow. *JAPCA, 32*:176-178.

Robbins, R.C., L.A. Cavanagh, L.J. Salas. 1973. Analysis of ancient atmospheres. *J. Geophys. Res., 78*:5,341-5,344.

Singer, S.F. 1971. Stratospheric water vapor increase due to human activities. *Nature, 233*:543-545.

Steele, L.P., et al. 1992. Slowing down of the global accumulation of atmospheric methane during the 1980s. *Nature, 358*:313-316.

Sze, N.D. 1977. Anthropogenic CO emissions: Implications for the atmospheric CO-OH-CH_4 cycle. *Nature, 195*:673-675.

Chapter 2

Working Group Report

Measurement and Research Techniques

CHAIRPERSONS: R. CONRAD[1] AND R.A. RASMUSSEN[2]

[1] *Max-Planck Institut für Terrestrische Mikrobiologie Karl-von-Frische-Strasse, D-3550 Marburg/Lahn, Germany*

[2] *Global Change Research Center, Oregon Graduate Institute P.O. Box 91,000, Portland, Oregon, 97219 U.S.A.*

Members of the group:
M. Beran, P. Bergamaschi, R. Conrad, B. Deck, M. Keller, I. Levin, G. Mroz, J. Murrell, S. Piccot, R.A. Rasmussen, J. Rau, J. Ritter, M. Torn, R. Wassmann, M.J. Whiticar

1. Introduction

The geographical distribution and increasing trend of trace gas concentrations in the global atmosphere are determined by a complex interplay of the geographical distribution of natural and human sources, biogeochemical and ecosystem dynamics, human agricultural and industrial practices, and atmospheric chemistry and dynamics. For an understanding of the global methane cycle, it is necessary to develop and apply methods that are suitable for measuring methane and its isotopes and that can be used for research on the many facets of the budget of atmospheric methane.

This report summarizes the available observational strategies and measurement options for the study of atmospheric methane and to identify the areas with promising methodological developments. For this purpose we focus on the following questions.

2. How can we study the present and future trends in the global methane budget?

The basic prerequisite for this study is the precise and accurate sampling and analysis of atmospheric methane. The procedures and techniques for this task are basically available by flask or continuous sampling and by the use of gas chromatography with flame ionization detection. The application of these

NATO ASI Series, Vol. I 13
Atmospheric Methane: Sources, Sinks, and Role in Global Change
Edited by M. A. K. Khalil

techniques during the last few years has provided a rather detailed insight into the temporal trends and horizontal distributions of methane within the planetary boundary layer (Khalil et al., this volume; Raynaud, this volume).

2.1 Vertical sampling. Whereas the presently available horizontal sampling density is relatively satisfactory (except for continental sites), the vertical sampling density still needs improvement. Three-dimensional simulations of CFC-11 and CH_3CCl_3 indicate that the upper tropical troposphere is an active region of inter-hemispheric transport (Prather et al., 1987; Spivakovsky et al., 1990). Three-dimensional general circulation models suggest that the vertical gradient of methane is a product of competing source and sink terms coupled with atmospheric mixing processes (Fung et al., 1991).

At present, regular observations of vertical gradients of methane and other trace gases are limited to southern Australia (Fraser et al., 1986). The availability of measured vertical and meridional profiles of methane and other important trace gases are limited because of the expense of operating manned aircraft for atmospheric research. Recent technical advancements in the use of composite structures, computational aerodynamics, and miniaturization of electronics have raised the prospect of utilizing unmanned aircraft for atmospheric research at a significantly lower cost than presently available manned aircraft. NASA, Harvard University, and Aurora Flight Services are collaborating on building Perseus A, an unmanned vehicle capable of carrying 50 kg of instruments to an altitude of 25 km. The U.S. Department of Energy is assessing a collaboration with Aurora Flight Services to build Perseus B, which would carry 150 kg payload at 18 km altitude for 24 hours or at 11 km altitude for nearly 100 hours. Present cost estimates indicate that these unmanned aircraft can be operated at flight-hour costs of hundreds of dollars compared to a thousands of dollars for manned scientific research aircraft. This new unmanned aircraft technology may present the climate change community with an affordable platform and new opportunities for making measurements of trace gases throughout the troposphere and even into the lower stratosphere.

2.2 Isotope measurements. The knowledge of the distributions and trends of the isotopic composition of methane is not satisfactory. Research using $\delta^{13}C$, $\delta^{14}C$, and, to a lesser extent, δ^2H content of methane is useful in constraining the allocation of source strengths in the global methane budget (Stevens, this volume; Whiticar, this volume). Present analytical techniques for determining $\delta^{13}C$, $\delta^{14}C$, and δ^2H require the conversion of the methane to CO_2 and H_2. The isotopic ratios of these species are then measured by mass spectrometry for the stable isotopes or by accelerator mass spectrometry (AMS) or counting the decay of ^{14}C. For AMS analysis CO_2 has to be further converted to graphite (Wahlen et al., 1989; Lowe et al., 1991). This is a labor-intensive task that limits the amount of available data.

A recent development shows promise for decreasing the time required for analysis and therefore for improving sample throughput. Schupp et al. (1993) have demonstrated the feasibility of measuring $\delta^{13}C$ in methane directly by tunable diode laser absorption spectroscopy. This approach is attractive because the measurement is performed on the CH_4 molecule without need of prior combustion to CO_2. The total time for the analysis of one sample is just 10-15 minutes including calibration. The presently achieved accuracy versus the PDB scale is 0.8 %., which is sufficient for measurement of a wide variety of CH_4 sources as their typical $\delta^{13}C$ variabilities are much larger. The present lower concentration limit is 2.5 %v methane. However, the application of a 200 m pathlength white cell may decrease the detection limit down to about 50 ppmv methane.

2.3 Isotope calibration. In order to allow routine measurements of the isotopic composition of methane, it will also be necessary to improve calibration. For this purpose it will be necessary to establish an intercalibration network and to provide for commercially available isotope standards of the methane molecule. While most investigators making methane isotope determinations use a procedure similar to that described by Stevens and Rust (1982) or Wahlen et al. (1989), different laboratories will have slightly different fractionation effects in the separation/combustion process for converting CH_4 to CO_2 and H_2O. The mass spectrometric determination of ^{13}C and ^2H in the combusted methane can be quantified against recognized standards on the PDB and SMOW scale using standards commercially available, but recognized methane isotope standards to calibrate the sampling and preparatory procedures are lacking.

A result of this is the variability of atmospheric concentration values for ^{13}C in methane given by various investigators, which exceed the variability expected due to differing sampling times and locations (Tyler, 1986; Lowe et al., 1988; Wahlen et al., 1989; Quay et al., 1991; Stevens, this volume). To date, only preliminary intercalibrations of methane isotopic composition among some of the above investigators have been made. In establishing criteria for a suitable standard, it is necessary to remember the large variability in $\delta^{13}C$ and $\delta^{2}H$ between microbiological sources (Whiticar et al., 1986) and thermogenic sources (Schoell, 1980). Measurements from natural wetlands or rice, where the $\delta^{13}C$ and $\delta^{2}H$ are depleted, would not ideally use standards designed for atmospheric or soil depletion studies, which are heavier. In addition, studies involving mechanisms of methane production, or flux measurements, where the sampled methane concentrations are high would find pure gas standards of use, whereas investigators of atmospheric composition would like to have standards diluted closer to ambient methane concentrations. Since there is no ideal standard for all applications, it might be better to suggest that a set of methane isotope standards be manufactured. As a first step towards this direction, the International Atomic Energy Agency (IAEA, Vienna) has initiated a Coordinated Research Program (CRP) on "Isotope Variations of Carbon Dioxide and Other Trace Gases in the Atmosphere" (Consultants Group Meeting, IAEA, Vienna, 3-6 December 1991). One of the aims of this CRP is to support close collaboration among research groups working on isotope measurements of greenhouse gases. Within this CRP framework, the IAEA is also developing the necessary standards for intercalibrations. An earlier report of the Consultants Group Meeting on Stable Isotope Reference Samples for Geochemical and Hydrological Investigations, IAEA, Vienna, 16-18 September 1985 recommended the adoption of natural gas standards NGS 1, 2, and 3 as initially prepared by M. Schoell. Natural gas standards would span the large range of $\delta^{13}C$ and $\delta^{2}H$ values encountered in most systems and would have additional value of being "dead" in ^{14}C for those investigators needing a methane radiocarbon blank. The formulation of the dilute standard similar to atmospheric composition in isotopes and concentration is more problematic but deserves consideration.

3. How can we study the contribution of a region to the global methane budget?

The contribution of a particular region to the global methane budget is most interesting for decision makers. Also of interest is what kinds of sources and sinks are dominant in the region. Regions are areas that vary in scale, e.g., a large country or an urban area. Regions generally comprise many different ecosystems with different strengths and different rhythms for methane sources or sinks and comprise diverse anthropogenic sources such as coal mines and landfills. Therefore, the determination of the source strengths of an entire region is not an easy task.

In Europe, a multidisciplinary project was started to study the budgets of pollutants such as NO_x and hydrocarbons. This "European Experiment on the Transport and Transformation of Environmentally Relevant Trace Constituents in the Troposphere over Europe" (EUROTRAC) comprises all the different aspects of fluxes, transport, and air chemistry that are involved in constraining the regional budgets of trace gases, in the case of EUROTRAC of short-lived trace gases (Borrell et al., 1991).

The available strategies to estimate a regional methane budget are not entirely satisfactory. Regional budgets may be based on the following elements: (1) intensive flux studies at as many different sites as possible to cover the whole region; (2) a grid of measurement platforms including towers, aircraft, and balloons to monitor methane concentrations in the atmosphere; (3) geographical information systems; (4) modelling.

The usefulness and availability of geographical information systems has been reviewed in a recent Dahlem conference by Stewart et al. (1989). Intensity of flux studies is usually limited for practical reasons. The same is true for the establishment of a dense grid of measurement platforms, which is very expensive. These necessary limitations emphasize the importance of developing a useful sampling strategy to obtain data that are suitable for combination with geographical elements using models (see also questions (3) and (4)).

3.1 Upwind-downwind measurements. One strategy is the measurement of methane concentrations upwind and downwind of a particular region. Usually fluxes of methane are measured over small plots of rice fields or other such sources. The extrapolation of these measurements to the regional scale is not straightforward nor is it easy to establish the reliability of the extrapolation. An atmospheric sampling strategy coupled to the flux measurements may provide constraints that can be used to estimate the area-wide fluxes more accurately.

If the horizontal transport flux of methane into a region is measured near the ground upwind of major sources such as rice fields and simultaneously measured downwind, then an estimate of the (vertical) flux in between can be obtained using a mass balance model. For more reliable information, the upwind-downwind fluxes may be measured on a vertical surface (e.g., Khalil and Rasmussen, 1988). The model may be a simple box such as has been used for urban airsheds, or it may be a detailed regional scale transport model, depending on the measurements available. The regional flux obtained in this manner should agree with the extrapolation of the small area chamber studies. If the agreement is not good, the causes can be investigated. The method may also be used without relying on small scale direct flux measurements from the rice fields.

3.2 Isotopic fingerprinting. Another strategy for the assessment of regional budgets and, most important, of the attribution of the various sources within the region is to use isotopic fingerprints, i.e., by measurement of $\delta^{13}C$ and $\delta^{2}H$ of methane (Levin et al., 1993; Whiticar, 1993). In a recent study (Thom et al., 1993), continuous concentration and stable isotope observations at a non-background site in Germany have been used to determine the isotopic signature of the mean methane source in the catchment area of the sampling site. The atmospheric boundary layer was used as an integrator for emissions from soil-borne methane sources. With a two-component mixing approach (of background air and source methane), the isotopic signature of the mean methane source is derived from the correlation of the inverse concentration with the isotope ratios ($\delta^{13}C$, $\delta^{2}H$). Then, taking advantage of the significant differences in the isotope ratios between recent biogenic and fossil fuel associated sources, respectively (Whiticar et al., 1990; Levin et al., 1993), allows for an estimate of the relative importance of these two source groups in the catchment area of the sampling site. However, the isotopic fingerprinting of methane may be refined further. For example, $\delta^{13}C$ for methane from wetlands, rice paddies, ruminants, and termites is similar. Likewise, $\delta^{13}C$ for methane from natural gas and coal mines is also similar. Adding $\delta^{2}H$ information for these sources helps but still does not allow for a separation of all relevant methane sources (Whiticar et al., 1990; 1993). Present analytical techniques for determining $\delta^{13}C$, $\delta^{14}C$, and $\delta^{2}H$ require the conversion of methane to CO_2 and H_2. Breaking the C-H bond of the methane molecule destroys the molecular deuteration pattern and, consequently, any signature imposed by kinetic isotope effects during the diverse biogenic and abiogenic processes that produce atmospheric methane. There are ten isotopomers (combinations of ^{13}C, ^{12}C, ^{1}H, and ^{2}H) present in methane. If the abundance ratios of the ten isotopomers could be determined, then the "under determined" nature of the isotopic approach to source attribution might be reduced. Achieving this goal will require high mass spectral resolution of about 15,000 with measurement precision of <1%. (Mroz, 1993). This approach should be technologically feasible but has yet to be demonstrated.

^{14}C has been used as classical isotope to distinguish between the global recent (biogenic) and fossil methane source components (e.g., Ehhalt, 1974; Wahlen et al., 1989, Quay et al., 1991). Moreover, the $^{14}CO_2$ spike from the atmospheric nuclear weapon tests in the late fifties and early sixties, introduced into the biosphere and therewith into the methane cycle, could have provided an independent measure of the mean residence time of methane in the atmosphere (Wahlen et al., 1989). However, recent $^{14}CH_4$ emissions from the nuclear industry today are significantly disturbing the "natural" $^{14}CH_4$ level. The use of this isotope for global and particularly for local methane budget evaluations is, therefore, of limited value today (Levin et al., 1992).

The abundance of tritium in methane has been measured sporadically (Kaye, 1987). There are not enough data to detect a trend. If such data were available, they would be very helpful in elucidating trends in the non-fossil component of atmospheric methane. Tritiated methane would have an atmospheric lifetime of about 5.5 years due to the additive effects of 12.3-year half-life of tritium and the approximately 10-year life-time of the methane from reaction with OH. As tritiated methane would come predominantly from modern biogenic sources, its meridional distribution and inter-annual trend would track non-fossil sources of methane to the trend of atmospheric methane. One potential drawback is that interpretation could be confused by tritiated methane from the nuclear industry.

3.3 Large point sources. The estimation of the methane source strengths of large point sources such as coal mines or landfills also poses methodological problems, which, however, may soon be overcome by the further development of Fourier Transform Infrared (FTIR) long-path spectroscopy. Open-path infrared spectroscopy for FTIR techniques have been used successfully to measure gaseous pollutants in the atmosphere (Hanst et al., 1973; Hanst, 1982; Herget, 1982). In general, the open-path FTIR13 sensor transmits a probing collimated infrared beam through the atmosphere, which is reflected back to the detector via a remotely placed retroreflector. The reflected beam is then subjected to absorption analysis, which identifies and quantifies the gases that are present along the path of the beam. Based on the absorption analysis, a path-average concentration can be determined for a variety of different gases.

The U.S. EPA's Office of Research and Development (ORD) has been conducting research to improve the estimates of methane emissions from coal mining operations. After several sampling and analysis techniques were investigated, open-path remote sensing techniques were identified as a viable option for measuring emission fluxes from surface mines (Kirchgessner et al., 1993). Their potential applicability to other sources of methane emissions (e.g., landfills, rice paddies, and biomass combustion) was also a consideration in selecting open-path remote sensing techniques.

In the measurements technique developed by EPA, the total methane liberated from a large surface coal mine was estimated by measuring the average ground-level concentration of methane in the plume emanating from the surface coal mine. The ground-level concentration of the overall plume was determined be taking path-average measurements along a line perpendicular to the plume's direction of travel. Throughout the measurements, a known release rate of sulfur hexafluoride (SF_6) was maintained in the mine pit so the concentration of SF_6 and methane could be measured simultaneously with the FTIR. Using the known release rate and measured concentration of SF_6, the vertical dispersion characteristics of the mine plume were determined using a dispersion model developed for assessing the behavioral characteristics of plumes from large area sources. Emissions flux and total mine emissions were calculated from the modeling assessments and the measured concentration of methane from the mine.

The advantages of using open-path FTIR techniques to conduct field measurements instead of the more traditional point sampling techniques are: (1) open-path FTIR gives path-integrated gas concentrations over long distances (up to several hundred meters), which average the heterogeneity inherent in the plumes from large and complex area sources and reduce the number of measurements needed; (2) they provide a rapid and cost-effective means for characterizing emission rates for a wide range of gases from very large and complex emission sources; and (3) they provide real-time measurement results in the field. Evaluations of measurements data obtained from FTIR systems have demonstrated that some instruments are capable of performing accurate ambient measurements and can be used to estimate the emissions flux from large and complex sources (Kirchgessner et al., 1993).

4. How can we bridge the gap in scales from chamber techniques to global inventories?

Methane emissions are often highly variable in both time and space. Observational strategies that rely on chambers may require massive efforts in order to precisely determine area fluxes. Strategies exist to measure fluxes by integrating techniques that are operative on an increasing scale (Harriss, 1989; Stewart et al., 1989; Khalil et al., 1993; Khalil et al., this volume). These scales and the corresponding techniques are the following:

1 m	chamber techniques
100 m	towers applying micrometeorological techniques
1 km	long path spectrometry/flux gradient technique
100 km	aircraft eddy flux
>100 km	aircraft eddy flux combined with satellite imagery

These techniques can potentially bridge the gaps arising from measurements on different scales. The measured fluxes may be further constrained by isotope fingerprinting of the methane (compare Question #2).

Techniques exist for measuring methane fluxes without enclosures. The flux estimation techniques called "micrometeorological" rely on measurements of atmospheric concentrations of methane and other meteorological parameters to estimate flux using simple models. The approaches discussed below are normally used at towers constructed at ground sampling sites although they may be extended to aircraft measurements (see below).

4.1 Boundary layer accumulation technique. The simplest approach to estimating area flux uses the surface boundary layer of the atmosphere instead of a chamber to accumulate emissions. This approach normally requires special micrometeorological conditions, most commonly a strong nocturnal inversion. When an inversion occurs, emitted methane accumulates in the boundary layer; the emission (J) may be estimated by the following formula:

$$\int_0^{h_i} (C_h - C_o) dz = \int_0^t (J - E) dt \qquad (1)$$

Measurements of boundary layer concentration (C_h) from ground level to the inversion layer (h_i) are compared to the background concentration (C_0) at an earlier time. The methane accumulation depends upon the balance between the net emission of methane within the boundary layer (J) and the exchange of methane across the boundary layer (E) over the accumulation time (t). Investigators have used this technique in areas of high methane emissions to estimate fluxes (Crill et al., 1988; Delmas et al., 1992). Normally, exchange across the boundary layer is neglected. Trumbore et al. (1990) caution against neglecting the exchange term. They found that even during strong inversions at night in the forest of the Central Amazon, exchange could cross the boundary on a 5-hour time scale. Trumbore et al. (1990) recommend measuring the concentration of a tracer within the boundary layer whose emission is well characterized (e.g., ^{222}Rn).

The $^{222}Radon$ tracer technique has indeed already been used in combination with carbon dioxide measurements in the atmospheric boundary layer (Dörr et al., 1983; Levin, 1987; Levin et al., 1989). On a regional scale Thom et al. (1993) used continuous methane concentration observations in combination with parallel atmospheric ^{222}Rn measurements to determine mean methane flux densities. The radioactive noble gas ^{222}Rn (daughter of $^{226}Radium$, a trace element present in all soils) emanates from continental surfaces rather homogeneously and is constant with time (Dörr and Münnich, 1990). The ^{222}Rn flux from ocean surfaces, if compared to continental surfaces, is nearly zero. The only sink for atmospheric ^{222}Rn is the radioactive α-decay ($\tau \approx 130$ h). With these source or sink characteristics, ^{222}Rn is an ideal tracer to evaluate the vertical mixing conditions in the atmospheric boundary layer over continental areas, respectively the residence time of an air mass over the continent: With a constant flux from the soils, the main reason for a variation of the ^{222}Rn activity in the atmospheric boundary layer is either the variability of the vertical mixing intensity or a change in the origin of the air mass (continental or maritime) in question.

During strong night time inversions, pronounced concentration increases of methane and ^{222}Rn are observed, e.g., in methane source regions (Thom et al., 1993). In order to calculate the methane flux from the observed records, a constant mixing height h_i of the inversion layer was assumed. If the sources of ^{222}Rn and CH_4 are distributed similarly homogeneously at the surfaces, Eqn. (1)

can be used for both trace gases separately. Division of both equations then eliminates the unknown terms, namely the height of the inversion layer h_i and the transport term E. The mean CH_4 flux (J_{CH4}) can then be calculated from the well known ^{222}Rn flux ($J^{222}Rn$) and the simultaneously observed concentration increases ΔC_{CH4} and $\Delta C^{222}Rn$ of both gases in the inversion layer:

$$J_{CH_4} = J^{222}Rn \quad (\Delta C_{CH_4}/\Delta C^{222}Rn) \qquad (2)$$

Neglecting exchange usually leads to an underestimate of emissions. The boundary layer approach may be extended to other meteorological conditions aside from nocturnal inversion. Wofsy et al. (1988) used aircraft measurements of carbon dioxide concentration above and inside the growing morning boundary layer over the Amazon forest to estimate the balance of forest respiration and photosynthesis.

4.2 Flux gradient technique. Other relations exist that allow for the estimation of methane fluxes. The flux-gradient technique is a classic approach to flux estimation, which has been used for estimation of methane emission from rice paddies in Australia (Denmead, 1991). This approach relies on the equivalence of the eddy diffusivities for heat, water vapor, momentum, and trace gases. Eddy diffusivities are usually determined from a profile of wind speeds (the aerodynamic method) or from an energy balance of sensible heat, latent heat, and soil heat fluxes (the Bowen Ratio method). The flux of methane is the product of the determined diffusivity and the vertical gradient of methane concentration. The reader should consult Businger (1986) or Fowler and Duyzer (1989) for a more complete description of these approaches. Practical considerations normally limit these approaches to flat areas of low stature vegetation. Measurements of methane concentration for this method require high precision but may be made at low frequency, so common relatively inexpensive instrumentation such as FID gas chromatography or standard infrared techniques may be used. Businger and Delaney (1990) discuss in detail the requirements for chemical instrumentation of this method and of the eddy correlation, and conditional sampling methods discussed below.

4.3 Eddy correlation techniques. The eddy correlation technique has wide applicability for obtaining a direct measurement of biosphere-atmosphere exchanges. This approach may succeed in forest systems where flux gradient techniques are problematic. The details of this method are reviewed by Businger (1986) and Fowler and Duyzer (1989). Briefly, the flux (F) of methane (or another trace gas) is determined by the following equation:

$$F = \overline{wc} + \overline{w'c'} \tag{3}$$

The quantities (w) and (c) are respectively the mean vertical wind speed and trace gas density. The term (w c) encompasses the fluctuations about the means of the same quantities. Practically, this approach requires precise high frequency (10 s^{-1}) measurements of the vertical mass flux and methane concentration. This technique has been used to measure methane flux during the NASA ABLE-3A and -3B campaigns that occurred over the tundra areas of the Yukon-Kuskokwim delta of Alaska and over the Hudson Bay Lowlands and southern boreal forest regions of Canada, respectively. During these campaigns, methane concentrations were measured using a tunable diode laser system (Ritter et al., 1991; 1992).

It may still be possible to relax the requirements of the eddy correlation method and determine trace gas fluxes reliably. Recent studies (Lenschow, personal communication) have indicated a potential advancement in the area of airborne flux measurements via intermittent sampling. This approach differs from the eddy correlation method in that the component of interest is sampled at a slower rate (e.g., 1 s^{-1}). A constraint of this method, which is held in common with the eddy correlation method, is that the data must still be sampled almost instantaneously and simultaneously (i.e., in phase) with other data. If the sampling frequency is chosen correctly, the data stream will still contain statistically independent samples of the fluctuating signal. The resulting data will then contain the information necessary to obtain a flux value, but the data will not be oversampled as in the eddy correlation method. The advantage of this method over the eddy correlation method is that instruments that have a response time of greater than about 1 s can be used in an aircraft to determine the vertical flux.

Desjardin (1972) first suggested a technique called eddy accumulation. In theory, it should be possible to pump air into containers in proportion to the strength of up and down eddies. Separate up and down canisters could be collected and analyzed after suitable periods (10-30 minutes). Unfortunately, this technique has posed grave difficulties, and proportional pumping has never been used successfully. Businger and Oncley (1990) have suggested that the eddy correlation approach may be relaxed in another way. Rather than proportional pumping, the approach known as conditional sampling uses constant pumping rates and a simple valve arrangement to switch between up and down collection vessels. A third collection vessel may be added for collection of air when eddies (up or down) are below a certain velocity threshold. This approach allows for flux measurements of a wide variety of gases where high frequency precise measurements are either prohibitively expensive, logistically difficult, or impossible with current technology. The conditional sampling approach has recently undergone field testing (MacPherson and Desjardin, 1991). In time we should know if this approach is applicable to a wide variety of systems over a range of micrometeorological conditions (see also Desjardin, 1991).

The area coverage of the micrometeorological approaches discussed above (frequently called a "footprint") varies according to atmospheric stability conditions but generally falls in the range of 10 to 100 ha. As a result, in heterogeneous landscapes, a measurement from a given tower platform may result in very different fluxes, depending upon size and direction of the footprint. Micrometeorological measurements are simplest to interpret in relatively uncomplicated terrain and over homogeneous systems.

Airborne eddy correlation flux measurement is an essential tool for understanding regional source and sink distributions of trace gases and the transport of these gases between the planetary boundary layer and the free troposphere. Transport occurs over a continuum of eddy sizes from meters to kilometers. The range of eddy sizes constrains sampling strategies in two ways. First, it necessitates a sampling rate that is fast enough to capture the contribution of small scale fluctuations to the total flux. Second, the sampling path must be long enough to measure the long wavelength contribution with minimal uncertainty.

The measurement of the vertical turbulent methane flux from aircraft platforms can be obtained using the eddy correlation technique (Eqn. 3). The simplicity of the equation belies the complexity of the calculation. For example, Webb et al. (1980) have shown that unless the instrumentation is designed to measure the trace gas mixing ratio relative to dry air, corrections to the measured flux based on concurrent heat and moisture fluxes have to be made. The magnitude of the required corrections can be as large as or larger than the measured flux. In addition to the "Webb" correction mentioned above, Sachse et al. (1987) found that an additional correction is necessary due to an absorption line broadening effect.

Lenschow et al. (1980) point out that limited sensor response time will induce a phase delay and reduce the amplitude of the signal from its true values. This signal degradation can be accounted for if the time response of the system is known. Ritter et al. (personal communication) used the time response data for a second order system (e.g., a typical airborne total temperature probe) to account for a portion of the signal that may have been lost due to the finite time response. An analytical estimation of a sensor's response may not be adequate if the assumption of "piston flow" within the ducting is assumed (Sachse, personal communication). Results from laboratory measurements indicate measured response times being almost a factor of two larger than that calculated on a piston flow assumption. Flow distortion effects have also been noted as an important factor to consider in the process of instrument design and the subsequent analysis of its performance (Wyngaard, 1988a, 1988b, 1990).

Typically, the inlet sampling port for a chemical species is not co-located within the point at which the turbulent vertical velocity is measured. Computationally, this can be accounted for by shifting the time series of one signal relative to the other so that the coherence of the turbulent signals is preserved. An additional time shift may also be needed to account for ducting the air from the intake port into the sampling chamber. Lenschow et al. (1980) point out that since the peak in the turbulent cospectra of vertical velocity and the species of interests (e.g., methane) shifts to higher frequencies as the altitude of the measurement is reduced, the effects of instrumental time lag and response time become more important for lower flight altitudes. In practice, the determination of these various time shifts can be obtained through spectral techniques employing actual flight data (Lenschow et al., 1980, 1982).

Previous discussions (Lawson, 1980; Lenschow, 1986) have described the instrumentation required to obtain air motion measurements necessary for eddy correlation flux measurements from an airborne platform. The determination of the three-component ambient wind field is not an easy task. The basic problem is to accurately determine the difference between two large numbers, i.e., the velocity of the plane relative to the ground (typically measured by an inertial navigation system; Broxmeyer, 1964; Kayton and Fried, 1969) and the velocity of the air relative to the aircraft (determined from the pitot-static probe and temperature sensors). For further details see Lenschow (1984, 1986).

The relative error variance of a measured species flux has been determined by Lenschow and Kristensen (1985). Ritter et al. (1990) obtained relative error variances of O_3 and CO fluxes measured over the Amazon basin being within the range of 10% to 20% of the reported flux values. To date, airborne eddy correlation flux measurements have been made twice for methane, i.e., in the Yukon-Kuskokwim delta region of Alaska (Ritter et al., 1992) and in the Hudson Bay Lowlands and boreal forest regions of Canada (Ritter et al., 1991).

4.4 Extrapolation techniques. The problem of how to relate point measurements to mean values representative of larger areas is an issue that is germane not only to atmospheric chemistry community but also to those concerned with modeling the dynamical processes of the atmosphere. The surface fluxes of momentum, heat, and moisture are fundamental in defining not only the steady state of the atmosphere but also in determining the mean profiles of trace gases within the boundary layer (Monin and Yaglom, 1971; Tennekes, 1982). The theory of Monin and Yaglom (1971), now known as Monin-Obukhov (MO) similarity theory, is strictly valid in the surface layer (i.e., the lowest 10 percent of the boundary layer) over uniformly flat and homogeneous terrain. Since conditions such as these are not typically satisfied due to variations in surface characteristics, modelers have employed the idea of a "blending height." The "blending height" is the height above which the flow no longer responds to variations in the surface characteristics and essentially becomes horizontally homogeneous (Wieringa, 1976, 1986; Mason, 1988). This is based on the fact that the timescale for turbulent eddies in the boundary layer is finite and cannot immediately adjust to variations in surface characteristics (Beljaars, 1982).

The analogy from the "meteorological" perspective, discussed above, to that which is relevant to the atmospheric chemistry community is clear. Significant variations in methane flux have been observed over a very limited spatial domain (Bartlett et al., 1989; Morrissey and Livingston, 1992). For a localized study area Wilson et al. (1989) found sizeable variations in the magnitude of the flux during the period of active emissions. Such variability would, by itself, pose some concern on the expected limits of uncertainty that might result from large-scale extrapolation of the data. However, the first attempt to intercompare flux measurements made over a range of scales from 1 m^2 to hundreds of km^2 using ground-level enclosures, tower, and aircraft data has met with reasonable success (Bartlett et al., 1992; Fan et al., 1992; Ritter et al., 1992). As noted by Bartlett and Harriss (1993), although the flux values from the ground-level and tower flux data agree very well in an integrated sense, the correspondence between their individual estimates of CH_4 source strength for the various local vegetation types does not agree as well. This is probably a result of the "blending" phenomena mentioned above in that source signatures from different vegetation types are being blended or averaged by the time the resulting signal arrives at the tower. The averaged flux values obtained from both the enclosure and tower data were somewhat lower than that observed from the aircraft. This is most likely due to a larger proportion of the aircraft's source area being covered by wetter and therefore generally higher CH_4-producing land areas. The enclosure data were extrapolated to the spatial scale representative of the tower by using SPOT satellite imagery and an area-weighted scheme for the CH_4 source strengths from the habitats present. This exercise illustrates how important the operational sampling strategy needs to be in order to obtain representative flux estimates that can then be extrapolated to larger scales using satellite data.

An extrapolation of surface CH_4 emission strength to large scales (i.e., hundreds of km^2) might be accomplished in one of the following ways. First, an aircraft with a CH_4 flux measuring capability could make repeated passes along a number of completely separate flight tracks. The source area for the fluxes measured by the aircraft could be determined with an algorithm such as that proposed by Schmid and Oke (1990). The variation in the fluxes measured along the different flight tracks could then be explained by examining the variation in the distribution of pixel radiances as seen from satellite imagery for the various source areas associated with the flight tracks. This would only be possible if the pixel radiance observed from satellite imagery has a near one-to-one mapping onto a distribution of CH_4 source strength (Matson et al., 1989). If, then, several independent source areas were sampled, each having different frequency distributions for the occurrence of a range of pixel radiances, a unique assignment of source strengths could be made for various pixel classes. The solution or assignment of source strengths would be constrained in that an areal integration over the source areas would match the aircraft flux values within a specified error limit. The resulting map of pixel radiances onto CH_4 source strengths could then be used to do a simple areal integration of the component source strengths of the region from satellite imagery. These data would then need to be complemented with data on the seasonal and diurnal variation of the fluxes in order to obtain an estimate of the annual CH_4 source strength for the region of interest.

A second way in which a CH_4 source distribution map might be extrapolated to large scales would involve a surface classification of CH_4 source strengths utilizing an intensive ground enclosure flux measurement program. The location and distribution of the chamber measurements would be dictated by high resolution satellite imagery. A one-to-one map of CH_4 source strength onto pixel radiance would then be determined. Then, in view of the large spatial variability in flux measurements, which has been mentioned previously, aircraft CH_4 flux measurements could be made above the "blending height" so that the actual areally averaged CH_4 flux to the atmosphere from the area of interest could be determined. The mapping of source strength onto pixel radiance could then be tuned to match the values obtained from the aircraft, thus reducing the concern of extremely large limits of uncertainty in the resulting areally averaged value. The source strength for the region of interest could then be determined by an areal integration of the component source strengths from the pixel values as in the first method. Additional discussion of the problem of extrapolation is given by Khalil and Shearer, this volume.

5. How can we study the environmental controls of methane flux on an ecosystem and landscape level?

The present observed trends of methane potentially affect climate and other conditions in the global or regional atmosphere. Changing environmental conditions, on the other hand, affect the biosphere and will probably result in changes of the spatial extension and functioning of the various ecosystems. This again may feed back on the flux of methane between the biosphere and the atmosphere. In order to get some understanding of these interactions, investigation of methane fluxes on an ecosystem and landscape level is required.

Methane flux studies on an ecosystem or landscape level can apply to the whole range of techniques available for flux studies as described in the previous sections. The practicability of applying chamber methods or micrometeorological techniques is primarily a matter of the extension and the structure of the terrain and of the logistic support and availability of funds. The ABLE experiments in the Yukon delta area and in the Hudson Bay Lowlands and boreal forest areas of Canada are good examples for methane flux studies by integration of the various types of flux measurement techniques on a landscape level.

A more serious problem for this type of study is the development of the right strategy for measurement and sampling. Ecosystems are very complex. The thorough understanding of an ecosystem requires its investigation by a large group of researchers working for decades. It is certainly impractical to demand the combination of methane flux studies to such a sophisticated approach of ecosystem research. On the other hand, this renunciation bears the potential danger that information important for the understanding of the methane flux may be missed. Therefore, it is important to think about strategies that are most suitable for specific questions and for particular ecosystems and landscapes. In particular, it is important to define at which sites (where) and at which time periods (when) the methane flux measurements must be made and what kind of other parameters (what) must be measured in addition. For obvious reasons, flux studies on an ecosystem level are still in their infancy.

An important first step would be to know whether a particular source or sink strength (or range thereof) could be attributed to a particular ecosystem under present-day conditions. The only approach presently available uses empirical relationships between fluxes and classes of ecosystems: "Where distinct and easily identified differences in ecosystem structure and vegetation composition coincide with differences in ecosystem function and trace gas flux, delineation of ecosystems into functionally different types is a useful basis for measurement strategies and for extrapolation" (Matson et al., 1989).

An example for this approach was the study of methane fluxes from the Florida Everglades. This ecosystem was stratified into major wetland types classified by satellite imagery, and fluxes for representatives of each type were measured over the season (Bartlett et al., 1989). A similar approach was taken by Whalen and Reeburgh (1988) studying the annual cycle of the methane flux from elements of an Alaskan tundra ecosystem, such as Eriophorum tussocks, intertussock depressions, moss-covered areas, and wet meadow stands with Carex. Continuation of these studies over the next three years showed, however, that the relative contribution of these different tundra elements to the total methane flux changed dramatically from year to year (Whalen and Reeburgh, 1992).

A similar empirical approach was used to study methane emissions from

Hudson Bay Lowland peatlands across a 100 km transect covering a 5,000 years chronosequence (Moore et al., 1990; Roulet et al., 1993; Roulet and Moore, 1993). The results revealed a distinct increase of both integrated methane emission rates and spatial variability of methane fluxes from the young tundra ecosystems towards the later successional stages.

6. How can we study methane fluxes on a process-oriented level?

The study of methane fluxes on a process-oriented level is required to address typical problems such as to assess the role of plants, animals, and microorganisms in emitting methane; or to assess the role of physico-chemical versus biological mechanisms; or to address the question after the portion of the net emission of methane compared to the total methane produced. A review of the factors controlling production and emission of methane from terrestrial ecosystems is found in the proceedings of a recent Dahlem conference (Conrad, 1989; Galchenko et al., 1989; Rosswall et al., 1989).

The different techniques for measuring methane fluxes have recently been reviewed and described by Schütz and Seiler (1989). For process-oriented studies a chamber method is certainly the method of choice as it provides the opportunity to manipulate a particular field plot in a way to extract information on processes. The chamber techniques are also suitable to include the measurement of fluxes of other tracers, e.g., ^{222}Rn, or to determine the isotope ratio in the released methane. For example, ^{222}Rn is a suitable tracer for gas transport in the soil profile and has been used to assess methane deposition fluxes (Dörr and Münnich, 1990; Born et al., 1990; Dörr et al., 1993). ^{222}Rn can also be used as a tracer for in situ bubble equilibration with dissolved gases in pore water of sediments (Martens and Chanton, 1989). The measurement of methane isotope ratios has so far been exploited only to a small extent (e.g., Martens et al., 1986; Burke et al., 1988; Chanton et al., 1992) but could potentially provide invaluable information on the processes involved in the production, oxidation, and transport of methane from rice fields and other wetlands.

Field experiments, i.e., controlled manipulations of a field plot, can give the first valuable information on the processes involved in methane flux. These manipulations may include treatment of the field with fertilizers or application of other agricultural practices that also help to assess the impact of various types of field management on CH_4 flux. For example, the influence of nitrogen fertilizer application has been studied with respect to methane emission from wetland rice fields (Schütz et al., 1989a) and to methane deposition in fertilized forest soil (Steudler et al., 1989) and cultivated grassland (Mosier et al., 1991).

Other manipulations include the removal or clipping of vegetation, shading or darkening of a field plot, addition of inhibitors into the soil or onto the leaves of the plants. There are many other manipulations that are feasible for specific questions and may help to elucidate the processes involved in the methane flux. For example, clipping rice plants provides a tool to assess the role of plant vascular transport for the emission of methane out of wetland rice fields. This relatively simple manipulation helped to assess the seasonal change of the methane emission path via ebullition versus plant vascular transport in an Italian rice field (Schütz et al., 1989b). An alternative procedure is to enclose the emergent parts of the rice plants in bags or plexiglass tubes (Butterbach and Rennenberg, personal communication). Similar approaches may be used to elucidate the role of plants or animals for accomplishing a methane flux under other circumstances.

However, by their nature emission data as such give only limited information for the understanding of the processes involved. In order to extract as much information as possible, methane flux measurements should be supplemented by (1) observation of crucial soil parameters, (2) determination of internal methane gross fluxes, and (3) evaluation by mechanistic models. The realization of these approaches will vary according to the ecosystem studied. Here, we will concentrate on the application in wetland rice fields.

Most investigations of methane flux from rice fields included the measurement of soil temperature, sometimes in different soil depths. Detailed studies on a diurnal basis for a whole season may provide hints for the location and importance of methane-producing processes, especially when they are supplemented by laboratory studies (Schütz et al., 1990). However, substantially more information on soil characteristics is needed for intercomparison of different studies and evaluation of possibly diverging effects. Therefore, the soil type should be classified according to an acknowledged soil taxonomy (e.g., IRRI, 1978). Additional information on the hydrological conditions in the field could be obtained by methods described by IRRI (1987). Redox potential and pH in different soil layers should be recorded in a reasonable temporal resolution, i.e., daily intervals shortly after flooding and weekly intervals during stable water regimes. Periodic 24-hour flux measurements should be conducted throughout the vegetation period if significant diurnal flux variations are to be expected.

Besides the measurement of the net emission rate of methane, it would be important to know the ratio between emission and production of methane. In other words, how much of the methane produced in the soil is emitted and how much of it is oxidized. Methods have been developed for quantifying the effect of methane oxidation on the diffusive flux of methane through the oxic surface layer of a sediment or flooded rice soil. These methods are based on flux measurements in whole soil cores under oxic versus anoxic conditions or by adding inhibitors to the supernatant water (Frenzel et al., 1990; King, 1990; Conrad and Rothfuss, 1991). However, these methods are not suitable to assess the methane oxidation, which is potentially active in a shallow oxic layer around the rice roots. The importance of methane oxidation operating in the rhizosphere has been demonstrated by laboratory studies (Holzapfel-Pschorn et al., 1985; 1986) and field studies (Schütz et al., 1989b). The latter studies mostly relied on retrieving of soil samples from the field and thus may be biased by sampling artifacts. However, recent laboratory studies of rice microcosms again demonstrated that oxygen transport from the atmosphere into the rhizosphere is important for methane oxidation and may account for oxidation of 80-90% of the produced methane (Frenzel et al., 1992).

At present, no feasible technique is available to determine the gross fluxes of methane in rice soil under field conditions. Inhibitors of methane-oxidizing bacteria could be applied to the soil, but difficulties arise in distributing the inhibitor homogeneously into the soil and in assessing the effectivity and specificity of the inhibitor under field conditions. Similar restrictions apply to techniques using isotopically labelled substrates in the field. The only practicable methods for studying production and oxidation of methane require the retrieval of soil cores and of horizontal cutting or drilling of subsamples. Due to the rooting of rice plants these manipulations may cause cutting off the roots. Cutting the roots may severely bias the results obtained with the soil cores. First, cutting may stimulate methane production by releasing substrates for the anaerobic microbial community. Second, cutting interrupts the transport of oxygen from the atmosphere into the soil and thus may result in depletion of the oxygen that is required for methane oxidation. At the moment, it can only be recommended to retrieve the soil cores from within the rows between the rice plants.

Mechanistic models of methane emission from rice fields are lacking at the moment. A model simulating the emission of methane either can be developed completely from the beginning or could be tailored by integrating the processes important for methane emission into existing models developed for optimizing rice yields. The CERES-RICE model operates with relatively detailed soil compartments (Goodwin et al., 1990) and therefore should be kept in mind for an extension towards a mechanistic methane emission model.

For the development of mechanistic emission models it will also be necessary to develop a quantitative understanding of the various pathways leading from the degradation of soil organic matter or of rice plant material towards the production of methane, and to develop the knowledge of the metabolic types of microorganisms involved in production and oxidation of methane in the paddy soil. The measurement of methane isotope ratios, especially of both $\delta^{13}C$ and δ^2H, of the methane in samples from soil reservoirs, from the gas vascular system of the plants, and from the gas entrapped in flux chambers could provide invaluable information on the pathways operative for production and oxidation of methane under field conditions (Whiticar and Faber, 1985; Whiticar et al., 1986). Besides this potentially useful approach, which has to be further developed and tested in the future, the presently available techniques for studying microbial metabolism require the retrieval and incubation of soil samples. The techniques that are useful in studying methanogenic processes and bacterial populations, including the isolation and counting of anaerobic bacteria and the application of radiotracer techniques, have been reviewed by Conrad and Schütz (1988). The present boom in the development of biomolecular techniques, e.g., DNA probes, will in future probably help to describe the composition of microbial communities in nature. For example, the application of DNA probes for methane-oxidizing bacteria helps to quantify the role of these bacteria in soil samples (Stainthorpe et al., 1990; Tsien et al., 1990; Hanson and Wattenberg, 1991; Murrell et al., 1993).

7. Summary

The Experimental Methods Working Group sought to identify the frontiers for future measurements of atmospheric methane. The five questions determined to be most important are: 1. How do we best continue to study the trends in the global methane budget? 2. How can we study the contribution of a region (geographical or political) to the global methane budget? 3. How can we reconcile the differences in scales from chamber techniques to global inventories? 4. How can we study the environmental controls of methane flux on an ecosystem and landscape level? 5. How can we study methane fluxes on a process-oriented level?

1. Trends of methane in the background air, measured by GC/FID methods, are well characterized. Further work is necessary to identify trends for continental sites and vertical distribution of methane. Trends of individual sources may be measured by the trend of the isotopic composition of methane. To improve understanding of the isotopic composition of different sources, the group recommends intercalibration between laboratories, better standards designed for measurements of atmospheric methane, and a more complete distribution of measurements.

2. Present methods of studying regional emissions are not yet satisfactory, either because of the expense and other practical limitations of tower studies or intensive chamber sampling, or because information from geographical information systems is not always applicable to flux estimation. Both limitations affect the usefulness of models to estimate regional fluxes. The group considered three techniques for improving regional estimates. Upwind-downwind measurements may be used to constrain the extrapolation of chamber studies. Isotopic fingerprinting is potentially useful for allocating fluxes to various sources within a region. New measurement technologies such as FTIR long path spectroscopy can be used to assess large point sources.

3. Techniques for measurement of trace gases vary in scale from chamber techniques (1 m) to aircraft eddy flux measurement and satellite imagery (100 km or more). "Micrometeorological" techniques such as boundary layer accumulation, flux gradient, and eddy correlation techniques may be used to estimate fluxes over area sources. Satellite data may be used to extrapolate measurement data to larger regions.

4. Investigation of methane fluxes on an ecosystem level is required to understand the interactions between changing environmental conditions and methane fluxes. However, these studies are time-consuming and expensive. It is important to determine which sites, time periods, and associated parameters should be measured.

5. Methane fluxes should be studied on a process-oriented level using small scale field experiments supplemented by laboratory studies. Flux measurements should be supplemented by determination and observation of crucial associated parameters, determination of gross methane fluxes, and evaluation by mechanistic models. The realization of these approaches must vary according to the source studied. While some sources have been studied at the process level, for example rice fields, mechanistic models are still lacking.

References

Bartlett, K.B., R.C. Harriss. 1993. Review and assessment of methane emissions from wetlands. *Chemosphere, 26* (1-4):261-320.
Bartlett, D.S., K.B. Bartlett, J.M. Hartman, R.C. Harriss, D.I. Sebacher, R. Pelletier-Travis, D. D. Dow, D.P. Brannon. 1989. Methane emissions from the Florida Everglades: Patterns of variability in a regional wetland ecosystem. *Global Biogeochem. Cycles, 3*:363-374.
Bartlett, K.B., P.M. Crill, R.L. Sass, R.C. Harriss, N.B. Dise. 1992. Methane emissions from tundra environments in the Yukon-Kuskokwim delta, Alaska. *J. Geophys. Res., 97*:16,645-16,600.
Beljaars, A.C.M. 1982. The derivation of fluxes from profiles in perturbed areas. *Boundary-Layer Meteorology, 24*:35-56.
Born, M., H. Dörr, I. Levin. 1990. Methane consumption in aerated soils of the temperate zone. *Tellus, 42B*:2-8.
Borrell, P., P.M. Borrell, W. Seiler (eds.). 1991. *Transport and Transformation of Pollutants in the Troposphere*, SPB Academic Publishing, The Hague (The Netherlands), 586 pp.
Broxmeyer, C. 1964. *Inertial Navigation Systems*, McGraw-Hill, New York, 254 pp.
Burke, R.A., Jr., T.R. Barber, W.M. Sackett. 1988. Methane flux and stable hydrogen and carbon isotope composition of sedimentary methane from the Florida Everglades. *Global Biogeochem. Cycles, 2*:329-340.
Businger, J.A. 1986. Evaluation of the accuracy with which dry deposition can be measured with current micrometeorological techniques. *J. Climate Appl. Meteorol., 25*:1,100-1,124.
Businger, J.A., A.C. Delaney. 1990. Chemical sensor resolution required for measuring surface fluxes by three common micrometeorological techniques. *J. Atmos. Chem., 10*:399-410.

Businer, J.A., S.P. Oncley. 1990. Flux measurement with conditional sampling. *J. Atmos. Ocean Technol., 7*:349-352.

Chanton, J.P., C.S. Martens, C.A. Kelley, P.M. Crill, W.J. Showers. 1992. Methane transport mechanisms and isotopic fractionation in emergent macrophytes of an Alaskan tundra lake. *J. Geophys. Res., 97*:16,681-16,688.

Conrad, R. 1989. Control of methane production in terrestrial ecosystems. In: *Exchange of Trace Gases between Terrestrial Ecosystems and the Atmosphere* (M.O. Andreae and D.S. Schimel, eds.), Dahlem Konferenzen, Wiley, Chichester, pp 39-58.

Conrad, R., H. Schütz. 1988. Methods of studying methanogenic bacteria and methanogenic activities in aquatic environments. In: *Methods in Aquatic Bacteriology* (B. Austin, ed.), Wiley, Chichester, pp 301-343.

Conrad, R., F. Rothfuss. 1991. Methane oxidation in the soil surface layer of a flooded rice field and the effect of ammonium. *Biol. Fertil. Soils, 12*:28-32.

Crill, P.M., K.B. Bartlett, J.O. Wilson, D.I. Sebacher, R.C. Harriss, J.M. Melack, S. MacIntyre, L. Lesack, L. Smith-Morrill. 1988. Tropospheric methane from an Amazonian floodplain lake. *J. Geophys. Res., 93*:1,564-1,570.

Delmas, R.A., J.P. Tathy, M. Labat, J. Servant, B. Cros. 1992. Sources and sinks of methane and carbon dioxide exchanges in mountain forest in equatorial Africa. *J. Geophys. Res., 97*:6,169-6,179.

Denmead, O.T. 1991. Sources and sinks of greenhouse gases in the soil-plant environment. *Vegetatio, 91*:73-86.

Desjardin, R.L. 1972. A study of carbon dioxide and latent heat fluxes using the eddy correlation technique. Ph.D. dissertation, Cornell University, NY.

Desjardin, R.L. 1991. Review of techniques to measure CO_2 flux densities from surface and airborne sensors. In: *Advances in Bioclimatology* (G. Stanhill, ed.), Springer, Berlin, pp 1-41.

Dörr, H., K.O. Münnich. 1990. 222Rn flux and soil air concentration profiles in West-Germany. Soil ^{222}Rn as tracer for gas transport in the unsaturated soil zone. *Tellus, 42B*:20-28.

Dörr, H., B. Kromer, I. Levin, K.O. Münnich, J.J. Volpp. 1983. CO_2 and ^{222}Radon as tracers for atmospheric transport. *J. Geophys. Res., 88*:1,309-1,313.

Dörr, H., L. Katruff, I. Levin. 1993. Soil texture parametrization of the methane uptake in aerated soils. *Chemosphere, 26* (1-4):697-714.

Ehhalt, D. 1974. The atmospheric cycle of methane. *Tellus, 26*:1-2.

Fan, S.M., S.C. Wofsy, P.S. Bakwin, D.J. Jacob, S.M. Anderson, P.L. Kebabian, J.B. McManus, C.E. Kolb, D.R. Fitzjarrald. 1992. Micrometeorological measurements of CH_4 and CO exchange between the atmosphere and the arctic tundra. *J. Geophys. Res., 97*:16,627-16,643.

Fowler, D., J.H. Duyzer. 1989. Micrometeorological techniques for the measurement of trace gas exchange. In: *Exchange of Trace Gases between Terrestrial Ecosystems and the Atmosphere*. Dahlem Konferenzen (M.O. Andreae and D.S. Schimel, eds.), Wiley, Chichester, pp 189-207.

Fraser, P.J., P. Hyson, R.A. Rasmussen, M.A.K. Khalil. 1986. Methane, carbon monoxide and methylchloroform in the southern hemisphere. *J. Atmos. Chem.*, 4:3-42.

Frenzel, P., B. Thebrath, R. Conrad. 1990. Oxidation of methane in the oxic surface layer of a deep lake sediment (Lake Constance). *FEMS Microbiol. Ecol.*, 73:149-158.

Frenzel, P., F. Rothfuss, R. Conrad. 1992. Oxygen profiles and methane turnover in a flooded rice microcosm. *Biol. Fertil. Soils.*, 14:84-89.

Fung, I., J. John, J. Lerner, E. Matthews, M. Prather, L.P. Steele, P.J. Fraser. 1991. Three-dimensional model synthesis of the global methane cycle. *J. Geophys. Res.*, 96:13,033-13,065.

Galchenko, V.F., A. Lein, M. Ivanov. 1989. Biological sinks of methane. In: *Exchange of Trace Gases Between Terrestrial Ecosystems and the Atmosphere* (M.O. Andreae and D.S. Schimel, eds.), Dahlem Konferenzen, Wiley, Chichester, pp 59-71.

Goodwin, D.C., U. Singh, R.J. Buresh, S.K. DeDatta. 1990. Modelling of nitrogen dynamics in relation to rice growth and yield. In: *Proc. 14th Int. Congr. Soil Science, Kyoto, Japan, vol. 4*, ISSS, Kyoto (Japan), pp 320-325.

Hanson, R.S., E.V. Wattenberg. 1991. Ecology of methylotrophic bacteria. In: *Biology of Methylotrophs, vol. 18* (I. Goldberg and J.S. Rokem, eds.), Butterworth-Heinemann, Stoneham, pp 325-348.

Hanst, P.L. 1982. Air pollution measurements by Fourier transforms spectroscopy. *Appl. Opt.*, 17:1,360.

Hanst, P.L., A.S. Lefohn, B.W. Gay. 1973. Detection of atmospheric pollutants at parts-per-billion levels by infrared spectroscopy. *Appl. Spectroscopy*, 27:188.

Harriss, R.C. 1989. Experimental design for studying atmosphere-biosphere interactions. In: *Exchange of Trace Gases between Terrestrial Ecosystems and the Atmosphere* (M.O. Andreae and D.S. Schimel, eds.), Dahlem Konferenzen, Wiley, Chichester, pp 291-301.

Herget, W.F. 1982. Analysis of gaseous pollutants using a mobile FTIR system. *Amer. Labs.* 72.

Holzapfel-Pschorn, A., R. Conrad, W. Seiler. 1985. Production, oxidation and emission of methane in rice paddies. *FEMS Microbiol. Ecol.*, 31:343-351.

Holzapfel-Pschorn, A., R. Conrad, W. Seiler. 1986. Effects of vegetation on the emission of methane from submerged paddy soil. *Plant and Soil*, 92:223-233.

IRRI. 1978. *Soils and Rice*, International Rice Research Institute, Los Banos (Phillippines), 82 pp.

IRRI. 1987. *Physical Measurements in Flooded Rice Soils,* International Rice Research Institute, Los Banos (Philippines), 65 pp.
Kaye, J.A. 1987. Mechanisms and observations for isotope fractionations of molecular species in planetary atmospheres. *Rev. Geophysics,* 25:1,609-1,658.
Kayton, M., W.R. Fried. 1969. *Avionics Navigation Systems,* Wiley, New York, 666 pp.
Khalil, M.A.K., R.A. Rasmussen. 1988. Trace gases over the western Atlantic Ocean: Fluxes from the eastern United States and distributions in and above the planetary boundary layer. *Global Biogeochem. Cycles,* 2:63-71.
King, G.M. 1990. Dynamics and controls of methane oxidation in a Danish wetland sediment. *FEMS Microbiol. Ecol.,* 74:309-323.
Kirchgessner, D.A., S.D. Piccot, A. Chadha, T. Minnich. 1993. Estimation of methane emissions from a surface coal mine using open-path FTIR spectroscopy and modeling techniques. *Chemosphere,* 26 (1-4):23-44.
Lawson, R.P. 1980. On the airborne measurement of vertical air velocity. *J. Appl. Meteorol.,* 19:1,416-1,419.
Lenschow, D.H. 1984. Instrumentation development needs for use of mass-balance technique. In: *Global Tropospheric Chemistry: A Plan for Action,* National Academy Press, Washington, D.C., pp 141-143.
Lenschow, D.H. 1986. Aircraft measurements in the boundary layer. In: *Probing the Atmospheric Boundary Layer* (D.H. Lenschow, ed.), American Meteorol. Soc., Boston Mass., pp 39-55
Lenschow, D.H., L. Kirstensen. 1985. Uncorrelated noise in turbulence measurements. *J. Atmos. Ocean Technol.,* 2:68-81.
Lenschow, D.H., A.C. Delany, B.B. Stankov, D.H. Stedman. 1980. Airborne measurements of the vertical flux of ozone in the boundary layer. *Boundary-Layer Meteorol,* 19:249-265.
Lenschow, D.H., R. Pearson Jr., B.B. Stankov. 1982. Measurements of ozone vertical flux to ocean and forest. *J. Geophys. Res.,* 87:8,833-8,837.
Levin, I. 1987. Atmospheric CO_2 in continental Europe - an alternative approach to clean air data. *Tellus,* 39B:21-28.
Levin, I., J. Schuchard, B. Kromer, K.O. Münnich. 1989. The continental European Suess-effect. *Radiocarbon,* 31:431-440.
Levin, I., R. Bösinger, G. Bonani, R. Francey, B. Kromer, K.O. Münnich, M. Suter, N.B.A. Trivett, W. Wölfli. 1992. Radiocarbon in atmospheric carbon dioxide and methane: global distribution and trends. In: *Radiocarbon After Four Decades: An Interdisciplinary Perspective* (R.E.T.A. Long and R.S. Kra, eds.), Springer, Berlin, pp 503-518.
Levin, I., P. Bergamaschi, H. Dörr, D. Trapp. 1993. Stable isotopic signature of methane from major sources in Germany. *Chemosphere,* 26 (1-4):161-178.

Lowe, D.C., C.A.M. Brenninkmeijer, M.R. Manning, R. Sparks, G. Wallace. 1988. Radiocarbon determinations of atmospheric methane at Baring Head, New Zealand. *Nature, 332*:522-525.

Lowe, D.C., C.A.M. Brenninkmeijer, S.C. Tyler, E.J. Dlugkencky. 1991. Determination of the isotopic composition of atmospheric methane and its application in the Antarctic. *J. Geophys. Res., 96*:15,455-15,467.

MacPherson, J.I., R.L. Desjardin. 1991. Airborne tests of flux measurement by the relaxed eddy accumulation technique. In: *Proc. 7th Symp. Met. Obs. Instrum.*, New Orleans, LA, pp 6-11.

Martens, C.S., J.P. Chanton. 1989. Radon as a tracer of biogenic gas equilibration and transport from methane-saturated sediments. *J. Geophys. Res., 94*:3,451-3,459.

Martens, C.S., N.E. Blair, C.D. Green, D.J. DesMarais. 1986. Seasonal variations in the stable carbon isotopic signature of biogenic methane in a coastal sediment. *Science, 233*:1,300-1,303.

Mason, P.J. 1988. The formation of arealy-averaged roughness lengths. *QJR Meteorol. Soc., 114*:399-420.

Matson, P.A., P.M. Vitousek, D.S. Schimel. 1989. Regional extrapolation of trace gas flux based on soils and ecosystems. In: *Exchange of Trace Gases Between Terrestrial Ecosystems and the Atmosphere* (M.O. Andreae and D.S. Schimel, eds.), Dahlem Konferenzen, Wiley, Chichester, pp 97-108.

Monin, A.S., A.M. Yaglom. 1971. *Statistical Fluid Mechanics: Mechanics of Turbulence, 3rd ed., vol. 1*, MIT Press, London.

Moore, T., N. Roulet, R. Knowles. 1990. Spatial and temporal variations of methane flux from subarctic/northern boreal fens. *Global Biogeochem. Cycles, 4*:29-46.

Morrissey, L.A., G.P. Livingston. 1992. Methane Flux from tundra ecosystems in arctic Alaska: An assessment of local spatial variability. *J. Geophys. Res., 97*:16,661-16,670.

Mosier, A., D. Schimel, D. Valentine, K. Bronson, W. Parton. 1991. Methane and nitrous oxide fluxes in native, fertilized and cultivated grasslands. *Nature, 350*:330-332.

Mroz, E.J. 1993. Deuteromethanes: potential fingerprints of the sources of atmospheric methane. *Chemosphere, 26* (1-4):45-54.

Murrell, J.C., V. McGowan, D.L.N. Cardy. 1993. Detection of methylotrophic bacteria in natural samples by molecular probing techniques. *Chemosphere, 26* (1-4):1-12.

Prather, M., M.B. McElroy, S. Wofsy, G. Russell, D. Rind. 1987. Chemistry of the global troposphere: Fluorocarbons as tracers of air motion. *J. Geophys. Res., 92*:6,579-6,613.

Quay, P. D., S.L. King, J. Stutsman, D.O. Wilbur, L.P. Steele, I. Fung, R.H. Gammon, T.A. Brown, G.W. Farwell, P.M. Grootes, F.H. Schmidt. 1991. Carbon isotopic composition of atmospheric CH_4: fossil and biomass burning source strengths. *Global Biogeochem. Cycles*, 5:25-47.

Ritter, J.A., D.H. Lenschow, J.D.W. Barrick, G.L. Gregory, G.W. Sachse, G.F. Hill, M.A. Woerner. 1990. Airborne flux measurements and budget estimates of trace species over the Amazon Basin during the GTE/ABLE 2B expedition. *J. Geophys. Res.*, 95:16,875-16,886.

Ritter, J.A., C. Watson, J. Barrick, G. Sachse, J. Collins, G. Gregory, B. Anderson, M. Woerner. 1991. Airborne boundary-layer measurements of heat, moisture, CH_4, CO, and O_3 fluxes over Canadian boreal forest and northern wetland regions. *EOS*, 72:85.

Ritter, J.A., J.D.W. Barrick, G.W. Sachse, G.L. Gregory, M.A. Woerner, C.E. Watson, G.F. Hill, J.E. Collins. 1992. Airborne flux measurements of trace species in an arctic boundary layer. *J. Geophys. Res.*, 97:16,601-16,625.

Rosswall, T., F. Bak, D. Baldocchi, R.J. Cicerone, R. Conrad, D.H. Ehhalt, M.M.K. Firestone, I.E. Galbally, V.F. Galchenko, P.M. Groffman, H. Papen, W.S. Reeburgh, E. Sanhueza. 1989. What regulates production and consumption of trace gases in ecosystems: Biology or physicochemistry? In: *Exchange of Trace Gases Between Terrestrial Ecosystems and the Atmosphere* (M.O. Andreae and D.S. Schimel, eds.), Dahlem Konferenzen, Wiley, Chichester, pp 73-95.

Roulet, N.T., R. Ash, T.R. Moore. 1992. Low boreal wetlands as a source of atmopsheric methane. *J. Geophys. Res.*, 97:3,739-3,749.

Sachse, G.W., G.F. Hill, L.O. Wade, M.G. Perry. 1987. Fast response, high-precision carbon monoxide sensor using a tunable diode laser absorption technique. *J. Geophys. Res.*, 92:2,071-2,081.

Schmid, H.P., T.R. Oke. 1990. A model to estimate the source area contributing to turbulent exchange in the surface layer over patchy terrain. *QJR Meteorol. Soc., 116*:965-988.

Schoell, M. 1980. The hydrogen and carbon isotopic composition of methane from natural gases of various origins. *Geochim. Cosmochim. Acta*, 44:649-661.

Schupp, M., P. Bergamaschi, G.W. Harris, P.J. Crutzen. 1993. Development of a tunable diode laser absorption spectrometer for measurements of the $^{13}C/^{12}C$ ratio in methane. *Chemosphere, 26* (1-4):13-22.

Schütz, H., W. Seiler. 1989. Methane flux measurements: methods and results. In: *Exchange of Trace Gases Between Terrestrial Ecosystems and the Atmosphere* (M.O. Andreae and D.S. Schimel, eds.), Dahlem Konferenzen, Wiley, Chichester, pp 209-228.

Schütz, H., A. Holzapfel-Pschorn, R. Conrad, H. Rennenberg, W. Seiler. 1989a. A 3-year continuous record on the influence of daytime, season, and fertilizer treatment on methane emission rates from an Italian rice paddy. *J. Geophys. Res., 94*:16,405-16,416.

Schütz, H., W. Seiler, R. Conrad. 1989b. Processes involved in formation and emission of methane in rice paddies. *Biogeochem., 7*:33-53.

Schütz, H., W. Seiler, R. Conrad. 1990. Influence of soil temperature on methane emission from rice paddy fields. *Biogeochem., 11*:77-95.

Spivakovsky, C.M., R. Yevich, J.A. Logan, S.C. Wofsy, M.B. McElroy, M.J. Prather. 1990. Troposheric OH in a three dimensional chemical tracer model: An assessment based on observations of CH_3CCl_3. *J. Geophys. Res., 95*:18,441-18,471.

Stainthorpe, A.C., G.P.C. Salmond, H. Dalton, J.C. Murrell. 1990. Screening of obligate methanotrophs for soluble methane monooxygenase genes. *FEMS Microbiol Lett., 70*:211-216.

Steudler, P.A., R.D. Bowden, J.M. Melillo, J.D. Aber. 1989. Influence of nitrogen fertilization on methane uptake in temperate forest soils. *Nature, 341*:314-316.

Stevens, C.M., R.E. Rust. 1982. The carbon isotopic composition of atmospheric methane. *J. Geophys. Res., 87*:725-733.

Stewart, J.W.B., I. Aselmann, A.F. Bouwman, R.L. Desjardin, B.B. Hicks, P.A. Matson, H. Rodhe, D.S. Schimel, B.H. Svensson, R. Wassmann, M.J. Whiticar, M.X. Yang. 1989. Extrapolation of flux measurements to regional and global scales. In: *Exchange of Trace Gases Between Terrestrial Ecosystems and the Atmosphere* (M.O. Andreae and D.S. Schimel, eds.), Dahlem Konferenzen, Wiley, Chichester, pp 155-174.

Tennekes, H. 1982. Similarity relations, scaling laws and spectral dynamics. In: *Atmospheric Turbulence and Air Pollution Modelling* (F.T.M. Nieuwstadt and H. vanDop, eds.), D. Reidel, Dordrecht (The Netherlands), pp 37-68.

Thom, M., R. Bösinger, M. Schmidt, I. Levin. 1993. The regional budget of atmospheric methane of a highly populated area. *Chemosphere, 26* (1-4):143-160.

Trumbore, S.E., M. Keller, S.C. Wofsy, J.M. daCosta. 1990. Measurement of soil and canopy exchange rates in the Amazon rain forest using ^{222}Rn. *J. Geophys. Res., 95*:16,865-16,873.

Tsien, H.C., B.J. Bratine, K. Tsuji, R.S. Hanson. 1990. Use of oligodeoxynucleotide signature probes for identification of physiological groups of methylotrophic bacteria. *Appl. Environ Microbiol, 56*:2,858-2,865.

Tyler, S.C. 1986. Stable carbon isotope ratios in atmospheric methane and some of its sources. *J. Geophys. Res., 91*:13,232-13,238.

Wahlen, M., N. Tanaka, R. Henry, B. Deck, J. Zeglen, J.S. Vogel, J. Southon, A. Shemesh, R. Fairbanks, W. Broecker. 1989. Carbon-14 in methane sources and in atmospheric methane: The contribution from fossil carbon. *Science, 245*:286-290.

Webb, E.K., G.E. Pearman, R. Leuning. 1980. Correction of flux measurements for density effects due to heat and water vapour transfer. *QJR Meteorol. Soc., 106*:85-100.

Whalen, S.C., R.S. Reeburgh. 1988. A methane flux time series for tundra environments. *Global Biogeochem. Cycles, 2*:399-409.

Whalen, S.C., W.S. Reeburgh. 1992. Interannual variations in tundra methane emission: a four-year time series at fixed sites. *Global Biogeochem Cycles, 6*:139-159.

Whiticar, M.J., E. Faber. 1985. Methane oxidation in sediment and water column environments - isotope evidence. *Adv. Org. Geochem., 10*:759-768.

Whiticar, M.J., E. Faber, M. Schoell. 1986. Biogenic methane formation in marine and freshwater environments: CO_2 reduction vs. acetate fermentation - isotopic evidence. *Geochim. Cosmochim. Acta, 50*:693-709.

Whiticar, M.J., E. Faber, M. Schoell. 1990. A geochemical perspective of natural gas and atmospheric methane. *Org. Geochem., 16*:531-547.

Wieringa, J. 1976. An objective exposure correction method for average wind speeds measured at a sheltered location. *QJR Meteorol. Soc., 102*:241-253.

Wieringa, J. 1986. Roughness-dependent geographical interpolation of surface wind speed averages. *QJR Meteorol. Soc., 112*:867-889.

Wilson, J.O., P.M. Crill, B.B. Bartlett, D.L. Sebacher, R.C. Harriss, R.L. Sass. 1989. Seasonal variation of methane emissions from a temperate swamp. *Biogeochem., 8*:55-71.

Wofsy, S.C., R.C. Harriss, W.A. Kaplan. 1988. Carbon dioxide in the atmosphere over the Amazon Basin. *J. Geophys. Res., 93*:1,377-1,387.

Wyngaard, J.C. 1988a. Flow-distortion effects on scalar flux measurements in the surface layer: Implications for sensor design. *Boundary-Layer Meteorol., 42*:19-26.

Wyngaard, J.C. 1988b. The effects of probe-induced flow distortion on atmospheric turbulence measurements: Extension to scalars. *J. Atmos. Sci., 22*:3,400-3,412.

Wyngaard, J.C. 1990. Scalar fluxes in the planetary boundary layer - theory, modeling and measurement. *Boundary-Layer Meteor., 50*:49-75.

Chapter 3

The Record of Atmospheric Methane

D. RAYNAUD AND J. CHAPPELLAZ

Laboratoire de Glaciologie et Géophysique de l'Environnement
B.P. 96 38402 St. Martin d'Heres Cedex (France)

1. *Introduction*

The global evolution of atmospheric CH_4 has been documented by sporadic direct measurements in the atmosphere during the 1960s and 1970s and by systematic survey only since 1979. The data from this time up to 1983 indicate an increasing trend at a rate of about 1% per year (Rasmussen and Khalil, 1986; Steele et al., 1987; Blake and Rowland, 1988). The most recent measurements indicate a slowing of the global accumulation of atmospheric CH_4 (Khalil and Rasmussen, 1990, 1993; Steele et al., 1992; see editor's note at the end of this chapter). The analysis of infrared solar absorption spectra (Rinsland et al., 1985; Zander et al., 1989) provides additional data of global concentrations for a few specific years (1951, 1975, 1981, 1984-87) and shows a mean increase of about 30% over the past 40 years. This long-term accumulation of CH_4 in the atmosphere is related to human activities, particularly from agricultural activities.

The analysis of gases trapped in ice makes possible the documentation of the anthropogenic influence back in time and the extension of the CH_4 record over periods such as the little ice age, medieval warming, and the glacial-interglacial cycle, encompassing climatic change with little or no human disturbance.

In this chapter we will first discuss the method itself. In particular, we will discuss why and how an atmospheric record of CH_4 from ice core analysis is obtained and what kind of accuracy can be expected from such a reconstruction. We will then review the existing ice records for CH_4 and discuss what we can learn in terms of the CH_4 cycle and its relation to climate.

NATO ASI Series, Vol. I 13
Atmospheric Methane: Sources, Sinks, and Role in Global Change
Edited by M. A. K. Khalil

2. How accurate is the atmospheric CH_4 record measured on the air trapped in polar ice?

2.1 How air becomes trapped in polar ice and dating problems. The air trapping process is illustrated on Figure 1. Snow deposited on ice sheets in areas free of summer surface melting is compressed and sintered as a result of water vapor diffusion and plastic deformation of the solid H_2O grains under the weight of subsequently fallen snow (Schwander, 1989). During this stage when snow is transformed into ice, called firn, the air found in the open pores becomes trapped in bubbles, which close off during the last step of the sintering process, mainly in the bottom of the firn layer (Schwander and Stauffer, 1984). When all the pores are finally closed, the resulting material, ice, contains, in volume, about 90% solid water and 10% air. Later, the trapped gases begin to associate with water molecules, creating hydrate structures, and the bubbles progressively disappear. There are no visible bubbles when the ice layers reach about 1000 m depth in the ice sheet.

In order to date the air trapped in an ice core, one must first find the age-depth relationship for the ice itself. This can be obtained by various techniques with different orders of precision (see for instance Paterson and Hammer, 1987) including: (1) counting annual layers in the ice using seasonal variations of various indicators (isotopic composition of the ice, acidity, and chemical compounds); (2) using horizons of known age, such as acid layers from volcanic eruptions; (3) radioactive dating (^{14}C from CO_2 in air bubbles or ^{32}Si); (4) matching features of an ice record with those contained in another dated record ($\delta^{18}O$ or dust); and (5) using an ice-flow model.

A second step is needed to date the air itself. The air is generally isolated from the surrounding atmosphere during the last stage of the transformation of snow into ice when air molecules can no longer diffuse through the open pores. This takes place after the snow deposition, and therefore the age of the air in the bubbles is younger than the age of the surrounding ice (see Table 1). The difference of age between the air and the ice depends on the isolation depth (a function of the accumulation rate and temperature) and on the mean accumulation rate between the surface and the close-off depth. Models of various complexity have been developed to take into account changes of these parameters along an ice core and to calculate the air-ice difference in age for each site (Schwander and Stauffer, 1984; Barnola et al., 1987; Barnola et al.,

1991). In the sites with high accumulation rates it may also be necessary to take into account the time required for the air to be in diffusive equilibrium between the surface and the depth of isolation (Etheridge et al., 1992; Schwander et al., 1993).

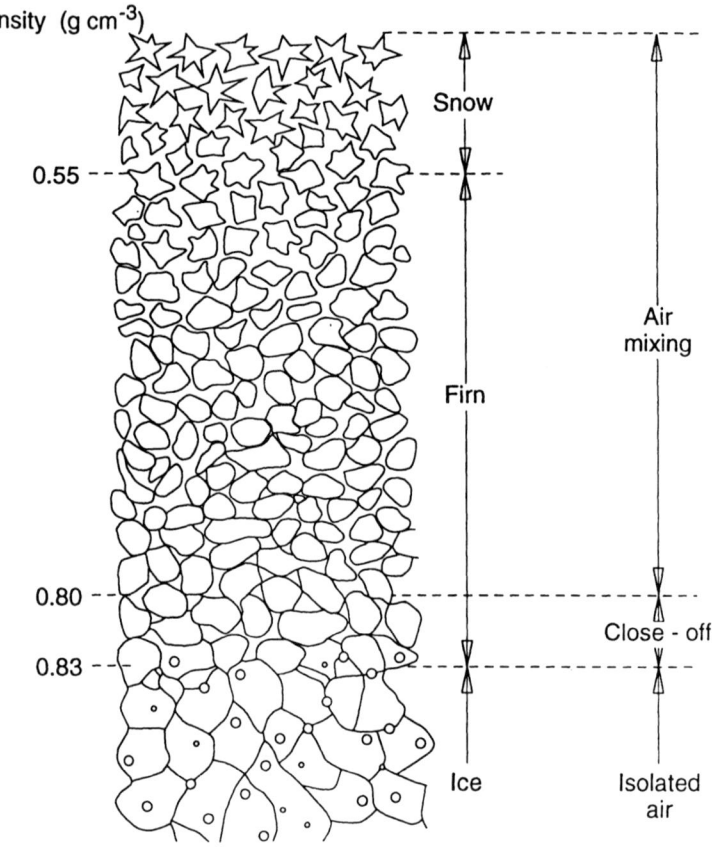

Figure 1. Diagram of the snow-ice transformation illustrating the trapping process for air bubbles. The depth at which the pore close-off occurs is typically in the range: 50-100 m below the surface, depending on the site characteristics (temperature and accumulation rate).

Also, because all the pores found in an ice layer do not close at the same time, the age of the air trapped in an ice sample analyzed for CH_4 typically results from the close-off of several thousands of bubbles, causing the measured concentration to be integrated over time intervals ranging from a few years to several hundreds of years (see Table 1).

Table 1. Antarctica - range of the chronological parameters used for evaluating the age of the air trapped in ice under current conditions. At station DEO 8 on the Law Dome the snow transforms very quickly in ice, mainly because of the particularly high snow accumulation rate (data from Etheridge et al., 1992). In contrast, the station Vostok in the central part of East Antarctica has a very low accumulation rate of snow and extremely cold temperatures, leading to a very slow firn densification. Moreover, changes in temperature and accumulation rate between glacial and interglacial conditions lead to large variations of the chronological parameters. Estimates indicate an age difference between air and surrounding ice of about 6000 years and a trapping time interval of about 900 years for glacial conditions at Vostok (Barnola et al., 1991).

Ice core	Law Dome DE08	Vostok
Age difference between snow deposition and air enclosure	35 years	2500 years
Time interval integrated by the air extracted from an ice sample	14 years	300 years

The accuracy of the mean age of the air depends upon the uncertainties in estimating both the age of the ice and the difference between the mean age of the air and the age of the ice. This accuracy varies, depending generally on the site and the age (the accuracy is lower for older ice). It ranges from a few years, for young ice in sites showing exceptionally high accumulation rates, to several thousands of years for old ice recovered in the central part of the ice sheets.

2.2 Parameters that could affect the record of atmospheric CH_4 trapped in polar ice. It is necessary to ascertain the extent to which the CH_4 concentration in the air extracted from the ice is the same as the original atmospheric CH_4 concentration prevailing at the time when the air was trapped. As illustrated on Figure 2, there are several parameters that could make the gas record measured in ice samples different from the real atmospheric concentration. However, the impact of some of these mechanisms can be experimentally tested or calculated and the gas measurements can be corrected.

2.3 Physical and chemical mechanisms involved in the air trapping. A recent thorough discussion of these mechanisms can be found in Schwander (1989).

They include:

(1) Physi- and chemi-sorption of gases on the surface of ice and snow crystals: in the case of methane, however, no adsorption on the snow crystals has been found (Rasmussen et al., 1982).
(2) Solid phase reactions between gases and ice on a long time scale: the absence of a regular trend in the records going back in time suggests that such a phenomenon does not occur.
(3) Separation by gravity in the firn column: this effect can be calculated from the barometric equation and depends on the height of the firn column where air mixing by diffusion occurs and on the temperature; in the extreme conditions of Antarctica, the maximum gravitational effect leads to CH_4 concentrations at the base of the firn, which are 0.4 to 0.9% larger than those at the surface of the ice sheet (Sowers et al., 1989).
(4) Alteration of the gas composition due to formation of air hydrates at great depths in the ice: as shown later in this chapter, similar glacial-interglacial CH_4 changes are found in different ice core sections that have experienced various amounts of clathrate formation (Stauffer et al., 1988; Chappellaz et al., 1990).
(5) Modified composition of the trapped air by fractures and thermal cracks in the ice: ice samples for gas measurements are generally selected to be free of fractures or thermal cracks. Nevertheless, we cannot assume that some CH_4 measurements are affected by contamination due to a poor ice quality (Chappellaz et al., 1990).

There is still fundamental work to be done to quantitatively evaluate the effects of these chemical and physical processes but current knowledge suggests that, on the whole, the CH_4 record is well preserved (within a few per cent) in ice.

The main reason for confidence arises from the ice core record itself. Australian Antarctic Expeditions work in a unique area (Law Dome, East Antarctica) where the snow transforms very quickly in ice (mainly because of the exceptionally high snow accumulation rate) and consequently can sample the very recent atmosphere. The analysis of ice retrieved on Law Dome indicates that the CH_4 record obtained from ice core measurements connects well with the atmospheric record measured since 1978 (Pearman and Fraser, 1988; Etheridge et al., 1992). Furthermore, the good agreement obtained for the last glacial-interglacial change between the records of ice cores taken at different places on both the Antarctic and Greenland ice sheets (Stauffer et al., 1988; Chappellaz et al., 1990) and the fact that methane concentrations remained quite constant for the past thousands of years (Rasmussen and Khalil, 1984) support the accuracy of the atmospheric CH_4 record based on ice core data.

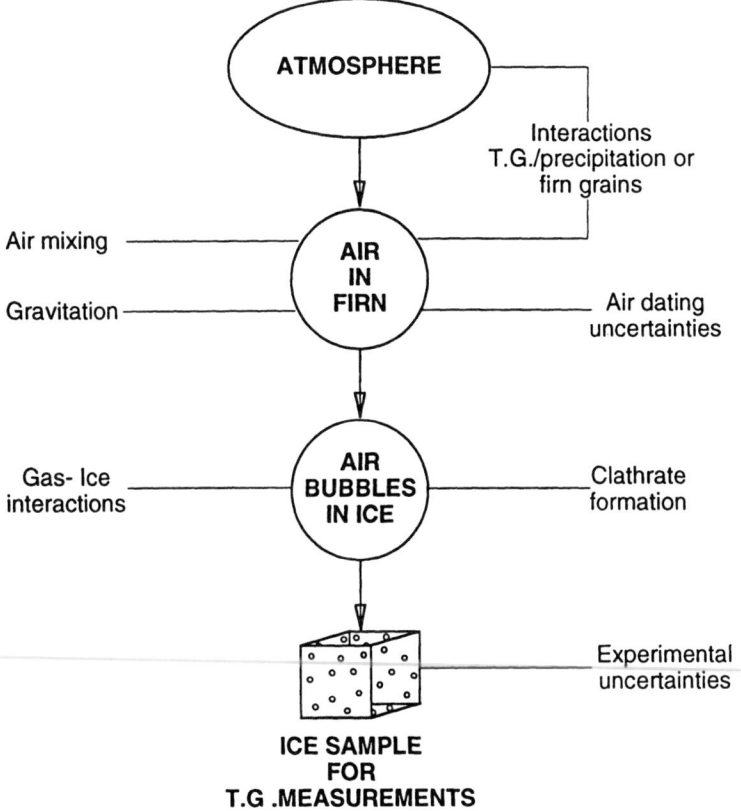

Figure 2. Diagram illustrating the physical processes involved in air trapping by ice and the sources of uncertainty when measuring the atmospheric record from ice core analysis. T.G. stands for Trace Gas.

Table 2. Experimental procedures for gas extraction and CH_4 measurements in polar ice cores. The contamination values refer to the concentration of CH_4 added by the experimental procedure to the original one in the ice. They are obtained from blank tests.

Authors	Conditions before extraction	Extraction method	Transfer to chromatograph	Amount of ice (g)	Contamination (ppbv)
Robbins et al. (1973)	vacuum	melting	He pressurization sampling with syringe	~1100	NA
Khalil & Rasmussen (1982) Rasmussen & Khalil (1984)	He pressurization	melting	syringe	NA	NA
Craig & Chou (1982)	vacuum	melting-refreezing repeated	condensation with liquid N_2	~40	50 ± 25
Stauffer et al. (1985)	vacuum	melting	pumping with Toepler pump	~400	70 ± 20
	vacuum	crushing with milling cutter	condensation with liquid He	~600	255 ± 45
Pearman et al. (1986)	vacuum	crushing with rasp	condensation with liquid He	500-1400	about 400

Table 2. Experimental procedures for gas extraction and CH_4 measurements in polar ice cores. The contamination values refer to the concentration of CH_4 added by the experimental procedure to the original one in the ice. They are obtained from blank tests.

Authors	Conditions before extraction	Extraction method	Transfer to chromatograph	Amount of ice (g)	Contamination (ppbv)
Zardini (1987)	N_2 pressurization	crushing with knifes	fluid pressurization	40	about 350
Stauffer et al. (1988)	vacuum	melting	condensation with liquid He	500-800	40 ± 10 (1 σ)
Etheridge et al. (1988)	vacuum	crushing with grater	condensation with liquid He	900-1400	26 ± 19
Raynaud et al. (1988) Chappellaz et al. (1990)	vacuum	melting-refreezing	expansion under vacuum	~40	36 ± 25
Tohjima et al. (1991) Nakazawa et al. (1992)	vacuum	melting-refreezing	pumped	~100	NA
Etheridge et al. (1992)	vacuum	crushing with grater	condensation with liquid He	900	6 ± 22

2.4 Experimental procedures used for extracting and measuring CH_4 in ice cores. Experimental procedures may constitute the major source of uncertainty concerning the reliability of the atmospheric CH_4 records from polar ice. So far, all laboratories that performed CH_4 measurements in ice cores measured the CH_4 concentration with a gas chromatograph equipped with a flame-ionization detector while the methods used to extract the gas from the ice and to transfer it into the chromatograph have differed considerably (Table 2).

The extraction is performed either by melting or by crushing the ice. In order to determine the potential impact of the extraction method on the original CH_4 concentration in the ice, blank tests are performed, generally using artificial bubble-free ice together with a calibration gas. An additional test consists of measuring halocarbons, which are indicators of contamination by modern air. The dry extraction method has created a large contamination (e.g., comparable to the low atmospheric CH_4 levels observed during full glacial conditions) due to friction between the metal container, the crushing system, and the ice. Recent efforts to decrease the friction in this method have greatly decreased the amount of CH_4 contamination (Etheridge et al., 1988, 1992; B. Stauffer, personal communication). Melting the ice limits the contamination to about 50 ppbv, but the origin of it is currently unknown. The amplitude of the contamination is not by itself a crucial point as it can be subtracted from the initial CH_4 measurement; the factor to consider is its reproducibility. In this regard, the difference between the two methods of extraction is small now (Table 2); with the new dry extraction techniques, both methods appear suitable, with results that are reproducible in a 10-50 ppbv range (2σ).

When comparing the results obtained by different groups, one should also keep in mind that the gas standards used are generally different, introducing an additional uncertainty in terms of absolute value. However, the good agreement existing on the pre-industrial level of atmospheric methane indicates that this uncertainty is within the range of experimental accuracies (Etheridge et al., 1992).

Future improvements of the experimental procedures should include modifications of the extraction methods in order to improve reproducibility and reduce contamination. Also, specific methods must be developed for the analysis of methane isotopes, which are used to constrain the methane cycle, with the objective of decreasing the amount of ice (about 25 kg) currently needed (Craig et al., 1988).

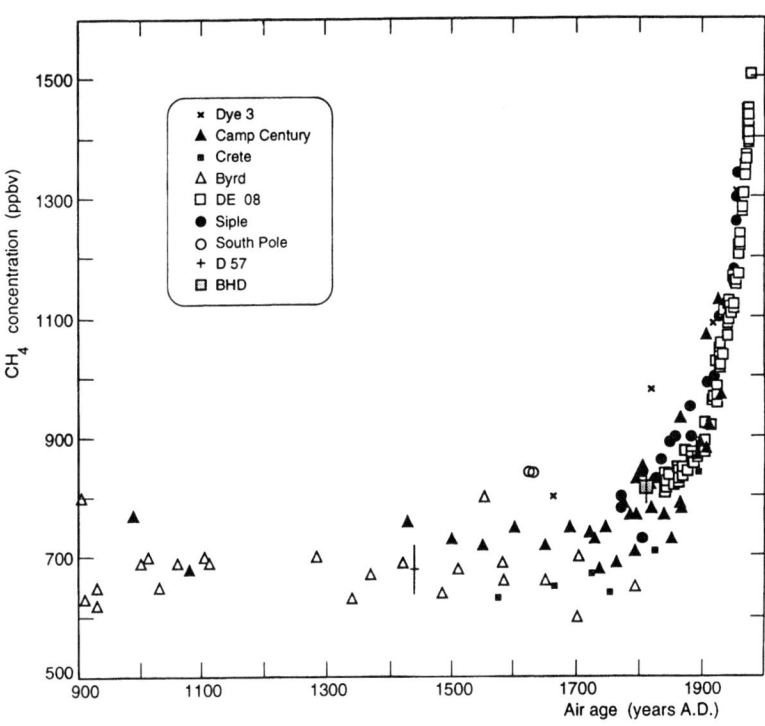

Figure 3. Atmospheric methane variations over the past 1000 years from ice core data. Greenland ice cores: Dye 3 (Craig and Chou, 1982), Camp Century and Crete (Rasmussen and Khalil, 1984). Antarctic ice cores: Byrd (Rasmussen and Khalil, 1984); DE 08 (Etheridge et al., 1992), Siple and South Pole (Stauffer et al., 1985), D 57 (Raynaud et al., 1988), BHD (Etheridge et al., 1988).

3. The industrial and pre-industrial ice record of CH_4

Apart from the historical work of Robbins et al. (1973), ice core data for CH_4 cover reliably the last few thousands years, with most details on the past two hundred years (Craig and Chou, 1982; Khalil and Rasmussen, 1982; Rasmussen and Khalil, 1984; Stauffer et al., 1985; Pearman et al., 1986; Pearman and Fraser, 1988; Etheridge et al., 1988; Raynaud et al., 1988; Tohjima et al., 1991; Nakazawa et al., 1992; Etheridge et al., 1992). A compilation of published measurements covering the past 1000 years is shown in Figure 3. These have been obtained by various groups from both Greenland and Antarctic ice cores over the past decade. From this composite profile, one may immediately note the remarkable agreement between all the results; this agreement further supports the reliability of ice cores and experimental methods for the reconstruction of past atmospheric CH_4 concentrations. Considering the temporal trend over the past 1000 years, atmospheric CH_4 remained essentially constant at a mean level of roughly 700 ppbv until at least 1700 AD. The CH_4 mixing ratio then increased progressively during the nineteenth and twentieth centuries until it reached the present-day level of roughly 1700 ppbv. The start of the increase is quite difficult to define from the published data; in Figure 4, it seems to be situated between 1750 and 1800 AD, but the scattering of results and uncertainties on the relative dating of the various ice cores used preclude a more precise estimate.

The important increase of atmospheric CH_4 over the most recent period, especially over the past century, is highly correlated with the evolution of human population, as first noted by Rasmussen and Khalil (1984). The increase results primarily from increased anthropogenic CH_4 sources, including domestic ruminants, rice cultivation, human-induced fires, landfills, and fossil fuel exploitation (Khalil and Rasmussen, 1985; Ehhalt, 1988). Although the increase of sources is thought to be primarily responsible for the increase in CH_4 concentrations, it is believed that some smaller part is due to a decreased sink resulting from enhanced anthropogenic sources of methane, carbon monoxide and other trace gases (for a recent review, see Thompson, 1992).

In addition to its unique overlap with direct atmospheric CH_4 measurements, the record from the DEO 8 ice core (Etheridge et al., 1992) provides the only estimate of changes in the growth rate of atmospheric methane since the onset of the industrial revolution (Figure 3). The observed growth rate generally increases from 1841 (oldest measurement) to 1978 (youngest measurement) except between about 1920-1945 when it stabilized. Etheridge et al. (1992) propose that a stabilization and occasional reduction in the fossil CH_4 sources (coal, gas, and oil exploitation) during World Wars I and II and the Great Depression would have induced the stabilized atmospheric growth rate, although other explanations cannot be ruled out.

The observation of possible CH_4 changes related to the climatic events of the past 1000 years (Little Ice Age, Medieval Warm Period) is currently supported only by measurements of Rasmussen and Khalil (1984). These data were re-analyzed and suggest the possibility of a drop of about 40 ppbv compared to the mean pre-industrial level, during the time of the Little Ice Age (Khalil and Rasmussen, 1989). If true, this observation implies that the natural CH_4 cycle can be influenced by climatic changes on a secular time scale. But here again, taking into account the scattering of the results and the dating uncertainties, more precise measurements are needed on well-dated ice cores to confirm this feature and to determine its temporal relationship with climatic conditions on the continents.

One final piece of information, the CH_4 interhemispheric gradient, has been detected in ice core measurements. This gradient is, as a first approximation, a function of the ratio between CH_4 sources of the northern and southern hemispheres. Therefore, it provides an important constraint on the global distribution of the natural CH_4 sources in the past. Ice cores from both Greenland and Antarctica for the period 3000 to 250 years BP show a concentration difference (north - south) of 70 ± 30 ppbv (Rasmussen and Khalil, 1984). A higher concentration in the Greenland cores is expected because natural CH_4 sources are essentially land-based and there is twice as much land in the northern hemisphere as in the southern hemisphere. New analyses are in progress that will help to constrain this important information of the natural CH_4 cycle.

4. The last climatic cycle

4.1. The ice record of CH_4. Observations of ice from both Greenland (Stauffer et al., 1988) and Antarctica (Raynaud et al., 1988) show that during the past glacial-interglacial climatic transition atmospheric concentrations of CH_4 nearly doubled, from 350 to 650 ppbv. Subsequently the CH_4 record has been documented over the complete sequence of the last climatic cycle (Chappellaz et al., 1990). Figure 4 shows strong variations of past atmospheric concentrations in the 350-650 ppbv range, well below the present mean atmospheric concentration of roughly 1700 ppbv. As for CO_2 (Figure 4), the overall correlation is remarkable ($r^2 = 0.78$) between the Vostok climate and CH_4 records. Nevertheless, the phase relationship with climate (at least as shown in measurements of the isotopic composition of the ice; see Figure 4) is, on the whole, closer than for CO_2. In particular, the following characteristics appear:

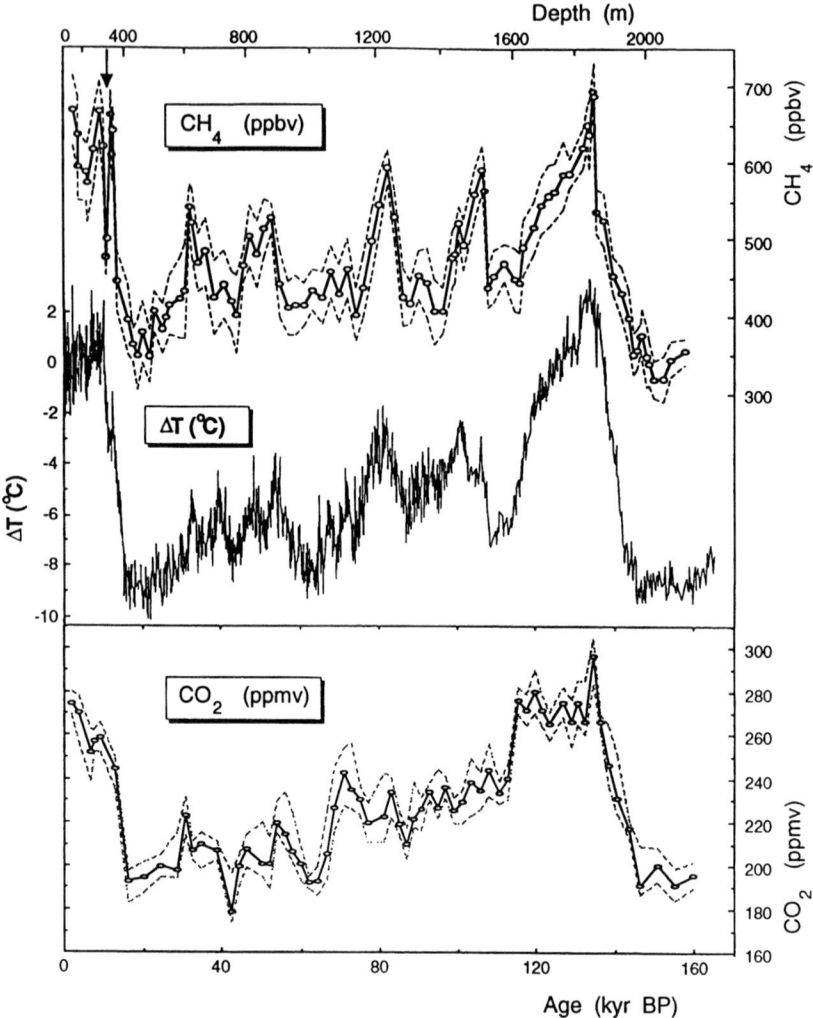

Figure 4. From Chappellaz et al., 1990.
 Top: CH_4 and surface temperature (as deduced from measurements of the ice isotopic composition) variations recorded in the Vostok ice core over the last climatic cycle. The envelope of the CH_4 curve corresponds to the measurement uncertainties. The arrow indicates the CH_4 oscillation attributed to the Younger Dryas.
 Bottom: the CO_2 record is plotted for comparison.
For details concerning the chronology used see Chappellaz et al., 1990.

(1) The decrease of CH_4 centered around 120 kyr BP in Figure 4 is roughly in phase when entering the glacial conditions shown by the climatic record of the Vostok core.

(2) The CH_4 fluctuations found during the glacial period between 100 and 20 kyr BP in Figure 4 are well defined as 4 peaks with the highest values only slightly less than the concentrations observed during the Holocene and the previous interglacial periods. This part of the record shows much greater similarities than for CO_2 with the temporal variations of Antarctic temperature.

A striking feature of the CH_4 record is the large oscillation (marked with an arrow in Figure 4), which has an amplitude of approximately 60% of the full glacial-interglacial CH_4 variation, at the end of the last glacial-interglacial transition. The CH_4 drop was possibly also recorded in the Dye 3 ice core but at only one level (Stauffer et al., 1988). This oscillation, which has been confirmed by replicate analyses (Chappellaz et al., 1990), is suspected to be associated with the large and abrupt but short-lived cooling experienced in the northern hemisphere between roughly 11-10 kyr BP known as the Younger Dryas event.

4.2. CH_4 budget during the last climatic cycle. The fundamental link between atmospheric CH_4 and climate variations during the last climatic cycle suggests a direct impact of climate on the methane budget for the past 160 kyr. Up to now, efforts to estimate the mechanisms involved in this relationship have been essentially qualitative and were concentrated on the main constituents of the natural CH_4 budget: the atmospheric sink by oxidation with OH radicals and the emissions from wetlands. The lack of direct and global evidence related to the natural CH_4 budget on the glacial-interglacial time scale, as well as the large uncertainties in the source strengths today, limit the reliability of quantitative approaches.

5. Sources

The main natural source of atmospheric methane is the decomposition of organic matter in soil wetlands. Changes in the strength of this source linked to changes in arid/humid conditions on the continents have been hypothesized as the main trigger of the glacial-interglacial variations of the atmospheric CH_4

mixing ratio (Stauffer et al., 1988; Raynaud et al., 1988; Khalil and Rasmussen, 1989; Chappellaz et al., 1990). As a large part of the present-day wetlands lie between 50° and 70°N, a direct impact of the growth and decay of continental ice sheets on the wetland extent at these latitudes was suspected. However, from the comparison between the signal and spectrum of both continental ice volume and methane, Chappellaz et al. (1990) suggested that this was not the main mechanism controlling past atmospheric CH_4 over the full climate cycle. By comparing the methane record with proxy data of the hydrological cycle in low-latitudes (associated with the monsoon intensity), they proposed a principal role of monsoon in controlling CH_4 concentration through changes of low-latitude wetland areas. A comparison of the Vostok CH_4 record with paleo-environmental indicators in north tropical Africa supports this hypothesis (Petit-Maire et al., 1991). Recently, a quantitative estimate of the terrestrial emission and consumption of methane under glacial and interglacial climate regimes has been proposed (Chappellaz et al., 1993). Based on a parameterization between wetlands, vegetation, and topography, and by relating other source elements to the vegetation distribution, they conclude that a large decrease of CH_4 emission from wetlands at the LGM time is responsible for most of the decrease in the global flux.

Other potential sources that may have participated in the CH_4 changes observed during the climatic cycle include termites, wild animals, wildfires, oceans and methane hydrates. Nisbet (1990, 1992), following an introductory study by MacDonald (1990), has proposed an original scenario to explain at least part of the CH_4 increase during the last deglaciation. This scenario involves an extra input of old CH_4 from destabilized hydrates in permafrost. A different evolution of hydrate reservoirs in the course of the climate cycle is also proposed (Paull et al., 1991). In all cases, the hydrate explanation requires large and time-limited bursts of the atmospheric CH_4 level, which are not so far revealed by the available ice core data. The questions concerning this potential contribution to the past CH_4 variability are numerous, and deeper investigations are needed, particularly through CH_4 analyses on high resolution ice cores.

6. Sinks

Warmer conditions during the interglacial periods could lead to an enhanced production of OH radicals and consequently reinforce the methane sink strength. Conversely, an increased CH_4 concentration can lead to an increased sink for OH and in turn to an even larger increase of methane. Photochemical models are the primary means available for studying changes in the CH_4 sink strength.

Several one- and two-dimensional models have attempted to simulate the oxidizing capacity of the atmosphere for the Last Glacial Maximum (McElroy, 1989; Valentin, 1990; Lu and Khalil, 1991; Pinto and Khalil, 1991; Thompson et al., 1993). A review and discussion of the results has been published by Thompson (1992). All models suggest that OH concentrations were higher than present-day levels but not much greater than OH concentrations during the pre-industrial time. This means that the lifetime of methane under glacial conditions was slightly shorter than during interglacial times, and that a small part of the glacial-interglacial doubling of the CH_4 mixing ratios is due to a decrease in the sink strength. However, the reliability of model estimates is strongly restricted by the lack of other chemical constraints from the ice cores and by uncertain assessments of the fluxes of carbon monoxide, nitrogen oxides, and non-methane hydrocarbons (Thompson, 1992).

The CH_4 consumption by bacterial activity in the soils is at present a small additional sink of atmospheric methane. It may also have contributed (in a minor way) to the glacial-interglacial changes of the CH_4 sink. The uptake rate essentially depends on the gas transport resistance and the degree of humidity of the soils (Dörr et al., 1993). Based on a parameterization of the soil uptake and the vegetation distribution, Chappellaz et al., (1993) suggest little change in the LGM soil sink compared to the pre-industrial time.

4.3. Climatic impact of the CH_4 changes over the climatic cycle. The changes of atmospheric CH_4 over the past 160,000 years appear to have been initially driven by the climatic changes. The effect of CH_4 changes on the variability of past climate is still unclear. These changes can affect the Earth's temperature in three ways: the direct radiative effect, chemical feedbacks, and climatic feedbacks.

The direct radiative effect of CH_4 results in a global surface equilibrium warming of about 0.08°C for the glacial-interglacial increase of about 300 ppbv (Chappellaz et al., 1993).

The chemical feedbacks are due to possible modifications by CH_4 of tropospheric ozone and stratospheric water vapor, two other radiatively active gases. For stratospheric water vapor, it has been estimated that the amount produced by the oxidation of increasing CH_4 would increase the direct radiative effect of atmospheric CH_4 by about 30% (Wuebbles et al., 1989). So the addition of the direct radiative effect with the effect of the chemical feedback would lead to a mean global warming of about 0.10°C in the case of the glacial-interglacial transitions. Although such an effect appears small compared to the 4-5°C warming of the deglaciation, it represents about 20% of the CO_2 contribution.

Because of the various climatic feedbacks essentially related to tropospheric water vapor, sea ice, and clouds, the climate warming due to the increase of CO_2 and CH_4 could have been larger, accounting for about half of the 4-5°C mean global warming during the glacial-interglacial transitions (Broccoli and Manabe, 1987; Rind et al., 1989; Lorius et al., 1990).

5. Summary

Ice cores are the only means available to reconstruct atmospheric concentrations of methane over long time periods. Several arguments support their overall good reliability for such purposes.

Past atmospheric CH_4 trends have been reconstructed from various ice cores. The main features revealed by these records are: (1) a relatively constant CH_4 mixing ratio around 700 ppbv during the Holocene up to approximately 200 years ago; (2) a large increase over the past 200 years up to the present-day level of roughly 1700 ppbv; (3) variations in the 350-700 ppbv range during the last glacial-interglacial cycle, remarkably correlated with climate changes.

Ice cores will continue to provide still more detailed information about atmospheric CH_4 and its trends. In particular, an effort will be made to measure the changes linked to climate variability at various time scales and to measure the interhemispheric gradient and its changes. Future work will also include isotopic measurements (^{13}C and D of CH_4), which provide an additional constraint on sources and sinks in the natural CH_4 cycle.

Acknowledgments. Thanks are due to the working group of the Oregon workshop on CH_4 records and to B. Stauffer and M.A.K. Khalil for further discussions and comments. This work was supported by CNRS Program on Environment and by the European Community.

[**Editor's Note on recent trends:** At the NATO meeting the editor had agreed to write a section on recent trends for this chapter. Delays in putting the book together, however, make it impractical to integrate the section into the main chapter, so it is being added here to provide readers with a synopsis of recent developments including the slowing in atmospheric methane trends, a finding of considerable current interest.]

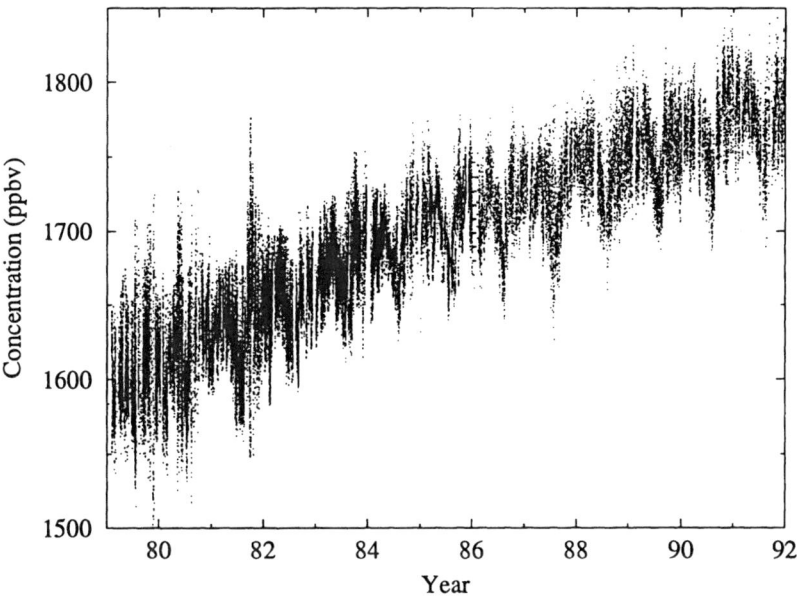

Figure A. Concentration of methane at Cape Meares, Oregon, between 1979 and 1992. Cape Meares is located in the quadrant of the earth with the largest emissions, particularly from human activities. The data show trends, seasonal variations, and the slowdown in the rate of accumulation (from Khalil, Rasmussen, and Moraes, 1993). Other features, not evident to the eye, include both shorter and longer term fluctuations, such as diurnal effects in some seasons, sub-monthly variabilities, and inter-annual cycles.

Concentrations.

The nature of the current atmospheric data is dramatically different from the ice core measurements described earlier in this chapter. Ice core measurements represent concentrations only in the polar regions and integrated over periods of a few years to hundreds of years. But the measurements go back thousands of years. In contrast, current systematic measurements extend over only the past 15 years but have been taken all over the world. There are some 28 sites in the NOAA/CMDL network and no fewer than four sites in other networks (for monthly average concentrations of methane at a six-site network, see Khalil and Shearer on methane sinks and distribution in this volume). Similarly, in time, measurements are usually taken every week, but sometimes the measurement frequency could be an hour or less. The Cape Meares record extends over 15 years and consists of more than 120,000 measurements (Figure A). As the space and time scales of the ice core data are so different from the current data, so are the applications of the two data sets in understanding the methane cycle. The current data are the foundation for understanding the effect of human activities on the global methane cycle, including present emission rates and their trends (see also Khalil and Shearer and Stauffer et al., both in this volume). Moreover, current data provide details of seasonal variations and even sub-monthly and diurnal variability, all of which are useful in validating the understanding of sources and sinks. Beyond the few-year time scale of the recent trends, the current data are not long enough to show how slower environmental changes affect atmospheric methane.

Trends.

When measurements of a trace gas are taken far from sources, it is accepted that the concentrations represent some usually unknown but large spatial scale. That the same idea applies to trends in time is not commonly recognized. When a trend is measured, it usually has an associated time scale over which the trend is expected to be valid. In longer time series, the trend changes, fluctuates, or undergoes cyclic variations over many time scales. Trends measured over short periods, therefore, do not reliably represent trends over longer periods extending either forward or backward in time from the period of the measurements. Of special interest are the longer-term changes of observed trends such as increases or decreases in the buildup of the gas in the atmosphere.

As mentioned earlier, detailed systematic measurements of methane have been taken only since 1979. Even over the relatively short period between then and now, the trends of methane have changed. The main feature of this change is the dramatic slowdown of trend between the early 1980s and recent times.

This change in global trend has been apparent for some time (Khalil and Rasmussen, 1990) and has recently been documented for the large number of sites in the NOAA/CMDL network (Steele et al., 1992). The decrease of trend over a longer period than the NOAA/CMDL measurements but at fewer sites has been reported by Khalil and Rasmussen (1993).

In Figure B we show the decline of trend as the change in the rate of increase at various sites between 1980 and the present, represented as "moving slopes" of the measured concentrations over 3-year periods. Global average trends in the first 3 years (1980-83) were about 21 ppbv/yr and in the last 3 years (1989-92) are about 11 ppbv/yr with uncertainties of about ±1 ppbv/yr. Changes of the same magnitude are seen in the Cape Meares data shown in Figure A. Smaller changes have been reported by Steele et al. (1992) between 1983 and 1990.

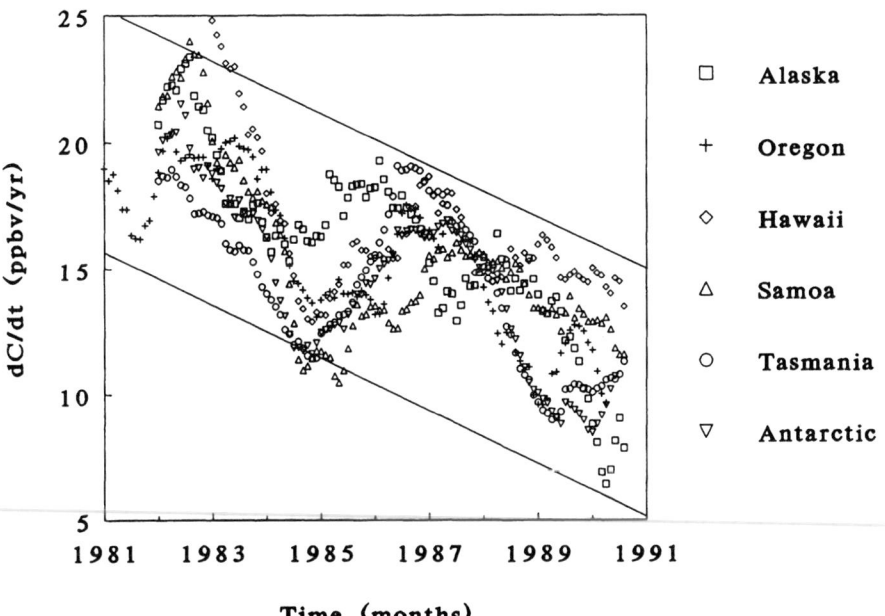

Figure B. The trends of methane over 3-year overlapping periods of time calculated by linear regression. (Adapted from Khalil and Rasmussen, 1993).

The causes of the slowdown of methane increase may come from increases of OH or a slowdown in the increases of anthropogenic emissions. There is evidence that the change of sources is perhaps the larger effect. Increases in historic anthropogenic sources such as rice agriculture and cattle have slowed, perhaps even stopped altogether during the last decade. Yet other, newer anthropogenic sources such as landfills, natural gas leakages, low-temperature coal burning, and the like may be increasing and eventually may cause methane trends to increase also. This represents a shift from agricultural sources to sources related to energy and waste disposal.

References

Barnola, J.M., D. Raynaud, Y.S. Korotkevich, C. Lorius. 1987. Vostok ice core provides 160,000-year record of atmospheric CO_2, *Nature, 329* (6138):408-414.
Barnola, J.M., P. Pimienta, D. Raynaud, Y.S. Korotkevich. 1991. CO_2-climate relationship as deduced from the Vostok ice core: a re-examination based on new measurements and on a re-evaluation of the air dating. *Tellus, 43B*:83-90.
Blake, D.R., F.S. Rowland. 1988. Continuing worldwide increase in tropospheric methane, 1978 to 1987. *Science, 239*:1129-1131.
Broccoli, A.J., S. Manabe. 1987. The influence of continental ice, atmospheric CO_2, land albedo on the climate of the Last Glacial Maximum. *Climate Dynamics, 1*:87-99.
Chappellaz, J., J.M. Barnola, D. Raynaud, Y.S. Korotkevich, C. Lorius. 1990. Ice-core record of atmospheric methane over the past 160,000 years, *Nature, 345*:127-131.
Chappellaz, J.A., I.Y. Fung, A.M. Thompson. 1993. The atmospheric CH_4 increase since the Last Glacial Maximum: 1. Source estimates. *Tellus, 45B* (in press).
Craig, H., C.C. Chou. 1982. Methane: the record in polar ice cores. *Geoph. Res. Lett., 9*:1221-1224.
Craig, H., C.C. Chou, J.A. Welhan, C.M. Stevens, A. Engelkemeir. 1988. The isotopic composition of methane in polar ice cores. *Science, 242*:1535-1539.
Dörr, H., L. Katruff, I. Levin. 1993. Soil texture parameterization of the methane uptake in aerated soils. *Chemosphere, 26* (1-4):697-713.
Ehhalt, D.H. 1988. How has the atmospheric concentration of CH_4 changed? In: *The Changing Atmosphere: Report of the Dahlem Workshop on the Changing Atmosphere* (F.S. Rowland and I.S.A. Isaksen, eds.), Berlin, J.

Wiley and Sons, New York, 25-32.
Etheridge, D.M., G.I. Pearman, F. De Silva. 1988. Atmospheric trace-gas variations as revealed by air trapped in an ice core from Law Dome, Antarctica. *Ann. Glaciol., 10*:28-33.
Etheridge, D.M., G.I. Pearman, P.J. Fraser. 1992. Changes in tropospheric methane between 1841 and 1978 from a high accumulation-rate Antarctic ice core. *Tellus, 44B*:282-294.
Khalil, M.A.K., R.A. Rasmussen. 1982. Secular trends of atmospheric methane (CH_4). *Chemosphere, 11*:877-883.
Khalil, M.A.K., R.A. Rasmussen. 1985. Causes of increasing methane: depletion of hydroxyl radicals and the rise of emissions. *Atmos. Environ., 19*:397-407.
Khalil, M.A.K., R.A. Rasmussen. 1989. Climate-induced feedbacks for the global cycles of methane and nitrous oxide. *Tellus, 41B*:554-559.
Khalil, M.A.K., R.A. Rasmussen. 1990. Atmospheric methane: recent global trends. *Environ. Sci. Technol., 24*:549-553.
Khalil, M.A.K., R.A. Rasmussen. 1993. Decreasing trend of methane: unpredictability of future concentrations. *Chemosphere, 26* (1-4):803-814.
Lorius, C., J. Jouzel, D. Raynaud, J. Hansen, H. Le Treut. 1990. The ice-core record: climate sensitivity and future greenhouse warming. *Nature, 347*:139-145.
Lu, Y., M.A.K. Khalil. 1991. Tropospheric OH: model calculations of spatial, temporal, and secular variations. *Chemosphere, 23*:397.
MacDonald, G.J. 1990. Role of methane clathrates in past and future climates, *Climatic Change, 16*:247-281.
McElroy, M.B. 1989. Studies of polar ice: insights for atmospheric chemistry. In: *The Environmental Record in Glaciers and Ice Sheets* (H. Oeschger and C.C. Langway, eds.), John Wiley and Sons, 363-377.
Nakazawa, T., T. Machida, K. Esumi, M. Tanaka, Y. Fujii, S. Aoki, O. Watanabe. 1992. Air extraction systems for ice core and CO_2 and CH_4 concentrations in the past atmosphere (abstract). *Proc. NIPR Symp. Polar Meteorol. Glaciol., 5*:178.
Nisbet, E.G. 1990. The end of the ice age. *Can. J. Earth Sci., 27*:148-157.
Nisbet, E.G. 1992. Sources of atmospheric CH_4 in early postglacial time. *J. Geophys. Res., 97*:12,859-12,867.
Paterson, W.S.B., C.U. Hammer. 1987. Ice core and other glaciological data. In: *The Geology of North America, Vol. K-3* (Ruddiman W.F. and H.E. Wright, Jr. eds.), Geological Society of America, Boulder, Colorado, 91-109.
Paull, C.K., W.III Ussler, W.P. Dillon. 1991. Is the extent of glaciation limited by marine gas-hydrates? *Geophys. Res. Lett., 18*:432-434.
Pearman, G.I., P.J. Fraser. 1988. Sources of increased methane. *Nature, 332*:489-490.

Pearman, G.I., D. Etheridge, F. De Silva, P.J. Fraser. 1986. Evidence of changing concentrations of atmospheric CO_2, N_2O and CH_4 from air bubbles in Antarctic ice. *Nature, 320*:248-250.
Petit-Maire, N., M. Fontugne, C. Rouland. 1991. Atmospheric methane ratio and environmental changes in the Sahara and Sahel during the last 130 kyrs. *Palaeogeography, Palaeoclimatology, Palaeoecology 86*:197-204.
Pinto, J.P., M.A.K. Khalil. 1991. The stability of tropospheric OH during ice ages, inter-glacial epochs and modern times. *Tellus, 43B*:347-352.
Rasmussen, R.A., M.A.K. Khalil. 1984. Atmospheric methane in the recent and ancient atmospheres: concentrations, trends and interhemispheric gradient. *J. Geoph. Res., 89*:11,599-11,605.
Rasmussen, R.A., M.A.K. Khalil. 1986. Atmospheric trace gases: trends and distributions over the last decade. *Science, 232*:1,623-1,624.
Rasmussen R.A., M.A.K. Khalil, S.D. Hoyt. 1982. Methane and carbon monoxide in snow. *J. Air. Poll. Cont. Asso., 32*:176-178.
Raynaud, D., J. Chappellaz, J.M. Barnola, Y.S. Korotkevich, C. Lorius. 1988. Climatic and CH_4 cycle implications of glacial-interglacial CH_4 change in the Vostok ice core. *Nature, 333*:655-657.
Rind, D., D. Peteet, G. Kukla. 1989. Can Milankovitch orbital variation initiate the growth of ice sheets in a General Circulation Model? *J. Geophys. Res., 41*:12,851-12,871.
Rinsland, C.P., J.S. Levine, T. Miles. 1985. Concentration of methane in the troposphere deduced from 1951 infrared solar spectra. *Nature, 330*:245-249.
Robbins, R.C., L.A. Cavanagh, L.J. Salas, E. Robinson. 1973. Analysis of ancient atmosphere. *J. Geophys. Res., 78*:5,341-5,344.
Schwander, J., B. Stauffer. 1984. Age difference between polar ice and the air trapped in its bubbles. *Nature, 311*:45-47.
Schwander, J. 1989. The transformation of snow to ice and the occlusion of gases. In: *The Environmental Record in Glaciers and Ice Sheets*, Report of the Dahlem Workshop held in Berlin 1988, March 13-18, John Wiley and Sons, Chichester, 51-67.
Schwander, J., J.M. Barnola, C. Andrié, M. Leuenberger, A. Ludin, D. Raynaud, B. Stauffer. 1993. The Age of the Air in the Firn and the Ice at Summit, Greenland. *J. Geophys. Res. 98* (D2):2,831-2,838.
Sowers, T., M. Bender, D. Raynaud. 1989. Elemental and isotopic composition of occluded O_2 and N_2 in polar ice. *J. Geophys. Res., 94*:5,137-5,150.
Stauffer, B., G. Fischer, A. Neftel, H. Oeschger. 1985. Increase of atmospheric methane in Antarctic ice core. *Science, 229*:1,386-1,388.
Stauffer, B., E. Lochbronner, H. Oeschger, J. Schwander. 1988. Methane concentrations in the glacial atmosphere was only half that of the preindustrial Holocene. *Nature, 332*:812-814.
Steele, L.P., P.J. Fraser, R.A. Rasmussen, M.A.K. Khalil, T.J. Conway, A.J.

Crawford, R.H. Gammon, K.A. Masarie, K.W. Thoning. 1987. The global distribution of methane in the troposphere. *J. Atmos. Chem.,* 5:125-171.
Steele, L.P., E.J. Dlugokencky, P.M. Lang, P.P. Tans, R.C. Martin, K.A. Masarie. 1992. Slowing down of the global accumulation of atmospheric methane during the 1980s. *Nature, 358*:313-316.
Thompson, A.M. 1992. The oxidizing capacity of the Earth's atmosphere: probable past and future changes. *Science, 256*:1,157-1,168.
Thompson, A.M., J.A. Chappellaz, I.Y. Fung, T.L. Kucsera. 1993. The atmospheric CH_4 increase since the Last Glacial Maximum: 2. Interactions with oxidants, *Tellus, 45B* (in press).
Tohjima, Y., T. Tominaga, Y. Makide, Y. Fujii. 1991. Extraction of air bubbles in Antarctic ice core samples and determination of methane concentration (abstract). *Proc. NIPR Symp. Polar Meteorol. Glaciol., 4*:134.
Valentin, K.M. 1990. Numerical modelling of the climatological and anthropogenic influences on the chemical composition of the troposphere since the Last Glacial Maximum, Ph.D. dissertation, University of Mainz, Germany, 238 p.
Wuebbles, D.J., K.E. Grant, P.S. Connell, J.E. Penner. 1989. The role of atmospheric chemistry in climatic change, *The Journal of the Air & Waste Management Association, 39* (1):22-28.
Zander, R., P. Demoulin, D.H. Ehhalt, U. Schmidt. 1989. Secular increases of the vertical abundance of methane derived from IR solar spectra recorded at the Jungfraujoch Station. *J. Geophys. Res., 94*:1,129-1,139.
Zardini, D.S. 1987. Analyse du protoxyde d'azote de l'air emprisonné dans la glace. Mise au point d'une méthode expérimentale, application à l'analyse des variations au cours du passé. Ph.D. dissertation, University of Paris VII, France, 205 p.

Chapter 4

Isotopic Abundances in the Atmosphere and Sources

C.M. STEVENS

Chemical Technology Division, Argonne National Laboratory, Argonne, IL 60439

1. Introduction

The ultimate goal of isotopic studies of atmospheric CH_4 is to contribute to the understanding of the atmospheric CH_4 cycle by determining the relative fluxes from various categories of sources and the causes of the increasing concentration (Stevens and Engelkemeir, 1988; Quay et al.,1988; Wahlen et al., 1989). Because the large number of generic anthropogenic source types makes it impossible to determine their relative strengths based on carbon-13 data alone, Stevens and Engelkemeir (1988) and Craig et al. (1988) used the isotopic data to calculate the flux of the source with the greatest uncertainty, namely biomass burning, making use of the estimated fluxes for the other sources from emission inventories. This method determined the flux and isotopic composition of the natural sources from the concentration and isotopic composition of CH_4 in old polar ice cores assuming the same lifetime as now. The lifetime was mostly determined by the fluxes based on the emissions inventories. This approach does not use the lifetime as a constraint nor contribute to the knowledge of the major sources, which have significant uncertainties in the estimates based on emissions inventories. A better approach is to start with the constraint of the lifetime value based on the methyl chloroform cycle (see Mayer et al., 1982; Khalil and Rasmussen, 1983; Prinn et al., 1987; Cicerone and Oremland, 1988). Then it is possible to calculate the fluxes of the two most isotopically different sources, providing an estimate based on emissions inventories for one of the anthropogenic sources is used as a constraint. A source is chosen that introduces the least error, namely landfills, which is one of the smallest and has an isotopic composition closest to the average.

Whiticar (this volume) discusses the overall cycles of CH_4 formation from the fundamental standpoint of the biogenic and thermogenic pathways and the isotopic fractionation effects at all stages from source generation to loss processes. The diverse and variable conditions for these fractionation effects hinder the isotopic characterization of the various generic sources and therefore the determination of a budget by this method. Examination of the measured values of the CH_4 produced by the various generic sources except fossil fuel sources (see Table 2) shows that they seem to fall in a narrow enough range to permit solving the budget with the practical approach of using these values directly, bypassing the fractionation effects of the formation pathways.

The following sections will summarize the knowledge about the flux strengths of the major anthropogenic sources using only the carbon isotopic results. Use is made of the carbon-14 results for the flux of "dead" CH_4 from fossil fuel sources as reported by others; the application of deuterium isotopic data to the atmospheric CH_4 is mentioned but not discussed in detail. The average measured isotopic compositions of the CH_4 produced by the various generic sources and released to the atmosphere are used to make up a budget whose isotopic composition is constrained to agree with the value determined from the atmospheric composition modified by the fractionation effect of the atmospheric removal process. The most recent determination of the kinetic isotope effect of the reaction $CH_4 + OH$ (Cantrell et al., 1990) is used. The results of calculations for the determination of the relative fluxes of the generic sources are shown subject to conditions for the lifetime. Measurements of the isotopic temporal trends, when considered with the constraints for the unequal distribution between hemispheres of the anthropogenic fluxes, lead to estimates of large increasing fluxes from biomass burning in the southern hemisphere and from natural fluxes in the northern hemisphere.

2. *Isotopic composition of the sources*

Table 1 lists the categories of the major sources of atmospheric CH_4 and the averages of the measured values of $\delta^{13}C$ reported by several laboratories.

Table 2 lists the measurements of $\delta^{13}C$ for the anthropogenic sources and the average values and 1 standard deviation (SD). These uncertainties were generally about ± 0.2 ‰, except in the case of rice (± 0.05). The uncertainty for the value for rice has been made greater, ± 0.2 ‰, to reflect an uncertainty from the lack of samples in the principal global rice growing areas. The uncertainty for the $\delta^{13}C$ of CH_4 from fossil fuels has been increased (± 0.4 ‰) from the indicated SD (± 0.2 ‰) of the measured values because of the very wide spread in the values and the uncertain distribution of a global emissions inventory. The non-anthropogenic source, sometimes called the "natural source," is not directly related to man's activities and consists of natural wetlands of all types, forest and savanna fires caused by lightning, oceans and lakes, any possible leakages from natural gas deposits, and termites. Natural wetlands constitute the major natural source.

Table 1. The carbon-13 isotopic composition of the sources of atmospheric CH_4. The uncertainty is estimated at ± 0.2 ‰, except (a), which is ± 0.4 ‰.

Source	$\delta^{13}C$
Natural (all non-anthropogenic)	−56.7
Rice Paddies	−64
Herbivores	−60
Landfills	−52
Natural Gas	−43 (a)
Coal Mining	−37 (a)
Biomass Burning	−22

Table 2. The carbon isotopic compositions of the anthropogenic sources of atmospheric CH_4.

Source	Method and number of samples	Mean $\delta^{13}C$ and range (‰)	Reference
RICE PADDIES			
California	Inversion (4)	-67 (-66 to -68)	Stevens and Engelkemeir, 1988
Louisiana	Flux Chamber (8)	-63.2 ± 2.9	Wahlen et al., 1989
Kenya	Flux Chamber (10)	-59.4 (-57 to -63)	Tyler et al., 1988
Vercelli, Italy	Flux Chamber (7)	-65.4 ± 1.6	
Average		-63.8 ± 1.5	Levin et al., 1993

Table 2. The carbon isotopic compositions of the anthropogenic sources of atmospheric CH_4.

Source	Method and number of samples	Mean $\delta^{13}C$ and range (‰)	Reference
RUMINANTS C_3 Diet			
Cattle	Fistula (5)	-63.7 (-61 to -76)	Rust, 1981
Cattle	Barn (1)	-61.1	Rust, 1981
Sheep	Fistula (2)	-68.6 (-67 to -70)	Rust, 1981
Cattle	Barn (4)	-71.3 ± 4	Wahlen et al., 1989
Cattle	Bag (1)	-65.1 ± 1.7	Levin et al., 1993
Sheep	Bag (1)	-70.6	Levin et al., 1993
Goat	Bag (1)	-65.2	Levin et al., 1993
Average		-6.3 ± 1.0	
RUMINANTS C_4 Diet			
Cattle	Fistula (3)	-50.3 (-47 to -51)	Rust, 1981
Cattle	Barn (1)	-45.4	Rust, 1981
Cattle (60-80% C_4 diet)	Bag (3)	-55.6 ± 1.4	Levin et al., 1993
Average		-50 ± 3	
LANDFILLS			
Indiana		-50 (-48 to -52)	Games & Hayes, 1976
Colorado	Flux chamber (2)	-53 (-51 to -55Z)	Tyler, 1986
Heidelberg	Upper layers (1)	-52	Levin et al., 1993
Average		-52 ± 2	
NATURAL GAS			
Thermogenic 20%		-38 (-25 to -52)	Schoell, 1980
Biogenic 80%		-65 (-60 to -70)	Rice & Claypool, 1981
Average		-43 ± 4(a)	
COAL MINING		-37 ± 4 (−14 to −60)	Deines, 1980
BIOMASS BURNING			
Wood fire	Plume (3)	-27 (−24 to −32)	Stevens & Engelkemeir, 1988
Grass fire	Plume (1)	-32	"
Brush fire	Plume (2)	-26.6	Wahlen et al., 1989
Wood fire	Plume (1)	-26.4	Levin et al., 1993

Table 3 lists the measurements of the isotopic composition of CH_4 emitted from or contained in the sediment gases of the various natural wetlands, such as the arctic tundra, Amazon floodplain, peatlands, Everglades marsh and temperate marshes, lakes and ponds, and estuaries. However, it would be difficult to determine the global average isotopic composition of all of the natural sources from these data because the relative emission inventories of the contributions from the various natural wetlands types, as well as other natural sources such as forest fires, cannot be easily ascertained. A direct measure of this average value can be taken from the isotopic composition of atmospheric CH_4 for a time when anthropogenic emissions were negligible, assuming the makeup of the natural sources has remained the same. Methane in the air contained in 100- to 300-year-old ice cores has $\delta^{13}C$ = -49.6 ± 0.7 ‰ (Craig et al., 1988); this leads to the average value for all the natural sources at that time of -55.7 ‰, after applying the correction $\Theta = -6.0$ for the fractionating factor $\alpha = 0.994$ (Cantrell et al., 1990) of the atmospheric loss processes, where $\Theta = (\alpha -1)(1 + 10^{-3}\delta^{13}C_{atm})$. The fractionation factor of the atmospheric loss processes is the ratio of the loss rates of $^{13}CH_4$ and $^{12}CH_4$, mainly the relative oxidation rates of the reaction of CH_4 by OH radicals. Assuming that the fluxes of all the component natural sources have not changed in the interim, the value at the present time would be -56.7 ‰ due to a small correction of -1 ‰ for the decreasing carbon-13 abundance of atmospheric CO_2 caused by contributions from fossil fuel burning (Keeling et al., 1979; Friedli et al., 1986).

Table 3. The carbon isotopic composition of CH_4 emitted from or contained in sediments of various natural wetlands.

Wetlands	Sampling Method	Mean $\delta^{13}C$ and Range (‰)	Reference
Alaska tundra	Flux chamber	-66 (-55 to -73)	Quay et al., 1988
"	Bubbles	-62 (-57 to -72)	Quay et al., 1988
Canadian tundra	Surface	-63 ± 1.9	Wahlen et al., 1989
USSR marshes	Not specified	-64 (-52 to -69)	Ovsyannikov & Lebedev, 1967
New England Lakes	Mud gases	-69 (-57 to -80)	Oona & Deevey, 1960

Table 3. The carbon isotopic composition of CH_4 emitted from or contained in sediments of various natural wetlands.

Wetlands	Sampling Method	Mean $\delta^{13}C$ and Range (‰)	Reference
New York wetlands	Surface	-58 ± 2.4	Wahlen et al., 1989
Minnesota peat bog	Inversion	-67 (-64 to -71)	Stevens & Engelkemeir, 1988
"	Flux chamber	-66 (-57 to -77)	Quay et al., 1988
"	Bubbles	-66 (-50 to -86)	Quay et al., 1988
Amazon River	Surface	-64	Tyler, 1986
Kenya papyrus swamp	Flux chamber	-51 (-31 to -62)	Tyler et al., 1988
Kenya river	Flux chamber	-54	Tyler et al., 1988
Kenya lake	Flux chamber	-48 (-44 to -50)	Tyler et al., 1988
Amazon flood plain	Bubbles, 1985	-62 (-47 to -73)	Quay et al., 1988
"	Bubbles, 1987	-52 (-42 to -60)	"
"	Dissolved in lakes	-56 (-41 to -66)	"
"	Flux chamber	-51 (-42 to -73)	"
"	Inversion	-52 (-49 to -56)	"
Florida Everglades	Inversion	-55 (-53 to -58)	Stevens & Engelkemeir, 1988
"	Sediment bubbles	-65 (-63 to -70)	Chanton et al., 1988
Florida, Crescent Lake	Flux chamber	-64 (-79 to -71)	Burke & Sackett, 1986
Florida, Mirror Lake	"	-55 (-52 to -56)	"
Florida, Lake Dias	"	-63 (-60 to -67)	"
Miss. River Delta	"	-60 (-59 to -60)	"
South Carolina pond	"	-53 (-51 to -55)	"

Table 3. The carbon isotopic composition of CH_4 emitted from or contained in sediments of various natural wetlands.

Wetlands	Sampling Method	Mean $\delta^{13}C$ and Range (‰)	Reference
Florida, Tampa Bay estuary	"	-66 (-63 to -71)	"
Colorado pond	"	-53 (-52 to -54)	Tyler, 1986
North Carolina marine basin	Bubbles	-60 + 2 (summer)	Chanton & Martens, 1988
"	Flux chamber	-67 (-66 to -70)	"
Illinois slough	Flux chamber	-50 (-49 to -51)	Stevens & Engelkemeir, 1988
"	Sediment gases	-56 (-55 to -58)	"
Swamp, Heidelberg, Germany	Flux chamber	-57 ± 2	Levin et al., 1993

The major natural source is the emission from the natural wetlands. The isotopic composition of the northern wetlands, including the regions of the arctic tundra and peatlands and some of the Amazon flood plains and Florida Everglades, falls in the range −60 to −70 ‰ with an average of about −65 ‰ (Table 3). More enriched values of CH_4, up to −50 ‰, are observed for some northern mid-latitude lakes and ponds and some regions of the Amazon and the Florida Everglades. It is likely that the difference between −57 ‰, the average for all natural sources derived from the atmospheric CH_4 in old polar ice cores, and −65 ‰, for the major natural wetlands, is at least partially accounted for by isotopically heavy CH_4 emissions from natural forest and savanna fires.

The value of −64 ‰ shown for rice is the average of the measurements by Stevens and Engelkemeir (1988), Tyler et al. (1988), Wahlen et al. (1989), and Levin et al. (1993) and is very close to the average value for CH_4 from the majority of natural wetlands. For CH_4 from herbivores the value −60 ‰ is used, which is based on the comprehensive study by Rust (1981). The values of −43 ‰ for natural gas and −37 ‰ for coal mining are based on the means of the extensive compilations of Schoell (1980), Deines (1980), and Rice and Claypool

(1981). These values are averages of available data and not weighted for the geographic flux distributions except for an estimated ratio of thermogenic and biogenic gas sources. For CH_4 from biomass burning a value of $\delta^{13}C = -22$ ‰ is used. This value is the weighted average of 80% due to C_3 plants and trees with an average isotopic composition of -25 ‰ (Craig, 1953) and 20% due to C_4 plants of tropical savanna fires with $\delta^{13}C = -12$ ‰. The measured values show little fractionation by the burning process from the isotopic composition of the biomass (Stevens and Engelkemeir, 1988; Wahlen et al., 1989). In order to reduce the number of independent variables, the CH_4 from rice paddies and herbivores are combined as a single generic source with an average $\delta^{13}C$ of -62 ‰, based on the emissions inventories of the relative fluxes. The fluxes from natural gas and coal mining can be combined into one generic fossil fuel source with an average value of -41 ‰, since the isotopic compositions are similar and both contain 100% "dead" carbon, i.e., no carbon-14.

3. Isotopic composition of atmospheric methane

Measurements of $\delta^{13}C$ of atmospheric CH_4 that have been done by various groups are listed in Table 4 and plotted in Figure 1. They agree well enough to determine the average value of the sources but not well enough to compare with each other for temporal trends or latitude gradients. There may be small real differences among these values due to different locations of sample collection, but the main differences are most likely caused by differences in sample preparation and calibration of the working isotopic standards used in the mass spectrometric analysis. Accurate calibration is a difficulty inherent in trying to establish to ± 0.1 ‰ the absolute accuracy of samples that differ by the large amount of -47 ‰ from the Peedee belemnite carbonate used as the common reference.

Table 4. Comparison of measurements of the carbon-13 isotopic composition of atmospheric CH_4; errors are 1 SD with numbers of samples in parentheses.

Laboratory	Year	$\delta^{13}C$ ‰ Southern Hemisphere	$\delta^{13}C$ ‰ Northern Hemisphere
A	1978.5	-47.84 ± .05 (4)	
"	1978.8		-47.88 ± .14 (7)
"	1980.5	-47.52 ± .09 (3)	-47.78 ± .08 (5)
"	1981.5	-47.49 ± .09 (3)	-47.69 ± .02 (7)
"	1982.5	-47.35 ± .04 (4)	-47.74 ± .26 (4)
"	1983.25		-47.60 ± .05 (20)
"	1983.5		-47.61 ± .05 (7)
"	1983.75		-47.61 ± .04 (20)
"	1984.2	-47.15 ± .15 (1)	
"	1984.5		-47.53 ± .10 (5)
"	1985.5		-47.66 ± .08 (5)
"	1986.25		-47.28 ± .06 (11)
"	1986.75		-47.25 ± .06 (10)
"	1986.9	-46.71 ± .08 (6)	
"	1987.5		-47.21 ± .10 (6)
"	1988.25		-46.98 ± .05 (15)
"	1988.3	-46.49 ± .09 (6)	
"	1989.25		-47.00 ± .10 (7)
"	1989.3	-46.32 ± .10 (2)	
B	1976.5		-47.3 ± .2 (1)
"	1977.5		-46.9 ± .2 (1)
"	1978.5		-46.5 ± .2 (1)
"	1986.5		-46.7 ± .3 (27)
"	1986.8	-46.0 ± .2 (5)	
"	1987.5	-45.2 ± .3 (2)	-46.5 ± .2 (11)
"	1988.0	-46.6 ± .3 (4)	
C	1987 to	-47.04 ± .2 (5)	-47.35 to
"	1989		-47.20 ± .13 (208)
D	1985.5		-46.5
E	1989.6 to 1991.8	-47.14 ± .03 (90)	

* A: Stevens, 1988. B: Wahlen et al., 1989. C: Quay et al., 1988. D: Tyler, 1986. E: Lassey et al., 1993.

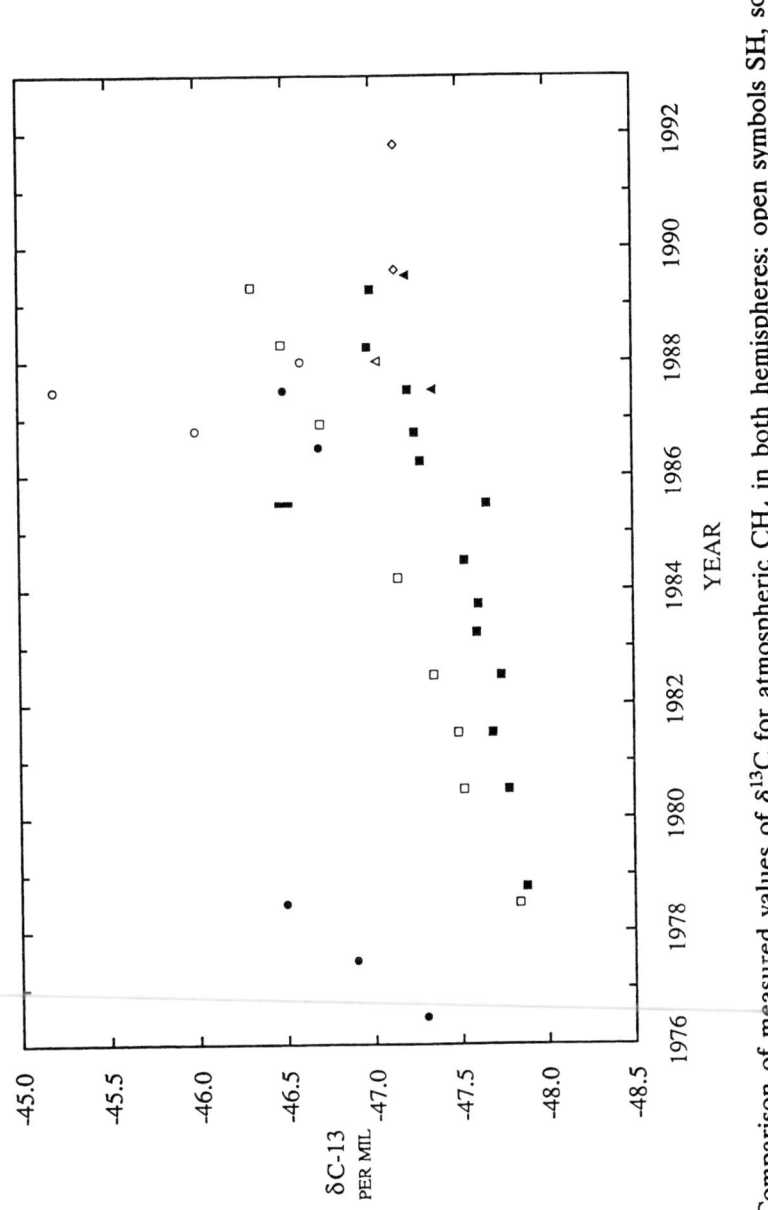

Figure 1. Comparison of measured values of $\delta^{13}C$ for atmospheric CH_4 in both hemispheres; open symbols SH, solid symbols NH, squares lab A, circles lab B, triangles lab C, upside down triangles lab D, diamonds lab E; Laboratory reference same as in Table 3. Error bars are 1 SD; numbers of samples are in parentheses. For location and source of samples see references cited in Table 4.

4. Calculation of source budgets

The overall objective is to determine the relative, if not absolute, fluxes of the five generic types of sources of atmospheric CH_4. There are only three constraints based on measurable quantities of the overall CH_4 cycle; two of them are the two mass balance equations for $^{12}CH_4$ and $^{13}CH_4$ for the current (1978) condition of atmospheric CH_4.

$$dC/dt = \Sigma F_i - C/\tau \tag{1}$$

$$d(C\delta_A)/dt = \Sigma F_i\delta_i - C(\delta_a + \Theta)/\tau \tag{2}$$

where C is the atmospheric burden in teragrams, F_i the individual fluxes, δ_i the isotopic composition of the sources, δ_A the average atmospheric isotopic composition, τ the lifetime and Θ the average fractionation factor of the atmospheric loss processes (-6.0 ‰). A third constraint is the abundance of $^{14}CH_4$ in atmospheric CH_4, which determines the fraction of CH_4 from fossil fuel sources at 16 to 21% of the total flux in 1987 (Wahlen et al., 1989; Quay et al., 1991; Lowe et al., 1991). The reference year for these calculations is 1978; back correcting for the growth rates of 2.75 %/yr of both global natural gas and coal production (see Figure 2) exceeding the concentration growth by 1.75 %/yr, the flux from these combined sources in 1978 would have been 13 to 17 % of the total flux or 60 to 80 Tg/yr if the total flux was 460 Tg/yr for a lifetime of 10 years. The mass balance equations for $^{12}CH_4$ and $^{13}CH_4$ 100 to 300 years ago, based on measurements of CH_4 in old polar ice cores when the concentration was 45% of that in 1978 and representing mainly natural sources (Craig and Chou, 1982; Craig et al., 1988), are important for determining the average isotopic composition of the natural sources, $\delta^{13}C = -56.7$ ‰. The $^{12}CH_4$ mass balance might seem to supply another constraint for the natural flux, but the lifetime at that time is another unknown, making for no net change in the number of degrees of freedom. Using the constraints cited above, the solution for the unknown fluxes is under-determined with an excess of three degrees of freedom, the unknown lifetime for the ice-core CH_4 and the fluxes of the following sources: combined rice and herbivores, landfills, and biomass burning. Assuming the lifetime at the time the methane was trapped in the polar ice was the same as now, then the fluxes of two of the sources can be expressed in terms of a third. This is as far as the calculation of a budget can be done based on carbon isotopic data alone. To proceed further the emissions inventories estimate for one of the sources is used. Using the estimate for landfills introduces the least uncertainty. This estimate is 50 ± 20 Tg/yr (Bingemer and Crutzen, 1987). On this basis Table 5 lists the fluxes calculated for the fluxes of all the sources. These results show that for a lifetime of ten years the calculated combined fluxes from rice and herbivores of 110 Tg/yr disagrees by a large amount with the emissions inventory estimate of 180 ± 60 Tg/yr. Also, the values for the natural fluxes based on the CH_4 in polar ice cores are much greater than the emissions inventory estimate by Matthews and Fung (1987). With a lifetime of 8 years, the discrepancy between the calculated values and the emissions inventory estimate becomes smaller for the combined rice and cattle flux and much larger for the natural fluxes. Figure 3 shows the calculated flux from the combined rice and herbivores versus the ratio of the lifetime in 1978 to the lifetime in pre-industrial times for lifetimes of 8 and 10 years in 1978. If the emissions inventory estimates for the combined fluxes of rice and herbivores are correct, then the lifetime has been decreasing since pre-industrial times by as much as 33%. The existing uncertainties for the lifetime as well as for the emissions inventory estimates of fluxes from rice and herbivores are the limiting factors in determining the natural fluxes, or the lifetime in pre-industrial times. There is other evidence that the concentration of tropospheric ozone, the precursor of OH radicals that scavenge CH_4, has been increasing in recent decades as well as in the past century (Logan, 1985; Hough and Derwent, 1990). In the following section on isotopic trends, other evidence is presented that shows that the sink rate in the northern hemisphere has been increasing in the past decade.

The flux from biomass burning that is calculated from the isotopic data in the above treatment falls in the narrow range of 37 to 50 Tg/yr if the flux from rice and herbivores is in the range of 110 to 178 Tg/yr. This is illustrated in Figure 4, which gives the fluxes from rice and herbivores versus the flux from biomass burning for a lifetime of ten years. A higher upper limit for the latter would seem unlikely since this would lead to unreasonably small fluxes for the natural sources.

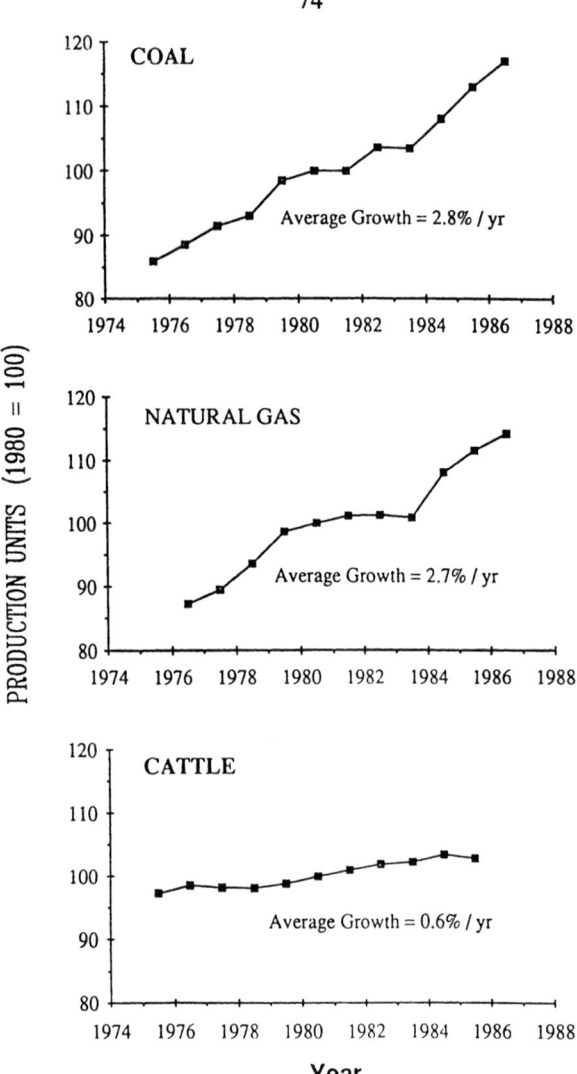

Figure 2. The growth rate for the global production of hard coal, natural gas, and cattle. Production units are based on setting the values for the year 1980 equal to 100 (from United Nations Statistical Yearbooks).

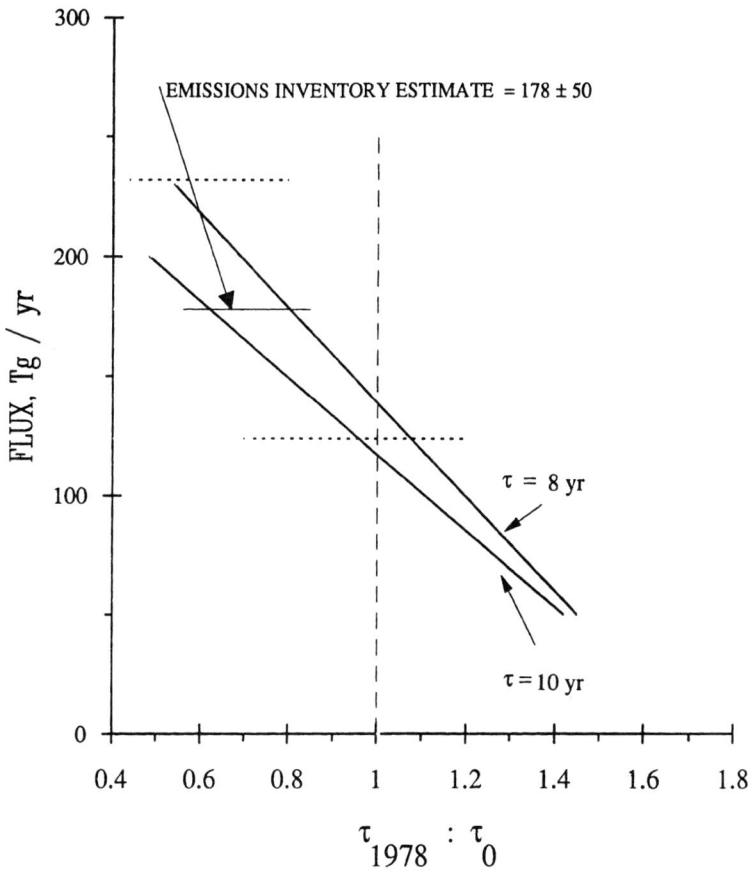

Figure 3. The calculated combined fluxes from rice and herbivores versus the ratio of the lifetime in 1978 to the lifetime in pre-industrial times for lifetimes of 8 and 10 years in 1978.

Table 5. The fluxes of the sources of atmospheric CH_4 in 1978, assuming a constant lifetime over the age of the ice cores in which the concentration and isotopic composition were measured and based on the flux from landfills of 50 Tg/yr.

Source	% of total	Flux Tg/yr		
		10-year lifetime	8-year lifetime	Estimate by emissions inventory
Natural	41	189 (a)	236 (a)	100 (b)
Rice + Herbivores	24	110	146	178 ± 54 (c) ranges
Landfills	10.8	50	50	50 ± 20 (d) ranges
Fossil Fuel	17 ± 3	78	85	60 (e)
Biomass Burning	7.5	35	50	15 to 71 (f)

a. Lifetime assumed constant. b. Matthews and Fung, 1987. c. Schütz et al., 1989; Khalil and Rasmussen, 1991; Crutzen et al., 1986. d. Bingemer and Crutzen, 1987. e. Holzapfel-Pschorn and Seiler, 1986; Hitchcock and Wechsler, 1972. f. Crutzen and Andreae, 1990.

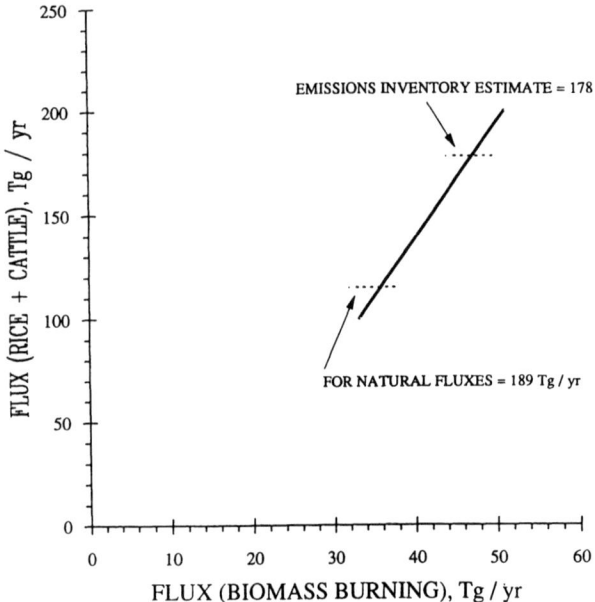

Figure 4. The calculated relationship between the combined fluxes from rice and herbivores and the fluxes from biomass burning for a lifetime of 10 years and the fluxes of the natural sources undetermined.

The isotopic abundance of deuterium in atmospheric CH_4 and its sources provide an additional constraint for a global CH_4 budget as shown by Wahlen et al. (1990). Their measurements show the global average δD for atmospheric CH_4 is -82 ‰, while δD for the biogenic CH_4 from the major wet environments sources is -290 to -360 ‰. The deuterium data provide additional constraints for calculating a budget of fluxes: first, the δD of the CH_4 from wet environments correlates with the δD of the local precipitation, which is latitude-dependent. This makes it possible to distinguish CH_4 from tropical and arctic wetlands; secondly, the δD of fossil fuel CH_4 has a different relative composition compared to the value for CH_4 from wetlands than in the case for the carbon isotopes. The fractionation factor of $(\alpha -1) = -0.330$ measured by Gordon and Mulac (1975) for the oxidation of CH_3D/CH_4 by OH radicals seems too large, leading to an average value for the sources more depleted than the value of the most depleted source. Wahlen et al. (1990) derives a value of -0.150 to -0.170 based on the enrichment of CH_3D in the stratosphere where the concentration decreases with altitude because of OH oxidation.

5. Temporal trends of the $\delta^{13}C$ ratio in atmospheric methane

The ice core data showed that the isotopic composition remained constant for 200 years while the concentration had increased from 0.65 to 1.2 ppm 3 to 5 decades ago (Craig et al., 1988). The 2 ‰ enrichment since then is attributed mainly to the heavy CH_4 from biomass burning and fossil fuel sources that must be increasing more rapidly than the increase in light CH_4 from rice and cattle.

Figure 5 shows the temporal trend of the carbon-13 isotopic composition of atmospheric CH_4 from 1978 to 1989 in both hemispheres. These plots are an update of earlier data reported by Stevens et al. (1985) and Stevens (1988). These data consist of analyses of 29 samples in the southern hemisphere (SH) and 129 in the northern hemisphere (NH), which are presented in this plot as annual averages. The slopes of these trends are greater by 0.04 ‰/yr than shown in an earlier publication of the same data up to 1987 (Stevens, 1988) because of a correction factor applied to the results of analyses done in 1983 of samples collected as far back as 1978 and stored. The oldest of these samples from 1978 collections were analyzed again five years later in 1988 and showed a small increase of $\delta^{13}C$ compared to the 1983 value, averaging 0.2 ‰, as well as slightly higher concentrations. The differences are interpreted as contamination

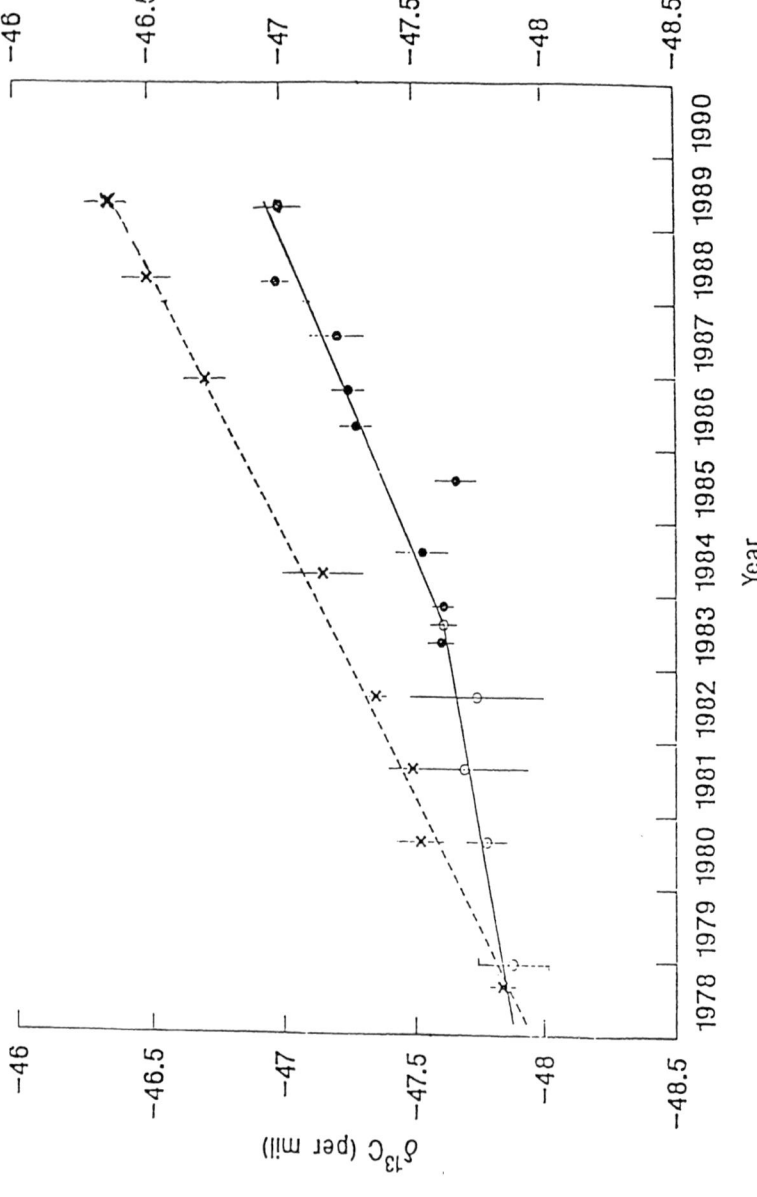

Figure 5. The stable carbon isotopic composition of atmospheric methane for the northern hemisphere (solid line) and southern hemisphere (dashed line) from 1978 to 1989. The points are the average values of annual or semi-annual sets of 4 to 15 measurements.

with isotopically heavier CH_4 from the walls of the storage cylinders accumulating over the 5-year interval between analyses, amounting to a correction of -0.04 ‰ per year of the interval between collection and analysis of the stored air samples. This correction makes only a small change in the calculated trends of the isotopic composition of the sources in each hemisphere because the principal term in this calculation is the rate of change of the difference between the atmospheric $\delta^{13}C$ trends, which is not altered by this correction (see equations 3 and 4 below). The correction increases the slopes of the trends in both hemispheres somewhat, but the difference between them remains the same.

The results in Table 4 and Figure 1 show that laboratory B measured a difference in the isotopic values between hemispheres of 0.7 ± 0.3 ‰ in 1986-87 versus 0.5 ±0.2 ‰ for laboratory A for the same years. The results of Lassey et al. (1993) (E) showed no change in the $\delta^{13}C$ of the SH from late 1989 to late 1991 with much smaller uncertainties and many more analyses per year (40 to 50) than any other group. This does not necessarily conflict with the increasing trend observed by lab A since the two studies do not overlap, and it is possible that there could have been a change in the trend as happened after 1983 in the NH data of laboratory A in Figure 5. A decrease in the trend of $\delta^{13}C$ in the SH would result if there was a significant reduction in biomass burning in the Amazon in recent years.

During the decade 1978-88 the trend of $\delta^{13}C$ for atmospheric CH_4 was 0.14 ‰/yr in the SH and averaged 0.08 ‰/yr in the NH; there was a change in the slope in the NH after 1983 from 0.054 ±0.017 to 0.112 ±0.008 ‰/yr. The difference of 0.058 ±0.019 ‰/yr between these trends is more than three times the standard deviation.

The trends of the average isotopic composition of the source fluxes in each hemisphere can be derived using a two box model of the atmosphere, with the mass balance equations for $^{13}CH_4$ and $^{12}CH_4$ taking into account the interhemispheric exchange. They are

$$d\delta_{NS}/dt = d\delta_{NA}/dt - \frac{\eta_{ex}(C_S/C_N)[d(\delta_{SA}-\delta_{NA})/dt]}{\eta + dC/Cdt + \eta_{ex}C_N/C_S - 1]} \qquad (3)$$

$$d\delta_{SS}/dt = d\delta_{SA}/dt + \frac{\eta_{ex}(C_N/C_S)[d(\delta_{SA} - \delta_{NA})/dt]}{\eta + dC/Cdt - \eta_{ex}[1 - C_S/C_N]} \quad (4)$$

where δ_{NA} and δ_{SA} are $\delta^{13}C$ of atmospheric CH_4 in the NH and SH respectively, δ_{SS} and δ_{NS} are $\delta^{13}C$ of the average CH_4 sources, and C_N and C_S are the concentrations in the respective hemispheres, η and η_{ex} are the loss rate and interhemispheric exchange rate. The $^{13}C/^{12}C$ ratio of the sources in the SH increased at an average rate of 0.63 ‰/yr for the decade (Figure 6). In that hemisphere, biomass burning is the major anthropogenic source; rice and cattle production are estimated at only 20 and 25%, respectively (Khalil and Rasmussen, 1983), of the global production, while losses from natural gas systems, coal mining, and landfills are negligible. The increasing trend can most plausibly be attributed to the increasing fluxes from the isotopically very heavy CH_4 from biomass burning associated with the rapid deforestation in that hemisphere in recent decades. The magnitude of the isotopic trend gives a quantitative measure of the rate of increase of biomass burning in the southern hemisphere from the relationship

$$dF_{BB}/dt = \frac{F_{S.H.}[d(\delta_{S.H.})/dt]}{(\delta_{BB} - \delta_{AVG})} \quad (5)$$
$$= 3.5 \text{ Tg/yr}^2$$

where F_{BB} is the flux from biomass burning in the southern hemisphere, $F_{S.H.}$ the total flux in the southern hemisphere of ca. 180 Tg/yr, and δ_{BB}, the isotopic composition for CH_4 from biomass burning, -22 ‰. This rate would be 0.2 Tg/yr^2 greater taking into account the increasing rates (1 to 2%/yr) of the small fluxes of light CH_4 from rice and cattle in this hemisphere. Since the increasing global concentration of atmospheric CH_4 is 1 %/yr (Blake and Rowland, 1988; Khalil and Rasmussen, 1990) and the global flux is about 500 Tg/yr, the increasing fluxes from biomass burning in the southern hemisphere account for 50% of the increasing concentration and must be its leading cause, assuming the other 50% is due to increasing fluxes of all the anthropogenic sources in the northern hemisphere and the sink rate and fluxes of the natural sources are constant. Analysis of the trends in the northern hemisphere will show that both the fluxes of the natural sources and the loss rate are probably increasing;

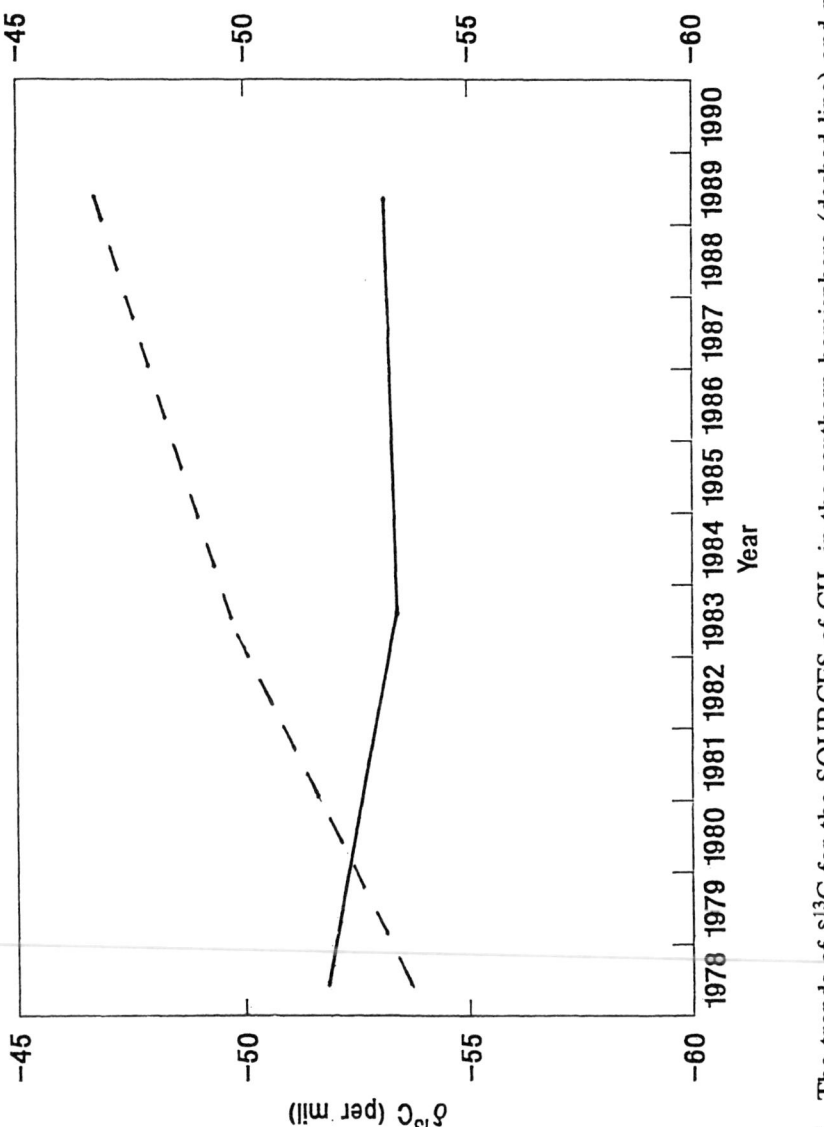

Figure 6. The trends of $\delta^{13}C$ for the SOURCES of CH_4 in the southern hemisphere (dashed line) and northern hemisphere (solid line) from 1978 to 1988, calculated for a lifetime of 10 years.

however, biomass burning in the southern hemisphere is the major anthropogenic source contributing to the increase in the concentration.

The trend of the $\delta^{13}C$ of the sources in the northern hemisphere averaged -0.17 ± 0.03 ‰/yr over the decade. The slope was -0.36 ‰/yr until 1983, when it changed to +0.02 ‰/yr. Table 6 lists the calculated average rate of change for the decade based on estimates of the rates of growth of the isotopically light and heavy CH_4 fluxes of the anthropogenic sources in this hemisphere as +0.02 to +0.08 ‰/yr; the measured trend is significantly different by -0.19 to -0.25 ‰/yr. The disparity with the average measured trend is too large to be explained by uncertainties in the factors involved in the calculation, namely, rates of growth and fluxes. The difference implies that there was a much greater increase in the fluxes of isotopically light CH_4 than can be reasonably accounted for by the increases in the production of rice and cattle, the only light anthropogenic sources. A plausible explanation is that the fluxes of the natural sources were changing (i.e., some combination of increasing light CH_4 or decreasing heavy CH_4 fluxes). The former possibility seems more likely because the temperate zone wetlands in the subarctic regions of the NH account for a major fraction of the global natural sources (Matthews and Fung, 1987) and produce light CH_4 (Quay et al., 1988). There was an increase in global temperature from 1975 to 1982 that could have caused increases in fluxes from these wetlands (Harriss, 1989), as well as increasing fluxes from rice paddies that were not accounted for by increasing acreage. The change in slope for the NH in 1983 coincides with a decrease in the slope of the global concentration trend after 1983 (Khalil and Rasmussen, 1990); thus, for the two phenomena to be caused by the same event would require a relatively rapid change over a few years of either a decrease of isotopically light fluxes or an increase in the loss rate. Based on the ratio of the change in slope of the isotopic trend to the change in slope of the concentration trend, the former possibility is more likely. This is further supported by the evidence of a pronounced cooling of 1° C during 1984-85 for the northern temperate zone (Angell, personal communication), which might have caused decreased fluxes from these northern wetlands and the change in the slope of the isotopic trend observed after 1983. The trend of the isotopic composition of atmospheric CH_4 might be a simple indicator of average temperature trends in the regions of the northern temperate zone wetlands. There is evidence that the arctic regions have undergone amplified warming (Lachenbruch and Marshall,

1986), which is predicted in models of climate change. Continuous monitoring of the isotopic trends in both hemispheres over the coming decades might be a means of indicating these changes. The results of the trend data show increasing fluxes of biomass burning in the SH of 3.5 Tg/yr and increasing light CH_4 fluxes in the NH of 4.2 Tg/yr. Adding these to the increasing fluxes from the other anthropogenic sources from increasing production of rice (1.0 Tg/yr), cattle (0.4 Tg/yr), and fossil fuel (1.7 Tg/yr) gives a total of 10.8 to 15.1 Tg/yr, or two to three times the total based on the increasing concentration of 1 %/yr for an annual flux of 460 Tg/yr in 1978. From this it is deduced that the loss rate must be increasing by about 1 to 2 %/yr.

Table 6. The calculated contribution of the increasing fluxes of the anthropogenic sources to the trend of the average isotopic composition of the sources for the northern hemisphere from 1978 to 1989 and comparison with the measured change.

Source	Flux (Tg/yr)	Growth Rate (%/yr)	$(\delta^{13}C_i - \delta^{13}C_{avg})$ (‰)	$d(\delta^{13}C)/dt$ (‰/yr)
Rice	80	0.5 to 1.0?	−12	−0.016 to −0.032?
Cattle	68	0.6	− 8	
Landfills	50	2.0 (a)	+ 2	−0.009
Fossil Fuel	69	2.75	+12	+0.003
Biomass Burning	20 (b)	0 to 2 (a)	+22	+0.06 +0 to 0.04

Calculated Total = +0.02 to +0.08 ‰/yr   Measured rate = -0.17 ‰/yr

DIFFERENCE = -0.19 to -0.25 ‰/yr

The indicated difference corresponds to 4 to 6 Tg/yr^2 in the rate of increase of a flux of CH_4 having $\delta^{13}C$ = -65 ‰.
a. Assumed rate equals population growth.
b. Flux in the northern hemisphere assumed to be half of the lower limit of the range calculated for biomass burning shown in Figure 3, 40 Tg/yr.

6. Summary

Carbon isotopic data have been used to elucidate three features of the atmospheric CH_4 cycle: (1) the magnitude of the fluxes from the combined rice and cattle source and from biomass burning, (2) annual rates of change for the fluxes of biomass burning in the SH and natural wetlands in the NH, and (3) indirect evidence of changes in the loss rate.

The sources of atmospheric CH_4 have been divided into five categories (Table 5) based on a combination of similar isotopic composition or source characteristics. The fraction of the annual flux due to fossil fuel sources can be determined from carbon-14 measurements. The flux of the natural source is based on the concentration of CH_4 in pre-industrial times as measured in polar ice cores with the caveat of possible changes in the lifetime. Taking the flux of the less important source of landfills based on emissions inventory data, then the carbon-13 data are limited to the determination of the ratio of the fluxes of the two remaining and isotopically different sources, namely those of heavy CH_4 from biomass burning and light CH_4 from the combined sources of rice paddies and ruminants. This ratio is subject to the possibility of changes in the lifetime (increasing loss rate) as shown in Figures 3 and 4.

The different isotopic trends for the two hemispheres are analyzed with the following conclusions: The isotopic composition of the overall sources in the southern hemisphere is becoming heavier and is most likely due to heavy CH_4 from the only important anthropogenic source in this hemisphere, burning of biomass, which has been increasing rapidly in recent decades in this hemisphere. The northern hemisphere showed an increasing net depletion of $^{13}CH_4$ of the source CH_4 during the past decade after taking account of increasing anthropogenic fluxes, as well as an abrupt increase of $^{13}CH_4$ in 1984-85. These changes are interpreted as caused by changing emissions of the isotopically light CH_4 from the natural northern wetlands as well as rice paddies due to climate changes.

Finally, there are indications from both the budget calculations and trend results that the sink rate has been increasing. Because of these important findings, especially the possibility of the natural fluxes changing with climate change, there should be continuing measurements of the isotopic trends in both hemispheres. It would be best to have these analyses for both hemispheres done by the same laboratory using the same processing techniques and isotopic measurement standards in order to avoid the difficult calibration problems between different laboratories.

Acknowledgements. The preparation of the graphics was by Carter Lewis. Work performed under the auspices of the Interdisciplinary Research Program in Earth Sciences, National Aeronautics and Space Administration, Order No. W-16188, and the Office of Basic Energy Sciences, Division of Mathematical and Geosciences, U. S. Department of Energy, under Contract No. W-31-109-Eng-38, and the U.S. Environmental Protection Agency through Interagency Agreement, Reference No. DW 89934989-01-0.

References

Bingemer, H.G., P.J. Crutzen. 1987. The production of methane from solid wastes. *J. Geophys. Res.,* 92:2,181-2,187.
Blake, D.R., F.S. Rowland. 1988. Continuing worldwide increase in tropospheric methane. *Science, 239*:1,129-1,131.
Burke, R.A., W.M. Sackett. 1986. Stable hydrogen and carbon isotopic compositions of biogenic methane from several shallow aquatic environments. In: *Organic Marine Geochemistry* (M.L. Sohn, ed.,) American Chemical Society, Washington, D.C., p. 297.
Cantrell, C.A., R.E. Shetter, A.H. McDaniel, J.G. Calvert, J.A. Davidson, D.C. Lowe, S.C. Tyler, R.J. Cicerone, J.P. Greenberg. 1990. Carbon kinetic isotope effect in the oxidation of methane by hydroxyl radicals. *J. Geophys. Res.,* 95:22,455-22,462.
Chanton, J.P., C.S. Martens. 1988. Seasonal variations in ebullitive flux and carbon isotopic composition of methane in a tidal freshwater estuary. *Global Biogeochem. Cycles,* 2:289.
Chanton, J.P., G.G. Pauly, C.S. Martens, N.E. Blair. 1988. Carbon isotopic composition of methane in Florida Everglades soils and fractionation during its transport to the troposphere. *Global Biogeochem. Cycles,* 2:245.
Cicerone, R.J., R.S. Oremland. 1988. Biogeochemical aspects of atmospheric methane. *Global Biogeochemical Cycles,* 2:299-327.
Craig, H. 1953. The geochemistry of the stable carbon isotopes. *Geochim. Cosmochim. Acta,* 3:53-92.
Craig, H., C.C. Chou. 1982. Methane: The record in polar ice cores. *Geophys. Res. Lett.,* 9:1,212-1,224.
Craig, H., C.C. Chou, C.M. Stevens, A. Engelkemeir. 1988. Isotopic composition of methane in polar ice cores. *Science, 242*:1,535-1,539.
Crutzen, P.J., M.O. Andreae. 1990. Biomass burning in the tropics: Impact on atmospheric chemistry and biogeochemical cycles. *Science, 250*:1,669.
Crutzen, P.J., I. Aselmann, W. Seiler. 1986. Methane production by domestic animals, wild ruminants, other herbivorous fauna, and humans. *Tellus, 38B*:271-284.

Deines, P. 1980. The isotopic composition of reduced organic carbon. In: *Handbook of Environmental Isotope Geochemistry, Vol. 1* (P. Fritz and J.C. Fontes, eds.), Elsevier Scientific, Chapter 9, p. 329-406.

Friedli, H., H. Lotscher, H. Oeschger, U. Siegenthaler, B. Stauffer. 1986. Ice core record of the $^{13}C/^{12}C$ ratio of Atmospheric CO_2 in the past two centuries. *Nature, 324*:237.

Games, L.M., J.M. Hayes. 1976. On the mechanisms of CO_2 and CH_4 production in natural anaerobic environments. In: *Proc. of the 2nd International Conference on Environmental Biogeochemistry, Vol. 1* (J.O. Nriague, ed.), Butterworth, Stoneham, Mass., p 51.

Gordon, S., W.A. Mulac. 1975. Reactions of the OH (X^2II radical produced by the pulse radiolysis of water vapor. *Int. J. Chem. Kinet., 7*:289.

Harriss, R.C. 1989. Historical trends in atmospheric methane concentration and the temperature sensitivity of methane outgassing from boreal and polar regions. In: *Proceedings of a Joint Symposium by the Board on Atmospheric Sciences and Climate and the Committee on Global Change Commission on Physical Sciences, Mathematics and Resources*. National Academic Press, Washington D.C., p. 79.

Hitchcock, D.R., A.E. Wechsler. 1972. Biological cycling of atmospheric trace gases. Contr. Rep. *NASA-CR 126663*. Natl. Aeron. Space Adm., Washington D.C., p. 117.

Holzapfel-Pschorn, A., W. Seiler. 1986. Methane during a cultivation period from an Italian rice paddy. *J. Geophy. Res., 91*:11,803.

Hough, A., R.G. Derwent. 1990. Changes in the global concentration of tropospheric ozone due to human activities. *Nature, 344*:645.

Keeling, C.D., W.G. Mook, P.P. Tans. 1979. Recent trends in the $^{13}C/^{12}C$ ratio of atmospheric carbon dioxide. *Nature, 277*:121.

Khalil, M.A.K., R.A. Rasmussen. 1983. Sources, sinks, and seasonal cycles of atmospheric methane. *J. Geophys. Res., 88*:5,131-5,144.

Khalil, M.A.K., R.A. Rasmussen. 1990. Atmospheric methane: Recent global trends. *Environ. Sci. Technol., 24*:549.

Khalil, M.A.K., R.A. Rasmussen. 1991. Methane emissions from rice fields in China. *Environ. Sci. Technol., 25*:979.

Lachenbruch, A.H., B.V. Marshall. 1986. Changing climate: Geothermal evidence from permafrost in the Alaskan Arctic. *Science, 234*:689.

Lassey, K.R., D.C. Lowe, C.A.M. Brenninkmeijer, A.J. Gomez. 1993. Atmospheric methane and its carbon isotopes in the southern hemisphere: Their time series and an instructive model. *Chemosphere, 26* (1-4):95-110.

Levin, I., P. Bergamaschi, H. Dörr, D. Trapp. 1993. Stable isotopic signature of methane from different sources in western Europe. *Chemosphere, 26* (1-4):161-178.

Logan, J.A. 1985. Tropospheric ozone: Seasonal behavior, trends and anthropogenic influence. *J. Geophys. Res., 90*:10,463.

Lowe, D.C., C.A.M. Brenninkmeijer, S.C. Tyler, E.J. Dlugokencky. 1991. Determination of the isotopic composition of atmospheric methane and its application in the antarctic. *J. Geophys. Res., 96*:15,455-15,467.

Matthews, E., I. Fung. 1987. Methane emissions from natural wetlands: Global distribution area, and environmental characteristics of sources. *Global Biogeochem. Cycles, 1*:61.

Mayer, E.W., D.R. Blake, S.C. Tyler, Y. Makide, D.C. Montague, F.S. Rowland. 1982. Methane: interhemispheric concentration gradient and atmospheric residence time. *Proc,. Natl. Acad. Sci., 79*:1,366-1,370.

Oona, S., E.S. Deevey. 1960. Carbon 13 in lake waters and its possible bearing on paleolimnology. *Am. J. Sci., 258A*:253.

Ovsyannikov, V.M., V.S. Lebedev. 1967. Isotopic composition of carbon in gases of biogenic origin. *Geochem. Int., 4*:453.

Prinn, R., D. Cunnold, R.A. Rasmussen, P. Simmonds, F. Alyea, A. Crawford, P. Fraser, R. Rosen. 1987. Atmospheric trends in methylchloroform and the global average for the hydroxyl radical. *Science, 238*:945.

Quay, P., S.L. King, J.M. Lansdown, D.O. Wilbur. 1988. Isotopic composition of methane released from wetlands: Implications for the increase in atmospheric methane. *Global Biogeochem. Cycles, 2*:385.

Quay, P., S.L. King, J. Stutsman, D.O. Wilbur, L.P. Steele, I. Fung, R.H. Gammon, T.A. Brown, G.W. Farwell, P.M. Grootes, F.H. Schmidt. 1991. Carbon isotopic composition of atmospheric CH_4: fossil and biomass burning source strengths. *Global Biogeochem. Cycles, 5*:25.

Rice, D.D., G.E. Claypool. 1981. Generation, accumulation and resource potential of biogenic gas. *Bull. Am. Assoc. Pet. Geol., 65*:5.

Rust, F.E. 1981. Ruminant methane $\delta(^{13}C/^{12}C)$ values: Relationship to atmospheric methane. *Science, 211*:1,044-1,046.

Schoell, M. 1980. The hydrogen and carbon isotopic composition of methane from natural gases of various origins. *Geochem. Cosmochim. Acta, 44*:649.

Schütz, H., A. Holzapfel-Pschorn, R. Conrad, H. Rennenberg, W. Seiler. 1989. A 3-year continuous record on the influence of daytime, season, and fertilizer treatment om methane emission rates from an Italian rice paddy. *J. Geophys. Res., 94*:16,405.

Stevens, C. 1988. Atmospheric methane. *Chem. Geol., 71*:11.

Stevens, C., A. Engelkemeir. 1988. Stable carbon isotopic composition of methane from some natural and anthropogenic sources. *J. Geophys Res., 93*:725.

Stevens, C., A. Engelkemeir, R. Rasmussen. 1985. Causes of increasing methane fluxes based on carbon isotope studies. In: *Special Environmental Report No. 16 (WMO-No.647)*; Lectures presented at the WMO Technical Conference on Observations and Measurement of Atmospheric Contaminants. World Meterol. Org., Geneva, Switzerland, p. 237.

Tyler, S.C. 1986. Stable carbon isotope ratios in atmospheric methane and some of its sources. *J. Geophys. Res., 91*:13,232.

Tyler, S.C., P.R. Zimmerman, C. Cumberbatch, J. Greenberg, C. Westberg, J.P.E.C. Darlington. 1988. Measurements and interpretation of δ13C of methane from termites, rice paddies, and wetlands in Kenya. *Global Biogeochem. Cycles, 2*:341.

United Nations Statistical Yearbook. 1988.

Wahlen, M., N. Tanaka, R. Henry, B. Deck, Zeglan, J.S. Vogel, J. Southon, A. Shemesh, R. Fairbanks, W. Broecker. 1989. Carbon-14 in methane sources and in atmospheric methane: The contribution from fossil carbon. *Science, 245*:286.

Wahlen, M., N. Tanaka, B. Deck, R. Henry. 1990. δD in CH_4: Additional constraints for a global budget. *EOS, 71* (43):1,249.

Chapter 5

Working Group Report

Atmospheric Methane Concentrations

B. STAUFFER[1], M. WAHLEN[2], AND F. MORAES[3]

[1]*Physikalisches Institut, Universität Bern, Sidlerstrasse 5, CH-3012 Bern, Switzerland*

[2]*Scripps Institution of Oceanography, University of California, San Diego, La Jolla, CA 92093*

[3]*Oregon Graduate Institute, Global Change Research Center, P.O. Box 91,000, Portland, OR 97291 USA*

Panel participants:
M.A.K. Khalil, K.R. Lassey, I. Levin, F. Moraes, D. Raynaud, M. Shearer, B. Stauffer, R. Sepanski, C.M. Stevens, M. Wahlen

1. Introduction

It was first discovered that the atmospheric concentration of methane was increasing based upon samples collected in the late seventies and early eighties at Cape Meares, Oregon (Rasmussen and Khalil, 1981). Since that time, more detailed and extensive data sets have allowed estimation of sources and sinks of methane on the global scale. As attempts are made to refine these estimates, the value of the current databases will be greatly increased, as will the need for more current data.

Direct atmospheric measurements of methane are the most accurate means available for determining concentrations, but they have the distinct disadvantage that precise measurements are available only since the early 1960s (Khalil et al., 1989), with continuous data starting in 1978 (Rasmussen and Khalil, 1981; Blake and Rowland, 1988). Ice core measurements have thus been used to determine methane concentrations in the past. This method is most important in that it allows researchers to look at methane levels at times when humans had little impact on the methane cycle. Because natural sources of methane make up a sizable portion of the yearly methane input into the atmosphere, these data can greatly constrain the overall budget.

NATO ASI Series, Vol. I 13
Atmospheric Methane: Sources, Sinks, and Role in Global Change
Edited by M. A. K. Khalil

The isotopic composition of atmospheric methane is an increasingly common part of databases from both direct atmospheric samples and ice core measurements. These data provide greater constraints on methane sources than do concentrations alone (see Raynaud and Chappellaz, this volume; Stevens, this volume). Sporadic isotope measurements from ambient air samples have been available since 1978, but continuous records are available only since 1987 (see Stevens, this volume). Obtaining isotopic data from ice cores has no constraints other than those mentioned above for obtaining concentration data alone except that they require a larger amount of air for analysis.

Other records of atmospheric methane (e.g., solar spectra) are available but are not discussed here since they are of limited value. This chapter discusses the current issues and frontiers of the three atmospheric methane measurement procedures presented above as well as the measurements of other gases in ice cores that can yield information on the methane cycle.

2. Concentration measurements on directly collected atmospheric air

A vast array of surface-based methane measurement sites has been developed over the past 15 years (Rasmussen and Khalil, 1981; Steele et al., 1987; Blake and Rowland, 1988). These sites allow us to measure methane concentrations with a great deal of spatial resolution, which is helpful in constraining the methane budget. In addition to these measurements, high time resolution data (as high as one measurement every 20 minutes) are available for the past 13 years at the station in Cape Meares, Oregon.

Recently it has been shown that the increase of tropospheric methane has been slowing (Khalil and Rasmussen, 1990; Steele et al., 1992; Khalil and Rasmussen, 1993). Alone, increases in the uptake rate do not adequately explain this decreasing trend; some slowdown in the sources is required. In addition to this, the size of the decrease is not agreed upon. This is an important issue to be resolved because of its ramifications on our understanding of the global sources and sinks of methane.

It has been well documented that there is little variation of methane concentration with height in the troposphere. We thus believe that the surface measurements are adequate for the entire troposphere. The concentration of methane in the stratosphere has been studied (e.g., WMO, 1985) due to the role methane plays in the removal of Cl from the stratosphere; however, the time and spatial resolution of these data is much less than the tropospheric measurements.

The direct methane record could be significantly improved in the area of flux measurements. We are concerned that chamber methods of flux determination may yield inaccurate representations. In addition, remote sensing instruments allow researchers to collect many more measurements than chamber methods do. This area of research, perhaps more than all other areas, is dependent upon improvements in measurement technologies.

3. Concentration measurements on air extracted from ice cores

For many gases, the concentration in air bubbles of ice is a good approximation of the concentration in the atmosphere at the time the ice was formed when the region the ice is from is cold enough that there is little surface melting. This has been independently verified for various gases including methane (Stauffer et al., 1988). The accuracy is still a major issue, however. Inaccuracies may occur in the derived concentrations by errors in calibration and dating. In addition to these experimental errors, inaccuracies may occur due to natural processes of fractionation, the gravitational enrichment of a gas, and air-ice reactions, already in the ice sheet or in the ice cores during transport and storage.

Fractionation is the only known cause of error in the derived concentrations of atmospheric methane from ice cores (Raynaud and Chappellaz, this volume). The upper limit of all systematic errors can be estimated based on the overlapping records of ice core methane with direct atmospheric measurements. Etheridge et al. (1992) found that the two records agree quite well. There are, however, complicating factors because of the uncertainty in the time estimates. The minimum uncertainty in dating is 5 years. Because the concentration has been increasing by about 15 ppbv per year for the past few decades, there is a 75 ppbv range of concentration values for each measurement. This yields a worst-case error of 75 ppbv. By traditional statistical methods this would be approximately 40, or roughly double the measurement precision of 22 ppbv.

The congruity of ice core records from different drilling sites with different temperatures and accumulation rates shows that it is unlikely that there are large unknown systematic errors resulting from temperature, accumulation rate, and impurity concentration variations of the ice. More data from different sites will further limit the range of inaccuracy associated with systematic errors.

Rasmussen and Khalil (1984) first discovered the pre-industrial interhemispheric concentration difference from ice core measurements of Greenland and the South Pole. They found that the northern hemisphere concentrations were, on average, 70 ppbv greater than the southern hemisphere concentrations. In general, a difference of more than 40 ppbv should be considered significant and reliable, assuming that both records have been measured with the same procedure and the same equipment. However, this emphasizes the need to increase the accuracy of ice core methane measurements.

All results reported until now have been obtained by gas chromatographic analysis of air samples extracted from ice cores. Analytical methods based on laser absorption spectroscopy are under development, however. Their main advantage over the traditional method is that they allow several gas components to be measured simultaneously.

The melt extraction method has provided the most reproducible way to acquire methane data. However, a melt extraction excludes the measurement of some gases, most notably carbon dioxide and nitrous oxide, because the contamination for these gases is too large with this extraction method. It is therefore important to develop dry extraction systems that allow CO_2, CH_4, and N_2O measurements with a low and reproducible contamination level. Fuchs et al. (1992) are currently working on a promising method that uses a milling tool with bearings flushed with helium.

Based on the previous discussion we recommend the following topics for future research. Analytical equipment should be modified to reach a precision of ±10 ppbv or better. Standards used by different laboratories should be compared and calibrated. Ice core measurements from different sites should be collected to check for systematic errors that are dependent on temperature and accumulation rates.

It is likely that more records reconstructed with improved analytical techniques will give more insight into the behavior of methane in the atmosphere. There are particular time "windows" that, we feel, will be particularly fruitful in this endeavor. The industrial period (the past 200 years) provides information about the anthropogenic effect on the methane cycle as well as serving as a method of calibrating ice core measurements. In order to investigate the influence of climate variability on the methane cycle, the Little Ice Age (1500 to 1800 A.D.) should be studied. The climatic optimum (6,000 years BP) could furnish valuable information on the effects of a warmer climate for both natural and anthropogenic sources such as rice agriculture. Finally, the transition from the last glaciation to the Holocene (14,500 to 10,000 years BP) might provide information on the effects of large climatic variations on the methane cycle.

4. Additional ice core records

In addition to methane, there are several other gases available from ice cores that could greatly expand our understanding of the global methane cycle. The comparison of ice core methane and HCHO records could allow us to examine changes in the sources of methane and in the oxidation capacity of the atmosphere (Staffelbach et al., 1992). Reactive components such as H_2O_2 and HCHO can be measured in ice cores. They are incorporated, however, not in the air bubbles but in the ice itself. The concentration measured in the ice would therefore be only a measure of the atmospheric concentration at the time of the corresponding snowfall. But this is true only if the transfer functions for these components are constant and known; it is also possible that reactive components may be depleted by slow chemical reaction. Sigg (1990) demonstrated that H_2O_2 can react with impurities in ice. We think that the methods based on reactive components are very interesting but that more tests are needed in order to obtain reliable atmospheric records.

Other than methane itself, the most valuable gas to have data on would be carbon monoxide. Early efforts to measure CO from ice cores demonstrated that it would be very difficult, if possible at all (Robbins et al., 1973; Rasmussen et al., 1982). It is possible that the amount of CO adsorbed onto the surface of snow flakes and aerosols is comparable to or greater than the amount of CO enclosed in bubbles and therefore makes a reconstruction impossible. In order to test whether CO in snow, firn, and ice can be measured, serious contamination problems that occur while extracting the air must be solved.

Ice core records of other gases, particularly CH_3Cl and O_3, would also provide valuable information in understanding the methane cycle.

5. Records of stable isotope ratios in methane

There are three isotopic ratios of methane that have been used to investigate the methane cycle: $\delta^{13}C$ ($^{13}C/^{12}C$), δD (D/H), and the abundance of the cosmogenic radionuclide ^{14}C. The isotopic composition of CH_4 places constraints on the different CH_4 sources because the isotopic composition of CH_4 from the various sources differs; specifically, methane from biogenic sources (natural wetlands and tundra, ruminants, rice production, landfills, and possibly termites) and abiogenic sources (biomass burning and fossil methane) differ, particularly with regard to ^{13}C (Schoell, 1980; Rust, 1981; Stevens and Rust, 1982; Tyler, 1986; Wahlen et al., 1987; Stevens and Engelkemeir, 1988; Quay et al., 1988; Wahlen et al., 1989a; Levin et al., 1993). From a comparison of the isotopic composition of CH_4 sources to that of the atmospheric CH_4 concentrations, estimates of the various source strengths can be obtained. In addition to CH_4 concentration and flux data, this process requires knowledge of the loss rate of atmospheric CH_4 to the stratosphere and the isotopic fractionation occurring in the tropospheric CH_4 destruction process. With all of these data, the sources of atmospheric CH_4 may be constrained further than with concentration data alone.

5.1 Carbon-14 in methane. The primary use of $^{14}CH_4$ data is to derive the magnitude of the fossil CH_4 source strength. This methane, which is thought to come from natural gas and coal acquisition and distribution, contains no ^{14}C. In contrast, biomass burning and biogenic CH_4 contain ^{14}C at levels close to that of contemporary atmospheric $^{14}CO_2$. Thus, from a comparison of the ^{14}C content of atmospheric CH_4 to that of biogenic and biomass burning CH_4 one can deduce the source strengths of the fossil CH_4 sources. This situation is somewhat complicated, however, by the time-dependent atmospheric disturbance of $^{14}CO_2$ by nuclear weapons testing, where atmospheric $^{14}CO_2$ gets incorporated into organic matter from which $^{14}CH_4$ is produced bacterially, as well as that caused directly by pressurized light water reactors (Kunz, 1985).

Wahlen et al. (1989a) tried to reconstruct the history of atmospheric $^{14}CH_4$ from historical samples and from more recent clean air samples. By modelling this reconstruction, the source contribution due to fossil CH_4 was found to be 21 percent of the global annual CH_4 production. A small interhemispheric gradient of $^{14}CH_4$ was also detected. Other studies based on the time trend observed over several years (Quay et al., 1991) deduced 16 percent, and modelling data from the southern hemisphere resulted in 26 percent (Manning et al., 1990). Independent estimates for the source contribution of fossil methane from natural gas and coal mining have been made, but no experimental data are available for fluxes from natural seepage.

Several records of atmospheric $^{14}CH_4$ exist (Quay et al., 1989; Wahlen et al., 1989a; Levin et al., 1991; Lassey et al., 1993) extending over the last few years. They all show annual increases of 1 to 2 percent, which is attributed to the nuclear reactor sources. Today, this source is substantial, almost completely offsetting the dilution caused by the addition of fossil methane into the atmosphere (Wahlen et al., 1989a). This source is expected to increase at an accelerated rate. From future accelerated increases in atmospheric $^{14}CH_4$ one could directly determine the magnitude of this source and reduce the uncertainties in the magnitude of the fossil CH_4 contribution to the entire CH_4 budget. We therefore feel that it is important to continue these measurements.

5.2 *Carbon-13 in methane.* The longest record of $\delta^{13}CH_4$ in the atmosphere is that established by Stevens (this volume) for both hemispheres, starting in 1978. It provides information on the interhemispheric gradient and the rate of change in both hemispheres for $\delta^{13}CH_4$ (becoming heavier). Details of the changing strengths of different sources (and perhaps sinks) can be deduced from this information. It is important that these records be continued.

The existing database of isotopic methane from the various sources is still small because these measurements are very labor intensive. Proposed in situ measurements, based on laser spectroscopic techniques would help to increase the database (Schupp et al., 1993) from which global extrapolations could be obtained.

Shorter records from remote locations were measured by Wahlen et al. (1989a) and, with much higher time resolution, by Quay et al. (1991) and Lassey et al. (1993). Time trends in $\delta^{13}CH_4$ are less discernible from these short-term records. The latter records show seasonal variations in $\delta^{13}CH_4$ (and methane concentration), with heavier values in summer than in winter. This is attributed to the destruction of methane by OH, being more intense in the summer because of isotopic fractionation. This seasonality seems to be more pronounced in northern high latitudes (Quay et al., 1991) but is unexpectedly strong in New Zealand (Lassey et al., 1993). It is possible that the observed seasonality is caused by a combination of factors including local and regional sources, long range transport, and varying atmospheric mixing intensity. The latter factor is demonstrated by the correlation of atmospheric CH_4 concentrations with that of ^{222}Rn (Thom et al., 1993). We believe that, in addition to obtaining high resolution records from remote sites, continental sites in proximity to various methane sources should also be investigated to gain more insight into these mechanisms.

As pointed out by Stevens (this volume), there are small but considerable differences in the assessment of $\delta^{13}C$ in atmospheric methane among different laboratories. This difference is attributed to the different techniques of collecting, separating, and converting methane to CO_2. Clearly there is a need for appropriate standards and inter-laboratory comparisons. Secondary standards for ^{13}C analyses by the National Institute of Standards and Technology (NIST, USA) and the International Atomic Energy Agency (IAEA, Austria) are being depleted and are becoming difficult to acquire. Methane standards from natural gases, covering the natural range of isotopic compositions in both ^{13}C and D, which test separation, conversion, and measuring techniques, are available from M. Schoell, Chevron Oil Field Research Station, La Habra, CA, USA. These standards are currently being evaluated by IAEA in a laboratory comparison. The IAEA is trying to establish a new international standards program including isotopic standards for methane.

Since pre-industrial times, the atmospheric methane concentration has more than doubled. In order to obtain constraints on the evolution of anthropogenically influenced sources, it is necessary to know the pre-industrial

$\delta^{13}CH_4$ value. This quantity was determined in an ice core from Dye 3 by Craig et al. (1988). Since that time, the pre-industrial $\delta^{13}CH_4$ value has not been reproduced based upon other ice core measurements, and we think it is important to confirm this number.

The isotopic fractionation for the methane destruction reaction initiated by OH has been variously determined in the past; the latest results were reported by Cantrell et al. (1990). Measurements of $\delta^{13}CH_4$ profiles from the lower stratosphere appear to show larger fractionation (Wahlen et al., 1989b). This discrepancy should be resolved.

The ^{13}C composition of methane from major sources has been variously assessed and is summarized in Stevens and in Whiticar (both this volume). The ^{13}C flux abundances are controlled by creation and destruction mechanisms. The creation mechanisms appear to be influenced by the composition of substrate carbon and the different methanogenic pathways (Whiticar et al., 1986). The isotopic composition of fluxes from inundated ecosystems are altered by bacterial methane oxidation (King et al., 1989). We feel that the isotopic consequences of CH_4 production and destruction should be explored in more detail.

Quay et al. (1992) demonstrated the average latitudinal isotopic trends from three stations in the northern and one station in the southern hemisphere. $\delta^{13}CH_4$ gets progressively enriched in going north to south. The biomass burning source of methane is by far the most enriched in $\delta^{13}CH_4$, and this trend is attributed to the geographical distribution of the source. Most measurements of the isotopic composition of CH_4 from this source involve burning of carbon derived from C-3 plants only. Burning of C-4 plants would make this source substantially heavier and influence the global isotopic source mix. More studies should be undertaken to evaluate the C-3 versus C-4 mix in the biomass burning source.

5.3 Deuterium in methane. There are considerably fewer data for δD in methane than for $\delta^{13}C$. Atmospheric CH_4 has a δD of about -85‰ (SMOW) (Ehhalt and Volz, 1976; Wahlen et al., 1987, 1990). A number of methane sources have been studied to determine the δD value (Schoell, 1980; Burke et al., 1988; Wahlen et al., 1987, 1990; Levin et al., 1993). The results cover a large range, with biogenic and fossil CH_4 more depleted, and CH_4 from biomass burning probably more enriched than δD of atmospheric CH_4. More measurements are needed to fully exploit deuterium in methane for refining the global budget.

No atmospheric time series has been developed for δD in methane. Such measurements would be useful. In addition, no laboratory measurements have been made to determine the deuterium fractionation due to CH_4 destruction by reaction with OH; only an estimate of the fractionation has been obtained from δD profiles of methane from the lower stratosphere (Wahlen et al., 1989b). As the fractionation is estimated to be large, time series would exhibit considerable seasonal variations. These seasonal variations would provide another means of determining the fractionation, if fractionation is the main cause of the seasonality in δD.

δD of biogenic CH_4 from wet ecosystems (wetlands, tundra, rice) is linked to δD of precipitation (Whiticar et al., 1986; Wahlen et al., 1990). This relationship could be used to reliably determine the global average δD of methane for the combination of these sources, as areal extent as well as fluxes, and δD in precipitation have been explored globally.

6. Summary

A continuous record of atmospheric methane concentrations in the future is essential for a better understanding of the global methane cycle and changes to it. Important additions to this record would be greater spatial and temporal information on the stratospheric distribution of methane and more flux measurements.

Records from the past, reconstructed based on ice core results, are necessary to refine source estimates and disentangle anthropogenic influences from the natural variability. The quantity and accuracy of these measurements should be increased.

Isotopic measurements of atmospheric methane give constraints on the size of different methane sources and sinks, which are not available by other methods. These measurements should be increased for both atmospheric and ice core measurements.

References

Blake, D.R., F.S. Rowland. 1988. Continuing worldwide increase in tropospheric methane, 1978 to 1987. *Science, 239*:1,129-1,131.

Burke, R.A., T.R. Barber, W.M. Sackett. 1988. Methane flux and stable hydrogen and carbon composition of sedimentary methane from the Florida Everglades. *Global Biogeochemical Cycles, 2*:329-340.

Cantrell, C.A., R.E. Shetter, A.H. McDaniel, J.G. Calvert, J.A. Davidson, D.C. Lowe, S.C. Tyler, R.J. Cicerone, J.P. Greenberg. 1990. Carbon kinetic isotope effect in the oxidation of methane by hydroxyl radicals. *J. Geophys. Res., 95*:22,455-22,462.

Craig, H., C.C. Chou, J.A. Welhan, C.M. Stevens, A. Engelkemeir. 1988. The isotopic composition of methane in polar ice cores. *Science, 242*:1,535-1,538.

Ehhalt, D.H., A. Volz. 1976. Coupling of the CH_4 with the H_2 and CO cycle: Isotopic evidence. In: *Microbial production and utilization of gases.* Akademie der Wissenschaften zu Goettingen, FRG, 23-33.

Etheridge, D.M., G.I. Pearman, P.F. Fraser. 1993. Changes of tropospheric methane between 1841 and 1978 from a high accumulation-rate antarctic ice core. *Tellus* (in press).

Fuchs, A., J. Schwander, B. Stauffer. 1993. A new ice mill allows precise concentration-determination of methane and most probably also other trace gases in the bubble-air of very small ice samples. *Journal of Glaciology* (in press).

Khalil, M.A.K., R.A. Rasmussen. 1990. Atmospheric methane: Recent global trends. *Environ. Sci. Technol., 24*:549-553.

Khalil, M.A.K., R.A. Rasmussen. 1993. Decreasing trend of methane: Unpredictability of future concentrations. *Chemosphere, 26* (1-4):803-814.

Khalil, M.A.K., R.A. Rasmussen, M.J. Shearer. 1989. Trends of Atmospheric Methane During the 1960s and 1970s. *J. Geophys. Res., 94*:18,279-18,288.

King, S.L., P.D. Quay, J.M. Lansdown. 1989. The $^{13}C/^{12}C$ Kinetic Isotope Effect for Soil Oxidation of Methane at Ambient Atmospheric Concentrations. *J. Geophys. Res., 94*:18,273-18,277.

Lassey, K.R., D.C. Lowe, C.A.M. Brenninkmeijer, A.J. Gomez. 1993. Atmospheric methane and its carbon isotopes in the southern hemisphere: Their time series and an instructive model. *Chemosphere, 26* (1-4):95-110.

Levin, I., R. Boesinger, G. Bonani, R. Francey, B. Kromer, K.O. Muennich, M. Suter, N.A.B. Trivett, W. Woelfli. 1991. Radiocarbon in atmospheric carbon dioxide and methane: Global distributions and trends. In: *Radiocarbon after four decades: An interdisciplinary perspective* (R.E. Taylor, A. Long, and R.S. Kra, eds.), Springer Verlag, New York, pp. 506-518.

Levin, I., P. Bergamaschi, H. Doerr, D. Trapp. 1993. Stable isotope signature of methane from different sources in western Europe. *Chemosphere, 26* (1–4):161-174.
Manning, M.R., D.C. Lowe, W.H. Melhuish, R.J. Sparks, G. Wallace, C.A.M. Brenningkmeijr, R.C. McGill. 1990. The use of radiocarbon measurements in atmospheric studies. *Radiocarbon, 32/1*:37-58.
Quay, P.D., S.L. King, J. Stutsman, D.O. Wilbur, L.P. Steele, I. Fung, R.H. Gammon, T.A. Brown, G.W. Farwell, P.M. Grootes and F.H. Schmidt. 1991. Carbon isotopic composition of atmospheric CH_4: Fossil and biomass burning source strength. *Global Biogeochemical Cycles, 5/1*:25-45.
Rasmussen, R.A., M.A.K. Khalil. 1981. Atmospheric methane (CH_4): Trends and seasonal cycles. *J. Geophys. Res., 86*:883-886.
Rasmussen, R.A., M.A.K. Khalil. 1984. Atmospheric Methane in the Recent and Ancient Atmospheres: Concentrations, Trends, and Interhemispheric Gradient. *J. Geophys. Res., 89*:11,599-11,605.
Rasmussen, R.A., M.A.K. Khalil, S.T. Hoyt. 1982. Methane and Carbon monoxide in snow. *J. Air Pollut. Control Assn., 32*:176-177.
Robbins, R.C., L.A. Cavanagh, L.J. Salas, E. Robinson. 1973. Analysis of ancient atmosphere. *J. Geophys. Res., 78*:5,341-5,344.
Rust, F.E. 1981. Delta 13C of ruminant methane and its relationship to atmospheric methane. *Science, 211*:1,044-1,046.
Schoell, M. 1980. The hydrogen and carbon isotopic composition from natural gases of various origin. *Geochim. Cosmochim. Acta, 44*:649-661.
Schupp, M., P. Bergamaschi, G.W. Harris, P.J. Crutzen. 1992. Measurement of the $^{13}C/^{12}C$ ratio in methane by tunable diode laser spectroscopy. *Chemosphere, 26* (1-4):13-22.
Sigg, A. 1990. Wasserstoffperoxid-Messungen an Eisbohrkernen aus Grönland und der Antarktis und ihre atmosphärenchemische Bedeutung. Ph.D. Dissertation, University Bern, Switzerland.
Staffelbach, T., A. Neftel, B. Stauffer, D. Jacob. 1991. A record of atmospheric methane: sink from formaldehyde in polar ice cores. *Nature, 349*:603-605.
Stauffer, B., E. Lochbronner, H. Oeschger, J. Schwander. 1988. Methane concentration in the glacial atmosphere was only half that of the preindustrial Holocene. *Nature, 332*:812-814.
Steele, L.P., P.J. Fraser, R.A. Rasmussen, M.A.K. Khalil, T.J. Conway, A.J. Crawford, R.H. Gammon, K.A. Masarie, K.W. Thoning. 1987. The Global Distribution of Methane in the Troposphere. *J. Atmos. Chem., 5*:125-171.
Steele, L.P., E.J. Dlugokencky, P.M. Lang, P.P. Tans, R.C. Martin, K.A. Masarie. 1992. Slowing down of the global accumulation of atmospheric methane during the 1980s. *Nature, 358*:313-316.

Stevens, C.M., F.E. Rust. 1982. The carbon isotopic composition of atmospheric methane. *J. Geophys. Res.*, 87:4,879-4,882.
Stevens, C.M., A. Engelkemeir. 1988. The carbon isotopic composition of methane from some natural and anthropogenic sources. *J. Geophys. Res.*, 93:725-733.
Thom, M., R. Boesinger, M. Schmidt, I. Levin. 1993. The regional budget of atmospheric methane in a highly populated area. *Chemosphere,* 26 (1–4):143-160.
Tyler, S.C. 1986. Stable carbon isotope ratios in atmospheric methane and some of its sources. *J. Geophys. Res.*, 91:13,232-13,238.
Wahlen, M., N. Tanaka, R. Henry, B. Deck, W. Broecker, A. Shemesh, R. Fairbanks, J.S. Vogel, J. Southon. 1987. ^{13}C, D and ^{14}C in methane. In: *Report to Congress and EPA on NASA Upper Atmosphere Research Program*, NASA, Washington, D.C., 324-325.
Wahlen, M., N. Tanaka, R. Henry, B. Deck, J. Zeglen, J.S. Vogel, J. Southon, A. Shemesh, R. Fairbanks, W. Broecker. 1989a. Carbon-14 in methane sources and in the atmosphere: The contribution from fossil carbon. *Science,* 245:286-290.
Wahlen, M., B. Deck, R. Henry, N. Tanaka, A. Shemesh, R. Fairbanks, W. Broecker and H. Weyer. 1989b. Profiles of $\delta^{13}C$ and δD in CH_4 from the lower stratosphere. *Trans. Am. Geophys. Union,* 71/43:1249.
Wahlen, M., N. Tanaka, B. Deck, R. Henry, A. Shemesh, R. Fairbanks, W. Broecker. 1990. δD in CH_4: Additional constraints for a global CH_4 budget. *Trans. Am. Geophys. U.,* 71/43:1249.
Whiticar, M.J., E. Faber, M. Schoell. 1986. Biogenic methane formation in marine and freshwater environments: CO_2 reduction versus acetate fermentation isotope evidence. *Geochim. Cosmochim. Acta,* 50:693-709.

Chapter 6

Biological Formation and Consumption of Methane

DAVID R. BOONE

Environmental Science and Engineering, Oregon Graduate Institute of Science and Technology
20,000 N.W. Walker Road, Portland, Oregon 97291-1000, USA

1. Introduction

Methane is an important product formed during the bacterial degradation of organic matter in environments such as flooded soils, wetlands, estuaries, marine and freshwater sediments, and the gastrointestinal tract of animals (Whitman et al., 1992). This chapter describes the conditions that lead to biogenic methane formation in natural environments, the metabolic pathways and interactions that lead to methanogenesis, and the implications of these factors on the biogeochemistry of methane.

Methane-producing bacteria (methanogens) are a specialized group of microbes that catabolize a small number of molecules and produce methane as the major catabolic product. They are the only known life-forms that produce a hydrocarbon as a major catabolic product. The known substrates that methanogens can catabolize is very limited: H_2 + CO_2, formate, acetate, methanol, methylamines, and methylsulfides. In addition to these, a small number of alcohols can be oxidized by some strains. Because the substrate range of methanogens is so limited, these microbes depend on other bacteria to provide them from a wide range of more complex molecules. Methanogenic decomposition of complex organic matter is possible because many non-methanogenic bacteria convert organic compounds into the few molecules that methanogens can use.

2. Ecology of methanogenic decomposition

The decomposition process in anoxic environments may be divided into three major steps, fermentation (or acidogenesis), syntrophic acetigenesis, and methanogenesis (Figure 1). Each step is catalyzed by a separate group of

NATO ASI Series, Vol. I 13
Atmospheric Methane: Sources, Sinks, and Role
in Global Change
Edited by M. A. K. Khalil

bacteria, but all the reactions occur simultaneously so the concentrations of extracellular intermediates often remain small. For instance, H_2 concentration in such ecosystems is typically less than 100 nM (with a partial pressure of about 5 to 10 Pa, or <0.01% of the gas phase), yet its rate of production and consumption is very rapid (Bryant et al., 1967; Boone, 1982; Wolin, 1982; Conrad et al., 1986; Dwyer et al., 1988).

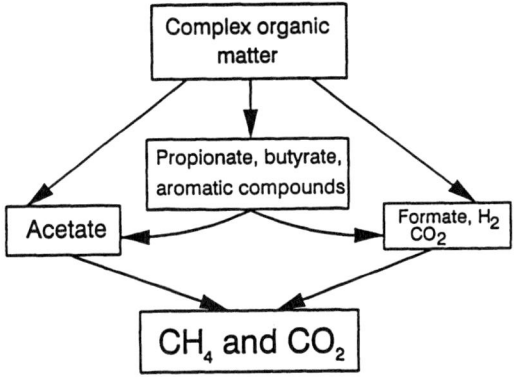

Figure 1.   Organic matter decomposition in methanogenic ecosystems.

The fermentative step converts a wide array of complex organic matter in to a relatively small number of products, mainly volatile fatty acids, carbon dioxide, and hydrogen (Figure 1). The conversion of most of the components of biomass (such as proteins, carbohydrates, nucleic acids, and fats) is typically rapid. Other organic molecules such as highly polymerized lignin, hydrocarbons, and some man-made chemicals are degraded slowly or not at all in anaerobic environments (Leisinger and Brunner, 1986).

The methanogens convert some of the products of the fermentative step to methane and carbon dioxide, the terminal carbon-containing products formed during the complete degradation of organic matter. One group of molecules produced during fermentation, the volatile fatty acids longer than acetic acid, cannot be broken down by the methanogens. These acids are degraded by a specialized group of bacteria called "obligate proton-reducing acetigens," which produce acetic acid by β-oxidation of the fatty acid (McInerney, 1986; McInerney et al., 1979, 1981). The acetigenic oxidation of odd numbered fatty acids produces, in addition to acetic acid, propionic acid (a 3-carbon monocarboxylic acid). Propionate is oxidized to acetic acid and carbon dioxide, probably by the succinate pathway (Houwen et al., 1987). The oxidation of volatile fatty acids is coupled to the production of H_2 (by proton reduction) or formate (by bicarbonate reduction), and is thermodynamically favorable only when H_2 and formate concentrations are very low (Bryant, 1979; Boone et al., 1989). Low concentrations are maintained by the rapid uptake of H_2 and formate by methanogens, which requires a tight coupling of fatty acid oxidation with methane formation. This process has been called "interspecies hydrogen transfer" (McInerney et al., 1981; Wolin, 1982), but the term "interspecies electron transfer" may be more appropriate because, in addition to H_2, formate may act as an extracellular electron transporter from fatty acid oxidizers to methanogens (Thiele and Zeikus, 1988; Boone et al., 1989).

Organic matter such as lignin monomers and pectin contain methoxyl groups that may be degraded in anoxic environments by a newly discovered pathway. The removal of methoxyl groups by acetigenic bacteria was documented during the 1980s (Barik et al., 1985). These bacteria convert the methoxyl groups and carbon dioxide to acetic acid, which is their catabolic product (Ljungdahl, 1986). Recently, other bacteria were shown instead to transfer successively the methyl groups to sulfide to form methane thiol and then dimethyl sulfide. The newly discovered ability of some methanogens to catabolize dimethyl sulfide (Oremland et al., 1989; Ni and Boone, 1991, 1993; Bak and Finster, 1992) and methane thiol (Bak and Finster, 1991; Ni and Boone, 1991, 1993) could allow the methoxyl groups to be converted ultimately to methane, regardless whether they were first combined with carbon dioxide (to form acetic acid) or with sulfide (to form methane thiol or dimethyl sulfide).

3. Partial methanogenic fermentations, including the rumen fermentation

In some methanogenic ecosystems all three decomposition steps (Figure 1) do not occur. For instance, when geothermal H_2 enters sediments, it may be converted directly to methane, so that only the methanogenic step is required (Bryant, 1979). In some environments, bacteria may be lost more rapidly than they are replaced by growth. This may be one of the factors that limit colonization of digestive tracts of man and animals by slow-growing methanogens (Bryant, 1979; Wolin, 1982; Miller, 1991). About half of the human population harbors substantial numbers of methanogens in their colons. Ungulates, or ruminants (such as cattle, sheep, deer, moose, elk) have a special forestomach (the rumen) that harbors an intense microbial fermentation. The function of this fermentation, from the standpoint of the animal, is to convert organic matter in the diet into a form that can be used by the animal. Cellulose is an important part of the diet of most of these animals, yet the animal itself cannot decompose cellulose. However, the rumen is colonized by anaerobic cellulose-fermenting microbes (Hungate, 1966), so cellulose and other biomass is fermented to compounds (mainly volatile fatty acids) which provide energy to the animal. This fermentation is not complete, however, because the slow-growing microbes responsible for volatile fatty acid degradation are not present. The methanogens which degrade acetate and the obligate proton-reducing acetigens are not active in the rumen, probably because they cannot grow fast enough to maintain large populations. The methanogens, which use H_2 and formate, do occur in the rumen, so the products of this fermentation are mainly volatile fatty acids longer than formate, methane and carbon dioxide. The gases, which are eructated, are an important source of global methane (see, for example, Miller, 1991). The volatile fatty acids are absorbed by the animal as its major source of carbon and energy. An important by-product of the rumen fermentation (again, from the standpoint of the animal) is microbial cells, which serve as a source of protein.

4. Alternate electron acceptors

In many environments conditions prevail that allow other catabolic pathways than methanogenesis to occur. When organic matter is completely oxidized, carbon dioxide is always the product. Thus, carbon dioxide is always available as an electron acceptor for methanogens, which can use the reducing equivalents derived from that oxidation to reduce a portion of the carbon dioxide to methane. When waters contain electron acceptors other than carbon dioxide, such as O_2, nitrate, sulfate, manganese(IV), or iron(III), the electrons generated from organic matter oxidation are preferentially passed to one of these electron acceptors. For instance, when SO_4^{2-} is present, sulfate-reducing bacteria may pass the electrons stripped from organic matter to SO_4^{2-}, making H_2S. Methanogenesis is thought to be inhibited by the presence of alternate electron acceptors because bacteria using those electron acceptors out-compete methanogens for biodegradable molecules.

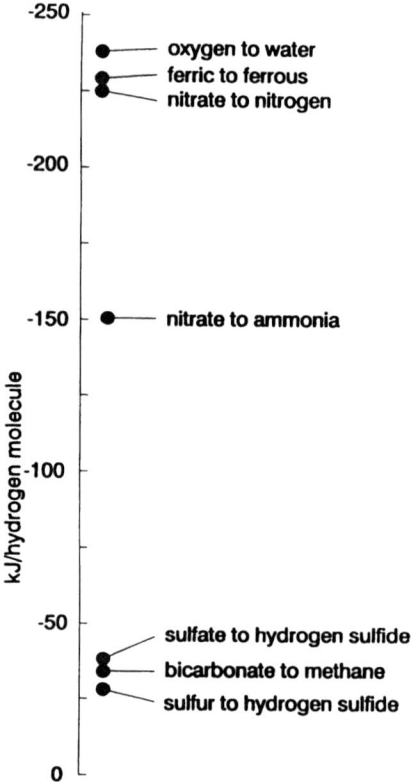

Figure 2. Energy released by coupling H_2 oxidation to the reduction of various catabolic electron acceptors.

The partial pressure of H_2 in sediments or groundwater can be used as an indicator of the dominant electron acceptor present, because it reflects the degree of competition for reduced substrates (Lovley and Goodwin, 1988). H_2 partial pressure is much lower when electron acceptors other than carbon dioxide are present, presumably because bacteria using those alternate electron acceptors have more energy available to scavenge reduced substrates at lower concentrations. This ability may derive from the extra energy that is released when electrons are passed to acceptors other than carbon dioxide. For instance, aerobic bacteria release approximately 10-fold more energy by using O_2 as electron acceptor than can anaerobic bacteria, which carry out the methanogenic decomposition of organic matter. This allows aerobes in the presence of O_2 to maintain concentrations of catabolic substrates at concentrations so low that methanogens cannot compete. In addition to this indirect inhibition, O_2 itself inhibits methanogens. In general, when organic matter is degraded in the presence of alternate electron acceptors, the electron acceptors whose reduction yield the most energy disappear first, and the others are used sequentially (Figure 2 shows the energy released by the reduction of various electron acceptors by H_2). Thus, methanogenesis generally begins only after all the alternate electron acceptors are depleted.

There are at least two exceptions to the general rule of sequential use of electron acceptors. The first is that heterogeneities in concentrations may lead to microenvironments in which different reactions predominate. For instance, in flocs or biofilms the outer surface may be exposed to sulfate-containing water, but the activities of sulfate-reducing bacteria on the surface may lower sulfate concentrations so it is not available deeper in the floc. Likewise, the use of insoluble electron acceptors may be kinetically limited by their availability. Fe(III) reduction may be limited by its solubilization. A second exception to the sequential use of electron acceptors is that specific organic molecules may be used by methanogens in preference to sulfate-reducing bacteria regardless of whether sulfate is present. These molecules are methyl amines and perhaps methyl sulfides and methanol, which can be used by methanogens to make methane without complete oxidation to carbon dioxide. Such compounds, which may be converted to methane even in the presence of excess sulfate, are called "non-competitive" substrates (Oremland et al., 1982). These "non-competitive" substrates" can also be catabolized by sulfate reducers, and it has not been established what environmental conditions determine whether they are degraded by methanogens or sulfate reducers.

5. Stoichiometry of methanogenesis

When microbial cells extract all the available free energy from organic matter, decomposition is carried to its ultimate conclusion and the organic matter is in its most stable form. For organic matter composed of carbon, hydrogen, and oxygen, the most stable form (at standard temperature and pressure) is a mixture of carbon dioxide (or bicarbonate), methane and water, in relative proportions dictated by stoichiometry (Buswell and Hatfield, 1939):

$$C_aH_bO_c + \left(a - \frac{1}{4}b - \frac{1}{2}c\right)H_2O \rightarrow \left(\frac{1}{2}a - \frac{1}{8}b + \frac{1}{4}c\right)CO_2 \\ + \left(\frac{1}{2}a + \frac{1}{8}b - \frac{1}{4}c\right)CH_4 \quad (1)$$

Other atoms, which may also be present in organic matter, are usually present in smaller concentrations and do not greatly affect stoichiometry. Their fates can be predicted by thermodynamics: transition metals are generally in a reduced state (Fe^{2+}, Ni^{1+}, Mn^{2+}), nitrogen, which may be found (NH_3 or N_2), sulfur as H_2S, and phosphorus as PO_4^{2-}. Their effect on the stoichiometry of methanogenesis can be calculated (McCarty, 1964), for instance for sulfur and nitrogen:

$$C_aH_bO_cS_dN_e + \left(a - \frac{1}{4}b - \frac{1}{2}c + \frac{1}{2}d + \frac{3}{4}e\right)H_2O \rightarrow \\ \left(\frac{1}{2}a - \frac{1}{8}b + \frac{1}{4}c + \frac{1}{4}d + \frac{3}{8}e\right)CO_2 + \left(\frac{1}{2}a + \frac{1}{8}b - \frac{1}{4}c - \frac{1}{4}d - \frac{3}{8}e\right)CH_4 \quad (2) \\ + dH_2S + eNH_3$$

These theoretical calculations reflect the products of anaerobic decomposition fairly accurately in environments where organic matter is the major energy source available for the microbial community, where the degradation of organic matter is complete, and where electron acceptors (such as O_2, NO_3^-, SO_4^{2-} Fe(III), and Mn(IV)) are not present in substantial concentrations. Adding such electron acceptors reduces the yield of methane and increases that of carbon dioxide. When excess quantities of electron acceptors are added (½a + ⅛b − ¼c − ¼d − ⅜e ≤ 0), methane production ceases (except perhaps for non-competitive substrates), and essentially all organic carbon is released as carbon dioxide.

6. The methanogenic bacteria

Reviews of the biology, ecology, and biochemistry of methanogens (Whitman, 1985; Vogels et al., 1988; DiMarco et al., 1990) indicate that all methanogens are very strictly anaerobic bacteria that all form methane as a major terminal catabolic product. However, methanogens may be found in environments that are grossly oxic, either because they inhabit microenvironments that are free of O_2 or because they can grow in the presence of O_2 when the inhibitory reaction products are removed by other bacteria within their environment.

Beginning in the late 1970s, evolutionary studies of ribosomal RNA indicated that methanogens are an evolution-coherent group of microbes that are distinct from most known bacteria (Jones et al., 1987; Woese, 1987). Together with some sulfur-dependent thermophiles and extremely halophilic bacteria, methanogens make up the "urkingdom," *Archaeobacteria*, separate from the urkingdoms of "normal" bacteria (prokaryotes) and of eukaryotes (plants, animals, fungi, and protozoa) in a number of important ways. The name "urkingdom" was coined to refer to a taxonomic grouping higher than the kingdom (such as *Planta*, *Animalia*). The three urkingdoms are distinct cell lines, which have evolved along different pathways since a time early in the evolution of life on Earth. Thus, the characteristics that methanogens share with prokaryotes and eukaryotes likely were developed early in the evolution of life on Earth, so studies of methanogen biochemistry and comparisons with that of prokaryote and eukaryotes may help us to understand how early life evolved. This is one of the reasons for the keen interest in the biochemistry of methanogens, including their physiology, genetics, biosynthetic pathways, structure of cell walls (Kandler and König, 1985) and membranes (Langworthy, 1985), and evolution. Table 1 lists the genera of methanogenic bacteria and their habitats.

7. Biochemistry of methanogenesis

The biochemistry of methanogenesis summarized here has been reviewed elsewhere (Whitman, 1985; Vogels et al., 1988; DiMarco et al., 1990; Ferry, 1993). The ultimate biochemical step in the formation of methane by methanogens is a reductive demethylation of methyl carrier coenzyme M (2-mercaptoethanesulfonate [Taylor and Wolfe, 1974]) (Figure 3). This reduction is coupled to ATP synthesis, but the small amount of free energy available under in situ conditions (Boone et al., 1989) implies that the stoichiometry must be less than one ATP per mole of methane formed. Barker (1956) correctly predicted that the mechanism of methane formation from carbon dioxide reduction was a step-wise 2-electron reduction of one or more C_1 carriers, with other methyl-group donors (e.g., methanol, acetate, methyl sulfides) feeding into the pathway at various points.

Table 1. Genera of methanogenic and methanotrophic bacteria.

	Shape	Substrates[1]	Habitat
Methanogens			
Methanobacterium	rod	$H_2 + CO_2$, formate	Anaerobic digestors, animal feces and rumen contents, ciliate endosymbiont, compost soil of rice paddies, oil wells, and aquifers, and sediments of rivers, hot springs, and the ocean.
Methanobrevibacter	short rod	$H_2 + CO_2$, formate	Anaerobic digestors, human and animal feces, rumen, human vagina, termite gut.
Methanococcoides	coccoid	methyl amines and methanol	Marine sediments and production water from an oil well.
Methanococcus	coccus	$H_2 + CO_2$, formate	Anaerobic digestors, and thermal and non-thermal sediments of oceans estuaries, and salt marshes.
Methanocorpusculum	coccoid	$H_2 + CO_2$, formate	Anaerobic digestors, saline and freshwater sediments of lakes and rivers.
Methanohalobium	coccoid	methyl amines[2]	Hypersaline sediments of lakes and salterns.
Methanohalophilus	coccoid	methyl amines, methanol, methyl sulfides[3]	Hypersaline lake sediments and aquifer solids.

Table 1. Genera of methanogenic and methanotrophic bacteria.

	Shape	Substrates[1]	Habitat
Methanolacinia	rod	$H_2 + CO_2$, formate	Marine sediments.
Methanolobus	coccoid	methyl amines, methanol, methyl sulfides	Thermal and nonthermal marine sediments, oil wells.
Methanomicrobium	rod	$H_2 + CO_2$, formate	Bovine rumen.
Methanoplanus	coccoid	$H_2 + CO_2$, formate	Sediments, saline swamp of oil-drilling wastes, ciliate endosymbiont.
"*Methanopyrus*"	rod	$H_2 + CO_2$	Thermal marine sediments.
Methanosarcina	coccoid, psudosarcina grape-like aggrevates	acetate, methyl amines, methanol, methyl sulfides, $H_2 + CO_2$	Anaerobic digestors, wastewaters, thermal and nonthermal sediments of lakes, rivers, estuaries, and ocean, anaerobic soils, bovine rumen, and animal manure.
Methanosphaera	rod	H_2 + methanol[4]	Feces of humans and rabbit.
Methanospirillum	spiral rod	$H_2 + CO_2$, formate	Anaerobic digestors.

Table 1. Genera of methanogenic and methanotrophic bacteria.

	Shape	Substrates[1]	Habitat
Methanothermus	rod	$H_2 + CO_2$	Thermal soils and sediments.
Methanothrix	sheathed rod	acetate	Anaerobic digestors.
Methanotrophs			
Methylobacter	rod	CH_4, CH_3OH	Soils, sediments, and waters.
Methylococcus	coccus	CH_4, CH_3OH	Soils, sediments, and waters.
Methylocystis	coccus	CH_4, CH_3OH	Soils, sediments, and waters.
Methylomonas	rod	CH_4, CH_3OH	Soils, sediments, and waters.
Methylosinus	rod or pear	CH_4, CH_3OH	Soils, sediments, and waters.

[1] Other substrates that may be used by methanogens are carbon monoxide and alcohols; the importance of these as methanogenic precursors in natural environments has not been established.
[2] Mono-, di- and trimethalamine.
[3] Methane thiol and dimethyl sulfide.
[4] Requires both H_2 and methanol and cannot grow on methanol alone as can other methanol-using methanogens.
[5] Other substrates that may be used by methanogens are carbon monoxide and alcohols; the importance of these as methanogenic precursors in natural environments has not been established.
[6] Mono-, di-, and trimethylamine.
[7] Methane thiol and dimethyl sulfide.
[8] Requires both H^2 and methanol and cannot grow on methanol alone as can other methanol-using methanogens.

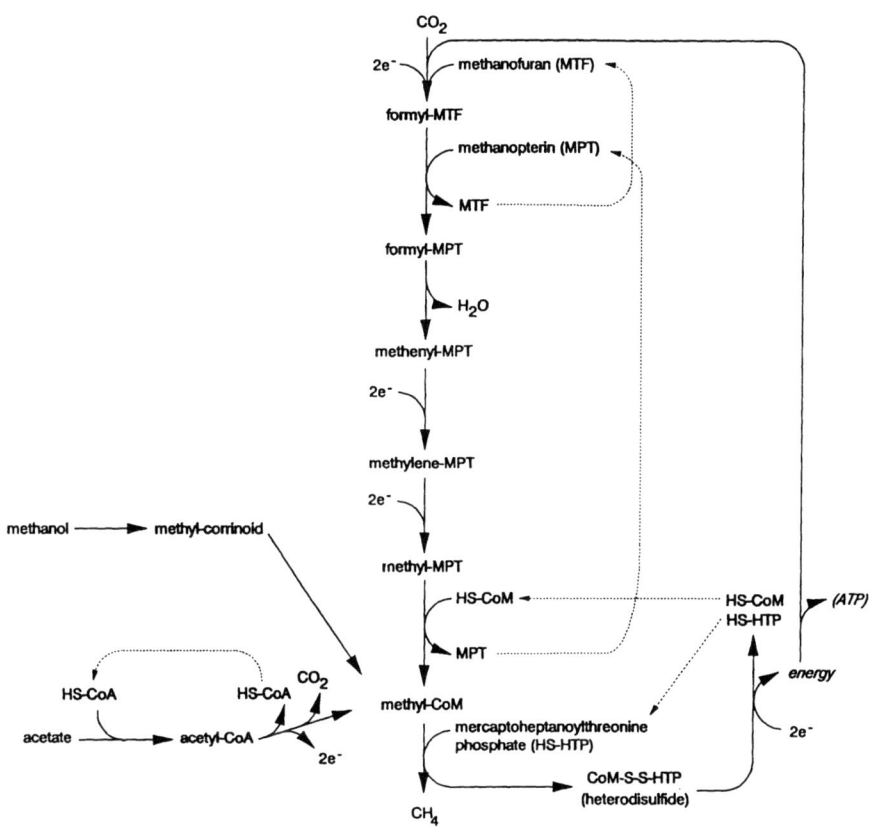

Figure 3. Pathway of methane formation by methanogens. Carbon dioxide is reduced to methane by four 2-electron reductions; the electrons are carried on the electron carrier F_{420} for at least two of these reduction. The C_1 carriers in methanogenesis (methanofuran, methanopterin, and coenzyme M) are recycled, as shown by the dotted arrows. During methanogenesis from methyl groups of acetate, methanol, methyl amines, and methyl sulfides, C_1 groups are transferred to coenzyme M (forming methyl-CoM), as shown in the left half of the diagram.

The first step of methanogenesis from H_2 plus carbon dioxide is perhaps the least-well understood (Figure 3). Carbon dioxide is reductively condensed with the unique coenzyme methanofuran in a reaction that is driven in some way by energy released in by later reactions. Of all of the reductive steps in methanogenesis (Figure 3) only the final cleavage of methane from methyl-coenzyme M can yield substantial energy. After the carbon dioxide is fixed as a C_1 group on methanofuran at the formyl oxidation state, this group is transferred to another coenzyme unique to methanogens, methanopterin. The formyl group is then dehydrated and reduced by two 2-electron reductions to the methyl level. This reduction is analogous to C_1 metabolism in folate biochemistry. For at least two of these reductions, the electron carrier that directly supplies the electrons is another coenzyme unique to methanogens, factor F_{420} (Cheeseman et al., 1972). The methyl group is then transferred from methanopterin to coenzyme M (HS-CoM) (Poirot et al., 1987), and finally it is reductively cleaved to yield methane. This reaction is complex, requiring several proteins and cofactors (Gunsalus and Wolfe, 1978, 1980; Nagle and Wolfe, 1983) and is in some way linked to the carbon dioxide fixing step. Methane is apparently reductively cleaved from methyl-CoM by reaction with 7-mercaptoheptanoylthreonine phosphate (HS-HTP), and resulting in the disulfide HTP-CoM (see Blaut et al., 1990). The regeneration of reduced HS-HTP and HS-CoM is likely the reaction coupled to energy generation for the cell.

When methyl compounds (acetate, methanol, trimethylamine, dimethyl sulfide) are the catabolic substrates, the methyl group of these compounds is transferred to one of the intermediates in the pathway for carbon dioxide reduction. For instance, the methyl group of methanol (and perhaps also dimethyl sulfide) is transferred to HS-CoM by a corrinoid methyltransferase; this methyltransferase is substrate specific. The methyl-CoM is then reductively demethylated to yield methane, as in carbon dioxide reduction. The source of the electrons for this reduction is not well known. When H_2 is available, methanogens of the genus *Methanosarcina* or *Methanosphaera* use it as the electron donor for methyl-CoM reduction. In fact, H_2 is the only source of these reducing equivalents that *Methanosphaera* can use. *Methanosarcina* can, when no H_2 is available, disproportionate methanol or trimethylamine. That is, it can generate reducing equivalents from (for instance) methanol by oxidizing some methanol. For every molecule of methanol oxidized to carbon dioxide, 3 pairs of electrons are generated, and these electron pairs can each be used to reduce one methyl-coenzyme M to methane. Thus, the stoichiometry of methanol oxidation is that four molecules of methanol are degraded to three of methane and one of carbon dioxide (plus two of water). The mechanism of methanol oxidation is not well understood, but another novel cofactor (sarcinopterin) may be the C_1 carrier for the oxidation of methanol (van Beelen et al., 1984).

In ecosystems other than intestinal tracts and marine or estuarine sediments, acetate is the major precursor to methane. However, acetate can be catabolized by only two genera of methanogens, *Methanosarcina* and *Methanothrix* (= "*Methanosaeta*"), via the aceticlastic reaction. These organisms transfer the methyl group of acetate to HS-CoM (making methyl-CoM) and oxidize the carboxyl group to carbon dioxide. The oxidation of the carboxyl group to carbon dioxide provides electrons for the reductive release of methane from methyl-CoM. Lysates of aceticlastic methanogens form methane from acetate (Krzycki and Zeikus, 1984) but at rates much lower than whole cells and then only when H_2 is added. However, when the membrane fraction is included in cell-free lysates, the rate of aceticlastic methanogenesis approaches that of whole cells, even without added H_2 (Baresi, 1984). In fact, H_2 inhibits methane production from acetate by these particulate fractions (Baresi, 1984).

8. Influence of environmental factors on methane production

The presence of alternate electron acceptors (O_2, nitrate, Fe(III), sulfate, etc.) is a major factor affecting methane production. Methane is rarely found in oxic environments, and the main reason is probably that O_2-using bacteria out-compete methanogens for catabolic substrates, as described above. In addition, methanogens are very strict anaerobes, being poisoned by even very tiny concentrations of O_2. This extreme sensitivity to O_2 is readily apparent in pure cultures of methanogens, but it may not be important in natural environments where other bacteria remove either O_2 itself or toxic reaction products of O_2. Methanogens may thrive in microenvironments, such as dead-end pores in soil, where O_2 does not penetrate, and methane that is formed there may be oxidized by methanotrophic bacteria as the methane diffuses out into oxic zones. Like O_2, nitrate, ferric, sulfate, and other electron acceptors tend to block the formation of methane by supporting the activities of bacteria, which out-compete methanogens for their substrates.

Like those of most bacteria, the activities of methanogens increase approximately linearly with temperature except within a few degrees centigrade of the maximum temperature of that organism. The maximum growth temperature of most methanogens is about 37° to 45°C, although many grow up to 65°C, and some even grow near the boiling point of water. These latter organisms may be found at hydrothermal vents, where they grow by using the H_2 emanating from the vent. Some methanogens have maximum temperatures as low as 20° to 25°C (Romesser et al., 1979), but these organisms can adapt to higher temperatures (Maestrojuán et al., 1990). Thus far, no psychrophilic methanogens have been isolated, although many important methanogenic environments are permanently at low temperature (Conrad et al., 1989; Conrad and Wetter, 1990).

The optimum pH for methanogens is generally near neutral (6 to 8), although some grow well at pH values as low as 4 (Patel et al., 1990; Maestrojuán and Boone, 1991) and as high as 10 (Worakit et al., 1985; Boone et al., 1986; Mathrani et al., 1988; Liu et al., 1990).

9. Factors affecting methane transport to the atmosphere

Probably the most important factor affecting the transport of methane to the atmosphere is the intervention of aerobic methane-oxidizing bacteria (Kiene, 1991). These bacteria can grow on methane as the sole carbon source in oxic environments, obtaining their energy by oxidizing methane to carbon dioxide. Table 1 shows the major genera of methane-oxidizing bacteria (see review: Lidstrom, 1992; Hansen et al., 1992). The importance of methane-oxidizing bacteria in diminishing methane inputs to the atmosphere may be seen in soils above sanitary landfills, where methane diffusing from underlying anaerobic zones enter aerobic zones. The high rates of methane oxidation in soils such as these may intercept most or all of the methane before it can reach the surface (Whalen and Reeburgh, 1988; Striegl and Ishii, 1989; Whalen et al., 1990).

Some mechanisms of transport avoid the activities of methanotrophs almost completely. When methane is formed slowly in the deep sediments of an oxic lake, the methane that diffuses upwards into the oxic upper sediments may be partly or completely oxidized before it reaches the surface. On the other hand, if methane is formed in the sediments more rapidly than it can leave via diffusion, the methane accumulates until its partial pressure is greater than the ambient pressure, and large bubbles can form and rise through the sediments. This methane passes through the oxic, methanotrophic sediments rapidly, and because methane is poorly soluble, most of the methane in the bubbles is released into the atmosphere. It is difficult to quantify methane loss to the atmosphere via bubbling, and it has been estimated that 80% of the methane formed in rice paddies is oxidized before it reaches the atmosphere (Craig, 1957). A more important mechanism by which methane can elude oxidation in rice paddies is by diffusion through plants (Dacey and Klug, 1979).

Another way methane may pass from the anoxic zones where it is formed into the atmosphere without exposure to methanotrophic environments is by eructation (belching) of ungulates. The methane accumulated in the anoxic rumen is discharged into the atmosphere without passage through an oxic environment that can support the activities of methanotrophs. Methane formed in groundwater or in landfills must pass through the soil before it enters the atmosphere, so some or most of the methane may be oxidized there (Striegl and Ishii, 1989; Whalen et al., 1990). Methane formed in ocean sediments must often pass through an anoxic, sulfate-containing zone and then an oxic zone before it reaches the atmosphere. The abilities of O_2-using methanotrophs to oxidize methane is well documented, but cultures of anaerobic (sulfate-reducing) methane-oxidizers has not been confirmed. However, geochemical evidence strongly suggests that methane oxidation occurs in the anaerobic waters of the ocean (Reeburgh, 1976; Reeburgh and Heggie, 1977). Catabolic oxidation of alkane was recently thought to require a monooxygenase as the first step, making alkane oxidation impossible in anoxic ecosystems, but Aeckersberg et al. (1991) demonstrated that higher molecular-weight alkanes are catabolized by anaerobic sulfate reducers. The possibility that low-molecular-weight alkanes such as methane may be similarly catabolized, or whether it is co-metabolized, remains to be proven.

10. Signatures of biological methane formation

Two types of signatures are used to distinguish the source of methane: the ratio of methane to ethane plus propane; and isotopic ratios.

10.1 Ratio of methane to ethane plus propane. This ratio of methane to light hydrocarbons is very high for biologically-produced methane (ratios much greater than 100 [Cicerone and Oremland, 1988]) because the biochemical mechanisms for methanogenesis are very specific (Oremland et al., 1988). In contrast, thermogenic reactions that produce methane always form substantial amounts ethane and propane as well (ratios typically less than 50 [Oremland et al., 1988]). These ratios may be modified by differential rates of microbial alkane oxidation. Thus, some results may be ambiguous, especially when the methane is derived from multiple sources and the ratio is between the normal values for biogenic and abiogenic methane.

10.2 Stable isotope ratios.

10.2.1 $^{13}C/^{12}C$ ratio. Two stable isotopic ratios give useful information of the source of methane, $^{13}C/^{12}C$ and D/H. The isotopic ratio of $^{13}C/^{12}C$ (recently reviewed by Tyler [1991]) is the most commonly used evidence of methane source, normally calculated relative to the standard, carbonate in Pee Dee Belemnite (Craig, 1957), expressed as:

$$\delta^{13}C\ (\text{‰}) = [(^{13}C/^{12}C)_{sample}/(^{13}C/^{12}C)_{standard} - 1] \times 1000 \qquad (3)$$

Enzymatic reactions select for lighter isotopes, so biologically-produced methane tends to have less ^{13}C (i.e., a lower $\delta^{13}C$). Divergence of the $\delta^{13}C$ of biologically produced methane could be due to one of two factors. The first factor is discrimination by methanogenic enzymes between the two carbon isotopes, which is referred to as α_C (Whiticar et al., 1986):

$$\alpha_C = (\delta^{13}C_{CO_2} + 10^3)/(\delta^{13}C_{CH_4} + 10^3) \qquad (4)$$

The second factor that can lead to divergence of $\delta^{13}C$ is formation of methane from a precursor pool whose $\delta^{13}C$ has been modified. Probably both of these factors are at work, because the biomass from which methane is produced is depleted in ^{13}C with respect to geological CO_2, having $\delta^{13}C$ values of -10‰ to -35‰. Thermogenic methane, which is derived from biomass, is slightly lighter than biomass ($\delta^{13}C$ values of -25‰ to -50‰), but biogenic methane is lighter still ($\delta^{13}C$ values of -45‰ to -80‰ [Tyler, 1991]).

This divergence of the $\delta^{13}C$ of the produced methane relative to the biomass from which it was produced is normally attributed to fractionation by enzymes of the methanogens themselves (α_C). However, there are at least two other mechanisms that may cause fractionation. The methane fermentation may selectively use certain carbons of the biomass (e.g., sediments preferentially convert the C-1 and C-6 carbons of glucose to methane [Krumböck and Conrad, 1991]), or the ultimate carbon substrates of methanogens may be preselected by other organisms, as illustrated by carbon flow in ruminants (see Hungate, 1966). In the rumen fermentation, biomass is fermented by bacteria to volatile fatty acids (mainly acetic, propionic, and butyric), bicarbonate/CO_2, and CH_4. The gases CO_2 and CH_4 (approximately 3:1) are released to the atmosphere by eructation. By swallowing saliva, the ruminant maintains a flow of liquids through the rumen, which buffers the rumen by its dissolved bicarbonate and transports the fatty acids to its next stomach (the abomasum) to facilitate absorption into the blood. The input of bicarbonate from the saliva to the rumen is a substantial source, and the isotopic ratio of that bicarbonate is affected by numerous enzymatic activities of the animal, most importantly by the carbonic anhydrase, which facilitates CO_2 removal in the lungs. Thus, the isotopic ratio of bicarbonate in the rumen may differ from that of the biomass that is eaten by the animal, and it is this bicarbonate that is the starting carbon for ruminal methane formation, but that may be further fractionated by the methanogens themselves.

Methane formation from CO_2 reduction accounts for about 30% of the methane from freshwater sediments and flooded soils, and nearly 100% of the methane from gastrointestinal tracts. The fraction of CH_4 in marine sediments that is due to CO_2 reduction is not well known; although most known halophilic methanogens catabolize methyl groups and do not reduce CO_2, the biogeochemical data of Whiticar et al. (1986) suggest that most methane comes from CO_2 reduction.

The precursor of about 70% or more of methane produced from flooded soils and freshwater sediments is acetate (Winfrey and Zeikus, 1979; Mackie and Bryant, 1981; Boone, 1982; Krumböck and Conrad, 1991). Whereas low $\delta^{13}C$ of methane from CO_2 reduction may be partially explained by isotopic discrimination by methanogens, such is not the case for methane formed from acetate. When methane is formed from a methyl group (such as the methyl group of acetate), methane formation requires an electron pair to reduce the methyl group. During methanogenesis from acetate, this electron pair is acquired by the oxidation of the carboxyl group to CO_2. Thus, the methyl group of acetate is converted intact to methane, and the carboxyl group is converted to carbon dioxide. Because these conversions are essentially complete, the $\delta^{13}C$ of the methane is the same as that of the methyl group. Thus, any measured differences between the $\delta^{13}C$ of the methane from acetate and that of the biomass are due to differences between the $\delta^{13}C$ of the methyl group of acetate and that of the biomass. In the most common fermentative pathway of hexoses, the Embden-Meyerhof-Parnas pathway, the C-1 and C-6 of glucose are converted to the methyl group of acetate, and hence to methane (Mah et al., 1978; Krumböck and Conrad, 1991).

During methanogenesis of other methyl compounds, such as methanol, trimethylamine, or dimethylsulfide, there is no carboxyl group to oxidize, so the electron pair needed to reduce the methyl group to methane must come from elsewhere. It may come from H_2 (Deppenmeier et al., 1988; Miller, 1991) or from oxidation of some of the methyl group to carbon dioxide (Deppenmeier et al., 1988). In the latter case, where there are two possible fates for the carbon of the methyl group, isotopic fractionation of the methane relative to that of the methyl group may occur. In addition, the $\delta^{13}C$ of the methyl group may be different than that of the biomass.

Thus, $\delta^{13}C$ of methane formed is determined in part by the isotopic make-up of the biomass from which it is produced and in part by isotopic discrimination of methanogenic enzymes. The $\delta^{13}C$ of methane released into the atmosphere may be affected by a third factor, isotopic discrimination by another group of bacteria, the methanotrophic bacteria, which oxidize methane. Aerobic methanotrophs selectively oxidize the lighter isotope, so their activity would raise the $\delta^{13}C$ of residual methane. It should be pointed out that this altered $\delta^{13}C$ would only be significant if a substantial fraction of produced methane were oxidized, therefore diminishing its impact on regional or global isotopic ratios. This may be the case in the ocean, where anaerobic methanotrophs oxidize methane. Anaerobic methane oxidation as a catabolic process has not been demonstrated in the laboratory, but there is convincing evidence in the shapes of methane gradients in the ocean. Also, the finding by Aeckersberg et al. (1991) of sulfate-reducing bacteria capable of oxidizing alkanes longer than methane suggests that slow-growing methane-oxidizing sulfate reducers may exist in the oceans.

10.2.2 D/H ratio. The ratio of deuterium to hydrogen in methane may also be measured. Analogous to $\delta^{13}C$ for carbon, $\delta(D/H)$ is calculated relative to ocean water (Hoefs, 1987, in Tyler, 1991). When CH_4 is formed by

CO_2 reduction, the hydrogens are derived from water (or protons). However, when methane is formed from methyl groups (acetate, methanol, etc.), the three hydrogens of the methyl group remain intact, and only the fourth hydrogen is derived from water. Thus, the $\delta(D/H)$ of methane from CO_2 reduction is determined by a methanogens's fractionation of the hydrogen of water, but most of the $\delta(D/H)$ of methane from methyl groups may be predetermined by the $\delta(D/H)$ of the methyl groups, based on the finding of Walther et al. (1981) that the methyl groups are reduced to methane with the hydrogens intact.

10.2.3 ^{14}C content of methane. Another isotopic measure of CH_4 which yields information about its source is the ratio of $^{14}C/^{12}C$. Because ^{14}C is radioactive, it is not present in very old carbon. Modern atmospheric carbon receives a constant input of ^{14}C which is derived from cosmic ray-derived neutrons, which are captured by ^{14}N which then ejects a proton to become ^{14}C. However, anomalies in atmospheric ^{14}C content caused by detonation of atomic weapons and escape of gases from nuclear power plants complicates ^{14}C dating of methane sources. Modern biomass, formed from atmospheric CO_2, has the same ^{14}C content as atmospheric CO_2, but old carbon loses its ^{14}C. The fraction of ^{14}C in CH_4 is expressed as a percentage of the fraction in modern carbon. Methane in geological formations can be dated by its ^{14}C content, but other measures (such as $\delta^{13}C$ or ratio of methane to ethane plus propane) must be used to differentiate old biogenic from old thermogenic methane (see also Whiticar, this volume, and Stevens, this volume).

11. Summary

Methane production occurs when organic matter is degraded in environments where light and inorganic electron acceptors such as O_2, Fe(III), Mn(IV), nitrate, sulfate, and sulfur are limiting. Under these conditions organic matter decomposition is catalyzed by consortia of bacteria including specialized bacteria that form methane. Methane and carbon dioxide are the terminal products of metabolism, and these compounds are stable (except at geological pressures). The contribution of this methane to the atmosphere may be moderated by the activities of methanotrophic bacteria, which can use O_2 to oxidize methane to carbon dioxide and water.

Acknowledgements. This work was supported by section 105 grant #14-08-001-G1636 from the U. S. Geological Survey.

References

Aeckersberg, F., F. Bak, F. Widdel. 1991. Anaerobic oxidation of saturated hydrocarbons to CO_2 by a new type of sulfate-reducing bacterium. *Arch. Microbiol., 156*:5-14.

Bak, F., K. Finster. 1992. Anaerobic formation and degradation of dimethyl sulfide and methane thiol by new types of acetogenic and methanogenic bacteria. In: *Proceedings of the 10th Int. Symp. on Environ. Biogeochem* (R.S. Oremland, ed.), San Francisco. [in press]

Baresi, L. 1984. Methanogenic cleavage of acetate by lysates of *Methanosarcina barkeri*. *J. Bacteriol., 160*:365-370.

Barik, S., W. J. Brulla, M. P. Bryant. 1985. PA-1, a versatile anaerobe obtained in pure culture, catabolizes benzenoids and other compounds in syntrophy with hydrogenotrophs, and P-2 plus *Wolinella* sp. degrades benzenoids. *Appl. Environ. Microbiol., 50*:304-310.

Barker, H. A. 1956. Bacterial fermentations, p. 1-27. Wiley, New York.

Blaut, M., V. Müller, G. Gottschalk. 1990. Energetics of methanogens. In: *The Bacteria, Vol. 12* (J.R. Sokatch and L. Nicholas Ornston, eds.), Academic Press, Inc., San Diego, 505-537.

Boone, D.R. 1982. Terminal reactions in the anaerobic digestion of animal waste. *Appl. Environ. Microbiol., 41*:57-61.

Boone, D.R., S. Worakit, I.M. Mathrani, R.A. Mah. 1986. Alkaliphilic methanogens from high-pH lake sediments. *J. Syst. Appl. Microbiol., 7*:230-234.

Boone, D.R., R.L. Johnson, Y. Liu. 1989. Diffusion of the interspecies electron carriers H_2 and formate in methanogenic ecosystems, and its implication in the measurement of K_m for H_2 or formate uptake. *Appl. Environ. Microbiol., 55:*,1735-1,741.

Bryant, M.P. 1979. Microbial methane production: theoretical aspects. *J. Anim. Sci., 48:*193-201.

Bryant, M.P., E.A. Wolin, M.J. Wolin, R.S. Wolfe. 1967. *Methanobacillus omelianskii*, a symbiotic association of two species of bacteria. *Arch. Mikrobiol., 59:*20-31.

Buswell, A.M., W.D. Hatfield. 1939. *Anaerobic fermentations.* Illinois State Water Survey, Urbana, Ill.

Cheeseman, P., A. Toms-Wood, R.S. Wolfe. 1972. Isolation and properties of a fluorescent compound, Factor F_{420}, from *Methanobacterium* strain M.o.H. *J. Bacteriol., 112:*527-531.

Cicerone, R.J., R.S. Oremland. 1988. Biogeochemical aspects of atmospheric methane. *Global Biogeochem. Cycles, 2:*299-327.
Conrad, R., B. Wetter. 1990. Influence of temperature on the energetics of hydrogen metabolism in homoacetogenic, methanogenic, and other bacteria. *Arch. Microbiol., 155:*94-98.
Conrad, R., B. Schink, T.J. Phelps. 1986. Thermodynamics of H_2-producing and H_2-consuming metabolic reactions in diverse methanogenic environments under in situ conditions. *FEMS Microbiol. Ecol., 38:*353-360.
Conrad, R., F. Bak, H.F. Seitz, B. Thebrath, H.P. Mayer, H. Schultz. 1989. Hydrogen turnover by psychrotrophic homoacetogenic and mesophilic methanogenic bcteria in anoxic paddy soil and lake sediment. *FEMS Microbiol. Ecol., 62:*285-294.
Craig, H. 1957. Isotopic standards for carbon and oxygen and correction factors for mass-spectroscopic analysis of carbon dioxide. *Geochim. Cosmochim. Acta, 12:*133-149.
Dacey, J.W.H., M.J. Klug. 1979. Methane efflux from lake sediments through water lilies. *Science 203:*1253-1255.
Deppenmeier, U., M. Blaut, A. Jussofie, G. Gottschalk. 1988. A methyl-coM methylreductase system from methanogenic bacterium strain Göl not requiring ATP for activity. *FEBS Lett., 241:*60-64.
DiMarco, A.A., T.A. Bobik, R.S. Wolfe. 1990. Unusual coenzymes of methanogenesis. *Annu. Rev. Biochem., 59:*355-394.
Dwyer, D.F., E. Weeg-Aessens, D.R. Shelton, J.M. Tiedje. 1988. Bioenergetic conditions of butyrate metabolism by a syntrophic, anaerobic bacterium in coculture with hydrogen-oxidizing methanogenic and sulfidogenic bacteria. *Appl. Environ. Microbiol., 54:*1,354-1,359.
Ferry, J.G. (ed.) 1993. *Methanogenesis.* Chapman & Hall. [in preparation].
Gunsalus, R.P., R.S. Wolfe. 1978. ATP activation and properties of the methyl coenzyme M reductase system in *Methanobacterium thermoautotrophicum*. *J. Bacteriol., 135:*851-857.
Gunsalus, R.P., R.S. Wolfe. 1980. Methyl coenzyme M reductase from *Methanobacterium thermoautotrophicum*: resolution and properties of the components. *J. Biol. Chem., 255:*1,891-1,895.
Hanson, R.S., A.I. Netrusov, K. Tsuji. 1992. The obligate methanotrophic bacteria: *Methylococcus, Methylomonas,* and *Methylosinus.* In: *The Prokaryotes, A Handbook on the Biology of Bacteria: Ecophysiology, Isolation, Identification, Applications* (A. Ballows, H.G. Trüper, M. Dworkin, W. Harder, and K.-H. Schleifer, eds.), second edition. Springer-Verlag, New York, p. 2,350-2,364.
Hoefs, J. 1987. *Stable Isotope Geochemistry*, 3rd edition, p. 22-24. Springer-Verlag, New York.

Houwen, F.P., C. Dijkema, C.C.H. Schoenmakers, A.J.M. Stams, A.J.B. Zehnder. 1987. ^{13}C-NMR study of propionate degradation by a methanogenic coculture. *FEMS Microbiol. Lett., 41:*269-274.
Hungate, R.E. 1966. *The Rumen and Its Microbes*, p. 1-533. Academic Press, New York.
Jones, W.J., D.P. Nagel Jr., W.B. Whitman. 1987. Methanogens and the diversity of archaebacteria. *Microbiol. Rev., 51:*135-177.
Kandler, O., and H. König. 1985. Cell envelopes of archaebacteria. In: *The Bacteria: Vol. VIII, Archaebacteria* (C.R. Woese and R.S. Wolfe, eds.), Academic Press, Orlando, Fla., p. 413-457.
Kiene, R.P. 1991. Production and consumption of methane in aquatic sediments. In: *Microbial Production and Consumption of Greenhouse Gases: Methane, Nitrogen Oxides, and Halomethanes* (J.E. Rogers and W.B. Whitman, eds.), American Society for Microbiology, Washington, D.C., p. 111-146.
Krumböck, M., R. Conrad. 1991. Metabolism of position-labelled glucose in anoxic methanogenic paddy soil and lake sediment. *FEMS Microbiol. Ecol., 85:*247-256.
Krzycki, J.A., J.G. Zeikus. 1984. Acetate catabolism by *Methanosarcina barkeri*: hydrogen-dependent methane production from acetate by a soluble cell protein fraction. *FEMS Microbiol. Lett., 25:*27-32.
Langworthy, T.A. 1985. Lipids of archaebacteria. In: *The Bacteria: Vol. VIII, Archaebacteria* (C.R. Woese and R.S. Wolfe, eds.), Academic Press, Orlando, Fla., p. 413-457.
Leisinger, T., W. Brunner. 1986. Poorly degradable substances. In: *Biotechnology: Microbial Degradations, Vol. 8* (W. Schönborn, ed.), VCH Verlagsgesellschaft, Weinheim, Germany, p. 475-513.
Lidstrom, M.E. 1992. The aerobic methylotrophic bacteria. In: *The Prokaryotes, a Handbook on the Biology of Bacteria: Ecophysiology, Isolation, Identification, Applications* (A. Ballows, H.G. Trüper, M. Dworkin, W. Harder, and K.-H. Schleifer, eds.), second edition. Springer-Verlag, New York, p. 432-445.
Liu, Y., D.R. Boone, C. Choy. 1990. *Methanohalophilus oregonense* sp. nov., a methylotrophic methanogen from an alkaline, saline aquifer. *Int. J. Syst. Bacteriol., 40:*111-116.
Ljungdahl, L.G. 1986. The autotrophic pathway of acetate synthesis in acetogenic bacteria. *Ann. Rev. Microbiol., 40:*415-450.
Lovley, D.R., S. Goodwin. 1988. Hydrogen concentration as an indicator of the predominant terminal electron acceptor reactions in aquatic sediments. *Geochim. Cosmochim. Acta, 52:*2,993-3,003.

Mackie, R.I., M.P. Bryant. 1981. Metabolic activity of fatty acid-oxidizing bacteria and the contribution of acetate, propionate, butyrate, and CO_2 to methanogenesis in cattle waste at 40 and 60°C. *Appl. Environ. Microbiol.,* *41:*1,363-1,373.

Maestrojuán, G.M., D.R. Boone. 1991. Characterization of *Methanosarcina barkeri* strains MS^T and 227, *Methanosarcina mazei* $S-6^T$, and *Methanosarcina vacuolata* $Z-761^T$. *Int. J. Syst. Bacteriol., 41:*267-274.

Maestrojuán, G.M., D.R. Boone, L. Xun, R.A. Mah, L. Zhang. 1990. Transfer of *Methanogenium bourgense, Methanogenium marisnigri, Methanogenium olentangyi,* and *Methanogenium thermophilicum* to the genus *Methanoculleus,* gen. nov., emendation of *Methanoculleus marisnigri* and *Methanogenium,* and description of new strains of *Methanoculleus bourgense* and *Methanoculleus marisnigri. Int. J. Syst. Bacteriol., 40:*117-122.

Mah, R.A., M.R. Smith, L. Baresi. 1978. Studies on an acetate-fermenting strain of *Methanosarcina. Appl. Environ. Microbiol., 35:*1,174-1,184.

Mathrani, I.M., D.R. Boone, R.A. Mah, G.E. Fox, P.P. Lau. 1988. *Methanohalobium zhilinae,* gen. nov. sp. nov., an alkaliphilic, halophilic, methylotrophic methanogen. *Int. J. Syst. Bacteriol., 38:*139-142.

McCarty, P.L. 1964. The methane fermentation. In: *Principles and Applications in Aquatic Microbiology* (H. Heukelekian and N.C. Dondero, eds.), John Wiley & Sons, New York, p. 314-343.

McInerney, M.J. 1986. Transient and persistent associations among prokaryotes. In: *Bacteria in Nature, Vol. 2* (E.R. Leadbetter and J.S. Poindexter, eds.), Plenum Publishing Corp., New York, p. 293-338.

McInerney, M.J., M.P. Bryant, N. Pfennig. 1979. Anaerobic bacterium that degrades fatty acids in syntrophic association with methanogens. *Arch. Microbiol., 122:*129-135.

McInerney, M.J., M.P. Bryant, R.B. Hespell, J.W. Costerton. 1981. *Syntrophomonas wolfei* gen. nov. sp. nov., an anaerobic, syntrophic, fatty acid-oxidizing bacterium. *Appl. Environ. Microbiol., 41:*1,029-1,039.

Miller, T.L. 1991. Biogenic sources of methane. In: *Microbial Production and Consumption of Greenhouse Gases: Methane, Nitrogen Oxides, and Halomethanes* (J.E. Rogers and W.B. Whitman, eds.), Amer. Soc. Microbiol., Washington, D.C., p. 175-187.

Nagle, D.P., Jr., R.S. Wolfe. 1983. Component A of the methyl coenzyme M methylreductase system of *Methanobacterium*: resolution into four components. *Proc. Nat. Acad. Sci. USA, 80:*2,151-2,155.

Ni, S., D.R. Boone. 1991. Isolation and characterization of a dimethylsulfide-degrading methanogen from an oil well, characterization of *Methanolobus siciliae* $T4/M^T$, and emendation of *M. siciliae. Int. J. Syst. Bacteriol., 41:*410-416.

Ni, S., D.R. Boone. 1993. Degradation of dimethyl sulfide and methane thiol by methanogenic bacteria. In: *Proceedings of the 10th Int. Symp. on Environ. Biogeochem* (R.S. Oremland, ed.), San Francisco. [in press]

Oremland, R.S., L.M. Marsh, S. Polcin. 1982. Methane production and simultaneous sulphate reduction in anoxic, salt marsh sediments. *Nature* (London), *296*:143-145.

Oremland, R.S., M.J. Whiticar, F.S. Strohmaier, R.P. Kiene. 1988. Bacterial ethane formation from reduced, ethylated sulfur compounds in anoxic sediments. *Geochim. Cosmochim. Acta, 51*:1,895-1,904.

Oremland, R.S., R.P. Kiene, I. Mathrani, M.J. Whiticar, D.R. Boone. 1989. Description of an estuarine methylotrophic methanogen which grows on dimethylsulfide. *Appl. Environ. Microbiol., 55*:994-1002.

Patel, G.B., G.D. Sprott, J.E. Fein. 1990. Isolation and characterization of *Methanobacterium espanolae* sp. nov., a mesophilic, moderately acidophilic methanogen. *Int. J. Syst. Bacteriol., 40:*12-18.

Poirot, C.M., S.W.M. Kengen, E. Valk, J.T. Keltjens, C. van der Drift, G.D. Vogels. 1987. Formation of methylcoenzyme M from formaldehyde by cell-free extracts of *Methanobacterium thermoautotrophicum*: evidence for involvement of a corrinoid-containing methyltransferase. *FEMS Microbiol. Lett., 40:*7-13.

Reeburgh, W.S. 1976. Methane consumption in Cariaco Trench waters and sediments. *Earth Planetary Sci. Lett., 28:*337-344.

Reeburgh, W.S., D.T. Heggie. 1977. Microbial methane consumption reactions and their effect on methane distributions in freshwater and marine environments. *Limnol. Oceanogr., 22*:1-9.

Romesser, J.A., R.S. Wolfe, F. Mayer, E. Spiess, A. Walther-Mauruschat. 1979. *Methanogenium*, a genus of marine methanogenic bacteria, and characterization of *Methanogenium cariaci* sp. nov. and *Methanogenium marisnigri* sp. nov. *Arch. Microbiol., 121:*147-153.

Striegl, R.G., A.L. Ishii. 1989. Diffusion and consumption of methane in an unsaturated zone in north-central Illinois, U.S.A. *J. Hydrology, 111*:133-143.

Taylor, C.D., R.S. Wolfe. 1974. Structure and methylation of coenzyme M ($HSCH_2CH_2SO_3$). *J. Biol. Chem., 249:*4,879-4,885.

Thiele, J.H., J.G. Zeikus. 1988. Control of interspecies electron flow during anaerobic digestion: significance of formate transfer versus hydrogen transfer during syntrophic methanogenesis in flocs. *Appl. Environ. Microbiol., 54:*20-29.

Tyler, S.C. 1991. The global methane budget. In: *Microbial Production and Consumption of Greenhouse Gases: Methane, Nitrogen Oxides, and Halomethanes* (J.E. Rogers and W.B. Whitman, eds.), Amer. Soc. Microbiol., Washington, D.C., p. 7-38.

Van Beelen, P., J.F.A. Labro, J.T. Keltjens, W.J. Geerts, G.D. Vogels, W.H. Laarhoven, W. Guijt, C.A.G. Haasnoot. 1984. Derivatives of methanopterin, a coenzyme involved in methanogenesis. *Eur. J. Biochem., 139*:359-365.
Vogels, G.D., J.T. Keltjens, van der Drift. 1988. Biochemistry of methane production. In: *Biology of Anaerobic Microorganisms* (A.J.B. Zehnder, ed.), John Wiley & Sons, New York, p. 707-770.
Walther, R., K. Fahlbusch, R. Sievert, G. Gottschalk. 1981. Formation of trideuteromethane from deuterated trimethylamine or methylamine by *Methanosarcina barkeri*. *J. Bacteriol., 148*:371-373.
Whalen, S.C., W.S. Reeburgh. 1988. A methane flux time series for tundra environments. *Global Geochem. Cycles, 2*:399-409.
Whalen, S.C., W.S. Reeburgh, K.A. Sandbeck. 1990. Rapid methane oxidation in a landfill cover soil. *Appl. Environ. Microbiol., 56*:3,405-3,411.
Whiticar, M.J., E. Faber, M. Schoell. 1986. Biogenic methane formation in marine and freshwater environments: CO_2 reduction vs. acetate fermentation--isotope evidence. *Geochim. Cosmochim. Acta, 50*:693-709.
Whitman, W.B. 1985. Methanogenic bacteria. In: *The Bacteria: Archaebacteria, Vol. 8* (C.R. Woese and R.S. Wolfe, eds.), Academic Press, Inc., New York, p. 3-84.
Whitman, W.B., T.L. Bowen, D.R. Boone. 1992. The methanogenic bacteria. In: *The Prokaryotes, a Handbook on the Biology of Bacteria: Ecophysiology, Isolation, Identification, Applications,* second edition (A. Ballows, H.G. Trüper, M. Dworkin, W. Harder, and K.-H. Schleifer, eds.), Springer-Verlag, New York, p. 719-767.
Winfrey, M.R., J.G. Zeikus. 1979. Anaerobic metabolism of immediate methane precursors in Lake Mendota. *Appl. Environ. Microbiol., 37*:244-253.
Woese, C.R. 1987. Bacterial evolution. *Microbiol. Rev., 51*:221-271.
Wolin, M.J. 1982. Hydrogen transfer in microbial communities. In: *Microbial Interactions and Communities, Vol. 1* (A.T. Bull and J.H. Slater, eds.), Academic Press, London, p. 323-356.
Worakit, S., D.R. Boone, R.A. Mah, M.-E. Abdel-Samie, M.M. El-Halwagi. 1985. *Methanobacterium alcaliphilum* sp. nov., an H_2-utilizing methanogen which grows at high pH values. *Int. J. Syst. Bacteriol., 36*:380-382.

Chapter 7

Working Group Report

Formation and Consumption of Methane

NIGEL T. ROULET[1] AND WILLIAM S. REEBURGH[2]

[1] *Department of Geography, York University, North York*
Ontario, CANADA M3J 1P3

[2] *Department of Geosciences, Physical Sciences Building, University of California*
Irvine, California 92717

Group Members:
N. Andronova, D. Boone, P. Glaser, D. Johnson, P. Steudler, R. Striegl,
D. Valentine, P. Westermann, M. Whiticar

1. Introduction

Laboratory and field studies of methanogenesis and methylotrophy are clearly at very different stages of development, yet information from both laboratory and field studies is needed to understand the global methane budget to the point of making predictions of the possible decrease or increase in methane emissions to the atmosphere.

Methanogenesis is understood at a high level in the laboratory, where organisms are in culture and their substrate requirements, biochemical pathways, physiology, and molecular biology are reasonably well known and under study (see Boone, this volume). Application of this knowledge to predictions of the rate and extent of methanogenesis in field situations, however, is a major challenge. Studies on enteric releases by ruminants are probably closest to bridging this gap, but application of the basic knowledge to field studies in wetland, rice paddy, and soil studies appears almost out of reach.

To date, most emphasis has been placed on the net atmospheric budget (Matthews and Fung, 1987; Aselmann and Crutzen, 1989; Fung et al., 1991), and soil oxidation and other consumption processes have been considered to be small.

This focus is entirely appropriate for understanding the atmospheric methane increase (Watson et al., IPCC, 1990), as the atmosphere is where the increase has been observed and where the radiative and reactive properties of methane lead to greenhouse warming. Clearly not only are the direct effects of methane important, but also the indirect, though the latter are more difficult to assess (Lelieveld and Crutzen, 1992). Further, most of the measurements used in assembling the atmospheric budget are measurements of net flux to the atmosphere.

The atmospheric methane budget is actually a net global budget, so understanding future changes and development of meaningful emission limiting strategies will require considering processes that precede emission to the atmosphere, production and consumption (Reeburgh et al., 1993). Unfortunately, we have no basis for estimating production and must estimate it by measuring consumption and emissions in a variety of environments. Several terms in the commonly used net methane budgets result from direct emission to the atmosphere, and we can assume that consumption does not occur in the biomass burning, ruminant, and gas leak budget terms.

Methane oxidation or methanotrophy is also well understood from laboratory studies (Beddard and Knowles, 1989), but its importance in field situations is an emerging issue (Conrad, 1984; Topp and Hanson, 1991). Recent field studies indicate not only that oxidation can reduce emission to the atmosphere (Yavitt et al., 1988; Whalen et al., 1992) but also that some soils are able to consume atmospheric methane, providing a negative feedback (Steudler et al., 1989; Keller et al., 1990; Khalil et al., 1990; Yavitt et al., 1990; Crill, 1991; Mosier et al., 1991; Whalen et al., 1991).

2. Methane production

Methane production rates can be measured in experiments where a system is isolated from oxidation and transport (Schütz et al., 1989; Sass et al., 1990). Methanogenesis is the terminal process in a sequence of fermentation reactions, and it might be possible to determine rates of hydrolysis (corrected for consumption) by other electron acceptors or by formation rates of hydrogen, formate, or acetate (Boone, this volume). Substrate production rates would be informative but would have little predictive value.

We also need to understand how populations limit production and consumption. An array of new techniques, such as membrane lipid compositions, 16S ribosomal RNA probes, and DNA probes (Tsien et al., 1990; Tsuji et al., 1991), are organism-specific and could be quantitative, but we need to determine their ability to discriminate among active, viable, and productive populations.

Substrate quality remains a major challenge and may be influenced by lignin content and refractory components.

3. Methane consumption

Methane oxidizers in moist soils have low thresholds (0.1 ppm) and a high capacity (10^5-fold over ambient; Whalen et al., 1990) to oxidize methane at much higher concentrations. They do so without a lag period (Whalen and Reeburgh, 1990). Actual oxidation of atmospheric methane in some soils appears to be limited by transport (Crill, 1991), which is a function of soil porosity, soil tortuosity, and diffusion lengths (Striegl, 1993).

In cases such as wetlands, where oxidation occurs in a moist surface zone caused by variations in water table level, some sites (generally those exposed to varying water levels) have a high capacity to oxidize methane without delay, while those that are generally waterlogged require an induction period. The history of water table level change, which influences the populations of methane-oxidizing bacteria, could be an important control that should be investigated with genetic probes and laboratory experiments similar to those performed by Moore and Knowles (1989).

Understanding consumption will also require an understanding of vascular transport by plants, which allows methane to bypass the oxidizing zone, and other transport processes like ebullition and diffusion (e.g., Chanton and Dacey, 1991; Schütz et al., 1991).

Studies of methane oxidation in temperate forest soils and grassland soils (Steudler et al., 1989; Mosier et al., 1991) show that disturbance by fertilization or cultivation can cause large (30%) reductions in methane oxidation rates. In contrast, boreal forest soils (Whalen et al., 1991, 1992) show unobservable disturbance effects following fertilization. These disturbance effects on methane oxidation appear to be unexpectedly long-lived. Understanding disturbance effects will require studies of the nitrogen cycle as well as land use history, types of disturbance, and the severity and duration of the disturbance. These studies may be confused by the presence of multiple disturbances.

Transport is a complication common to field studies of methanogenesis and methane oxidation (Born et al., 1990; Dörr and Munnich, 1990; Dörr et al., 1993). Transport mechanisms such as advection, diffusion, ebullition, and vascular transport by plants complicate studies of microbial processes and must be understood. We have limited knowledge of whether the plants responsible for vascular transport are active or passive transporters (Chanton and Dacey, 1991). Also, many of the field studies have limited information on lateral and vertical movement of stored methane by groundwater. Further, the large day to day scatter in fluxes suggests that the emissions can be episodic (Moore et al., 1990; Whalen and Reeburgh, 1992). We have limited knowledge of the frequency of emission episodes as a function of time, temperature and barometric pressure.

4. Transition from laboratory and field studies to predictive modelling

In an ideal situation the results of laboratory incubation studies on the production and consumption of methane would be transferrable to the field situation. In reality, the laboratory experiments are done in unrealistic settings where the physical and chemical environment of a soil are manipulated in isolation, when in reality they are synergistically related. Because of this, there are large differences between absolute methane production and consumption from laboratory experiments and emissions from in situ measurements. The in situ emissions measurements explicitly incorporate the synergy of the methane-producing and -consuming environments. At present, the real benefit of laboratory investigations is their ability to provide a qualitative understanding of the key controlling variables, which can help in the interpretation of field emissions data (Crill, 1991). For example, Moore et al. (1993) measured the methane emissions from wetlands of the Hudson Bay Lowland and also determined the production and consumption rate of methane for peat samples obtained from the same wetlands where emissions were measured. Laboratory consumption rates were greater than the production rates for all the peat samples, yet the field measurements indicate emissions to the atmosphere at most sites were low (< 5 mg CH_4 m^{-2} d^{-1}). Differences in production and consumption, however, followed a very similar trend to the trend in emissions: sites where consumption was significantly greater than production showed the lowest emissions, while sites where the consumption and production rates were nearly equal had the highest emissions. Valentine et al. (1993) experimented with the physical and chemical controls of methane production from other peat samples from the Hudson Bay Lowland. Their results indicate that, while temperature is a factor, the availability of the H^+ and an acetate precursor were more important. Methane production was clearly inhibited as pH dropped, and it increased several fold with the addition of acetate precursors.

At present it appears that the modelling of the in situ emissions from production and consumption rates derived in the laboratory is not possible for most situations. The in situ methane emissions E are a function of the production P, consumption C, and the change in stored methane, such that:

$$E = P - C \pm \int \frac{d[CH_4]}{dz} dt \qquad (1)$$

where t indicates the time interval under study. If functions were available for the in situ production and consumption and the soil gas storage was in steady state, the system could be modelled. On the basis of work of Whalen and Reeburgh (1990) it appears that methane consumption is high in almost all conditions, except where soils have been disturbed (Steudler et al., 1989). Once it has been established the methane storage is in steady state and a laboratory consumption rate, C_L, has been determined, then an expected in situ methane production, P_E, could be calculated, as a residual from measured methane emissions:

$$P_E = E - C_L \qquad (2)$$

Actual consumption in the field is unlikely to equal laboratory consumption, but the two rates should be correlated if the laboratory experiments bear any resemblance to reality (if they do not, one has to question to usefulness of them in the first place). In this case, to compare laboratory and field measurements, models of their relationship need to be established:

$$P_E = \alpha P_L \qquad (3)$$

where α is the slope of this relationship. Presumably α would implicitly contain all the differences between the field and laboratory situations. It is also very likely that α would be highly ecosystem-specific.

There are two situations where it would be desirable to model the methane flux from the primary processes. The first is the rate of methane consumption in soils that lack any significant methane production. The modelling of consumption in this case requires an estimation of the diffusive exchange of methane between the atmosphere and the ground, and the limiting control appears to be diffusion to the soil (Striegl, 1993). It is assumed that the rate of methane consumption will always exceed the rate of diffusion (Whalen and Reeburgh, 1990) except in disturbed soils (Steudler et al., 1989; Mosier et al., 1991). The flux of methane in a soil can be described by:

$$F_s = D_a (\epsilon - \theta) \frac{d[CH_4]}{dz} \qquad (4)$$

where D_a is the diffusion of methane in air, ϵ is the porosity of the soil, and θ is the volumetric soil moisture (Striegl, 1993). Using typical air-filled porosities, Striegl (personal communication) has estimated that the maximum uptake of CH_4 by soils should be $\approx$ -5 mg m^{-2} d^{-1}. These numbers compare well with most of the higher field observations of soil methane uptake. Using soil and ecosystem classification such as those developed by Matthews and Fung (1987) and remote sensing to map soil moisture, it should be possible to model regional soil methane uptake. Field experiments such as the Boreal Ecosystem Atmosphere Exchange Study (BORAES) offer an ideal opportunity to attempt such an approach.

The modelling of methane emissions from soils that both produce and consume methane is very difficult, if not impossible, at present. If methane production exceeds the consumption of methane, then emissions are positive. But many soils where methane is produced are overlain by a layer of aerobic activity. The thickness of the aerobic zone plays a key role in determining the size of the flux. Whalen et al. (1992) have called this thickness of the oxidation zone the oxidation path-length. Roulet et al. (1992a) have used a simple model to estimate the change in the oxidation path-length using the position of the water table. These approaches, however, do not independently model production and consumption; they simply correlate emissions with moisture conditions. It is clear from other field studies that have correlated methane emissions with such variables as moisture (e.g., Crill et al., 1988; Moore and Knowles, 1989; Roulet et al., 1992b) and temperature (Crill et al., 1988; Moore et al., 1990; Roulet et al., 1992b) that these relationships are very ecosystem-specific. But until the in situ methane production and consumption can be adequately modelled, they will remain the best predictive tools available. Process work, such as that of King (1990) and Valentine et al. (1993), is very useful in narrowing the gap between the laboratory process studies and the results of field experiments.

5. Future studies

As indicated above, our understanding of methanogenesis and methane oxidation at the cultural, biochemical, and molecular levels greatly exceeds our understanding at the ecological level. We are not optimistic about bridging this gap in time to provide useful input to global climate models.

An additional approach to providing information on controls and the sensitivity of various systems to global climate change involves manipulations at a variety of scales (laboratory, plot, watershed) involving variables such as temperature, moisture and water table level, and substrate quality. These studies should be focused on systems with a substantial data base (e.g., Harvard Forest and Bonanza Creek LTERs, Canadian and Alaskan wetland/tundra sites) and should also take advantage of long-term peatland drainage operations in Finland and Canada.

References

Aselmann, I., P.J. Crutzen. 1989. Global distribution of natural freshwater wetlands and rice paddies, their net primary productivity, seasonality and possible methane emissions. *J. Atmos. Chem., 8*:307-358.
Beddard, C., R. Knowles. 1989. Physiology, biochemistry and specific inhibitors of CH_4, NH_4^+ and CO oxidation by methylotrophs and nitrifiers. *Microbiol. Rev., 53*:68-84.
Born, M., H. Dörr, I. Levin. 1990. Methane consumption in aerated soils of the temperate zone. *Tellus, 42B*:2-8.
Chanton, J.P., J.W.H. Dacey. 1991. Effects of vegetation on methane flux. In: *Gas Emissions from Plants* (H. Mooney, E. Holland, and T. Sharkey, eds.), Academic Press, San Diego, p. 65.
Conrad, R. 1984. Capacity of aerobic microorganisms to utilize and grow on atmospheric trace gases. In: *Current Perspectives in Microbial Ecology* (M.J. Klug and C.A. Reddy, eds.), American Society for Microbiology, Washington, p. 461.
Crill, P.M. 1991. Seasonal patterns of methane uptake and carbon dioxide release by a temperate woodland soil. *Global Biogeochem. Cyc., 5*:319-334.
Crill, P.M., K.B. Bartlett, R.C. Harriss, E. Gorham, E.S. Verry, D.I. Sebacher, L. Madzar, W. Sanner. 1988. Methane flux from Minnesota peatlands. *Global Biogeochem. Cyc., 2*:371-384.
Dörr, H., K.O. Munnich. 1990. ^{222}Rn flux and soil air concentration profiles in West Germany: soil ^{222}Rn as tracer for gas transport in the unsaturated soil zone. *Tellus, 42B*:20-28.
Dörr, H., L. Katruff, I. Levin. 1993. Soil texture parameterization of the methane uptake in aerated soils. *Chemosphere, 26* (1-4):697-714.
Fung, I., J. John, J. Lerner, E. Matthews, M. Prather, L.P. Steele, P.J. Fraser. 1991. Three-dimensional model synthesis of the global methane cycle. *J. Geophys. Res., 96*:13,033-13,065.
Hogan, K.B., J.S. Hoffman, A.M. Thompson. 1991. Methane on the greenhouse agenda. *Nature, 354*:181-182.
Keller, M., M.E. Mitre, R.F. Stallard. 1990. Consumption of atmospheric methane in soils of central Panama. *Global Biogeochem. Cyc., 4*:21-47.
Khalil, M.A.K., R.A. Rasmussen, J.R. French, J. Holt. 1990. The influence of termites on atmospheric trace gases: CH_4, CO_2, $CHCl_3$, N_2O, CO, H_2, and light hydrocarbons. *J. Geophys. Res., 95*:3,619-3,634.
King, G.M. 1990. Regulation by light of methane emissions from a wetland. *Nature, 345*:513-515.
Lelieveld, J., P.J. Crutzen. 1992. Indirect effects of methane on climate warming. *Nature, 355*:339-342.

Matthews, E., I. Fung. 1987. Methane emissions from natural wetlands: global distribution, area, and environmental characteristics of sources. *Global Biogeochem. Cyc.,* 1:61-86.
Moore, T.R., R. Knowles. 1989. The influence of water table levels on methane and carbon dioxide emissions from peatland soil. *Can. J. Soil Sci.,* 69:33-38.
Moore, T.R., N.T. Roulet, R. Knowles. 1990. Spatial and temporal variations of methane flux from subarctic/northern boreal fens. *Global Biogeochem. Cyc.,* 4:29-46.
Moore, T.R., A. Heyes, N.T. Roulet. 1993. Methane emissions from wetlands, southern Hudson Bay Lowland. *J. Geophys. Res.* in press.
Mosier, A.R., D. Schimel, D. Valentine, K. Bronson, W. Paton. 1991. Methane and nitrous oxide fluxes in native, fertilized and cultivated grasslands. *Nature,* 350:330-332.
Reeburgh, W.S., S.C. Whalen, M.J. Alperin. 1993. The role of microbially-mediated oxidation in the global CH_4 budget. In: *Microbiology of C_1 Compounds* (J.C. Murrell and D.P. Kelly, eds.), Intercept, Andover, U.K., pp. 1-14.
Roulet, N.T., T.R. Moore, P. Lafleur, J. Bubier. 1992a. Northern fens, atmospheric methane, and climate change. *Tellus,* 44B:100-105.
Roulet, N.T., R. Ash, T.R. Moore. 1992b. Low boreal wetlands as a source of atmospheric methane. *J. Geophys. Res.,* 97:3,739-3,749.
Sass, R.L., F.M. Fisher, P.A. Harcombe, F.T. Turner. 1990. Methane production and emissions in a Texas rice field. *Global Biogeochem. Cyc.,* 4:47-68.
Schütz, H., W. Seiler, R. Conrad. 1989. Processes involved in formation and emission of methane in rice paddies. *Biogeochem.,* 7:33-53.
Schütz, H., P. Schroder, H. Renenburg. 1991. Role of plants in regulating the methane flux to the atmosphere. In: *Gas Emissions from Plants* (H. Mooney, E. Holland, and T. Sharkey, eds.), Academic Press, San Diego, p. 29.
Steudler, P.A., D.M. Bowden, G.E. Lang, J.D. Aber. 1989. Influence of nitrogen fertilization on methane uptake in temperate forest soils. *Nature,* 341:314-316.
Striegl, R.G. 1993. Diffusional limits to the consumption of atmospheric methane by soils. *Chemosphere,* 26 (1-4):715-720.
Topp, E., R.S. Hanson. 1991. Metabolism of radiatively important trace gases by methane-oxidizing bacteria. In: *Microbial Production and Consumption of Greenhouse Gases: Methane, Nitrogen Oxides and Halomethanes* (J.E. Rogers and W.E. Whitman, eds.), American Society of Microbiology, Washington, p 71.

Tsien, H.C., B.J. Bratina, K. Tsuji, R.S. Hanson. 1990. Use of olignucleotide signature probes for identification of physiological groups of methylotrophic bacteria. *Appl. Environ. Microbiol.*, 56:2,858-2,865.
Tsuji, K., H.C. Tsien, S.R. DePalma, R. Scholtz, S. LaRoche. 1991. 16S ribosomal RNA sequence analysis for determination of phylogenetic relationships among methylotrophs. *J. Gen. Microbiol.*, 13:1-10.
Valentine, D.W., E.A. Holland, D.S. Schimel. 1993. Methane production in northern wetlands: ecosystem and physiological controls. *J. Geophys. Res.* in press.
Watson, R.T., H. Rohde, H. Oeschger, U. Sigenthaler. 1990. Greenhouse gases and aerosols. In: *Climate Change: The IPCC Scientific Assessment* (J.T. Houghton, G.J. Jenkins, and J.J. Ephraums, eds.), Cambridge University Press, Cambridge, p. 1.
Whalen, S.C., W.S. Reeburgh. 1990. Consumption of atmospheric methane by tundra soils. *Nature*, 346:160-162.
Whalen, S.C., W.S. Reeburgh. 1992. Interannual variations in tundra methane flux: a 4 year time series at fixed sites. *Global Biogeochem. Cyc.*, 6:139-159.
Whalen, S.C., W.S. Reeburgh, K.A. Sandbeck. 1990. Rapid methane oxidation in a landfill cover soil. *Appl. Environ. Microbiol.*, 56:3,405-3,411.
Whalen, S.C., W.S. Reeburgh, K.S. Kizer. 1991. Methane consumption and emission from taiga soils. *Global Biogeochem. Cyc.*, 5:261-274.
Whalen, S.C., W.S. Reeburgh, V. Barber. 1992. Oxidation of methane by boreal forest soils: a comparison of seven measures. *Biogeochem.*, 16:181-211.
Yavitt, J.B., G.E. Lang, D.M. Downey. 1988. Potential methane production and methane oxidation rates in peatland ecosystems of the Appalachian Mountains, United Sates. *Global Biogeochem. Cyc.*, 2:253-268.
Yavitt, J.B., D.M. Downey, G.E. Lang, A.J. Sextone. 1990. Methane consumption in two temperate forest soils. *Biogeochem.*, 9:39-52.

Chapter 8

Stable Isotopes and Global Budgets

MICHAEL J. WHITICAR

School of Earth and Ocean Sciences, University of Victoria, P.O. Box 1700, Victoria, B.C. V8W 2Y2, Canada

1. Introduction

Global climate change associated with the increasing atmospheric methane burden is an important societal concern. Today we can monitor with good precision the yearly 1% rise in lower tropospheric methane mixing ratios (e.g., Blake and Rowland, 1988), and we have adequate, basic global coverage of atmospheric methane latitudinal variation. Mesoscopically, we are able to roughly estimate the various source strengths, e.g., from wetlands, agriculture, fossil fuels, but there is considerable uncertainty in the actual magnitudes of the various individual fluxes of methane across the geosphere-biosphere-atmosphere interface. This knowledge deficit includes our understanding of both release and uptake process-groups. Control of methane emissions to the atmosphere requires that we reliably characterize these source-sink relationships.

Stable isotope signatures of natural gases, and in particular those of methane, are potentially a key tool to track the movement of volatile carbon compounds in geosphere-biosphere-atmosphere systems. However, for this tool to be successful, three prerequisites must be satisfied:

1. *Classification* - Can we reliably characterize the stable carbon and hydrogen isotope ratios of the various methane sources? Are their isotope signatures relatively distinct, consistent, diagnostic, and measurable?

2. *C-, H-isotope effects* - Do we know or can we reliably predict the magnitude of the carbon and hydrogen isotope effects leading to isotope fractionation at every step of the way from formation to destruction?

3. *Uncertainty in input and output flux magnitudes, influences on isotope budgets* - Will or do changes in the different source strengths of methane cause a significant (distinct and measurable) shift in the stable carbon and hydrogen isotope signatures of the atmospheric methane input? What influence do changing sink functions have on atmospheric methane isotope ratios?

The objective of this paper is to directly address these three concerns.

2. Classification of methane sources

2.1 Natural gas types. The geosphere is the primary source of methane released to the atmosphere. Therefore, the isotope signal of the integrated source input to atmospheric methane will be determined by the relative proportions of the various types of methane leaving the geosphere. Natural gases are derived from biogenic and non-biogenic sources through diverse processes, including bacterial formation, catagenesis (thermogenic natural gas generation), hydrothermal and geothermal activity and, to an unknown degree, from primordial or mantle emissions. The relative sizes of most global natural gas reservoirs, shown in Figure 1, can generally be constrained to within an order of magnitude (Whiticar, 1990, 1992). Conventional subsurface natural gas accumulations (e.g., thermogenic, bacterial, coal gases) together are estimated to comprise around 0.16 million Tg hydrocarbons (120 Gt C), far larger than 3.6 Gt C of atmosphere methane and 3 Gt C terrestrial gas hydrates (5.6×10^{12} m^3). Biogas accumulations from anthropogenic sources, including rice paddies (5 Tg C, 1×10^{10} m^3) and land fill and from wetlands (75 Tg C, 14×10^{10} m^3), are the next largest reservoirs. But all of the above natural gas reservoirs are minute compared with the amount of hydrocarbons bound up as gas hydrates on continental margins (Table 1). Conservative estimates for ocean clathrates range from 3 to 11 million Tg hydrocarbons (2000 to 8000 Gt C, 4 to 15×10^{15} m^3), and projections of up to 5 billion Tg hydrocarbons (4×10^5 Gt C, 7×10^{18} m^3) have been made (see Kvenvolden, 1988). The large uncertainty regarding the extent of gas hydrates precludes any firm estimation of the total budget for global natural gas. Furthermore, our lack of knowledge concerning the amounts of hydrocarbons in geothermal, crystalline, and mantle systems complicates this budget estimation.

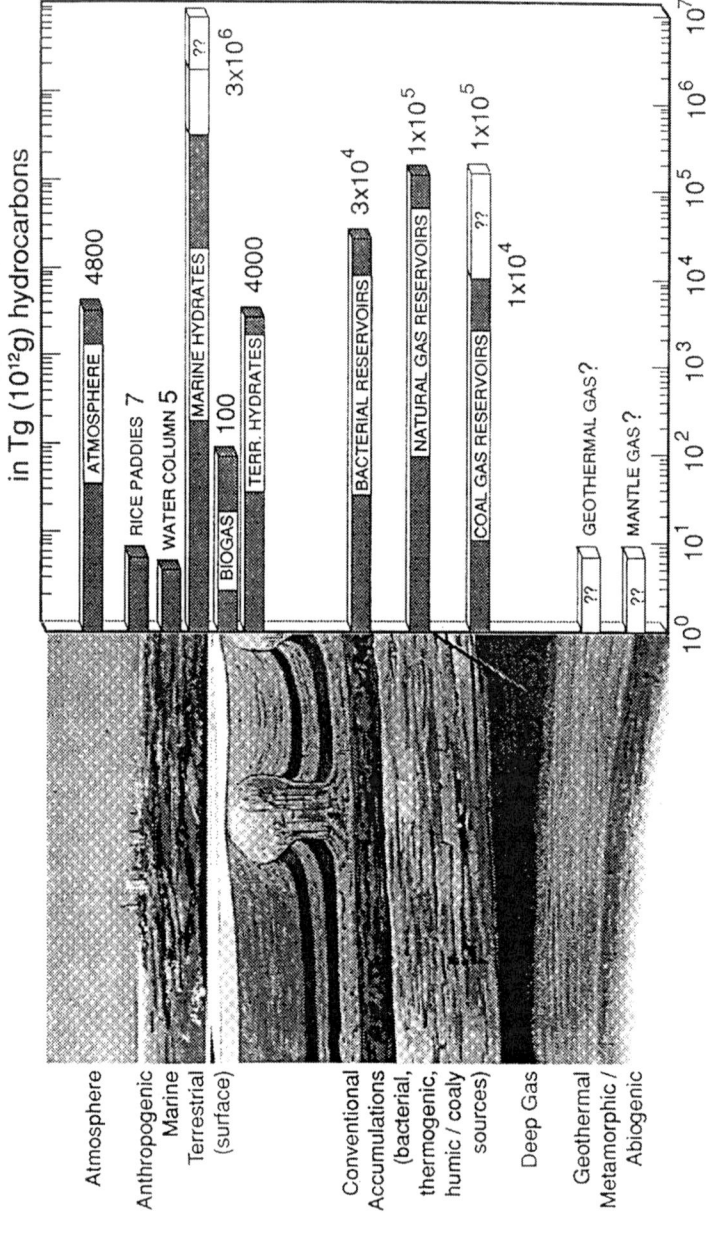

Figure 1. A geochemist's view of natural gas types and their respective reservoir sizes. Hydrates are by far the largest hydrocarbon source type but their contribution to atmospheric methane is questionably thought to be minimal (<1% of total influx).

2.1.1 Diagenetic gases. Trace hydrocarbons (<1 ppb CH_4 wt. gas/wt. sediment) are often encountered in near-surface soils or sediments. Their source can be autochthonous (in situ), low temperature diagenetic reactions (Hunt et al., 1980; Whelan et al., 1982) or allochthonous, e.g., carried in by migration of gases that were more deeply buried in sediments. These diagenetic gases are not necessarily mediated by microbial processes as compared with bacterial gases. In many instances, they are not primary gas types, rather they represent a mixture of residual, allochthonous, and diagenetic gases. As such, it is usually not possible to assign a specific source or history to these trace gases. Due to their low concentrations they are not a significant source of methane to the atmosphere.

Table 1. Magnitudes of methane fluxes to the lower troposphere from various sources and their associated mean carbon and hydrogen isotope ratios

Source	flux (Tg CH_4 y^{-1})	% total emission	$\delta^{13}C$-CH_4 (‰, PDB)	δD-CH_4 (‰, SMOW)
rice paddies	110	20	-63	-390
cattle/enteric	80	15	-60	-330
natural gas	45	8	-44	-180
coal	35	7	-37	-110
biomass burning	55	10	-25	-90
natural wetlands	115	21	-58	-380
termites	40	7	-70	-390
landfill	40	7	-55	-380
oceans (water column)	10	2	-60	-220
freshwater	5	1	-58	-385
hydrates	5	1	-60	-240
mean input value			-54	-307

2.1.2 Bacterial gases. Roughly 20% of the conventional natural gas reserves are of bacterial origin (Rice and Claypool, 1981; Rice, 1991). Methane is the major hydrocarbon constituent of this natural gas (usually <1% ΣC_{2+}, Figure 2) and have been formed, for the most part, by near-surface fermentation reactions by methanogenic bacteria as one of the final diagenetic remineralization stages. Methanogens are a broad consortium of obligate anaerobic microorganisms that can utilize a limited suite of precursor compounds such as 1) acetate and formate or 2) bicarbonate, i.e., $CO_2 + H_2$, to form methane. The carbonate reduction pathway, which predominates in marine environments (Whiticar et al., 1986), can be represented by the general reaction

$$CO_2 + 8H^+ + 8e^- \longrightarrow CH_4 + 2H_2O \qquad (1)$$

and the net reaction for the acetate fermentation pathway more common in freshwater settings is

$$*CH_3COOH \longrightarrow *CH_4 + CO_2 \qquad (2)$$

where the * indicates the intact transfer of the methyl position to CH_4. In addition to these so-called "competitive substrates," there are non-competitive substrates for methanogenesis including methanol, mono-, di- and tri-methylamines and certain organic sulphur compounds, i.e. dimethylsulphide, but their relative importance as the source for bacterial methane is uncertain. The microbiology and ecology of the various methanogenic pathways have been reviewed in several monographs (e.g., Zehnder, 1988; Boone, this volume).

2.1.3 Thermogenic natural gas. During the catagenic transformation and reorganization of organic matter, various short-chained hydrocarbons, such as methyl groups, are being cleaved off higher molecular weight organic compounds and saturated to form the light hydrocarbons of conventional natural gas. The degree and type of product formed is dependent on several key factors including source kerogen type (sapropelic vs. humic) and richness (e.g., Total Organic Carbon = TOC, and molar H/C ratio), maturity, expulsion efficiency, and the presence of catalysts such as clays.

Sapropelic, Type I/II kerogen sources generate over suitable time periods significant quantities of thermogenic hydrocarbon gases at temperatures over 70°C. The initial gases at low source maturity (Vitrinite Reflectance level: R_m < 0.5%) will be relatively dry (<5% ΣC_{2+}), but the proportion of higher hydrocarbons will increase with maturation into and through the petroleum window (peak natural gas generation at 150 - 160°C). Oil-associated natural gases can have over 80% ΣC_{2+} (e.g., Evans and Staplin, 1971). As maturation proceeds into late mature or condensate range (R_m <1.6%), subsequent kerogen transformation and cracking of hydrocarbons leads to a greater proportion of shorter-chained hydrocarbons and essentially a methane-rich gas at the base of the catagenic stage, roughly about 200°C.

Natural gas generation profiles from humic Type III kerogens and coals are quite different from Type I/II kerogens. Significant hydrocarbon generation is traditionally thought to occur at higher maturity levels, i.e., $R_m >0.7\%$, for humic kerogens than for sapropelic kerogens (Karweil, 1969; Tissot and Welte, 1978; Chung and Sackett, 1979). Throughout their maturation history, humic kerogens generate less C_{2+} hydrocarbons, and the total methane generative potential, including cracking of oil and condensates, is about 1 - 2 times lower than for a sapropelic source. Reliable characterization of humic kerogens is complicated by the high retention capacity of humic organic matter, which partitions gases, resulting in a time/maturity-dependent compositional fractionation of the gases released. These difficulties are compounded due to the high retention by humic organic matter of bacterial gases that were formed while the peat or fen deposit was at or near the surface.

2.1.4 Secondary gases. In addition to these "primary" gas sources, secondary alteration process can also influence the character of the natural gas emission. These alteration effects include microbial oxidation (aerobic and anaerobic), migration fractionation and mixing. Altered gases are an important type of natural gas and must be considered in the atmospheric methane budgets.

2.2 Natural gas classification tools

2.2.1 Essential parameters. Driven initially by the needs of the energy exploration sector, a suite of geochemical parameters has been developed to classify the various types of natural gas. Three categories of geochemical tools are commonly used to correlate natural gases to their sources:

    1) gas concentration,
    2) molecular composition, and
    3) stable isotope ratios.

Analytical routines devised to find hydrocarbon fuels are and have been applied widely to samples from near-surface soils and seepages, cuttings from drill wells, and natural gas reservoirs, and thus a well-defined data set is available for environmental applications. Through the combination of parameters from the three categories, i.e., carbon isotope ratio of methane ($\delta^{13}C\text{-}CH_4$), molecular ratio: methane to ethane + propane ($C_1/(C_2+C_3)$, %vol.), and hydrocarbon concentration (ΣC_1-C_4) an interpretative schema can be applied as illustrated in Figure 2.

Concentration and molecular composition information can broadly characterize gas types, i.e., bacterial vs. thermogenic, but stable isotope ratios, such as carbon and hydrogen for hydrocarbons, are more specific and distinctive. In addition to identifying the natural gas type, the carbon and hydrogen isotope signature of a reservoired natural gas can often provide more detailed information about its source(s), for example, kerogen type (e.g., sapropelic Type I/II or coal/humic Type III), the level of maturity of the source, and sometimes the degree of generation. In addition to source typing and maturity estimates, the combination of molecular and isotope composition of a gas can serve to delineate altered (secondary) gases or gas mixtures from different sources. Molecular and concentration information alone is generally insufficient to do this.

For the purposes of this paper, emphasis is on the stable carbon and hydrogen isotope ratios to describe and discriminate the different methane sources, and formation/destruction pathways.

2.2.2 Stable isotope effects. Stable isotope data are determined as ratios, e.g. $^{13}C/^{12}C$, rather than as absolute molecular abundances and are reported as the magnitude of excursion in per mil of the sample isotope ratio relative to a known standard isotope ratio. The usual d-notation generally used in the earth sciences is:

$$\delta R_X (\text{\textperthousand}) = B(F(R_a / R_b\text{-sample}, R_a/R_b\text{-standard}) - 1) \times 10^3 \qquad (3)$$

where R_a/R_b are the isotope ratios, e.g. $^{13}C/^{12}C$ or D/H, referenced relative to the PDB or SMOW standards, respectively.

Numerous factors control the distribution of isotopes in natural gas components. The principal ones include the isotope ratios of the precursor material, e.g., the $^{13}C/^{12}C$ or D/H ratios of the source organic compounds, and the isotope effects associated with the processes of formation, retention, and destruction of natural gas. For the purpose of distinguishing the different natural gases, it is important that the isotope signatures and/or the isotope effects are distinctive and diagnostic for the different types of natural gas and for the mechanisms governing them.

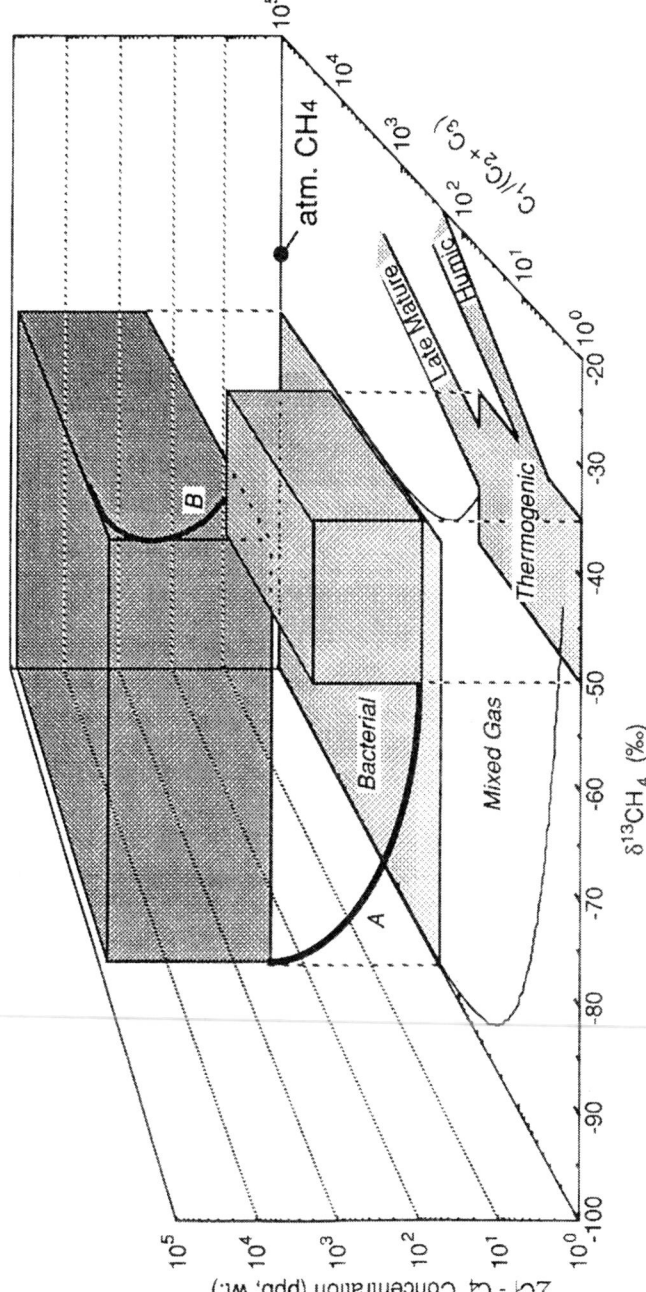

Figure 2. Carbon isotope ratios of methane are alone often sufficient to define a natural gas type (i.e., bacterial, thermogenic, abiogenic), but sub-grouping is severely limited. The combination of concentration and molecular composition can often help, particularly to identify secondary alteration effects such as (1) gas mixtures or (2) microbial oxidation

Typical methane carbon isotope values are given in Figure 3 for the various bacterial and thermogenic natural gases (see also Stevens, this volume). However, this can only serve as an initial orientation. For example, methanogenesis by carbonate reduction can have $\delta^{13}C_{CH4}$ values ranging from −20 ‰ to −110 ‰ due to (a) variations in the initial carbon isotope ratio of the substrate, (b) the degree of substrate utilization, or (c) the ambient temperature (Whiticar, 1992). In comparison with bacterial methane, thermogenic hydrocarbons have a much more constrained range of isotope compositions.

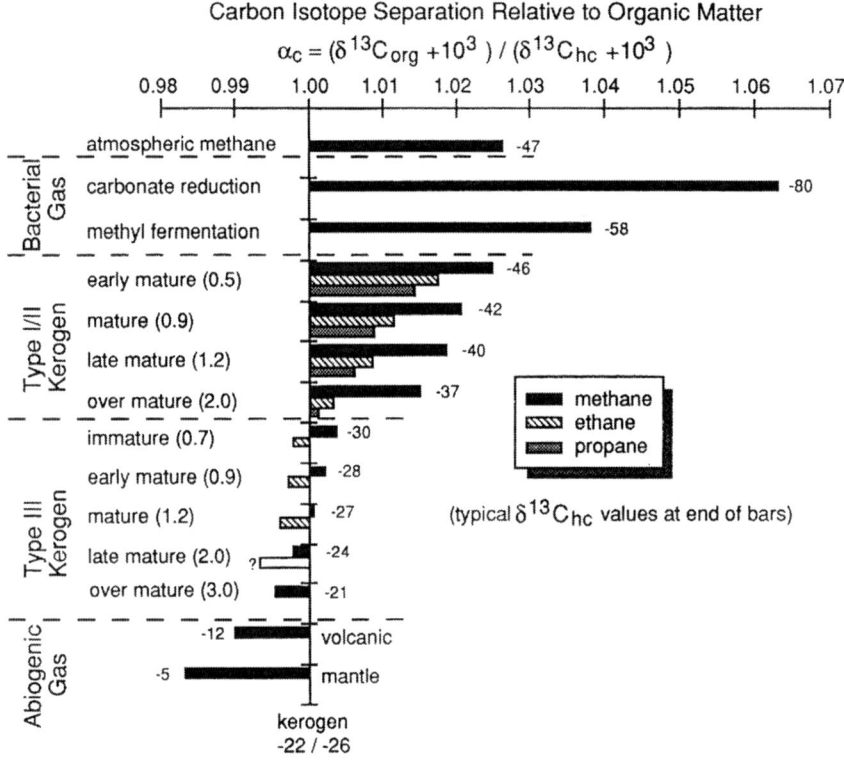

Figure 3. Natural gas formation is typically the result of various bacterial or thermocatalytic kinetic reactions, which have isotope effects associated with the A -> B. Our knowledge of the magnitude of these isotope effects is paramount to our tracking the changes in isotope signals. The isotope offset relative to general organic matter (δ^{13}C-Corg = -22‰ and -26‰)

Initially equilibrium isotope effects (EIEs) had been proposed to explain the distribution of carbon isotopes in thermogenic hydrocarbons, most notably by Galimov and Petersil'ye (1967), Galimov and Ivlev (1973), and Galimov (1973), but Galimov (1985) has modified this equilibrium effect to a "thermodynamically ordered distribution," which is more consistent with the accepted view that kinetic isotope effects (KIEs) control the redistribution of isotopes (see for carbon: Sackett, 1968; Stahl, 1973; McCarty and Felbeck, 1986; Chung et al., 1988; and for hydrogen: Frank, 1972). For thermogenic hydrocarbon formation, kinetic isotope theory predicts that the light hydrocarbon formed by the saturation of an alkyl group cleaved from the kerogen molecule will be depleted in the heavier isotope relative to the remaining reactive kerogen. Similar KIE considerations apply to methanogenesis. Methanogens preferentially utilize isotopically lighter substrates; thus, the methane formed by methanogens is depleted in the heavier isotope relative to the precursor material (Whiticar, 1992).

Although the KIEs fractionate reactant and products pools, and thus can potentially generate a considerable range in stable isotope values for methane in natural gases, in many cases their carbon isotope signatures are sufficiently unambiguous to permit classification.

Isotopic fractionation or discrimination resulting from KIEs can be described by Rayleigh distillation relationships. The isotope ratio of the remaining reactant pool (e.g., a generating kerogen), which is being depleted in the lighter isotope, can be approximated by

$$R_r/R_o = f^{(1/\alpha - 1)} \qquad (4)$$

and the progressive isotopic shift of the cumulative product pool (e.g., methane accumulation) by

$$R_\Sigma/R_o = (1 - f^{(1/\alpha 1)}) / (1-f) \qquad (5)$$

where R is the isotope ratio of the initial reactant (R_o), the residual reactant at a specified time (R_r), and the cumulative product (R_Σ), respectively (e.g., Claypool and Kaplan, 1974). The fraction of the reactant remaining is f, and α is the isotope fractionation factor for the conversion of the reactant to the product. Isotope shifts related to bacterial methane production (substrate depletion) and microbial methane oxidation, shown in Figures 4 and 6, are treated below. Although the KIEs partition isotopes between the reactant and products pools and can potentially generate a considerable range in stable isotope values for methane in natural gases, in many cases the carbon isotope signatures of different natural gas types are sufficiently unambiguous to permit classification.

Figure 3 illustrates the relative magnitudes of isotopic offset between various generalized types of natural gas ($\delta^{13}C_{hc}$) and organic matter ($\delta^{13}C_{org}$), according to the equation:

$$\alpha_{org\text{-}hc} = (\delta^{13}C_{org} + 10^3) / (\delta^{13}C_{hc} + 10^3) \qquad (6)$$

The greatest isotope fractionation is observed for bacterial methane formed by the carbonate reduction ($\alpha_{org\text{-}hc} > 1.055$) and then methyl fermentation pathways ($\alpha_{org\text{-}hc} \sim 1.04$) (Whiticar et al., 1986). In general, thermogenic hydrocarbons formed from Type I or II kerogen ($\alpha_{org\text{-}hc} \sim 1.02$) have a larger KIE expressed than for Type III kerogens ($\alpha_{org\text{-}hc} \sim 1.003$). In both cases, the magnitude of the isotope separation is less than for methanogenesis (Figure 3). It also decreases with increasing maturity of the organic matter and with carbon number, i.e., α_{org-hc} of methane > ethane > propane > butane (Figure 3). Hydrocarbons more enriched in ^{13}C than their precursor organic matter are rarely measured in Type I/II kerogens but is common in humic or Type III sources. In Type I/II cases, this could signal the presence of secondary effects such as mixing of inorganic methane (e.g., volcanogenic, mantle), oxidation of hydrocarbons, or sometimes sampling and production artifacts.

2.2.3 Carbon isotope ratios. Carbon isotope ratio of methane is the most common isotope measurement made to classify a natural gas (e.g., Colombo et al., 1965; Sackett, 1968; Silverman, 1971; Stahl, 1973; Schoell, 1980, 1988). Basic differences in carbon isotopes between bacterial and thermogenic natural gases are identified in Figures 2 and 3. As mentioned above, bacterial gases are commonly "lighter," i.e., more depleted in ^{13}C than thermogenic gases, and thermogenic gases become heavier with increasing maturity.

Owing to the multiple sources for methane and its susceptibility to secondary effects, the classification of thermogenic natural gases has recently placed more emphasis on the carbon isotope ratios of higher homologues, such as ethane, propane, and the butanes (e.g., James, 1983; Whiticar et al., 1984; Chung et al., 1988; Clayton, 1991). Furthermore, the differentiation of bacterial methane formation pathways can be improved by the carbon isotopes of co-existing species, including CO_2 or volatile organic acids (e.g., Whiticar et al., 1986). The latter is illustrated in Figure 4 (after Whiticar et al., 1986) where the combination of carbon isotope ratios co-existing CO_2 and CH_4 delineate the methanogenic formation pathways of carbonate reduction or methyl-fermentation. In addition, Figure 4 shows the $\delta^{13}C_{CO2}$-$\delta^{13}C_{CH4}$ trajectory of a microbially oxidized methane.

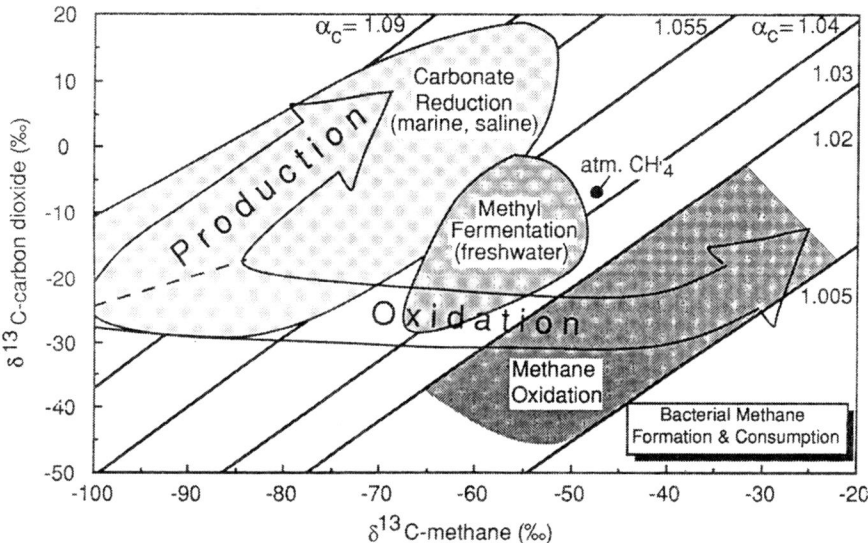

Figure 4. Classification can be enhanced by isotope measurements on co-existing species, such as dissolved CO_2 or formation water. Methanogenesis follows strict relationships between the methane formed and these co-existing compounds. The example of $\delta^{13}C$-CO_2 vs. $\delta^{13}C$-CH_4 is shown; similar interpretative schemes are available between δD-H_2O vs. δD-CH_4 and $d^{13}C$-substrate vs. $d^{13}C$-CH_4.

2.2.4 Hydrogen isotope ratios. Carbon isotopes, particularly those of methane, are the most common isotopic measurement made on natural gases, but hydrogen isotopes of methane are a diagnostic parameter in classifying its source and type. The work by Schoell (1980) provided one of the first rigorous treatments of carbon and hydrogen isotope variations of bacterial and thermogenic hydrocarbons. This work was extended later to include information on the different bacterial gas formation pathways, as shown in the CD-diagramme of Figure 5 (after Whiticar et al., 1986). The general zonation of C-, H-isotope signatures in the outlined areas of Figure 5 represent the regions occupied by the dominant natural gas types. The shaded regions of the diagramme outside the major gas types areas are less common or well defined gas signatures.

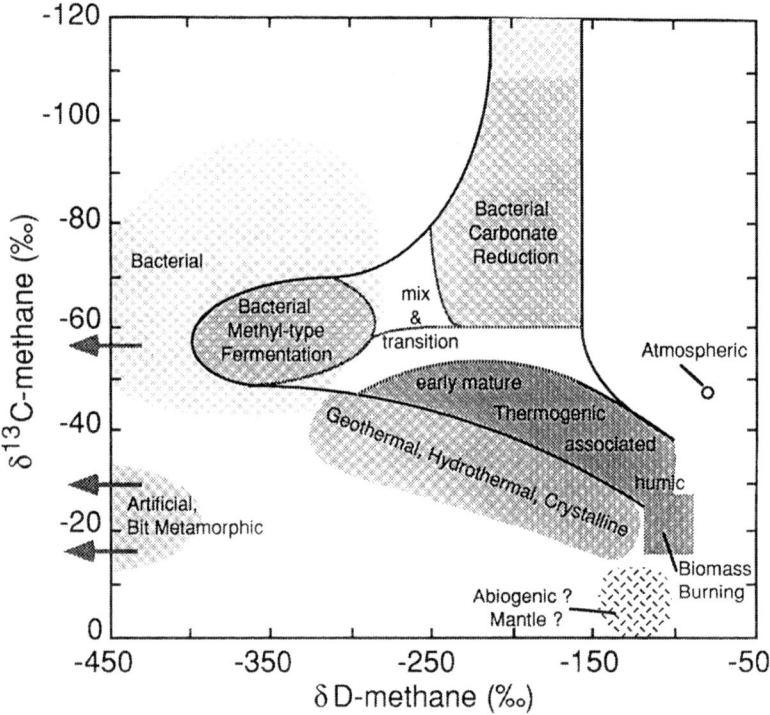

Figure 5. Combination of carbon and hydrogen isotope ratios greatly improves our classification precision. Although less frequently measured, $\delta D\text{-}CH_4$ is well suited to distinguish between different methanogenic pathways and environments, which are the major source of methane to the atmosphere. Note: the mean atmospheric methane CD-value does not directly correspond with any of the known natural gas types, and the isotope offset associated with the removal of atmospheric methane needs to be considered (see Figure 7 for explanation).

In contrast to carbon isotope ratios, the hydrogen isotope data of methane do not exhibit a clear dependency on maturity or oxidation effects; rather, they provide details on the depositional environment and pathway of formation. Hydrogen isotopes are particularly useful in distinguishing (1) methane from different methanogenic pathways, (2) bacterial from early mature thermogenic gas, (3) thermogenic from geothermal-hydrothermal gas, and (4) artificial or bit metamorphic sources. In these examples, methane carbon isotope ratios, if used alone, would deliver ambiguous results.

The geothermal-hydrothermal-crystalline zone in Figure 5 is defined, for example, by the gas data from open boreholes in the Canadian Shield crystalline rocks (Sherwood et al., 1988) and from the Gravberg-1 drill well (Schmidt, 1987; Laier, 1988; Jefferey and Kaplan, 1988). Examples of geothermal-hydrothermal data include Yellowstone and Lassen Parks (Welhan, 1988; Whiticar and Simoneit, 1992), New Zealand fields (Lyon and Hulston, 1984), Guaymas Basin (Welhan and Lupton, 1987, Simoneit et al., 1988) or Bransfield Strait (Whiticar and Suess, 1989).

Methane found at the East Pacific Rise, 21°N by Welhan (1988) and from the Zambales ophiolite (Abrajano et al., 1988) are maintained to be of abiogenic origin (Figure 5). Similarly, gas inclusions from south Greenland reported by Konnerup-Madsen et al. (1988) have ^{13}C-enriched and ^{2}H-depleted methane (δ^{13}C-CH$_4$ = -1.0‰ to -5.1 ‰, δD-CH$_4$ ca. -100 ‰ to -150 ‰, Figure 5). These three sample locations are used to define the region of Figure 5 for abiogenic or mantle methane. The free gases in dolerites from the Gravberg-1 well (Schmidt, 1987; Laier, 1988; Jefferey and Kaplan, 1988) also have carbon isotope signatures approaching those of abiogenic gas, but their origin is unsure.

Bit metamorphism hydrocarbons noted in Figure 5 are formed artificially by drilling hard lithologies have distinctive methane isotope signatures enriched in ^{13}C (δ^{13}C-CH$_4$ ca. -20 ‰) and strongly depleted in deuterium (δD-CH$_4$ ca. -750 ‰), (e.g., Gerling, 1985; Faber et al., 1987; Faber and Whiticar, 1989).

2.3 Resume. Stable carbon and hydrogen isotope signatures of methane are capable of reliably defining and distinguishing the various natural gas types and sources. In particular, C- and H-isotopes can clearly differentiate between thermogenic methane and methane from bacterial carbonate reduction and methyl-type fermentation pathways. The latter is important because these are the predominant sources of atmospheric methane. The isotopes are also capable of identifying secondarily altered methane, such as that subjected to microbial oxidation.

However, despite the promising usefulness of isotope signatures, two major difficulties complicate the situation. First, roughly 75% of atmospheric methane is derived from methyl-type fermentation, which limits the value knowing the isotope signatures of the other natural gases. Second, due to kinetic reactions, a range in values rather than unique carbon and hydrogen isotope ratios define bacterial methane. These caveats are treated in greater detail below.

3. C-, H-isotope effects

3.1 C-D isotope variation of atmospheric methane sources. In the course of the global carbon cycle, it is atmospheric carbon dioxide that sets the carbon isotope base for organic matter and ultimately the isotope ratios of atmospheric methane sources. However, kinetic isotope effects determine magnitude of the associated isotopic fractionation or offset of the methane from the precursor.

The CD-diagramme of Figure 6 shows the possible range of methane isotope pairs in response to (a) initial natural isotope variations in precursor substrates, (b) substrate depletion effects due to methanogenesis or thermocatalytic conversion, and (c) secondary, microbial methane oxidation. In addition, factors such as temperature can influence the magnitude of the isotope effects. While this confuses the issue of pegging unique isotope signatures to sources of atmospheric methane formed by similar processes, e.g., methyl-type fermentation, the isotope shifts provide key information on the degree to which the formation or oxidation process has proceeded.

Important in this context is that the CD-isotope ratio pair of the mean atmospheric methane does not fall within the CD-isotope signatures of any of the major natural gas types. On the face of it, the atmospheric $\delta^{13}C$-CH_4, δD-CH_4 pair of -47 ‰, -80 ‰ could not be created only by a direct mixture of these major natural gas types, and an isotope fractionation must be involved with the removal of atmospheric methane.

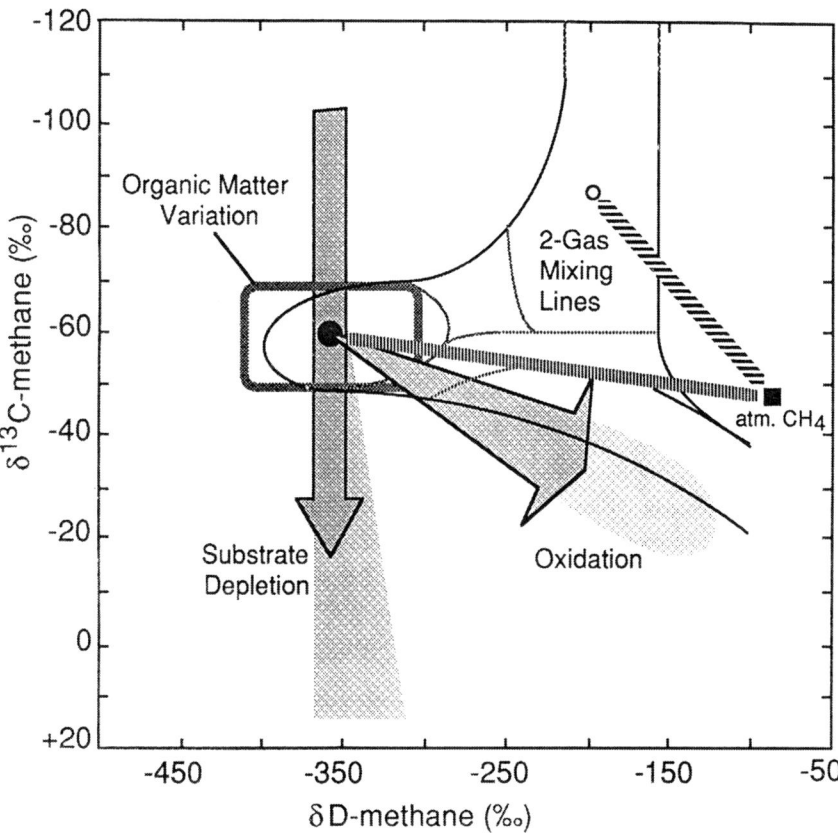

Figure 6. How these various isotope effects combine to control the isotope signature of a particular gas is demonstrated for bacterial methane. The potential degree of the isotope shifts is shown for methanogenesis (pathway, substrate depletion), microbial oxidation, and isotope variations in the precursor organic matter (e.g., acetate, TMA, DMS). To confuse the issue, two mixing lines (are not isotope effects) are drawn to illustrate the mixing trajectories between atmospheric methane and two bacterial gas end-members. This mixing often takes place at soil-air and sea-air interfaces.

For comparison, mixing lines, which are not isotope effects, are drawn in Figure 6 between the current mean C- and H-isotope ratios of lower tropospheric methane and bacterial methane from both carbonate reduction and methyl-type fermentation pathways. Again, recognizing that the majority of atmospheric methane is generated by methyl-type fermentation, severe or even combined isotope shifts could not provide a methane source with the isotope signature close to that of atmospheric methane.

It is the isotope effect associated with the dominant atmospheric methane sinks, atmospheric OH-abstraction and soil uptake, which are the final fractionation steps responsible for the current atmospheric methane CD-isotope ratios.

3.2 Atmospheric methane sinks (hydroxyl abstraction, microbial oxidation). The CD-isotope pair for atmospheric methane is not merely the averaged value of the gas entering the atmosphere; rather, it is the net atmospheric isotope value between the addition and removal terms. The removal of atmospheric methane is primarily by the hydroxyl abstraction reaction, although bacterial methane consumption in surface soils is thought to be an important atmospheric sink (see this volume). Isotope fractionation factors are known for both aerobic and anaerobic bacterial methane oxidation in soils and sediments (αC = 1.004 to 1.02, αD = 1.2, Whiticar and Faber, 1986; Alperin et al., 1988; Whiticar, 1992), but it is uncertain if these values are strictly applicable to this situation, or indeed to the role of soils as a regulator of atmospheric methane.

As mandated by isotope theory, the lighter isotope of methane species ($^{12}CH_4$) in the atmosphere react more rapidly with OH-radicals than the heavier species (e.g., $^{13}CH_4$ or $^{12}CH_3D$). As a result, the steady state CD-isotope signature of the residual atmospheric methane is enriched in ^{13}C and ^{2}H relative to the CD-isotope signature of the mean source input.

One of the greatest concerns in the application and suitability of stable isotopes to constrain estimates of atmospheric methane budgets is the uncertainty in the estimated magnitudes of carbon and hydrogen isotope effects associated with the hydroxyl abstraction reaction. Initial experimental studies by Rust and Stevens (1980) suggested that the carbon isotope fractionation factor for the OH-reaction (α_C) is approximately 1.003. Davidson et al. (1986, 1987) offered α_C values of 1.025 and 1.010. More recently, Cantrell et al. (1990) determined α_C to be 1.0054 ± 0.0009, which is comparable to the calculations of Lasaga and

Gibbs (1991). The only value published for the hydrogen fractionation factor associated with the hydroxyl abstraction reaction (α_D) is 1.5 (Gordon and Mulac, 1975). Figure 7 shows the combined carbon and hydrogen isotope shift of the mean methane input to the mean atmospheric methane. Our limited understanding of these isotope effects, particularly in view of their degree of offset, curtails our use of isotope signatures in atmospheric methane mass balances. We may expect the isotope effects to be consistent but not necessarily constant. For example, do the isotope effects increase in magnitude with greater elevation due to lower temperatures?

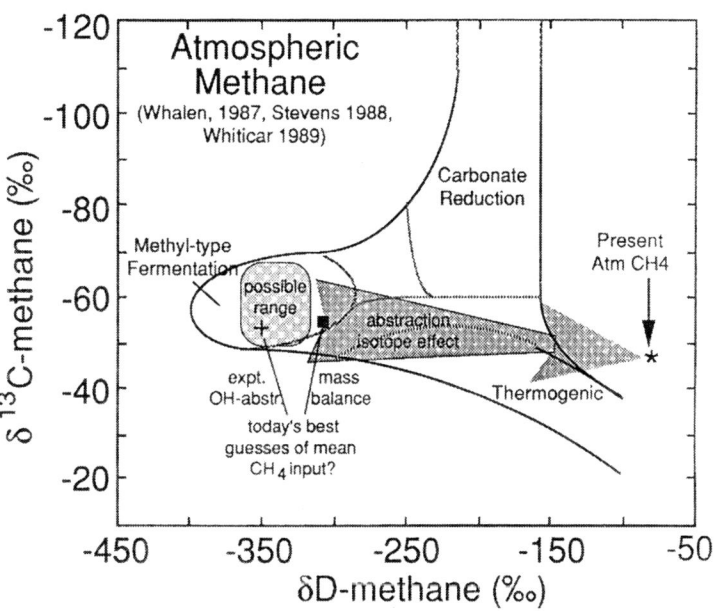

Figure 7. Perhaps the weakest link in our ability to constrain atmospheric methane budgets by stable isotopes is the correctness of the isotope effects associated with OH radical removal of atmospheric methane. The isotope experimental base on OH-abstraction is limited, and the "best guesses" of the CD shift (experimental and calculated) are indicated on Figure 7. The shaded box is the range of values cited in the literature. We do not know if the fractionation is consistent or uniform temporally and geographically. What roles do temperature effects or OH-concentration variations play?

Recognizing that several of the major atmospheric methane sources are of bacterial origin, i.e., by methyl-type fermentation, then their δ^{13}C-CH$_4$ values lie around -55 ‰ to -65 ‰ and the δD-CH$_4$ around -300 ‰ to -400 ‰. The present uncertainty in the estimates of the C,H-isotope effects associated with hydroxyl abstraction is similar to the isotope range experienced for these sources.

3.3 Resume. Carbon and hydrogen isotope effects are generally predictable for the process controlling the bacterial or thermogenic formation of thermogenic natural gas. Despite the potentially large range in isotope values possible, the actual methane CD-pairs of typical gases are quite restricted in excursion. Metabolic rates or thermal activation energies are possible restrictors in these systems. Similarly, we have some preliminary understanding of isotope effects associated with methane oxidation in sediments, soils and water columns, but our confidence is considerably less in the estimation of the C- ,H-isotope effects for the OH abstraction sink of atmospheric methane. However, if the estimated values of α_C = 1.005 and α_D = 1.5 for the OH-reaction are close to the actual values, then the globally averaged C-, H-isotope values of methane entering the atmosphere would be around δ^{13}C-CH$_4$ = -53 ‰, δD-CH$_4$ = -350 ‰. As plotted in Figure 7, this mean source signal lies within and is dominated by the CD-zone defining methanogenesis by methyl-fermentation, i.e., bacterial methane from terrestrial sources.

4. Uncertainty in flux magnitudes

Provided that we are able to assign a definitive ranges of carbon and hydrogen isotope ratios to specific sources of atmospheric methane, and provided that we can make reasonable assumptions regarding the magnitudes of the operative isotope effects, we could then proceed to calculate mass balances for carbon and hydrogen isotopes of atmospheric methane based on the source input fluxes. Conversely, if we are able to ascertain changes or variations in the isotopes' ratios atmospheric methane, we may be able to ascribe them to changes in the flux strengths from different sources.

The success of this mass balance approach lies on dependable flux and isotope signal estimations. Table 1 lists generally accepted values of methane fluxes to the atmosphere from the major sources. Accompanying the flux magnitudes in Table 1 are crude estimates of mean carbon and hydrogen isotope ratios for these individual sources (after Schoell, 1980; Whiticar, 1990; Bundesanstalt für Geowissenschaften und Rohstoffe, (BGR) data base). As discussed earlier (Figure 6), a certain latitude in the assigned CD values must be tolerated at this stage. Figure 8 illustrates the relative proportions of the various source strengths. The sources are arranged in Figure 8 such that they are progressively heavier isotopically (^{13}C enriched) in a clockwise direction. Even though the flux estimates may be in error up to a factor of 2, it is readily apparent from Table 1 and Figure 8 that bacterial methane from rice paddies, wetlands and landfills constitute close to 50% of the emissions. Furthermore, the addition of livestock and insect releases bring the bacterial contribution to atmospheric methane to around 75%, i.e., as predicted by the CD diagramme (Figure 7). Important in this context is that both the carbon and hydrogen isotope ratios of these various sources differ very little between these sources. Sources of methane from coal and thermogenic natural gas comprise around 15%, probably similar to biomass burning, whereas marine and freshwater environments are currently not significant sources <5%.

Methane from termites illustrates the difficulty in obtaining representative flux values, and they continue to be a topic of considerable debate since Zimmerman et al. (1982) first suggested that termites could generate between 75 Tg and 310 Tg methane yearly. Taking the total annual global influx of methane to be 540 Tg (Table 1), then the termite contribution could comprise over 50%. More conservative estimates of termite methane emissions strengths of 40 to 50 Tg CH_4 were suggested by Rasmussen and Khalil (1983), Seiler et al. (1984), and Zimmerman et al. (1987), and more recently downwardly revised flux values of about 12 Tg/yr have been cited by Fraser et al. (1986) and Khalil et al. (1990). The conservative estimates are used in Figure 8.

Accepting the uncertainties in the flux and isotope estimates, we calculate by mass balance a mean (integrated) input value of methane entering the atmosphere to have a $\delta^{13}C\text{-}CH_4 = -54.2$ ‰, and $\delta D\text{-}CH_4 = -307$ ‰ (Table 1, Figure 9). The carbon isotope value is similar to that of $\delta^{13}C\text{-}CH_4 = -53$ ‰ predicted by the OH-reaction, but the hydrogen isotope ratio is significantly heavier than $\delta D\text{-}CH_4 = -350$ ‰ estimated for the abstraction. Assuming steady state, the revised fractionation factors based on the mass balance approach would be values $\alpha_C = 1.0076$ and $\alpha_D = 1.33$. The discrepancy in methane isotope values between the calculated mass balance and the expected isotope input, particularly in D/H, points to our weakness in identifying, estimating, or characterizing the various sources and sinks for atmospheric methane.

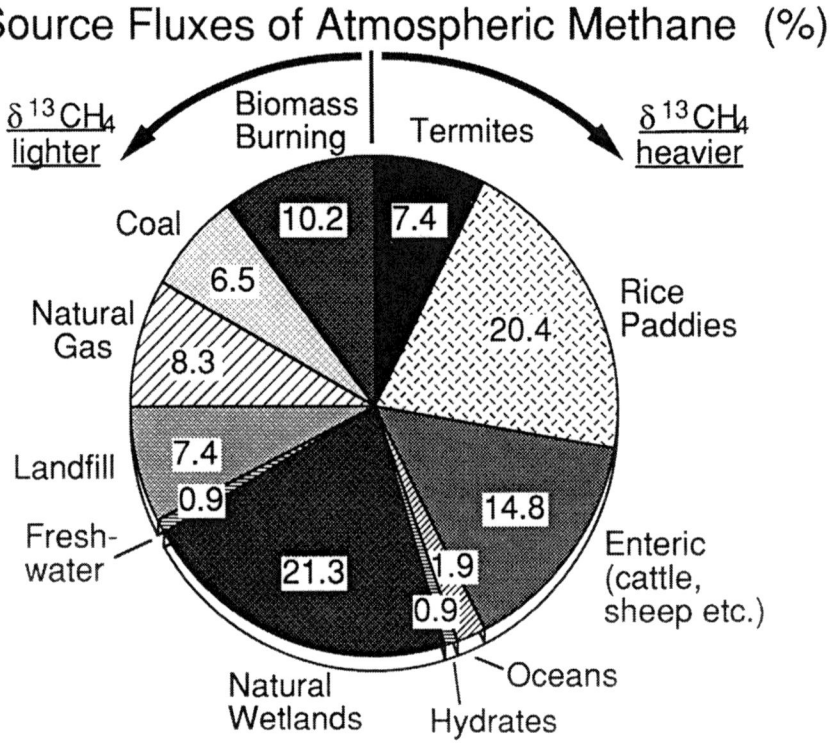

Figure 8. Commonly referred to as the "Pie-of-Culprits," we can use the information on the input sizes and stable isotope signature of the different sources to calculate a mass balance. Perhaps more than "fine-tuning" of the pie wedges is still required (e.g., termite or rice controversies).

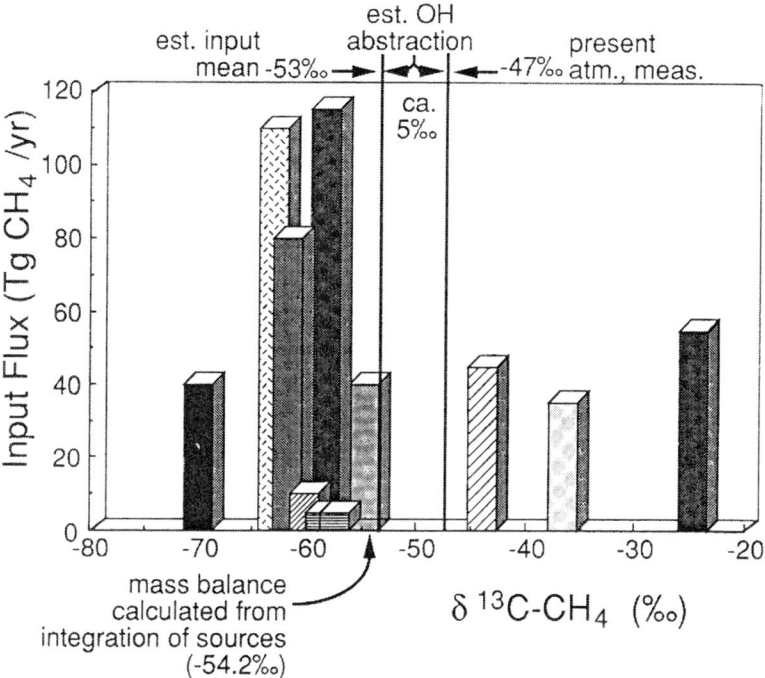

Figure 9. A re-representation of the flux and isotope data of Figure 8. Based on an isotope effect causing an approximate 6 per mil offset in δ^{13}C-CH$_4$ (α_C ca. 1.0054) from the steady state atmospheric methane value, a value of -53‰ is set for the mean δ^{13}C-CH$_4$ entering the atmosphere (refer to Figure 7). The mass balance presented here gives an integrated δ^{13}C-CH$_4$ input value of -54.2‰. The difference between the measured and calculated carbon isotope effect for the methane reaction is now much less than found previously, but the discrepancy for the hydrogen isotope effect remains significant.

Because the majority of the sources have similar isotope ratios, uncertainties in the estimates of the flux magnitudes of various sources may not significantly affect the mass balance, nor change the overall story. Figure 10 illustrates the sensitivity of the mean δ^{13}C-CH$_4$ value of methane entering the atmosphere on the correctness of the flux magnitude estimate for four different sources (rice, wetlands, natural gas and biomass burning). In Figure 10, the 540 Tg/yr mass balance of methane entering the atmosphere is maintained and the flux increase for any one species is proportionately reduced amongst the remaining sources.

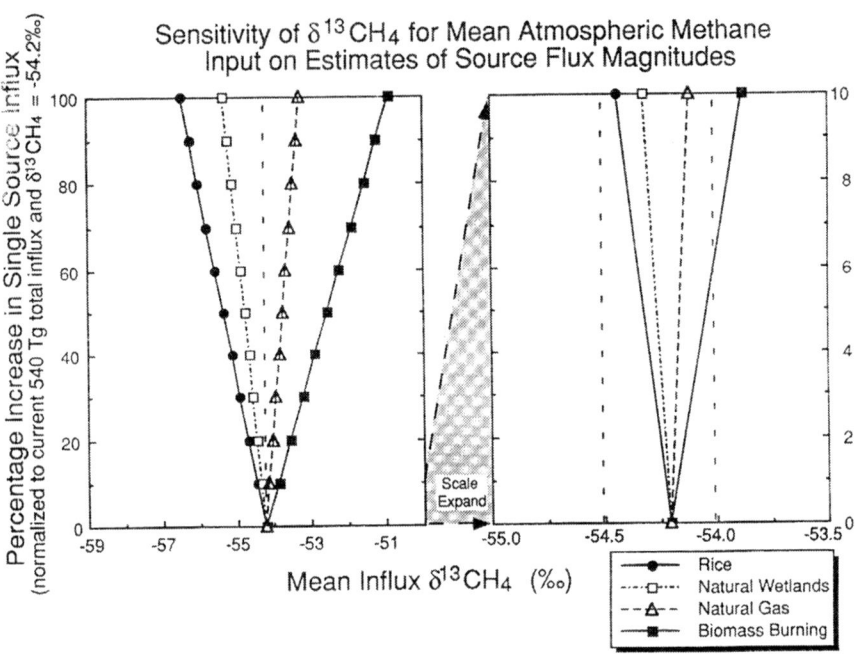

Figure 10. The sensitivity of the integrated δ^{13}C-CH$_4$ input value to adjustments in our source strengths is demonstrated. The left-hand figure shows a relative increase in the magnitude of the source flux from 0 - 100% (up to twice now estimated) for 4 candidates and the corresponding shift in integrated δ^{13}C-CH$_4$ input value (total emission held to 540 Tg, loss distributed proportionally amongst remaining sources). The right-hand figure is an expanded 0-10% scale. In all cases, the isotope offset is well within our measurement capability, but obviously the system is probably not so simple.

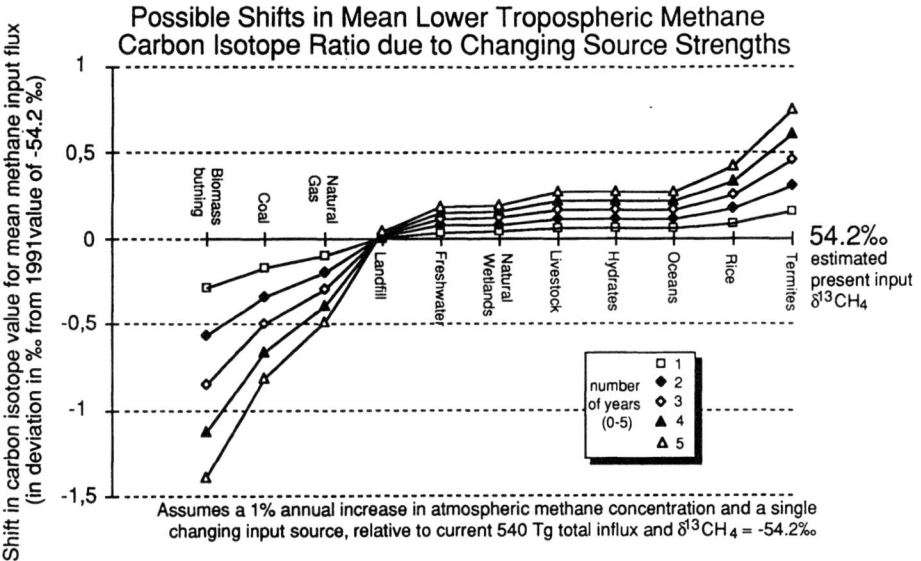

Figure 11. The shift of the integrated δ^{13}C-CH$_4$ input value, based on the present 1% annual increase in atmospheric methane, that could be experienced over the next 5 years. As for Figure 10, the relative isotope shift calculation only accounts for single source strength adjustments, e.g., only biomass burning or rice paddies are responsible for the atmospheric methane rise. Again, measurable isotope offsets result, but multiple source changes would severely complicate the story.

In the cases of 100% error, i.e., where the single source is actually twice that normally estimated, then for an extreme case a shift of over 4 ‰ (only for the ^{13}C-rich biomass burning) could be realized. Typically changes in ^{13}C-CH$_4$ for atmospheric methane for a 100% single source revision would be 1 ‰ or less. More conservative revisions of individual flux strengths of 10% (see expanded scale, Figure 10) would generally result in δ^{13}C-CH$_4$ shifts of less than 0.5‰.

Perhaps our greatest need of isotope constraints of atmospheric methane budgets is to recognize which source(s) is(are) contributing to the annual 1% increase in the atmospheric methane. Unfortunately, most of the important sources have similar or identical isotope ratios, so that "catching the culprit" may require further evidence, such as improved flux estimations, or $^{14}CH_4$ constraints. Figure 11 attempts to illustrate the point that if the entire annual 1% atmospheric methane increase were due to a single changing source (e.g., rice or biomass burning, etc.), then even after 5 years many of these sources would offset the current mean input $\delta^{13}C$-CH_4 value of -54.2 ‰ less than ± 0.5 ‰. Only biomass burning, coal-bed methane, or perhaps termites would generate $\delta^{13}C$-CH_4 changes > 0.5 ‰ after 5 years. If multiple sources are changing, which is likely, then these extreme excursions would be dampened considerably, and possibly below the analytical threshold for recognizing the isotope shift. The recent improvement in analytical precision by isotope ratio monitoring (GC-C-IRMS) may provide us with the improved sensitivity necessary.

5. Summary

The three prerequisites, which needed to be satisfied in order that we can use stable isotopes to constrain estimates of the atmospheric methane budgets, as set out at the start of the paper (i.e., 1. classification, 2. C-, H-isotope effects, and 3. uncertainty in input/output flux magnitudes) have been only partially met. We can discriminate the primary methanogenic and thermogenic (catagenic) formation and oxidation pathways. However, our knowledge of isotope fractionation during migration/transport is weak and our understanding of the isotope effects associated with the atmospheric OH-abstraction of methane insufficient. Based on mass balance calculations, the isotope effects for the OH-reaction are less than predicted by experimental data. Some of this discrepancy may be due to the impact of other methane sinks. Soils are now recognized as important sinks of atmospheric methane, and the possibility that the oceans may also be large scale sinks is currently being tested. The consequence of these new sinks for atmospheric methane, is that the magnitude of the total source fluxes may have to be adjusted upward to accommodate the annual 1% increase in atmospheric methane burden.

If the rise in atmospheric methane concentration is due to essentially a single source, and that source has an isotope signature dissimilar from the present mean $\delta^{13}C$-CH_4 (-54.2 ‰), then the isotope shift could constrain the budget estimates. Multiple sources definitely confuse the issue.

Stable isotopes are excellently suited to track the various process of formation, migration, and consumption, but in short time spans (i.e., next 5 years) they appear to lack sufficient resolution to constrain estimates of mean atmospheric methane budgets. Longer-term (30 - 50 years, or palaeo-atmospheric) changes in source strengths may be followed in the atmospheric methane isotope record, but again only if the changing components are significantly dissimilar from the isotope signature of mean methane input.

Finally, the intricate relationship between atmospheric methane and OH-radicals may permit us to estimate changes in atmospheric OH-radical abundance by monitoring the stable isotope ratios of methane, but more confidence in predicting and quantifying the operative isotope effects plays a key role in this development.

Acknowledgments. I would like to thank Keith Lassey for his incisive review, which improved the manuscript and alerted me to some recent investigations. This research effort is funded in Canada through the National Science and Engineering Research Council (NSERC) Strategic Research Grant No. 105389.

References

Abrajano, T.A., N.R. Sturchio, J.H. Bohlke, G.L. Lyon, R.J. Poreda, C.M. Stevens. 1988. Methane-hydrogen gas seeps, Zambales Ophiolite, Philippines: Deep or shallow origin? *Chem. Geol.,* 71:211-222.

Alperin, M.J., W.S. Reeburgh, M.J. Whiticar. 1988. Carbon and hydrogen isotope fractionation resulting from anaerobic methane oxidation. *Global Biogeochem. Cycles,* 2:279-288.

Blake, D.R., F.S. Rowland. 1988. Continuing worldwide increase in tropospheric methane, 1978 to 1987. *Science,* 239:1,129-1,131.

Cantrell, C.A., R.E. Shetter, A.H. McDaniel, J.G. Calvert, J.A. Davidson, D.C. Lowe, S.C. Tyler, R.J. Cicerone, J.P. Greenberg. 1990. Carbon kinetic isotope effect in the oxidation of methane by hydroxyl radicals. *J. Geophys. Res.,* 95:22,455-22,462.

Chung, H.M., W.M. Sackett. 1979. Use of stable isotope compositions of pyrolytically derived methane as a maturity indices for carbonaceous materials. *Geochim. Cosmochim. Acta,* 43:1,979-1,988.

Chung, H.M., J.R. Gromly, R.M. Squires. 1988. Origin of gaseous hydrocarbons in subsurface environments: Theoretical considerations of carbon isotope distribution. *Chem. Geol.,* 71:97-103.

Claypool, G.E., I.R. Kaplan. 1974. The origin and distribution of methane in marine sediments. In: *Natural Gases in Marine Sediments* (I.R. Kaplan, ed.), Plenum, New York, pp.99-139.

Clayton, C. 1991. Carbon isotope fractionation during natural gas generation from kerogen. *Mar. and Petrol. Geol., 8*:232-240.

Colombo, U., F. Gazzarini, G. Sironi, R. Gonfiantini, E. Tongiorni. 1965. Carbon isotope composition of individual hydrocarbons from Italian natural gases. *Nature, 205*:1,303-1,304.

Davidson, J.A., C.A. Cantrell, S.C. Tyler, J.G. Calvert, R.J. Cicerone, R.E. Shetter. 1986. The carbon kinetic isotope effect in the CH_4 + OH reaction. *EOS, 67*:245.

Davidson, J.A., C.A. Cantrell, S.C. Tyler, R.E. Shetter, R.J. Cicerone, J.G. Calvert. 1987. Carbon kinetic isotope effect in the reaction of CH_4 with HO. *J. Geophy. Res., 92*:2,195-2,199.

Evans, C.R., F.L. Staplin. 1971. Regional facies of organic metamorphism in geochemical exploration. In: *3rd Intl. Geochem. Explor. Symp.*, Proc. Can. Inst. Mining and Metallurgy, *Spec. Vol., 11*:517-520.

Faber, E., M.J. Whiticar. 1989. C- und -Isotope in leichtfluchtige Kohlenwasserstoffen der KTB. *KTB Reports*.

Faber, E., P. Gerling, I. Dumke. 1987. Gaseous hydrocarbons of unknown origin found while drilling. *Org. Geochem., 13*:875-879.

Frank, D.J. 1972. Deuterium variations in the Gulf of Mexico and selected organic materials. Ph.D. thesis, Texas A&M Univ.

Galimov, E.M. 1973. *Carbon Isotopes in Oil and Gas Geology* Nauka, Moscow, Engl. trans: *NASA TT-682*, Washington, D.C. 1975, 395 p.

Galimov, E.M. 1985. *The Biological Fractionation of Isotopes*, Academic Press, N.Y., 261 p.

Galimov, E.M., I.A. Petersil'ye. 1967. Isotopic composition of the carbon of methane isolated in the pores and cavities of some igneous minerals. *Doklady AN SSSR 176 (4)*.

Galimov, E.M., A.A. Ivlev. 1973. Thermodynamic isotope effects in organic compounds: 1. Carbon isotope effects in straight-chained alkanes. *Russian J. Phys. Chem. 47*:1564-1566.

Gerling, P. 1985. Isotopengeochemische Oberflächenprospektion Onshore. *BGR Internal Report No. 98576*, 36 pp.

Gordon, S., W.A. Mulac. 1975. Reaction of the OH ($X2\pi$) radical produced by the pulse radiolysis of water vapour. *Int. J. Chem. Kinetics*:289-299, Proc. Symp. on Chemical Kinetics Data for the Upper and Lower Atmosphere.

Hunt, J.M., A.Y. Huc, J.K. Whelan. 1980. Generation of light hydrocarbons in sedimentary rocks. *Nature, 288*:688-690.

James, A.T. 1983. Correlation of natural gas by use of carbon isotope distribution between hydrocarbon components. *Am. Assoc. Petrol. Geol., 67*:1,176-1,191.

Jefferey, A.W.A.; I.R. Kaplan. 1988. Hydrocarbons and inorganic gases in the Gravberg-1 well, Siljan Ring, Sweden. *Chem. Geol.*, 71:237-255.
Karweil, J. 1969. Aktuelle Probleme der Geochemie der Kohle. In: *Advances in Organic Geochemistry 1968* (P.A. Schenk and I. Havenaar, eds.), Pergamon Press, Oxford, pp. 59-84.
Khalil, M.A.K., R.A. Rasmussen, J.R.J. French, J.A. Holt. 1990. The influence of termites on atmospheric trace gases: CH_4, CO_2, $CHCl_3$, N_2O, CO, H_2, and light hydrocarbons. *J. Geophys. Res.*, 95:3,619-3,634.
Konnerup-Madsen, J., R. Kreulen, U. Rose-Hansen. 1988. Stable isotope characteristics of hydrocarbon gases in the alkaline Ilimaussaq complex, south Greenland. *Bull. Minéral.*, 111:567-576.
Kvenvolden, K.A. 1988. Methane hydrate - A major reservoir of carbon in the shallow geosphere. *Chem. Geol.*, 71:41-51.
Laier, T. 1988. Hydrocarbon gases in the crystalline rocks of the Gravberg-1 well, Swedish deep gas project. *Mar. and Petrol. Geol.*, 5:370-377.
Lasaga, T., G.V. Gibbs. 1991. Ab initio studies of the kinetic isotope effect of the CH_4 + OH• atmospheric reaction. *Geophys. Res. Lett.*, 18:1,217-1,220.
Lyon, G.L., J.R. Hulston. 1984. Carbon and hydrogen isotopic compositions of New Zealand geothermal gases. *Geochim. et Cosmochim. Acta*, 48:1,161-1,171.
McCarty, H.B., G.T. Felbeck, Jr. 1986. High temperature simulation of petroleum formation, IV. Stable carbon isotope studies of gaseous hydrocarbons. *Org. Geochem.*, 9:183-192.
Rasmussen, R.A., M.A.K. Khalil. 1983. Global production of methane by termites. *Nature*, 301:704-705.
Rice, D.D. 1993. Controls, habitat, and resource potential of ancient bacterial gas. In: *Biogenic Natural Gas* (in press).
Rice, D.D., G.E. Claypool. 1981. Generation, accumulation and resource potential of biogenic gas. *AAPG Bull.*, 67:1,199-1,218.
Rust, F.E., C.M. Stevens. 1980. Carbon kinetic isotope effect in the oxidation of methane by hydroxyl. *Int. J. Chem. Kinetics*, 12:371-377.
Sackett W.M. 1968. Carbon isotope composition of natural methane occurrences. *AAPG Bull.*, 52:853-857.
Schmidt, M. 1987. Isotope-geochemical analysis of dunk tank gases, headspace gases, desorbed gases of cuttings and cores. *Vattenfall Deep Gas Project, Init. Rpt., Nov. 1987*, 14p.
Schoell, M. 1980. The hydrogen and carbon isotopic composition of methane from natural gases of various origins. *Geochim. et Cosmochim. Acta*, 44:649-661.
Schoell, M. (ed.) 1988. Origins of Methane in the Earth. *Chem. Geol.*, 71, 265 p.

Seiler, W., R. Conrad, D. Scharffe. 1984. Field studies of methane emission from termite nests into the atmosphere and measurements of methane uptake by tropical soils. *J. Atmos. Chem.*, 1:171-186.

Sherwood, B., P. Fritz, S.K. Frape, S.A. Macko, S.M. Weise, J.A. Welhan. 1988. Methane occurrences in the Canadian Shield. *Chem. Geol.*, 71:223-236.

Silverman, S.R. 1971. Influence of petroleum origin and transformation on its distribution and redistribution in sedimentary rocks. *Proc., 8th World Petrol. Congr.* 2, 47-54.

Simoneit, B.R.T., O.E. Kawka, M. Brault. 1988. Origin of gases and condensates in the Guaymas Basin hydrothermal system (Gulf of California). *Chem. Geol.*, 71:169-182.

Stahl, W. 1973. Carbon isotope ratios of German natural gases in comparison with isotopic data of gaseous hydrocarbons from other parts of the world. In: *Advances in Organic Geochemistry 1973* (B. Tissot and F. Bienner, eds.), Pergamon Press, Oxford, pp. 453-462.

Tissot, B.P., D.H. Welte. 1978. *Petroleum Formation and Occurrence*. Springer Verlag, Berlin, 538 p.

Welhan, J.A. 1988. Origins of methane in hydrothermal systems. *Chem. Geol.*, 71:183-198.

Welhan, J.A., J.E. Lupton. 1987. Light hydrocarbon gases in Guaymas Basin hydrothermal fluids: Thermogenic versus abiogenic origin. *Bull. Am. Assoc. Petrol. Geol.*, 71:215-223.

Whelan, J.K., M.E. Tarafa, J.M. Hunt. 1982. Volatile C_1-C_8 organic compounds in macroalgae. *Nature*, 299:50-52.

Whiticar, M.J. 1990. A geochemical perspective of natural gas and atmospheric methane. In: *Advances in Organic Geochemistry* (B. Durand and F. Behar, eds.), Org. Geochem. 16, 531-547.

Whiticar, M.J. 1992. Isotope tracking of microbial methane formation and oxidation. In: *Cycling of Reduced Gases in the Hydrosphere* (D.D. Adams, P.M. Crill, and S.P. Seitzinger, eds.), Mitteilung (Communications) v. 23, Internationalen Vereinigung für Theoretische und Angewandte Limnlogie, E. Schweizerbart'sche Verlagsbuchhandlung (Nägele u. Obermiller), Stuttgart, Germany

Whiticar, M.J., E. Faber. 1986. Methane oxidation in sediment and water column environments - isotope evidence. *Org. Geochem.*, 10:759-768.

Whiticar, M.J., E. Suess. 1989. Hydrothermal hydrocarbon gases in the sediments of the King George Basin, Bransfield Strait, Antarctica. In: *Geochemistry of Hydrothermal Systems, Applied Geochemistry*, 5 (B.R.T. Simoneit, ed.), 135-147.

Whiticar, M.J., B.R.T. Simoneit. 1993. Carbon and hydrogen isotope systematics of hydrothermal hydrocarbons at Yellowstone Park, USA. (in press)

Whiticar, M.J., E. Faber, M. Schoell. 1984. Carbon and hydrogen isotopes of C_1-C_5 hydrocarbons in natural gases. *AAPG Research Conference on Natural Gases*, San Antonio TX.
Whiticar, M.J., E. Faber, M. Schoell. 1986. Biogenic methane formation in marine and freshwater environments: CO_2 reduction vs. acetate fermentation - isotope evidence. *Geochim. et Cosmochim. Acta, 50*:693-709.
Zehnder, A.J.B. (ed.) 1988. *Biology of Anaerobic Microorganisms*. Wiley, N.Y.
Zimmerman, P.R., J.P. Greenberg, S.O. Wandiga, P.J. Crutzen. 1982. Termites: A potentially large source of atmospheric methane, carbon dioxide and molecular hydrogen. *Science, 218*:563-565.
Zimmerman, P.R., C. Westberg, J. Darlington. 1987. Global methane production by termites. *Div. of Geochem., 193rd National Mtg., Am. Chem. Soc.* (abstract), p. 55.

Chapter 9

Methane Sinks and Distribution

M.A.K. KHALIL, M.J. SHEARER, AND R.A. RASMUSSEN

*Global Change Research Center
Department of Environmental Science and Engineering
Oregon Graduate Institute, Portland, Oregon 97291, USA*

1. Introduction

At present the amount of methane removed from the atmosphere each year is about 500 Tg/yr or more than 90% of that released into the atmosphere each year. Most of the methane is removed by reacting with tropospheric OH radicals; lesser amounts are removed by soils and stratospheric oxidation by OH, $O(^1D)$, and minor reactions. This chapter is on the removal rate of CH_4 and its variability in space and time.

The oxidation of methane in the troposphere depends on the concentrations of OH and thus varies greatly with latitude and altitude. Sixty-four percent of it is destroyed in the tropics where OH is more abundant throughout the year compared to any other part of the world. The removal rates vary seasonally, reflecting the variability of OH, and also may be affected by seasonal variations in soil conditions and of stratospheric processes, particularly at middle and higher latitudes. On still longer time scales, the destruction rate and hence the atmospheric lifetime of methane may vary because of secular changes in OH, the soil sink, and possibly stratospheric conditions.

The global mass balance of methane may be written as:

$$\frac{\partial C}{\partial t} = S - L - \frac{1}{n}\nabla \cdot n \,(CV - K \cdot \nabla C) \qquad (1)$$

where C is the mixing ratio, S is the source, L is the destruction or loss rate, ∇ is the spatial gradient operator, n is the number density of air molecules, and the last term represents transport by mean (**V**) and turbulent (**K**) processes.

The destruction rate L_R (molecules/yr or ML_R/No in gm/yr) and the average lifetime τ_R (yrs) in a region R, which includes a surface area A at ground level, are:

$$L_R = \int_R \sum_J \eta_J(x,t)\, \rho(x,t)\, C(x,t)\, dx \\ + \int_A v_d(x,t)\, \rho(x,t)\, C(x,t)\, dA \qquad (2)$$

$$\tau_R = \frac{\int_R C\rho\, dx}{L_R} \qquad (3)$$

$$\eta_J = K_J [J] \qquad (4)$$

M = Molecular weight (gm/mole).
No = Avogadro's number (molecules/mole).
ρ = Density of air at location x and time t (molecules/cm^3).
v_d = Deposition velocity (cm/sec).
K_J = Rate constant for reaction of CH_4 with chemical species J (cm^2/molecule-sec).
[J] = concentration of species J (molecules/cm^3).

To estimate the current annual and seasonal global removal rates, we use the measured concentrations of methane, deposition velocities, reaction rate constants, and calculated distributions of OH and other reactants in the equations mentioned above (with data from Crutzen and Schmailzl, 1983; Spivakovsky et al., 1990; Brasseur et al., 1990; NOAA/CMDL, 1990; Lu and Khalil, 1991; Vaghjiani and Ravishankara, 1991; DeMore et al., 1992; Weisenstein et al., 1992).

In the next section we discuss the global distribution of methane used in our calculations. Section 3 is on the vertical, latitudinal, and seasonal destruction of methane under current climatic and environmental conditions. In Section 4 we discuss possible long-term changes of the lifetime, and in Section 5 we consider some emerging issues in evaluating the sinks of methane.

2. Spatial and Seasonal Concentration Distribution of Methane

Recent measurements are used to construct latitudinal and vertical distributions of methane. We then adjust the distributions to 1990 as the base year and convert all data to a single calibration (Rasmussen scale as in Rasmussen and Khalil, 1986).

The vertical profile for methane at middle northern latitudes is a composite of several published data sets (Fabian et al. 1981; Schmidt et al. 1984, 1987; Taylor et al. 1989). It is shown in Figure 1 extending from the surface to nearly 70 Km. In the troposphere concentrations are nearly constant with height, but in the stratosphere methane concentrations fall rapidly. Above the tropopause, concentration can be described by $C = C_T \, e^{(\lambda z)}$ ($z_T < z < z_S$) where C_T is the concentration at the tropopause, z is the altitude, z_T is the tropopause height, and z_S is the stratopause height. Much higher up in the stratosphere, between 30-60 Km, methane concentrations have been measured by satellites and spacecraft (results reported by Ehhalt et al., 1972; Jones and Pyle, 1984; Gunson et al., 1990). Stratospheric concentrations fall off more rapidly in the higher latitudes than in the tropical latitudes (Bush et al., 1978).

The latitudinal variations are taken from the NOAA/GMCC (now CMDL) data and our measurements (Steele et al., 1987; NOAA/CMDL, 1990; Khalil and Rasmussen, 1990, 1993) and are shown in Figure 2. Concentrations are generally higher in the middle and higher latitudes of the northern hemisphere than elsewhere because of the similar global distribution of sources (Khalil and Shearer, this volume and references therein). The seasonal variations shown in Figures 3a and 3b are taken from our flask sampling network. Concentrations are lower in the summers compared to winters. The seasonal cycles of CH_4 concentrations at various latitudes are determined mostly by seasonal changes of OH, but substantial contributions from seasonal variations of sources are expected at some latitudes, particularly the middle northern latitudes (Khalil and Rasmussen, 1983, 1993; Fung et al., 1991).

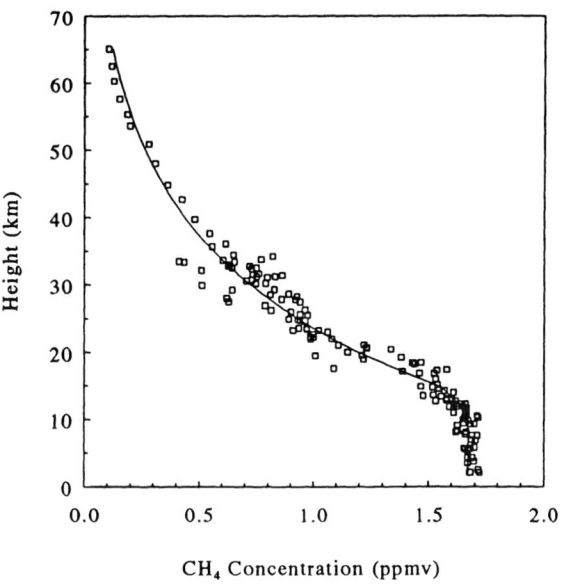

Figure 1. Vertical distribution of CH_4 at 44° N latitude. Data from Fabian et al., 1981, Schmidt et al., 1984 and 1987, and Taylor et al., 1989.

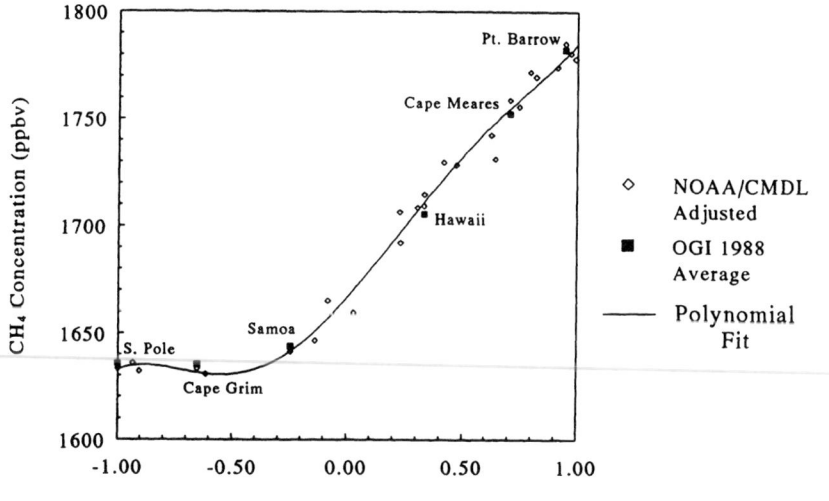

Figure 2. Latitudinal distribution of CH_4. The 1988 NOAA/CMDL data (NOAA/CMDL, 1990) are shown next to our 1988 data from six sites. (Calibration adjustment applied: C(NOAA) = C(ours) + 11 ppbv.)

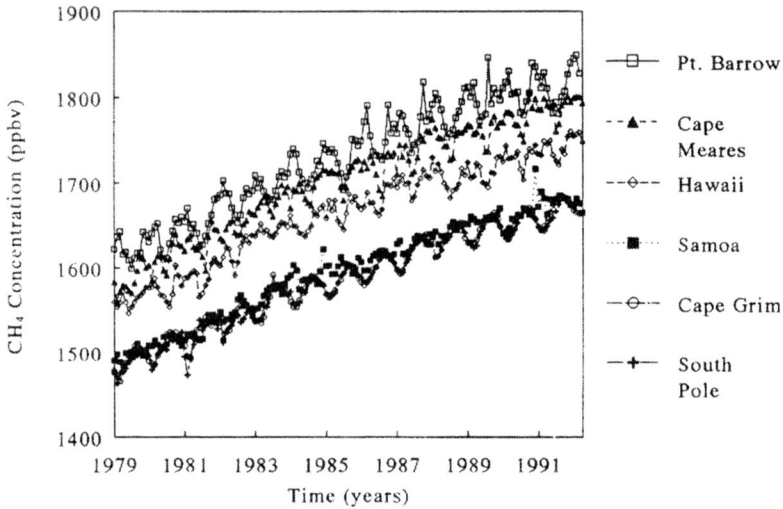

Figure 3a. Time series of monthly measurements of CH_4 at six sites.

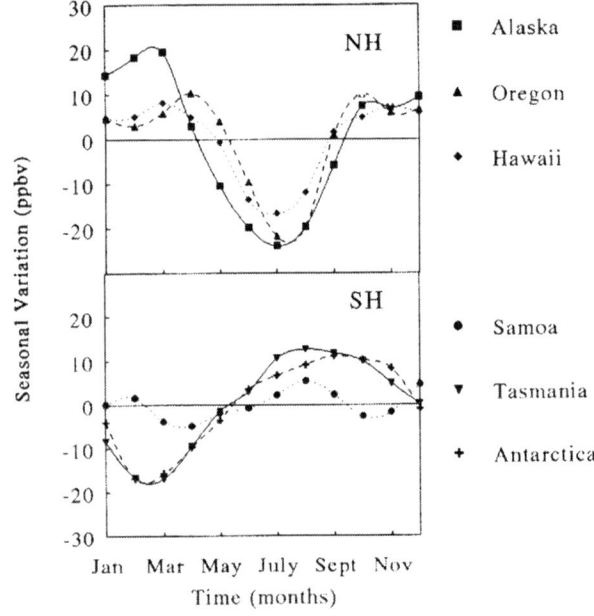

Figure 3b. Average seasonal variations at the same six sites.

3. Seasonal and Spatial Destruction Rates and Lifetimes

The hydroxyl radical is the principle sink of atmospheric methane through a series of reactions that begins with $CH_4 + OH \rightarrow CH_3 + H_2O$ (see Wuebbles and Tamaresis, this volume). The OH concentrations by latitude, altitude and season are taken from the tables published by Spivakovsky et al. (1990) with stratospheric data from Brasseur et al. (1990); concentrations in both studies are derived from model calculations. With the methane distribution described above, we estimate 440 Tg/yr of methane are destroyed in the troposphere by OH mostly during the daytime. Small amounts, around 5 Tg/yr of methane may be destroyed by nighttime OH (Lu and Khalil, 1992). At higher latitudes most of the annual destruction of methane occurs during summers compared to other seasons. An additional 10 Tg/yr is destroyed by OH in the stratosphere.

Other sinks believed to be important in the stratosphere are CH_4 reactions with $O(^1D)$ and Cl. We estimate about 5 Tg/yr of methane is destroyed by $O(^1D)$ and Cl destroys much less than 1 Tg/yr. (See Crutzen and Schmailzl (1983) and Weisenstein et al. (1992) for base data.)

The soil sink was estimated from a global vegetation database (Matthews, 1983; 1984), using methane consumption factors compiled by Bartlett and Harriss (1993). Methane uptake by soils is mainly controlled by soil inundation and soil porosity (see Bartlett and Harriss, 1993, for a review). The total soil sink is estimated to be 25-30 Tg/yr.

Table 1. Annual sinks of CH_4 and its mass balance for 1990.

	30-90° S	0-30° S	0-30° N	30-90° N	Global
C (Tg)	1166	1201	1243	1256	4865
dC/dt[§] (Tg/yr)	7.8	7.9	8.5	7.1	30
S (Tg/yr)	20	115	170	205	510
τ^* (yr)	24	7	7	17	10
L (Tg/yr)	48	178	180	74	480

Global values are rounded.
[§] Values for dC/dt are calculated from Khalil and Rasmussen (1993).
[*] Composite of destruction by oxidation and soil sink:
$$\tau = (\tau_{OH}^{-1} + \tau_{Soil}^{-1})^{-1}$$

The destruction rate by latitude in Figure 4 shows that most of the methane is destroyed in the tropics. A summary of the destruction rates and lifetimes is given in Table 1. Other estimates are given by Crutzen and Schmailzl (1983), Warneck (1988), Prather and Spivakovsky (1990), and Lelieveld et al. (1993).

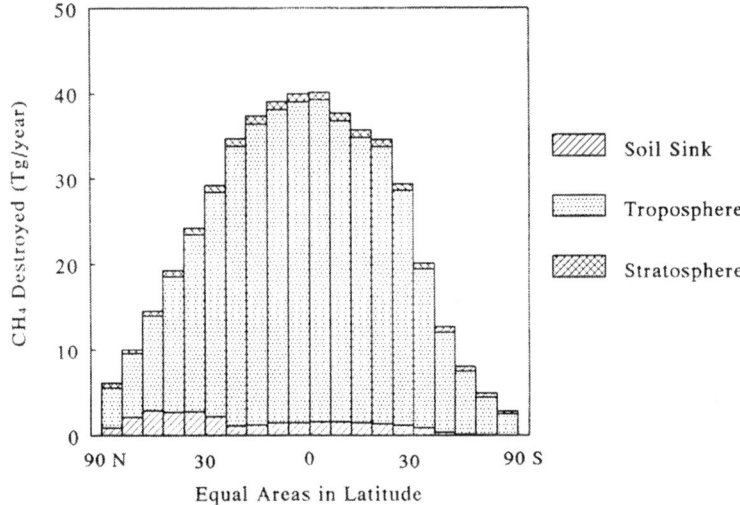

Figure 4. Latitudinal distribution of CH_4 sinks by equal area latitude bands (increments of 0.1 in sine of latitude).

4. Secular trends of lifetime

There has been much speculation on the long-term changes in atmospheric OH which is a principal oxidant of the global atmosphere. Such changes in the oxidation capacity affect not only the sinks of methane but of many other trace gases in the earth's atmosphere (for a current review see Thompson, 1992). If OH concentrations decrease over time, methane lifetime would increase and contribute to increasing trends. OH is produced mostly by the reaction of sunlight with O_3 ($O_3 + h\nu \rightarrow O_2 + O(^1D)$) followed by the reaction of the excited oxygen atoms with water vapor ($H_2O + O(^1D) \rightarrow 2\ OH$). It is destroyed mostly by reacting with CO and methane. There are other production and destruction processes as well, but they are individually of lesser importance.

It was thought that because CO and methane have been increasing, OH must be declining over the last century and particularly in recent decades (Levine et al. 1985; Khalil and Rasmussen, 1985; Thompson and Cicerone, 1986). Current research suggests however, that OH concentrations may be stabilized because at the same time that anthropogenic activities increase the removal rate of OH (by increasing CH_4 and CO), they also tend to increase production. This occurs from increasing tropospheric O_3 due to human activities and changes of water vapor in the transitions between climatic regimes (Pinto and Khalil, 1991; Lu and Khalil, 1991). In addition, as stratospheric O_3 is reduced it allows more uv to penetrate the troposphere thus increasing the production of $O(^1D)$ which also increases OH production (Madronich and Granier, 1992). Since it is possible that both the production and destruction of methane are increasing, the present concentrations of OH may be constant or even increasing slightly.

It is also possible that the soil sink may be slowing down due to human activities as suggested by Steudler et al. (1989) who showed that nitrogen deposition onto the soils reduces the uptake of methane. Ojima et al. (1993) estimate that land use and management changes have led to a 30% decrease in the methane soil sink in the northern temperate latitudes. The global significance of this mechanism is not yet understood.

5. Current Issues

The study of the OH sink is dependent mostly on laboratory studies of the reaction rate constants and model calculations of tropospheric OH. Since there is no global climatology of measured OH concentrations there are a number of theoretical factors that affect the calculation of the oxidation of methane, and many other gases. Two identified factors that affect OH but are not included in current models are the non-methane hydrocarbons and the effects of clouds.

The non-methane hydrocarbons have short lifetimes compared to methane and most have sources that are much greater in the northern hemisphere compared to southern latitudes. Over the vast expanses of the oceans the total non-methane hydrocarbon concentrations are practically nothing compared to concentrations over populated or forested land. Only the light non-methane hydrocarbons exist over most of the Earth's surface. These hydrocarbons are likely to reduce OH concentrations both by direct reactions and the reactions of OH with the organic by-products. Qualitatively the effect of nonmethane hydrocarbons is reduced on the global scale by their low concentrations that decline with altitude and latitude into the southern hemisphere and the lack of significant concentrations of the higher molecular weight hydrocarbons (see Khalil and Rasmussen, 1992). The magnitude of the effect of non-methane hydrocarbons on tropospheric OH is not well known but may well be important since a 10% reduction in calculated tropospheric OH would lead to a reduction of about 50 Tg/yr in the calculated destruction rate of methane.

The effect of clouds on OH may be substantial also. Actinic flux is larger above the clouds and smaller below compared to clear sky conditions. Recent calculations show that low clouds at 1.5 Km altitude tend to increase the average vertical column concentration of OH and middle clouds at 4.5 Km altitude reduce the OH concentration (Lu, 1993). In both cases the concentration of OH below the cloud is about an order of magnitude smaller than under clear sky conditions. In the case of low clouds, there is an increase of OH above the cloud due to increase of actinic flux and water vapor, that spans a large altitude and overcomes the effect of reduced OH below the clouds. For higher clouds, the concentration of OH is reduced below the cloud, but is not compensated by the slightly higher calculated OH concentration above the cloud.

For the oxidation of trace gases, and particularly methane these results translate into a net reduction of methane removal from the atmosphere compared to clear sky conditions. At present we are not able to put definitive numbers behind the cloud effect since many theoretical issues remain unresolved so we can only report the first approximations. OH concentrations were calculated by a detailed photochemical model with and without the effect of clouds. Taking the average annual global cloud cover to be 35% (Hahn et al.,1987), annual methane destruction with only low clouds (< 2 Km) would increase to 460 Tg/yr, and with only middle clouds (3-6 Km) methane destruction would decrease to 340 Tg/yr. This is a sizeable range compared to calculations for clear sky conditions and warrants further study.

It is generally believed that methane oxidation is greatest in the boundary layer. This may not be so because the effects of both clouds and nonmethane hydrocarbons are likely to be greatest in the boundary layer and both tend to reduce the sink of methane.

6. Summary

Most of the annual removal of methane from the atmosphere is by reactions with tropospheric OH radicals (440 Tg/yr). Methane is also destroyed by deposition onto dry soils (30 Tg/yr) and 15 Tg/yr is destroyed in the stratosphere by reactions with OH, O(^{1}D) and minor reactions. Non-methane hydrocarbons and clouds may substantially lower the total annual removal rate of methane by OH.

Acknowledgements. We have benefitted from discussions with Y. Lu and F. Moraes. Support for some portions of this work was provided in part a grant from the Department of Energy (DOE DE-FG06-85ER60313). Additional support was provided by the Andarz Co.

References

Bartlett, K.B., R.C. Harriss. 1993. Review and assessment of methane emissions from wetlands. *Chemosphere, 26*:261-320.
Brasseur, G., M.H. Hitchman, S. Walters, M. Dymek, E. Falise, M. Pirre. 1990. An interactive chemical dynamical radiative two-dimensional model of the middle atmosphere. *J. Geophys. Res., 95* (D5):5,639-5,655.
Bush, Y.A., A.L. Schmeltekopf, F.C. Fehsenfeld, D.L. Albritton, J.R. McAfee, P.D. Goldan, E.E. Ferguson. 1978. Stratospheric measurements of methane at several latitudes. *Geophys. Res. Lett., 5*:,1027-1,029.
Crutzen, P.J., U. Schmailzl. 1983. Chemical budgets of the stratosphere. *Planet. Space Sci., 31* (9):1,009-1,032.
DeMore, W.B., S.P. Sander, C.J. Howard, A.R. Ravishankara, D.M. Golden, C.E. Kolb, R.F. Hampson, M.J. Kurylo, M.J. Molina. 1992. *Chemical Kinetics and Photochemical Data for Use in Stratospheric Modeling.* NASA Evaluation No. 10.
Ehhalt, D.H., L.E. Heidt, E.A. Martell. 1972. The concentrations of atmospheric methane between 44 and 62 kilometers altitude. *J. Geophys. Res., 77*:2,193-2,196.
Fabian, P., R. Borchers, G. Flentje, W.A. Matthews, W. Seiler, H. Giehl, K. Bunse, F. Müller, U. Schmidt, A. Volz, A. Khedim, F.J. Johnen. 1981. The vertical distribution of stable trace gases at mid-latitudes. *J. Geophys. Res., 86* (C6):5,179-5,184.
Fung, I., J. John, J. Lerner, E. Matthews, M. Prather, L.P. Steele, P.J. Fraser. 1991. Three-dimensional model synthesis of the global methane cycle. *J. Geophys. Res., 96* (D7):13,033-13,065.
Gunson, M.R., C.B. Farmer, R.H. Norton, R. Zander, C.P. Rinsland, J.H. Shaw, B.-C. Gao. 1990. Measurements of CH_4, N_2O, CO, H_2O, and O_3 in the middle atmosphere by the Atmospheric Trace Molecule Spectroscopy

Experiment on Spacelab 3. *J. Geophys. Res., 95* (D9):13,867-13,882.
Hahn, C.J., S.G. Warren, J. London, R.L. Jenne, R.M. Chervin. 1987. *Climatological Data for Clouds over the Globe from Surface Observations.* Report NDP-026, Carbon Dioxide Information Center, Oak Ridge, TN.
Jones, R.L., J.A. Pyle. 1984. Observations of CH_4 and N_2O by the NIMBUS 7 SAMS: a comparison with in situ data and two-dimensional numerical model calculations. *J. Geophys. Res., 89* (D4):5,263-5,279.
Khalil, M.A.K., R.A. Rasmussen. 1983. Sources, sinks, and seasonal cycles of atmospheric methane. *J. Geophys. Res., 88* (C9):5,131-5,144.
Khalil, M.A.K., R.A. Rasmussen. 1985. Causes of increasing atmospheric methane: depletion of hydroxyl radicals and the rise of emissions. *Atmos. Environ., 19*:397-407.
Khalil, M.A.K., R.A. Rasmussen. 1990. Atmospheric methane: recent global trends. *Environ. Sci. Technol., 24*:549-553.
Khalil, M.A.K., R.A. Rasmussen. 1992. Forest hydrocarbon emissions: relationships between fluxes and ambient concentrations. *J. Air & Waste Manage. Assoc., 42*:810-813.
Khalil, M.A.K., R.A. Rasmussen. 1993. Decreasing trend of methane: unpredictability of future concentrations. *Chemosphere, 26*:803-814.
Lelieveld, J., P.J. Crutzen, C. Brühl. 1993. Climate effects of atmospheric methane. *Chemosphere, 26*:739-768.
Levine, J.S., C.P. Rinsland, G.M. Tennille. 1985. The photochemistry of methane and carbon monoxide in the troposphere in 1950 and 1985. *Nature, 318*:254-257.
Lu, Y. 1993. Model calculations of radiative transfer and tropospheric chemistry. Ph.D. dissertation, Oregon Graduate Institute, Beaverton, OR.
Lu, Y., M.A.K. Khalil. 1991. Tropospheric OH: model calculations of spatial, temporal, and secular variations. *Chemosphere, 23*:397-444.
Lu, Y., M.A.K. Khalil. 1992. Model calculation of night-time atmospheric OH. *Tellus, 44B*:106-113.
Madronich, S., C. Granier. 1992. Impact of recent total ozone changes on tropospheric ozone photodissociation, hydroxyl radicals and methane trends. *Geophys. Res. Lett., 19*:465-467.
Matthews, E. 1983. Global vegetation and land use: new high-resolution data bases for climate studies. *J. Climate & Appl. Met., 22*:474-487.
Matthews, E. 1984. Vegetation, land-use and seasonal albedo data sets: documentation of archived data tape. NASA Technical Memorandum 86107, Goddard Space Flight Center, New York, U.S.A.
NOAA/CMDL (National Oceanic and Atmospheric Administration, Climate Monitoring and Diagnostics Laboratory Flask Sampling Program). 1990. In: *Trends '90, A Compendium of Data on Global Change* (T.A. Boden, P. Kanciruk, and M.P. Farrell, eds.), 148-189. Carbon Dioxide Information Analysis Center, Oak Ridge, TN, USA, ORNL/CDIAC-36.

Ojima, D.S., D.W. Valentine, A.R. Mosier, W.J. Parton, D.S. Schimel. 1993. Effect of land use change on methane oxidation in temperate forest and grassland soils. *Chemosphere, 26* (1-4):675-685.

Pinto, J., M.A.K. Khalil. 1991. The stability of tropospheric OH during ice ages, inter-glacial epochs and modern times. *Tellus, 43B*:347-352.

Prather, M., C.M. Spivakovsky. 1990. Tropospheric OH and the lifetimes of hydrochlorofluorocarbons. *J. Geophys. Res., 95* (D11):18,723-18,729.

Rasmussen, R.A., M.A.K. Khalil. 1986. Atmospheric trace gases: trends and distributions over the last decade. *Science, 232*:1623-1624.

Schmidt, U., A. Khedim, D. Knapsa, G. Kulessa, F.J. Johnen. 1984. Stratospheric trace gas distributions observed in different seasons. *Adv. Space Res., 4* (4):131-134.

Schmidt, U., G. Kulessa, E. Klein, E.-P. Röth, P. Fabian, and R. Borchers. 1987. Intercomparison of balloon-borne cryogenic whole air samplers during the MAP/GLOBUS 1983 campaign. *Planet. Space Sci., 35*:647-656.

Spivakovsky, C.M., R. Yevich, J.A. Logan, S.C. Wofsy, M.B. McElroy, M.J. Prather. 1990. Tropospheric OH in a three-dimensional chemical tracer model: an assessment based on observations of CH_3CCl_3. *J. Geophys. Res., 95* (D11):18,441-18,471.

Steele, L.P., P.J. Fraser, R.A. Rasmussen, M.A.K. Khalil, T.J. Conway, A.J. Crawford, R.H. Gammon, K.A. Masarie, K.W. Thoning. 1987. The global distribution of methane in the troposphere. *J. Atmos. Chem., 5*:125-171.

Steudler, P.A., R.D. Bowden, J.M. Melilo, J.D. Aber. 1989. Influence of nitrogen fertilization on methane uptake in temperate forest soils. *Nature, 341*:314-316.

Taylor, F.W. A. Dudhia, C.D. Rodgers. 1989. Proposed reference models for nitrous oxide and methane in the middle atmosphere. In: *Handbook for MAP, Vol. 31.* (G.M. Keating, ed.), 67-79.

Thompson, A.M., R.J. Cicerone. 1986. Possible perturbations to atmospheric CO, CH_4, and OH. *J. Geophys. Res., 91* (D10):10,853-10,864.

Thompson, A.M. 1992. The oxidizing capacity of the Earth's atmosphere: probable past and future changes. *Science, 256*:1,157-1,165.

Vaghjiani, G.L., A.R. Ravishankara. 1991. New measurement of the rate coefficient for the reaction of OH with methane. *Nature, 350*:406-408.

Warneck, P. 1988. *Chemistry of the Natural Atmosphere.* Vol. 41, International Geophysics Series, Academic Press, Inc., San Diego, CA, USA.

Weisenstein, D.K., M.K.W. Ko, N.-D. Sze. 1992. The chlorine budget of the present-day atmosphere: a modeling study. *J. Geophys. Res., 97* (D2):2,547-2,559.

Chapter 10

Sources of Methane: An Overview

M.A.K. KHALIL AND M.J. SHEARER

*Global Change Research Center, Department of Environmental Science & Engineering
Oregon Graduate Institute, Portland, Oregon, 97291, USA*

1. Introduction

The sources of methane are the most complex and critical element in understanding the concentrations of atmospheric methane and their trends. For those who want to reduce methane in the atmosphere or prevent it from increasing, controlling the sources is perhaps the only practical approach. Accordingly, a significant portion of this book is devoted to estimating the global and regional emission rates. The purpose of this chapter is to introduce the subsequent chapters on individual sources and to lay the foundation for the common elements of determining global emission rates from the many and varied sources of methane.

There are three major sources (> 50 Tg/yr), all biogenic, namely rice agriculture, ruminants (particularly cattle), and the natural wetlands. There are many more minor sources that each emit between 10-50 Tg/yr but collectively are a significant fraction of the global budget. These sources include landfills, coal mines, biomass burning, urban areas, sewage disposal, natural gas leakages, lakes, oceans, termites, and tundra. Finally there are yet smaller sources including biogas pits, asphalt, several industrial sources, and possibly others that have not yet been identified (~ < 5 Tg/yr) (see Judd et al., this volume; Lacroix, 1993). The very small sources are thought to be a small fraction of the annual emissions even when taken together. There is still enough uncertainty in the estimates of emission rates from individual sources that some may go from minor to major or vice versa, but it is very unlikely that there are any unknown major sources. Research has concentrated more on the major sources so less is known about the minor sources, particularly the smaller of the minor sources.

2. Estimating global emissions: issues and procedures

The process of evaluating global emissions usually consists of two fundamental pieces of information. The first is data on fluxes or emission factors, often measured directly under field or laboratory conditions. The second is an extrapolating factor or extrapolant that, when associated (or multiplied) with the measured fluxes, results in the global emission rate.

$$S_{Global} = \phi \cdot G \qquad (1)$$

Here ϕ is the average emission rate (Tg/yr/Unit), G is the extrapolant (the number of Units in the world) and S_{Global} is the global emission rate. For instance, the emission factor may be the methane emitted per year by a global average cow, and the extrapolant may be the number of such cows in the world.

While this appears straightforward enough, in practice neither the flux from the standard unit nor the extrapolant is accurately known. At the same time, direct laboratory or field measurements can give rather precise (and probably accurate) estimates of emission factors from the source under the conditions of the experiment. This high precision and repeatability of the measurements in the field or laboratory can sometimes lead to a false sense of accuracy in the estimate of global emission rates. The conditions of any experiment, however, necessarily limit the representativeness of the measured emission factors because there usually are many variables that alter the flux under different environmental (or experimental) conditions. To approach the problem experimentally requires that we know the main factors that control emission rates from a particular source. For the major sources (rice agriculture, cattle, and wetlands) these factors are only now becoming known (see Johnson et al., Neue and Roger, Bachelet and Neue, and Matthews, this volume).

While the flux can be measured experimentally, the extrapolant is often obtained from geographical, economic, or political data; this introduces a new class of uncertainties and systematic biases (see Mitchell, 1982). The number of cattle, for instance, is obtained from country censuses, which are better for some regions and worse for others and more accurate in some years and less so in others. Finally, the correctness of associating a measured flux with an extrapolant is also not always apparent. For instance, when measurements of methane emissions are taken from rice fields in one area, how large an area of rice fields has similar emission rates? To obtain a global emission rate, fluxes measured in one region may be multiplied by the world-wide area of rice fields (extrapolant), as indeed had to be done when no other information was available. Such an estimate is not likely to be accurate, however. To go further, we must account for the factors that affect emission rates differently in different regions. For instance, rice agriculture may be defined as continuously flooded, intermittently flooded, or dryland (see Shearer and Khalil, this volume). Annual emission rates of methane from these regimes are very different. The global flux may then be calculated as:

$$S_{Global} = \sum_{i=1}^{N} \phi_i A_i \qquad (2)$$

where ϕ_i are the measured annual fluxes and A_i are the areas of rice harvested under the three regimes (N = 3). The annual flux ϕ_i may itself may be a product or function of other factors. For instance, it may consist of average (measured) emission rates per day times the number of days in the growing season. In this way measurements taken in regions with one length of the growing season may be extrapolated to regions with a longer or shorter growing season (and assuming all other factors are the same). This processes can be continued further. For each irrigation regime ϕ_i one may add the effects of characteristic soil types so that ϕ_{ij} is the measured emission rate (or determined by some other means) under irrigation regime i and soil type j. ϕ_{ij} may be further sub-divided to include the effect of fertilizers to obtain ϕ_{ijk} and extrapolant G_{ijk} (the area of rice harvested, under irrigation regime i, soil type j, and fertilizer type k) and so on. In general, then, the annual flux from any source may be represented as:

$$S_{Global} = \sum_{i_1=1}^{N_1} \sum_{i_2=1}^{N_2} \cdots \sum_{i_j=1}^{N_j} \phi_{i_1 i_2 \cdots i_j} G_{i_1 i_2 \cdots i_j} \qquad (3)$$

Progression to apparently more and more detailed global estimates does not come without a price, however; the amount of information necessary increases geometrically, and if that information is highly uncertain, a more complicated calculation may be less accurate than a simpler one!

This procedure is essentially a search to find regions over which the methane flux is constant and can be represented adequately by measured or estimated fluxes. There are other equivalent methods for calculating global emission rates. For some variables there are better ways of expressing the flux. This formulation is general and allows us a common foundation for assessing emissions from many different types of sources (and for many gases), but there are some complications. One is that the effect of two or more factors may be synergistic, and we then cannot represent the factors additively as in Eqn. (3). In order for readers to reproduce calculated global emissions, the factors mentioned above must be specified. We have attempted to include such factors in the chapters on individual sources.

3. Constraints

There are several constraints that work at different levels of the global budget of methane. For the total emission rates from all sources, there is a constraint imposed by the global mass balance:

$$S = dC/dt - C/\tau \qquad (4)$$

Here, S (Tg/yr) are the global emissions, τ is the average atmospheric lifetime (years) (see Khalil and Shearer, this volume), and C is the global burden (Tg). dC/dt and C can be determined quite accurately from atmospheric measurements. The atmospheric lifetime is dominated by reaction with tropospheric OH, with lesser removal by soils and other chemical processes in the stratosphere so that:

$$1/\tau = K_R[OH] + 1/\tau_{other} \qquad (5)$$

where K_R is a suitably averaged reaction rate constant (see, for example, Khalil and Rasmussen, 1985). If we can calculate τ accurately, we can estimate the total emissions of methane from Eqn. (4). Fortunately, the effective global average concentration of OH may be estimated by empirical models from which τ can be obtained using Eqn. (5). A common method is to use the mass balance of methylchloroform (CH_3CCl_3) to determine the effective average OH (Lovelock, 1977; Singh, 1977; Khalil and Rasmussen, 1984; Prinn et al., 1987; Butler et al., 1991). Methylchloroform is an industrial de-greasing solvent that has been used for many years. It is emitted to the atmosphere soon after production (and sale), so that the well-documented industrial production records can be used as the emission rate (Midgely, 1989). It is removed from the atmosphere primarily by reacting with OH, although there may also be other lesser sinks (Khalil and Rasmussen, 1984; Butler et al., 1991). Knowing the emission rate of CH_3CCl_3 from the industry records and the concentration and rate of change of CH_3CCl_3 from direct atmospheric measurements, we can determine average OH by:

$$<[OH]> \approx <1/K_R[S/C - dlnC/dt]> \qquad (6)$$

where $<\cdots>$ is an average over a suitable period. This average OH can be used in Eqns. (4)-(5) to determine the total annual emissions of CH_4. This method has many associated uncertainties, but it seems that it can limit the possible range of total methane emissions to within ± 15%. There are some who would argue that the method is more accurate than this, and others who believe that the OH from CH_3CCl_3 is but a very weak constraint. One of our early budgets relied on this constraint to produce the total emission of about 500 Tg/yr, which has persisted for a decade (see Khalil and Rasmussen, 1983).

The method we have discussed involves two parameters for estimating total annual removal. Both parameters are subject to systematic errors. In addition to knowing the effective global OH concentration discussed above, we also need to know the reaction rate constant between OH and CH_4. This has undergone a recent revision that reduced its value by 20%, resulting in a lower estimate for the total methane sink (Vaghjiani and Ravishankara, 1991). There are other uncertainties, particularly systematic errors in estimates of average OH, that may offset the adjustment to the rate constant (see Rasmussen and Khalil, 1981; Khalil and Rasmussen, 1984; Butler et al., 1991).

For many reasons we need to know the methane budget in more detail than just the global average balance. Measurements of methane in polar ice core have shown that concentrations now are 2.5 to 3 times higher than 200 to 300 years ago (Khalil and Rasmussen, 1987; Chappellaz et al., 1990; Raynaud and Chappellaz, and Stauffer et al., this volume). If we assume that several hundred years ago human activities had not greatly disturbed the global methane cycle, we can regard the measured concentrations as representative of natural sources. The change of concentrations between then and now can be attributed to anthropogenic sources. Using a mass balance model, this process tells us what fractions of the present emissions are natural and what fractions are anthropogenic (Khalil and Rasmussen, 1990b). The results show that 40%-70% may come from anthropogenic sources, although this fraction continues to increase in recent times. This constraint allows us to obtain important information about the effect of human activities on the global methane cycle.

Calculations based on the ice core data that separate anthropogenic and natural emissions raise questions about how we define what is natural and what is not. For a long time methane emissions from rice fields and cattle were not regarded as anthropogenic. Clearly, however, the number of cattle in the world and the hectares of land under rice agriculture are determined by human needs. Archived data show that cattle and rice areas have increased dramatically, at first in some proportion to increasing human population but much less so now. Nonetheless, the number of domestic cattle has increased almost three-fold over the past century, and the hectares of rice harvested have increased about two-fold (Mitchell, 1980, 1982, 1983; U.N. FAO Yearbook, various years). These increases are linked directly to human activities and population growth. The increase of these sources has probably caused major increases in the concentration of atmospheric methane (Khalil and Rasmussen, 1985). On the other hand, wetlands have been drained over the past century, and bison in North America have been nearly exterminated. These activities may have led to a decrease of methane emissions that can also be attributed to human factors. These decreases do not, however, come close to compensating for the increases in the many other sources. For some sources there is less doubt that they are anthropogenic -- sources such as leakage from natural gas lines or automobile exhaust.

At the next stage, we have constraints on emissions from individual sources (or source groups) or from specific regions. The stable isotopes of C and H provide constraints on specific sources or source types because each source has a characteristic isotopic composition. There are a number of factors that cause uncertainties and variabilities in the measured isotopic fractions in each source. Moreover, the available data provide many fewer constraints than there are sources. This situation (under-determined system) and the associated variability of isotopic composition from each source produce only weak constraints but hold considerable promise as measurement techniques are advanced and if isotopic combinations can be distinguished. More can be found on this subject in the chapters by Whiticar and Stevens, both in this volume, and articles by Mroz (1993), Levin et al. (1993), and Lassey et al. (1993) (see also Appendix 2). Constraints on regional emissions may be obtained by meteorological techniques where measurements at the boundaries of a region are used to evaluate the potential emissions inside the region (see Conrad and Rasmussen, this volume). These constraints may become more important in the future as measurement techniques advance further.

In this context it is worth mentioning that there is also a constraint that limits the uncertainties of the total global emissions or emissions from a collection of sources. Often, as emissions from each source are evaluated based on the type of procedures mentioned in Section 2, a large range of emissions is calculated from each source. This range may be justified based on the variability of measured emission rates or more likely based on the uncertainties in the knowledge of extrapolants. Frequently, all sources are added to arrive at the total emissions. Then the sum of the lower limits of all sources is used as the lower limit of the total emissions, and the sum of the upper limits from all the sources is taken as the upper limit of the total global emissions. This range often violates the first two constraints mentioned above, namely the OH-sink constraint and the natural and anthropogenic fractions constraint. In some cases the ranges may also violate isotopic constraints. Even a conservative statistical procedure shows that the range of the total global emissions is considerably smaller (by a factor of 2 or so) than the range obtained from the sum of minima and the sum of maxima. The procedure is described by Khalil (1992) and may lead to a reconciliation between the constraints on global emissions and the unreasonably large ranges of emissions that can be obtained by adding minima and maxima of individual sources.

4. Spatial and temporal aspects of emissions

4.1 Spatial variations. Methane emissions change appreciably from latitude to latitude and over oceans and land. Knowledge of the distribution of methane emissions on the Earth's surface is important to know both from a scientific perspective to understand the fundamental processes of emissions and removal and from a practical perspective for those who have an interest in mediating global climate change. The oceans turn out to be a relatively small (although there is some justifiable debate about this matter; see Judd et al., this volume); the majority of the emissions are confined to the biologically productive land, which excludes the polar regions, the deserts, mountains, and drylands.

Estimates of latitudinal distributions are based on the same equations as in Section 2 but applied to each latitude or regional area. For the major sources such as rice agriculture, wetlands, and cattle such latitudinal information is available and adequate estimates can be made. For other sources the information is sketchy, unreliable, or unavailable, thus requiring many assumptions to obtain estimates. The spatial emissions rate as a function of latitude is shown in Figure 1 from our own work, with total emissions for some of the sources taken from this volume and additional information from Marland et al. (1985), Lerner et al. (1988), Cofer et al. (1991), Stocks (1991), Taylor and Zimmerman (1991), Fung et al. (1991), and Bartlett and Harriss (1993).

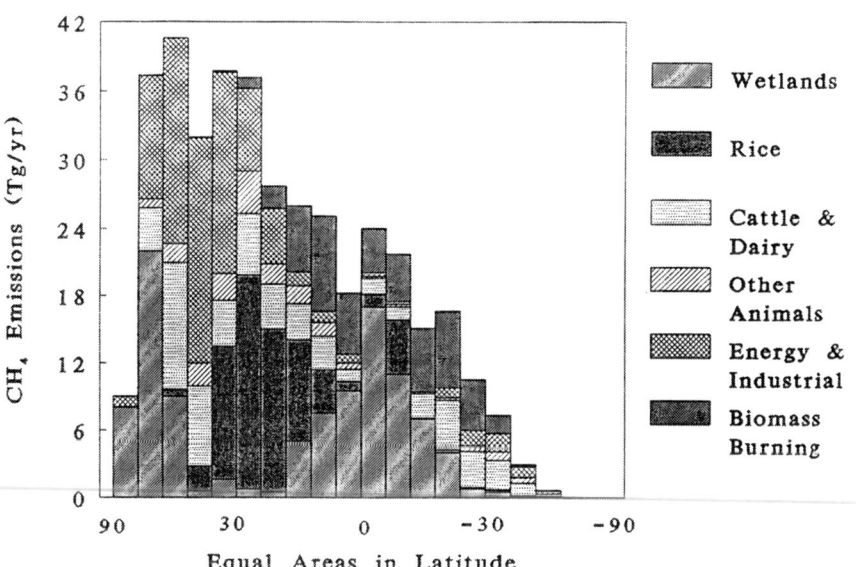

Figure 1: The current latitudinal distribution of methane sources. About 57% of the sources are in the tropical and sub-tropical latitudes (30°N-30°S) and another 40% are in the middle and higher northern latitudes (> 30°N).

Emissions inventories on smaller spatial scales are currently being developed. It is worth noting that on small spatial scales there are no constraints available at present, and estimates may be substantially in error even though when added over the whole world, they may be quite accurate. The summing procedure reduces uncertainties.

Regional estimates add another dimension leading to a matrix, S_{ij}, of emissions with one variable (i) representing the source type and the other the region (j). The region may be a country or a latitudinal and longitudinal box, depending on the intended use of the data. The sum $S_{*j} = \Sigma_i S_{ij}$ represent the total methane emissions from the all sources in a region, and $S_{i*} = \Sigma_j S_{ij}$ represents the global methane emissions from a given source (which is the focus of the subsequent chapters). The total methane emissions are $S = \Sigma_i \Sigma_j S_{ij}$. Uncertainties in S_{i*} or S_{*j} may be smaller than the average uncertainties in the S_{ij} that make up these sums because of cancellation of errors, provided the errors are not systematic. If there is systematic bias in the way emissions are calculated for a given source, then it would spread across all the regions and would persist in the sum representing the global emissions from that source. Constraints mentioned above can be brought to bear on the matrix S_{ij}, S_{*j}, S_{*j}, and S.

4.2 Temporal variations. The changes of emissions in time can occur on several time scales ranging from seasonal (annual), to inter-annual, to decadal, and to longer times. Observations have confirmed that there must be substantial seasonal variations of methane emissions to account for the atmospheric seasonal cycles, particularly at middle to higher northern latitudes (see Khalil and Rasmussen, 1983). Direct measurements of methane emissions from various sources such as rice fields, wetlands, or swamps have shown that emissions are confined to specific periods of the year (for example, Harriss et al., 1982; Seiler et al., 1984; Whalen and Reeburgh, 1992). At present there is no rigorous quantitative connection between known cycles of emissions from sources and the cycles of observed concentrations. A large fraction of the atmospheric seasonal cycle is driven by seasonal variations of OH (especially at higher latitudes), which complicates the evaluation of the effect of seasonal emissions on the atmospheric seasonal cycle.

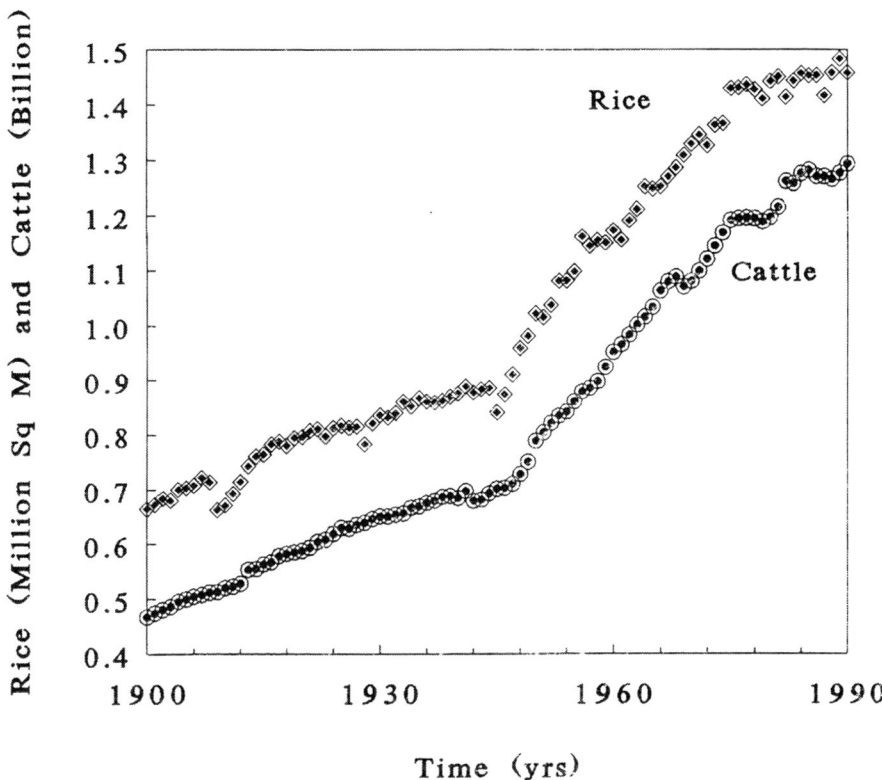

Figure 2: Increases in cattle populations and area of rice harvested for the turn of the century to the present. Cattle populations have increased by a factor of almost 3 and rice fields by 2 during this time. These agricultural sources represent a large role of human activities on the global methane cycle.

There is perhaps a greater interest in the long-term changes of the sources because they tell us which sources have contributed to the doubling of methane over the past century and indeed whether it is the sources that have caused global increases of methane. The probable changes of emissions from the major sources over the past 100 years can be evaluated based on available geographical and agricultural data (Chappellaz et al., 1993; Kammen and Marino, 1993; Shearer and Khalil, 1989 and this volume). Figure 2 shows the increases in cattle populations and hectares of rice harvested. The increases are so large that they are bound to have a significant effect on the emissions and hence the atmospheric concentrations, regardless of the uncertain extrapolants that have to be applied to these data to calculate methane emissions. The growth of these and other sources suggests that the major increases of methane over the past 100 to 300 years have been caused by increasing emissions as opposed to being caused by decreasing OH (see Khalil and Rasmussen, 1985; Levine et al., 1985; Thompson and Cicerone, 1986; Lu and Khalil, 1991; Pinto and Khalil, 1991). There are still other important sources such as biomass burning for which we have no readily available means of estimating trends. It is interesting that the major sources such as rice fields and cattle have not increased much over the last decade -- certainly not comparable to the rapid rates of increase in the 1950s. This slowdown of major sources may be affecting atmospheric trends, which also show a deceleration (Khalil and Rasmussen, 1990a, 1993, and elsewhere; Steele et al., 1992).

The connection between human population and methane emissions may be written as:

$$S_{ij} = e_{ij} P_j \qquad (7)$$

where e_{ij} is the per capita emission of methane from source i (such as rice fields) in region j (China, for example) and P_j is the population of region j, making S_{ij} the emissions from region j due to source i. To have a constant relationship between population and emissions from source i, $\Sigma_j e_{ij} P_j / \Sigma_j P_j$ must be constant in time; for a constant relationship between emissions and population from a certain region j, $\Sigma_i e_{ij} P_j / \Sigma_j P_j$ must be constant; and to have a constant relationship between emissions from all anthropogenic sources and population, $\Sigma_{ij} e_{ij} P_{ij} / \Sigma_j P_j$ must be constant.

In fact, the anthropogenic sources increase or decrease according to complex economic, social, and technological factors, which makes it difficult or perhaps impossible to predict future emissions. Generally, sources such as rice fields and cattle are related to human population since a larger population requires more food. This is not the only connection, however, since emissions are also related to "per capita" demand for the commodity. According to Eqn. (7), the emissions are a product of the per capita consumption rate (from which per capita emissions are calculated) and the population, but both the per capita emission rate and population can change in time. It is likely that population increases are more predictable than the per capita demand or per capita consumption rates. For instance, if people are generally poor and undernourished, there will be a demand to grow more rice, causing increases of methane emissions even if the population is not increasing. After a certain point, however, as people become richer, the per capita demand for rice may decline as other foods are substituted and preferred, so methane emissions may not increase even if the population does. (See, for example, Ito et al., 1989.) Also, populations that have the greatest contribution to one source or another rise at different rates. The shifting nature of the connection between population and agricultural emissions makes population an unreliable predictor of the future even if it works well to explain the past. For instance, rice fields and cattle rose steadily in the past, keeping pace with population. As mentioned earlier, methane from rice and cattle may not be increasing rapidly compared to the previous decades, and yet the population is continuing to rise at close to the same rates as a decade ago, thus breaking the link between population and methane emissions. This effect is demonstrated in Figure 3 where we have plotted the expected methane emissions from cattle and rice (as shown in Figure 2) divided by the world population. Per capita emissions from these sources have fallen from about 39 Tg/billion people around the turn of the century to about 25 Tg/billion people; this is a decrease of about 35% since 1900. Most of this change is very recent, about 19% over the last decade and 29% since 1960. Before 1960, population was a good indicator of agricultural CH_4 emissions, but now it is not.

Another way to evaluate trends in global emissions (possibly also over large regions such as semi-hemispheres) is to use Eqn. (4) and calculate the emissions necessary to balance the time series of atmospheric concentrations and trends (C and dC/dt from measurements). To do this requires a calculation not only of the mean OH but also its possible trends over decadal periods. One such calculation is shown in Figure 4. Here we assumed several different possible trends for OH during the period of measurements extending from decreases at −1.5 %/yr to increases at 2.0 %/yr. The results show that if OH is increasing at a rate faster than about 0.5%/yr, it requires a rapid increase of emissions, which appears untenable given the slowdown in the major sources mentioned earlier. We favor the scenario with no OH change over the period of measurements, which gives modest rises in sources during the past decade. These calculations also show that the average emissions are about 500 Tg/yr as expected from the constraints mentioned earlier. It is this 500 Tg/yr that has to be distributed among the various sources. These calculations also show that inter-annual variations are quite small at around ± 10 Tg/yr or ± 2% from the mean.

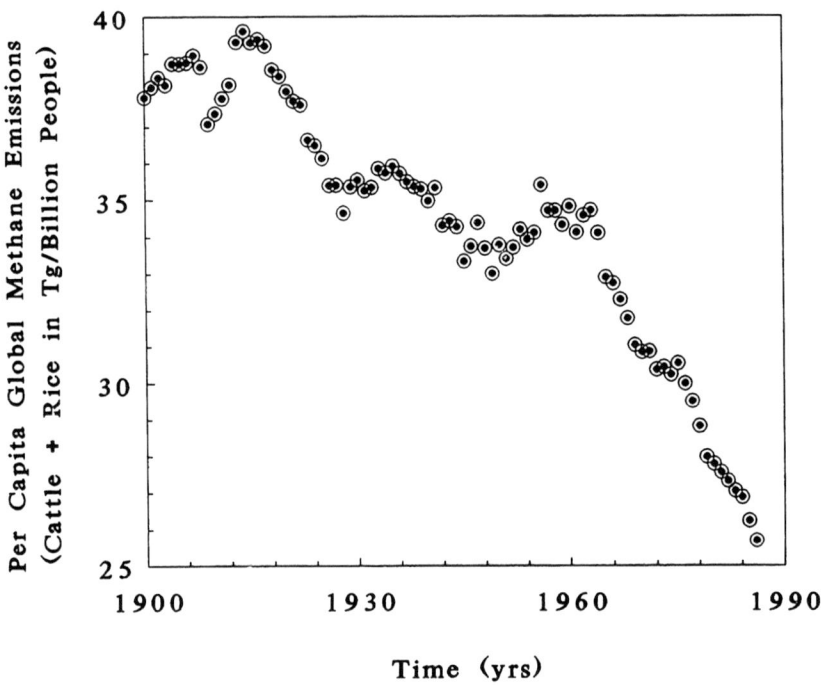

Figure 3: The global per capita emission rate of methane from rice and cattle since the turn of the century. The per capita emission rates are not constant and show substantial declines in the recent decade.

There is active research on the spatial and temporal emissions of methane, which will eventually lead to credible estimates. In most of the chapters on individual sources, spatial and temporal aspects have not been discussed.

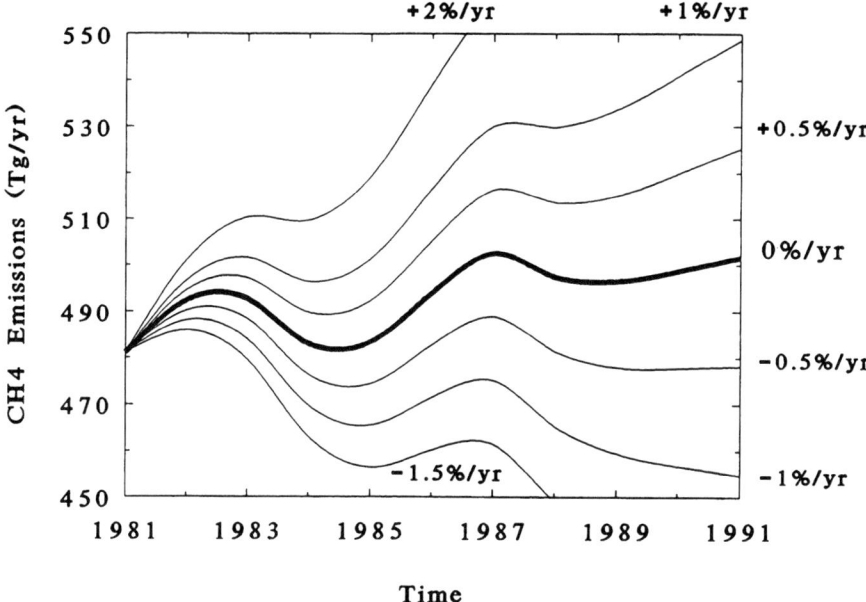

Figure 4: The global emissions of methane during the period of systematic atmospheric measurements spanning more than a decade. The emissions are calculated so as to balance the observed concentrations and the removal rates. The effect of changing removal rates (particularly OH) is shown by the various lines. The bold line is for the case when the sinks are constant (0% change).

5. The NATO-ARW budget

The estimated emissions from each of the sources as presented in the subsequent chapters were not subjected to any constraints. Once the work was together, we wanted to see if the resulting source budget would indeed satisfy the constraints mentioned earlier. Some of these estimates are based on very tenuous data (particularly emissions from wastewater treatment, low temperature fuels, and manure). Table 1 shows the estimates of emissions from various sources. The sum satisfies the constraint on total emissions (500 ± 75 Tg/yr), and the natural and anthropogenic components satisfy the constraint from the ice core data (anthropogenic fraction 50%-70%), although they are on the high side.

The fossil fuel fraction (23%) is within the range constrained by ^{14}C (21 ± 3%, Wahlen et al., 1989; see Stevens, this volume, for further discussion). Perhaps these agreements are fortuitous, but they suggest that budgets and methods for estimating emissions are converging. The detailed estimates of emissions from each source are given in the chapters that follow.

6. Summary

According to current assessments, the major sources of methane are rice agriculture, domestic animals, mostly cattle, and the wetlands, each emitting more than 50 Tg/yr. Cattle and rice are directly linked to population and agricultural productivity, and even the global wetland emissions are affected by human activities. Another group of sources each emit lesser amounts (10-50 Tg/yr), but there are so many that together they constitute a large fraction of the total emissions. These sources include landfills, coal mines, biomass burning, urban areas, sewage disposal, natural gas leakages, lakes, oceans, termites, and tundra. Finally, there are yet smaller sources including biogas pits, asphalt, several industrial sources, and possibly others that have not yet been identified (~ < 5 Tg/yr). In this chapter we discuss the procedures for calculating global emissions from various sources and the inherent uncertainties. It shows some of the common foundations of the calculations in subsequent chapters on individual sources. We also test the combined methane budget that emerged from the NATO-ARW to show that it meets the constraints on total emissions and the anthropogenic and isotopic fractions.

Acknowledgements. We have benefitted from discussions with R.A. Rasmussen, F. Moraes, E. Matthews, J. Pinto, S. Thorneloe, K. Smith, V. Aneja, and D. Johnson. This work was supported in part by a grant from the U.S. Department of Energy (DE-FG06-85ER60313) and the resources of Andarz Co.

Table 1. Estimates of methane emissions from various sources described in the following chapters.

NATO-ARW Methane Budget

Source	Methane (Tg)	Range (if available)
Natural Sources		
Wetlands[1]	110	
Termites[2]	20	15-35
Open ocean[2]	4	
Marine sediments[2]	**	8-65
Geological[2]	10	1-13
Wild fire[3]	2	2-5
Total*	150	
Anthropogenic Sources		
Rice[4]	65	55-90
Animals[5]	79	
Manure[6]	15	
Landfills[6]	22	
Wastewater treatment[6]	25	27-80
Biomass burning[3]	50	
Coal mining[7]	46	25-50
Natural gas†[7]	30	
Other anthropogenic[2,7]	13	7-30
Low temperature fuels[2]	17	
Total*	360	
^{14}C Depleted Sources‡*	120	
All Sources*	510	

* Rounded
** The amount of CH_4 generated was estimated, but no estimates of emissions were given.
† Beck et al. did not make a final estimate for natural gas emissions. This was inferred from Tables 5-8 in their chapter.
‡ ^{14}C depleted sources were considered to be the "geological" source, plus emissions from coal mining, natural gas, other anthropogenic (mainly industrial and transportation fossil fuel combustion), and low temperature fuel combustion.
[1] Matthews; [2] Judd et al.; [3] Levine et al.; [4] Shearer and Khalil; [5] Johnson et al.; [6] Thorneloe et al.; [7] Beck et al.

References

Aselmann, I., P.J. Crutzen. 1989. Global distribution of natural freshwater wetlands and rice paddies, their net primary productivity, seasonality and possible methane emissions. *J. Atmos. Chem.,* 8:307-358.
Bartlett, K., R.C. Harriss. 1993. Review and assessment of methane emissions from wetlands. *Chemosphere,* 26:261-320.
Butler, J.H., J.W. Elkins, T.M. Thompson, B.D. Hall, T.H. Swanson, V. Koropolov. 1991. Oceanic consumption of CH_3CCl_3: implications for tropospheric OH. *J. Geophys. Res.,* 96:22,347-22,355.
Chappellaz, J., J.M. Barnola, D. Raynaud, Y.S. Korotkevich, C. Lorius. 1990. *Nature,* 345:127-131.
Chappellaz, J.A, I.Y. Fung, A.M. Thompson. 1993. The atmospheric CH_4 increase since the Last Glacial Maximum, 1. Source estimates. *Tellus,* in press.
Cofer, W.R. III, J.S. Levine, E.L. Winstead, B.J. Stocks. 1991. Trace gas and particulate emissions from biomass burning in temperate ecosystems. In: *Global Biomass Burning, Atmospheric, Climatic, and Biospheric Implications* (J.S. Levine, ed.):203-208.
Fung, I., J. John, J. Lerner, E. Matthews, M. Prather, L.P. Steele, P.J. Fraser. 1991. Three-dimensional model synthesis of the global methane cycle. *J. Geophys. Res.,* 96 (D7):13,033-13,065.
Harriss, R.C., D.I. Sebacher, F.P. Day, Jr. 1982. Methane flux in the Great Dismal Swamp. *Nature,* 297:673-674.
Ito, S., E.W.F. Peterson, W.R. Grant. 1989. Rice in Asia: is it becoming an inferior good? *Amer. J. Agr. Econ.,* 71:32-42.
Kammen, D.M., B.D. Marino. 1993. On the origin and magnitude of pre-industrial anthropogenic CO_2 and CH_4 emissions. *Chemosphere,* 26:69-86.
Khalil, M.A.K. 1992. A statistical method for estimating uncertainties in the total global budget of trace gases. *J. Environ. Sci. Health, A27* (3):755-770.
Khalil, M.A.K., R.A. Rasmussen. 1983. Sources, sinks, and seasonal cycles of atmospheric methane. *J. Geophys. Res.,* 88:5,131-5,144.
Khalil, M.A.K., R.A. Rasmussen. 1984. The atmospheric lifetime of methylchloroform (CH_3CCl_3). *Tellus, 36B*:317-312.
Khalil, M.A.K., R.A. Rasmussen. 1985. Causes of increasing atmospheric methane: depletion of hydroxyl radicals and the rise of emissions. *Atmos. Environ., 19*:397-407.
Khalil, M.A.K., R.A. Rasmussen. 1987. Atmospheric methane: trends over the last 10,000 years. *Atmos. Environ., 21*:2,445-2,452.
Khalil, M.A.K., R.A. Rasmussen. 1990a. Atmospheric methane: recent global trends. *Environ. Sci. Tech.,* 24:549-553.
Khalil, M.A.K., R.A. Rasmussen. 1990b. Constraints on the global sources of methane and an analysis of recent budgets. *Tellus, 42B*:229-236.

Khalil, M.A.K., R.A. Rasmussen. 1993. Decreasing trend of methane: unpredictability of future concentrations. *Chemosphere, 26* (1-4):803-814.
Lacroix, A.V. 1993. Unaccounted-for ssources of fossil and isotopically-enriched methane and their contribution to the emissions inventory: a review and synthesis. *Chemosphere, 26* (1-4):507-557.
Lassey, K.R., D.C. Lowe, C.A.M. Brenninkmeijer, A.J. Gomez. 1993. Atmospheric methane and its carbon isotopes in the southern hemisphere: their time series and an instructive model. *Chemosphere, 26* (1-4):95-109.
Lerner, J., E. Matthews, I. Fung. 1988. Methane emission from animals: a global high-resolution data base. *Global Biogeochem. Cycles, 2*:139-156.
Levin, I., P. Bergamaschi, H. Dörr, D. Trapp. 1993. Stable isotopic signature of methane from major sources in Germany. *Chemosphere, 26* (1-4):161-178.
Levine, J.S., C.P. Rinsland, G.M. Tennille. 1985. The photochemistry of methane and carbon monoxide in the troposphere in 1950 and 1985. *Nature, 318*:254-257.
Lovelock, J.E. 1977. Methyl chloroform in the troposphere as an indicator of OH radical abundance. *Nature, 267*:32-33.
Lu, Y., M.A.K. Khalil. 1991. Tropospheric OH: model calculations of spatial, temporal, and secular variations. *Chemosphere, 23* (3):397-444.
Marland, G., R.M. Rotty, N.L. Treat. 1985. CO_2 from fossil fuel burning: global distribution of emissions. *Tellus, 37B*:243-258.
Matthews, E., I. Fung, J. Lerner. 1991. Methane emission from rice cultivation: geographic and seasonal distribution of cultivated areas and emissions. *Global Biogeochemical Cycles, 5* (1):3-24.
Mitchell, B.R. 1980. *European Historical Statistics*, 2nd Rev. Ed. Facts on File, New York, U.S.A.
Mitchell, B.R. 1982. *International Historical Statistics, Africa and Asia*. New York University Press.
Mitchell, B.R. 1983. *International Historical Statistics, The Americas and Australasia*. Gale Research Co., Detroit, Michigan.
Midgely, P.M. 1989. The production and release to the atmosphere of 1,1,1-trichloroethane (methyl chloroform). *Atmos. Environ., 23*:2,663-2,665.
Mroz, E.J. 1993. Deuteromethanes: potential fingerprints of the sources of atmospheric methane. *Chemosphere, 26* (1-4):45-53.
Pinto, J., M.A.K. Khalil. 1991. The stability of tropospheric OH during ice ages, inter-glacial epochs and modern times. *Tellus, 43B*:347-352.
Prinn, R.G., D. Cunnold, R.A. Rasmussen, P. Simmonds, F. Alyea, A. Crawford, P. Fraser, and R. Rosen. 1987. Atmospheric trends in methylchloroform and the global average for the hydroxyl radical. *Science, 238*:945-950.
Rasmussen, R.A., M.A.K. Khalil. 1981. Interlaboratory comparison of fluorocarbons 11, 12, methylchloroform, and nitrous oxidemeasurements. *Atmos. Environ., 15*:1,559-1,568.

Seiler, W., A. Holzapfel-Pschorn, R. Conrad, D. Scharffe. 1984. Methane emission from rice paddies. *J. Atmos. Chem.*, *1*:241-268.

Shearer, M.J. and M.A.K. Khalil. 1989. The global emissions of methane over the last century. *Eos Trans.*, *70* (43):1017.

Singh, H.B. 1977. Preliminary estimation of average tropospheric HO concentrations in the northern and southern hemispheres. *Geophys. Res. Let.*, *4*:453-456.

Steele, L.P., E.J. Dlugokencky, P.M. Lang, P.P. Tans, R.C. Martin, K.A. Masarie. 1992. Slowing down of the global accumulation of atmospheric methane during the 1980s. *Nature*, *358*:313-316.

Stocks, B.J. 1991. The extent and impact of forest fires in northern circumpolar countries. In: *Global Biomass Burning, Atmospheric, Climatic, and Biospheric Implications*, J.S. Levine, ed.:197-202.

Taylor, J.A., P.R. Zimmerman. 1991. Modeling trace gas emissions from biomass burning. In: *Global Biomass Burning, Atmospheric, Climatic, and Biospheric Implications* (J.S. Levine, ed.), 345-350.

Thompson, A.M., R.J. Cicerone. 1986. Possible perturbations to atmospheric CO, CH_4, and OH. *J. Geophys. Res.*, *91*:10,853-10,864.

United Nations. 1977, 1978, 1980, 1982, 1984, 1986, 1989, 1990, 1991. *FAO Production Yearbook, vols. 30, 31, 33, 35, 37, 39, 42, 43, 44*. Food and Agriculture Organization of the United Nations, Rome.

Vaghjiani, G.L., A.R. Ravishankara. 1991. New measurement of the rate coefficient for the reaction of OH with methane. *Nature*, *350*:406-408.

Wahlen, M., N. Tanaka, R. Henry, B. Deck, J. Zeglen, J.S. Vogel, J. Southon, A. Shemesh, R. Fairbanks, W. Broecker. 1989. Carbon-14 in methane sources and in atmospheric methane: the contribution from fossil carbon. *Science*, *245*:286-245.

Whalen, S.C., W.S. Reeburgh. 1992. Interannual variations in tundra methane emission: a 4-year time series at fixed sites. *Global Biogeochem. Cycles*, *6*:139-159.

Chapter 11

Ruminants and Other Animals

D.E. JOHNSON, T.M. HILL, G.M. WARD, K.A. JOHNSON, M.E. BRANINE,
B.R. CARMEAN, AND D.W. LODMAN

Department of Animal Sciences, Colorado State University, Fort Collins, CO 80523, U.S.A.

1. Introduction

Estimates of global animal methane emissions have ranged from 60 (Crutzen, 1983) to 160 Tg/yr (Ehhalt and Schmidt, 1978). The potential of a significant animal contribution to atmospheric methane has been recognized for some time. Approximately 95 percent of animal methane emissions are accounted for by ruminants. The emphasis in this paper will therefore be primarily focused on these animals.

2. Gastrointestinal fermentation

Many animals have a portion of their digestive tract developed to suit a fermentation that enables them to derive energy and other nutrients from materials that would otherwise be excreted. The ruminant digestive system is the most extensively studied and characterized of all of herbivore digestive tracts. It is known that many of the other animals that possess enteric fermentations have the same species of bacteria and protozoa involved in digestion of plant materials. Microbes in the gut grow and digest the diet into end products the host animal utilizes (Buchner, 1965). The numbers and importance of the symbiotic microflora are dependent on the diet and the conditions that exist within the stomach or intestine. The factors that influence microbial growth and their fermentation of the diet include: temperature, pH, nutrient and liquid flow rates, diet composition, osmotic pressure, and redox potential. Alterations in any one of these factors may alter fermentation efficiency by selecting for or against specific microbial species.

The hindgut and foregut are the two areas that have adapted to provide a suitable environment for a fermentation. The general characteristics of fermentations in either the fore or hind gut are similar. Plant structural carbohydrates such as cellulose or hemicellulose are broken down by bacteria and protozoa, which possess exoenzymes that hydrolyze beta bonds. Once the constituent sugars are released from the plant material by the action of these cellulolytic bacteria, these and other species of bacteria ferment the sugars to short chain fatty acids (volatile fatty acids - VFA). The fatty acids are then absorbed and utilized by the host animal. The other endproducts of the fermentation include CO_2, H_2, microbial protein, and methane. Methane results from the symbiotic relationship between those bacteria that produce H_2 as an end product and the methanogens who couple the H_2 with CO_2 or formate. The result of this interaction is that methanogens gain energy for their own growth and the metabolic hydrogen is removed from the environment, which benefits the other bacteria.

2.1 Ruminant fermentation. The ability of ruminant animals to derive energy from a variety of fibrous food sources is made possible by the symbiotic relationship with the diverse microbial community inhabiting the digestive tract. In order for the dietary energy and nutrients to be useful to the ruminant, it must ferment, digest, and absorb these from the gastrointestinal tract and metabolize them in the body. During these processes, some is never absorbed (excreted as feces or belched as methane) and additional losses occur in the urine. The balance of absorbed energy is metabolizable by body tissue, a portion of which is wasted as heat with the remainder used to maintain the animal or produce a product.

Ruminants ferment their diets extensively in a foregut compartment, the reticulorumen, and additionally, to a lesser degree, in the hindgut. There are several advantages to having a fermentation in the foregut. First, those fermentation products that are not directly absorbed (i.e., microbes, undigested organic matter) pass through the rest of the digestive tract and may be digested there rather than being excreted. This permits maximal digestion of all portions of the diet and fermentation endproducts (Bauchop, 1977).

The rumen is a large, multicompartmented portion of the ruminant stomach that is lined with papillae. It is maintained at a constant 39°C with heat from the animal's body as well as heat from the fermentation. The pH is maintained at a relatively constant 6.5 to 6.8 through copious salivary secretions and the absorption of volatile acids through the rumen wall. Constant mixing is provided by strong ruminal contractions. Rate of passage of liquid and solids out of the rumen is slow, which allows time for fermentation of cellulose. The microorganisms that exist within this environment consist of bacteria, protozoa, and fungi. Because of the large numbers, continuous turnover, wide variety of species and metabolism of the ruminal microflora, adaptations to different feed substrates is rapid, thus allowing the ruminant flexibility in diet selection. These adaptations result in population shifts in microorganisms within the rumen.

From the standpoint of the host animal's metabolism, formation of VFA and microbial protein represents usable nutrients, while formation of methane constitutes a net energetic loss to the animal. Therefore, understanding the factors influencing ruminal methane production is important for developing nutritional strategies and livestock management practices to improve production efficiency of food and fiber from ruminants in addition to providing information concerning the potential global warming consequences.

Table 1. Examples of measured methane yield by cattle and sheep across a range of diets and treatments

Reference/species	Basal diet and treatment	Methane yield (%CH_4)[a]
Birkelo et al., 1986 Cattle	wheat straw	5.90
	ammoniated straw	6.47
Blaxter and McGraham, 1956 Sheep	chopped grass	7.61
	medium ground/cubed grass	5.87
	fine ground/cubed grass	4.61
Armstrong, 1964 Sheep	first cutting of timothy hay	7.81
	second cutting of timothy hay	7.45
	third cutting of timothy hay	7.13
Blaxter and Wainman, 1964 Cattle	100% hay	7.45
	80% hay:20% flaked corn	7.84
	60% hay:40% flaked corn	8.17
	40% hay:60% flaked corn	8.08
	20% hay:80% flaked corn	5.74
	5% hay:95% flaked corn	3.40
Whitelaw et al., 1984 Cattle	faunated (85% barley diet)	11.46
	cilate-free (85% barley diet)	6.68

Table 1. Examples of measured methane yield by cattle and sheep across a range of diets and treatments

Reference/species	Basal diet and treatment	Methane yield ($\%CH_4$)[a]
Van der Honing et al., 1981	0% tallow or oil	5.9
Lactating dairy cows	5% animal tallow	5.3
	5% soybean oil	5.0
Johnson, 1974	control (90% concentrate)	6.19
Sheep	methane inhibitor (90% conc.)	3.70

[a] $\%CH_4$ = methane enthalpy ÷ total diet enthalpy.

3. Methane emissions as affected by diet

Methane losses are influenced by many factors, including intake, digestibility and diet type (some examples are shown in Table 1). In general, fractional methane loss ($\%CH_4$, expressed as a fraction of the animal's diet gross energy intake) will decrease as daily dietary intake increases. At restricted intakes $\%CH_4$ yield will increase as digestibility of the diet increases. However, in practice, highly digestible diets are seldom fed at restricted, maintenance levels. It should also be noted that, as amount of dietary energy ingested increases, the actual quantity of methane produced daily may increase due to the greater mass of potentially fermentable substrate within the rumen, although $\%CH_4$ declines.

Physical and chemical processing of low quality forages have been commonly used to improve nutritive value. Studies examining the effects of ammoniation and sodium hydroxide treatment of various cereal straws have shown that $\%CH_4$ increases primarily as a result of the increased digestibility (Robb et al., 1979; Sundstol, 1982; Birkelo et al., 1986). Physical alteration of low quality forages by chopping, grinding, or pelleting have also been shown to affect $\%CH_4$ (Blaxter and McGraham, 1956; Wieser and Wenk, 1970). Generally, $\%CH_4$ will be higher for more coarsely chopped diets compared to finely ground and pelleted diets. This effect is primarily due to increased intake and decreased digestibility resulting from an increased particulate passage rate and shorter ruminal retention times decreasing $\%CH_4$ on finely ground and pelleted diets. In studies examining the effects of grain processing methods on energy balance and $\%CH_4$, Johnson (1966) found greater methane energy loss when beef steers fed an 80% grain diet were provided cracked corn compared to a similar diet of steam-flaked corn. Moe and Tyrrell (1977) observed no differences among mixed diets consisting of either whole corn, cracked corn, or ground corn meal when fed at the same level of intake to non-lactating cows. However, for lactating cows, $\%CH_4$, when expressed as a percentage of digestible energy intake, was lower for the cracked corn and ground corn meal diets and contributed to the increased energy availability ratios observed for these diets. With pelleted diets or increased dietary intake, the reduction in energy loss attributable to methane may often at least partially compensate for the increased fecal energy loss observed under these conditions.

Only a few studies have investigated the effects of feed preservation method, forage species, or maturity on $\%CH_4$. Armstrong (1964) observed that, with sheep fed timothy grass hay harvested at different stages of maturity, $\%CH_4$ decreased from 7.8% to 7.1% as the digestibility of the forage decreased from 71% to 57% with advancing maturity. Similar results have been obtained with timothy grass that was harvested at three stages of maturity and fed fresh-frozen, ensiled, or as hay (Sundstol and Ekern, 1976) and for clover grass silage ensiled at an immature and a mature stage of development (Sundstol et al., 1979). Corbett et al. (1966) observed a greater $\%CH_4$ for late season compared to early season herbage. In this study, despite a negligible difference in digestible energy between early and late season herbages, the greater methane yield associated with the latter was related to a lower content of soluble carbohydrate and a greater proportion of cellulose. Varga et al. (1985) examined differences in energy balance between alfalfa silage and orchard grass silages and observed a decreased $\%CH_4$ (5.8 vs. 6.3) for cattle consuming alfalfa silage. Tyrrell and Varga (1985) indicated a slight, but significant decrease in methane energy loss from non-lactating dairy cows fed diets containing 40% shelled corn that had been dried as compared to ensiled (8.7 vs. 8.85 $\%CH_4$, respectively).

In many studies a common denominator that apparently exists for methane production is the relative proportions of soluble and structural carbohydrates within the plant tissue. The ratio of soluble to structural carbohydrate will decrease as a plant matures and likewise is generally lower for grass species compared to legumes (Van Soest, 1982). Varga et al. (1985) suggests that because a major portion of the digestible organic matter of orchard grass is derived from structural carbohydrate, whereas a major fraction of the digestible organic matter from alfalfa silage would be soluble carbohydrate, shifts in ruminal fermentation can occur that would either inhibit or enhance ruminal methanogenesis.

Under ruminal conditions that favor propionate-yielding fermentation pathways, propionate serves as a competitive hydrogen sink, decreasing methanogenesis. Conversely, fermentation pathways that favor acetate formation will generate hydrogen and stimulate methanogenesis. The latter type of fermentation is typically characteristic of bacterial degradation of structural carbohydrate such as cellulose. These observations agree with results of Moe and Tyrrell (1980) who indicated that when cows were consuming diets at high levels of intake, then digestion of cellulose yields 3 to 5 times more methane per unit hexose fermented as compared to soluble carbohydrate and starch. In summary, the results of these studies suggest that methane yield from forage-based diets cannot be generically evaluated but must take into account factors such as stage of maturity, species differences, and methods of processing and preservation.

3.1 Effects of high grain diets on methane yield. In countries with excess agricultural production, animal feeding and management optimization practices for some segments of the industry utilize high proportions of grain and other concentrate feeds to maximize energy intake required for maintaining high levels of production. Typically, U.S. beef feedlot diets contain about 10% roughage and 90% grain, and/or other concentrate feeds. In their classic study, Blaxter and Wainman (1964) indicated that as the percentage of steam-flaked corn in the diet increased, %CH_4 increased until the corn was incorporated into the diet at a level of 60-80%, at which point %CH_4 abruptly declined. For cattle and sheep fed at the higher intake level, %CH_4 ranged from 7.5% and 6.5% for the 100% hay diets to 3.4% and 3.7% for the all-grain diets, respectively. In general, where diets in excess of 80% grain (usually corn-based diets) have been fed at levels of intake substantially above maintenance, %CH_4 are typically below 5 (Johnson, 1966; Byers, 1974; Wedegaertner and Johnson, 1983). Recent experiments (Abo-Omar, 1989; Carmean, 1991), when extrapolated to industry intake levels, suggest three percent methane losses. The percentage of methane for high grain feedlot diets cannot be accurately predicted from general relationships such as the Blaxter and Clapperton equation.

Barley may be a special case for evaluating %CH_4. As indicated by the results of Hashizume et al. (1967) and Whitelaw et al. (1984), %CH_4 from high grain, barley based diets ranges from 6.5% to 12%. These values are higher than would be expected from feeding a comparable corn-based diet at similar levels of intake. A primary reason for this difference in %CH_4 might be the greater fermentability of the starch of barley. Theurer (1987) found ruminal starch digestibility coefficients of 93% and 73% for barley and corn, respectively.

Other factors probably associated with the low methane yields from high grain diets are factors that create a more hostile environment for hydrogen-producing microbes or the methanogenic bacteria which utilize the hydrogen. Van Soest (1982) has suggested some of these factors to be an increased rate of passage and digestion, depression in rumination, decreased ruminal pH, and reduction or elimination of certain populations of ruminal protozoa.

Rumen protozoa may have an important role in ruminal methane production. Krumholtz et al. (1983) observed that methanogens were directly adhered to protozoa and that protozoa supported methanogenic activity of the attached methanogens via an interspecies transfer of metabolic hydrogen. Whitelaw et al. (1984) observed reductions of approximately 50% in methane yield for defaunated steers fed a high grain, barley-based diet compared to faunated steers.

3.2 Effect of dietary fat on methane yield. Supplemental fats are often provided to high-producing dairy cows for the primary purpose of increasing the energy density of the diet. Swift et al. (1948) observed decreased %CH_4 for sheep when fed isocaloric diets containing from 3% to 8% fat. Van der Honing et al. (1981) likewise found decreases of 10 and 15 percent in methane yield for dairy cows receiving supplemental fat in the form of either animal tallow or soybean oil, as compared to non-supplemented controls. Haaland (1978) observed reductions in methane production for beef steers fed a protected tallow. The reduction in methane yield in this case, as in some others, was attributed to decreased fermentable substrate rather than a direct inhibition of methanogenesis.

The practice of providing polyunsaturated fatty acids in the diet for the express purpose of reducing methane has also been studied. Czerkawski et al. (1966) demonstrated that dietary additions of long chain, polyunsaturated fatty acids reduced ruminal methanogenesis by providing an alternative receptor of metabolic hydrogen, rather than the reduction of CO_2 to methane. In many plant tissues, particularly grasses, unsaturated fatty acids account for a large proportion of the total fatty acids. However, the percentage of the total metabolic hydrogen produced during the fermentation process that is actually used for the biohydrogenation of endogenous polyunsaturated fatty acids is only 1% to 2% (Czerkawski, 1972). This is a very small amount compared to the typical 48% used for reduction of CO_2 to methane, 33% used for VFA synthesis, and 12% used for bacterial cell synthesis (Czerkawski, 1988).

3.3 Effects of ionophores and other inhibitors on methane yield. The energetic loss represented by ruminal methane production has stimulated a considerable research effort directed at minimizing methanogenesis. This endeavor has been based on the premise that by decreasing methane energy loss, more energy will be available for productive purposes. Czerkawski (1988) has estimated the total annual methane energy produced by ruminants around the world is roughly equivalent to the gross energy contained in over 300 million tons of feed representing an economic value of 24 billion dollars being belched into the atmosphere each year (Czerkawski, 1988). Therefore, methods for decreasing methane production in ruminants could provide dividends in terms of both increased animal productivity and feed efficiency.

During the past several years many compounds have been developed that have shown a suppressive effect on ruminal methanogenesis. In general, these compounds can be subdivided into two groups: 1) polyhalogenated methane inhibitors and 2) polyether ionophores. The polyhalogenated inhibitors include the hemiacetal of chloral and starch (Johnson, 1972), chlorinated fatty acids (Czerkawski and Breckenridge, 1975 a,b), and bromochloromethane (Sawyer et al., 1974). All of these compounds have demonstrated a reduction in %CH_4 of 20% to 80% compared to untreated animals. Responses in animal performance to these compounds have been variable although improvements in digestibility and metabolizability (Sawyer et al., 1974), feed efficiency (Trei et al., 1971), and average daily gain (Johnson et al., 1972) have been reported. The use of polyether ionophores such as monensin and lasalocid have found widespread acceptance for improving growth and feed efficiency in cattle and sheep while the polyhalogenated methane inhibitors have not been commercially successful. The antimethanogenic properties of ionophores are not the result of a direct toxic effect on the methanogenic bacteria (Van Nevel and Demeyer, 1977); rather, methane losses are decreased in two ways. Firstly, 5% to 6% less feed is consumed and fermented per unit of production (Goodrich et al., 1984), thus reducing the corresponding methanogenesis. Secondly, several studies have demonstrated suppression of methane yield from feeding monensin (Joyner et al., 1979; Thorton and Owens, 1981; Benz and Johnson, 1982; Wedegaertner and Johnson, 1983) or lasalocid (Delfino et al., 1988) through alteration of ruminal fermentation patterns shifting H_2 away from methane. This suppression may not persist for long periods, however.

A problem common to using either the polyhalogenated methane inhibitors or ionophores for the suppression of methane production has been the phenomena of an apparent microbial adaptation and return to baseline methane levels after a relatively short period of exposure to these compounds. Johnson (1974), feeding sheep a pelleted diet with and without a halogenated methane inhibitor, observed (initial) methane depression to 36% of controls. However, on d 30, methane production by inhibitor-fed sheep had increased to 88% of control values. Considering the sum of both methane and hydrogen production, the inhibitor decreased total gas energy losses to 49% of control initially with a significant rise to 88% of control values by day 30. Similarly, in an ionophore study conducted by Rumpler et al. (1986), feeding a 70% cracked corn basal diet to beef steers, which supplied either no ionophore, monensin, or lasalocid, indicated methane levels for the monensin and lasalocid fed steers, while initially decreasing, had returned to control values by day 12 of the study. In an attempt to circumvent this apparent microbial adaptation to ionophores, Hubbert et al. (1987) initiated an experiment where monensin and lasalocid were fed on alternate weeks to growing-finishing heifers. The premise for this study being that if continuous feeding of a single ionophore throughout a feeding period does in fact lead to an attenuation of ionophore-induced methane suppression, the alternate provision of different ionophores at specified intervals could decrease microbial adaptation and maintain a level the methane suppression which is initially observed. Results from this and several subsequent studies with ionophore rotation programs (Johnson, 1987; Branine et al., 1989) have shown benefits in terms of improved animal productivity. The reasons, however, for these responses may not be related to a sustained inhibition of methane production. Recent research conducted at Colorado State University by Abo-Omar (1989) and Carmean and Johnson (1990) have indicated an initial 25% to

30% decrease in methane production in response to singly or alternatively fed ionophores with high-grain diets; however, as also shown by Rumpler et al. (1986), methane production had essentially returned to control levels after 16 d of feeding the ionophores. The results of these studies have indicated that long-term reduction in methane yield (%CH_4) by feeding ionophores to cattle consuming high grain diets may be small considering the overall approximately 120-day feeding period. The decrease in total methane production (i.e., l/d) will primarily result from the 5% to 6% lower feed consumption commonly observed for cattle fed ionophores and only secondarily from a methane yield savings.

4. Wild ruminant and non-ruminant

While domestic ruminants (sheep and cattle) are very similar in methane fractional loss from a wide range of common feeds, some of the other ruminant species may be unique. Deer consuming a browse and concentrate diet (Robbins, 1973) produced methane at rates that were 1/5 or less than expected from farm animals. It is unclear whether this is diet- or species-related; however, the relatively fast rate of passage of ingesta noted in deer (Milchunas et al., 1978) could contribute to lower methane loss. Recent measurements with llamas fed common feedstuffs (Carmean et al., 1991), however, showed methane losses of seven percent of gross diet energy, very similar to expected sheep or cattle rates. The hindgut fermentation in the ruminant animal is much less important than the foregut fermentation. The species of bacteria are the same as those found in the rumen, but the rate of fermentation may be lower (McBee, 1977).

Hindgut fermenters include: birds, fish, reptiles, rats, rabbits, pigs, horses, man, and elephants (McBee, 1977). In many of the species identified as hindgut fermenters, the evidence for the existence of the fermentation is indirect. In most cases, the identification of the VFAs and isolation of bacterial and protozoal species known to be involved in cellulolytic fermentations are the evidence of a fermentation.

Rats, beavers, voles, porcupine, guinea pigs, and rabbits are all animals that harbor important gut microbial populations to digest cellulose to VFAs and appear able to absorb and use the fatty acids for their energy needs. It has been estimated the 5% to 33% and 4% to 12% of the maintenance energy requirement of the beaver and rabbit, respectively, may be met by the use of these VFAs (McBee, 1977).

The omnivores (pig and man) and herbivores (horses, elephants, and ruminants) also have cecal fermentations and are dependent on the fermentation endpoints to varying degrees. Pigs on normal production diets (mostly grain) have little necessity for the hindgut fermentation end products, but when these animals are fed on high roughage diets, VFA from fermentation can supply up to 24% of their energy supply (Yen and Nienaber, 1991). The horse is much more dependent on the cecal fermentation than the pig. All of the fiber digestion occurs in the cecum or colon of the horse or pony (Hintz et al., 1971), regardless of the hay to grain ratio. Their efficiency of cellulose digestion has been estimated to be 39% to 50%.

5. Empirical predictions in ruminant methane production

Available data relating %CH_4 to variables such as level of intake (LOI) as multiples of maintenance, percent concentrate in the diet and percent DE in the diet for beef cattle and sheep were summarized. Extremes in %CH_4 ranged from 2% to 12% in beef cattle diets and both extremes occurred with diets of 90% or greater concentrates. These large variations for somewhat comparable diets cause difficulties in empirical prediction of %CH_4. Variations observed in our database were compared to variations predicted by the equation of Blaxter and Clapperton (1965), which relates %CH_4 to LOI and percent digestible energy (DE) of the diet at maintenance. This equation is not easy to apply as originally intended. The LOI depends on assessing the fasting heat production, which was loosely defined for animal age categories. Also, the DE of the diet was determined at maintenance level intake. Neither age nor diet DE at maintenance is commonly reported or determined in the methane production literature. In addition, the original publication contains a sign error preceding the LOI term. Blaxter and Clapperton's (1965) equation is:

$$\%CH_4 \text{ equals } 1.30 + 0.112\, DE + LOI\, (2.37 - 0.05\, DE).$$

The equation of Blaxter and Clapperton predicts a considerably narrower range of methane loss than those observed for beef cattle (Figure 1) and sheep, suggesting their equation is inaccurate or does not account for other important factors that influence methane losses. Certainly, their equation was based on a small subset of data that did not cover as wide a range of dietary conditions.

The relationship of %CH$_4$ to either LOI or both LOI and digestible energy is shown in Table 2 for both beef cattle and sheep. For beef cattle, difficulties in predicting %CH$_4$ increased with increasing percent concentrate, as reflected by r^2 values of one-third or less for high vs low concentrate diets (Table 2). Methane losses from high concentrate diets are highly variable and are frequently between 2 and 5 %CH$_4$ for beef cattle. Under most circumstances, increasing intake by one multiple of maintenance will decrease methane yield by 1.8 and 1.6 percentage units of GEI in cattle and sheep, respectively. Methane production appears influenced to a greater degree by variables other than LOI in sheep compared to cattle as reflected by the changing regression coefficients relating %CH$_4$ to LOI or to digestible energy (Table 2).

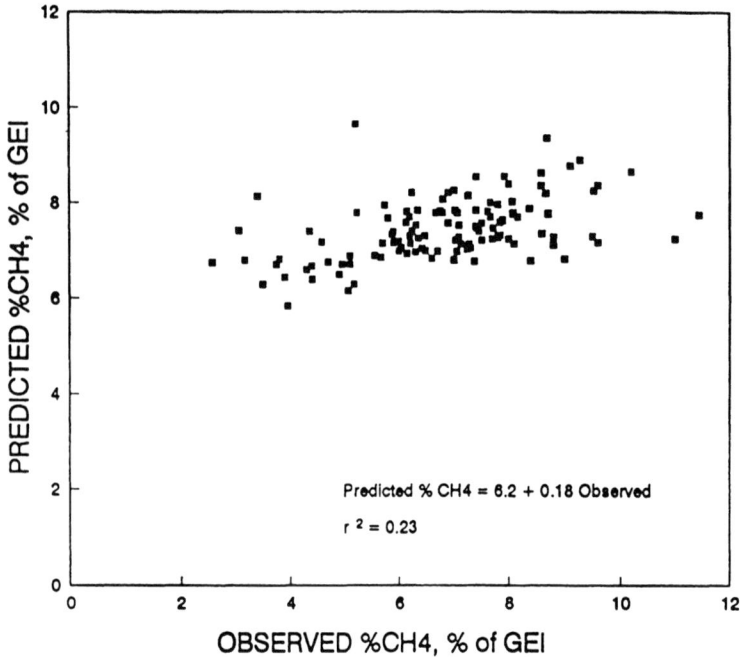

Figure 1. Predicted vs. observed methane production (using Blaxter and Clapperton, 1965, equation)

Table 2. Methane relationship to level of intake (LOI) and percent diet digestible energy (DE) for beef cattle and sheep[a].

Species/ % Forage in diet	%CH$_4$ vs LOI	r^2	Equation No.
Beef Cattle			
80-100	9.12 - 1.71 LOI	.57	1
20-80	9.80 - 1.84 LOI	.27	2
< 20	8.85 - 2.14 LOI	.17	3
all diets	9.49 - 1.84 LOI	.31	4
80-100	5.50 - 2.25 LOI + 0.06 DE	.72	5
20-80	11.33 - 1.81 LOI - 0.02 DE	.27	6
< 20	9.90 - 1.54 LOI - 0.02 DE	.18	7
all diets	10.32 - 1.79 LOI - 0.01 DE	.31	8
Sheep			
80-100	8.76 - 0.86 LOI	.11	9
20-80	10.92 - 2.29 LOI	.69	10
< 20	8.85 - 2.14 LOI	.29	11
all diets	9.40 - 1.39 LOI	.25	12
80-100	4.07 - 1.36 LOI + 0.07 DE	.47	13
20-80	4.59 - 2.74 LOI + 0.08 DE[b]	.72	14
< 20	28.3 - 2.74 LOI - 0.21 DE[c]	.43	15
all diets	5.29 - 1.58 LOI + 0.06 DE	.41	16

[a] Beef cattle, n = 118 experimental means; sheep, n = 159 experimental means.
[b] DE is not significant (P > 0.2) effect.
[c] Equation is not significant (P > 0.2).

Prediction of methane losses from typical diets fed to various classes of beef cattle and sheep are shown in Table 3. Predictions are made using relationships of %CH$_4$ to LOI, LOI, and digestible energy and the methane yield predicted by the Blaxter and Clapperton equation. Prediction from our equations for beef cattle indicate fattening cattle lose a smaller percentage of the gross energy intake as methane compared to other classes of beef cattle. This relationship is similar for sheep. The extremely low methane yield values projected for some categories (i.e., sheep, fatten, equation 11) are unrealistic and reflect the difficulty of using empirical equations outside of the range of their determination.

Table 3. Predictions of methane loss from typical diets fed to beef cattle and sheep using Table 2 or Blaxter equations[a]

Class	Gain (kg)	Daily ME intake (Mcal)	Diet DE %	ARC[b] LOI	%CH$_4$ predicted by[c]: LOI	DE-LOI	B-C
Beef Cattle							
Cow, maint.	0	15.5	48	1.1	7.2 (1)	5.9 (5)	6.6
Cow, lact.	0	20.3	56	1.5	6.5 (1)	5.5 (5)	6.9
Growing	0.7	17.2	63	2.0	6.2 (2)	6.5 (6)	6.8
Fatten	1.4	21.1	84	2.6	4.4 (3)	4.2 (7)	5.0
Sheep							
Ewe, maint.	0	2.4	55	1.4	7.6 (9)	6.0 (13)	6.9
Ewe, lact.	0	5.9	66	3.4	5.8 (9)	4.1 (13)	5.5
Growing	0.18	3.3	66	2.8	4.5 (10)	2.9 (14)	6.1
Fatten	0.28	4.4	75	3.8	0.7 (11)	2.1 (15)	4.5

[a] NRC (1984, beef) and NRC (1985, sheep) requirements by class with LOI corrected to be comparable with the equation of Blaxter and Clapperton (1965). ME = metabolizable energy and DE = digestible energy.

[b] Level of intake (LOI) calculated as per Blaxter and Clapperton (1965) but using the MAFF (1976) linear equation to predict fasting heat production.

[c] LOI and DE-LOI values as per Table 2. Numbers in ( ) refer to equations per Table 2 used in calculating %CH$_4$. B-C = as per equation of Blaxter and Clapperton, 1965, where %CH$_4$ = 1.30 + .112 DE + LOI (2.37 - .05 DE).

The relationship of %CH_4 to digestible energy is much more variable than the relationship of %CH_4 to LOI. The high variability in the relationship of %CH_4 to digestible energy reflects the conflict between increasing amount of fermentable substrate versus changing pattern of fermentation, i.e., decreasing acetate/propionate occurring with percent digestible energy increases.

6. Estimates of methane production by U.S. cattle by classes

Estimates of the number, diet and duration of time for the major classes of cattle in the United States was used as the basis for determining total methane output (Table 4). Methane losses were estimated to range from 5.5 to 6.5 %CH_4 for all classes, except for the feedlot phase, which was estimated at 3.5 %CH_4. The estimated yearly production for beef cattle is 3.8 million metric tons (Tg). If dairy beef are added into this total, it raises it to approximately 4.1, and if total U.S. cattle are summed, including dairy cattle (1.8 Tg), the estimate reaches 5.9 Tg per year. This estimate of beef cattle methane emissions is sixteen percent lower than the 4.9 Tg estimated by Crutzen et al. (1986) but higher than the 3.2 Tg estimated by Byers (1990). Our estimates suggest that Crutzen overestimated beef cattle emissions but underestimated contributions from dairy cattle, providing similar total amounts.

Table 4. Estimated methane production by U.S. cattle.

Class	Yearly mean herd[a]			Estimated methane liters per	
	Numbers x 10^6	Days fed	%[b]	Hd/day	Tg/yr[c]
Cows, beef	33.7	365	6.2	262	2.3
Cows, dairy	10.1	365	5.8	492	1.3
Replacements	9.9	365	6.5	220	0.6
Bulls	2.2	365	6.0	380	0.2
Calves[d]	38.6	210	6.0[d]	53	0.3
Stockers	37.9	150	6.5	202	0.8
Feedlot	26.6	140	3.5	153	0.4
All cattle total					5.9
Subtotals:					
Dairy cows with their replacements					1.8
Dairy beef					0.3
Beef cattle					3.8

[a] Taken or extrapolated from USDA cattle inventory data, August 1989 and January 1990.
[b] Percent of diet gross energy intake.
[c] Million metric ton (Tg) per year.
[d] Beef and dairy calves carried to 210 d of age, fraction of dry feed energy estimated from birth to 210 d.

7. Estimates of global methane production from livestock

The potential contribution of animals to atmospheric methane has been recognized for some time. Global animal methane emissions estimated during the 1970s were 100 and 160 Tg/yr (Ehhalt, 1974; Ehhalt and Schmidt, 1978). Estimates made in the early 1980s were similar (90 Tg, Sheppard et al., 1982; 120 Tg, Khalil and Rasmussen, 1983; and 115 Tg, Blake, 1984). The extensive investigation and summary of 1983 livestock emissions by Crutzen et al. (1986) resulted in lower estimates, even though animal numbers were increasing by about 1.2%/yr during this period. Crutzen et al. (1986) combined literature data and expert opinion to estimate the typical fraction of diet energy lost as methane (5.5% to 9.0%) and diet consumed by animal classes with FAO estimates of animal numbers to arrive at global sums. They conclude that domestic animals emit seventy-four million metric tons annually with an additional four tons contributed by wild animals. Of the domestic animals, cattle, with their large size, appetites, and numbers, combined with their extensive gastrointestinal tract fermentation, are the major contributors, producing seventy-four percent of animal emissions. Another bovine, the water buffalo, contributes another eight percent. Smaller ruminant species, sheep and goats, produce approximately thirteen percent, while camels produce about one percent of animal emissions. Horses and mules appear to be the principle nonruminant methane emitters (two percent), followed by pigs (one percent). These authors also estimated that wild ruminants globally produce about five percent additional animal methane.

A recent examination by largely independent methods (Gibbs and Johnson, 1993) estimated that the 1990 global livestock animal methane emissions were 79 Tg, which supports the Crutzen et al. (1986) estimates. While there are many differences in estimated emissions within groups of animals by the two investigations, total emissions are very similar. Increased emissions largely result from the 5% increase in global numbers of cattle and buffalo between 1983 and 1990.

Livestock methane emissions appear to have increased from 17 to 79 Tg, nearly five-fold over the last century (Table 5), although the increase has been slower during recent years. Livestock, particularly cattle, thus have contributed to the doubling of atmospheric methane during this time period, however, could not have solely caused the increase.

All of the global estimates suffer from lack of information about age distributions and body weights of animals, feed type, quality, and consumption: factors that are necessary to estimate methane emissions. Furthermore, almost no experimental data exists for methane production in the rumen of the extensive cattle and buffalo populations in developing countries of the world.

Table 5. Changes in animal population and methane emissions during the last century.

Item	1890[a]	1983[a]	1990[b]
Population ($\times 10^9$)			
Cattle and buffalo	0.34	1.35	1.42
Sheep and goats	0.84	1.61	1.75
Methane, Tg/yr			
Cattle and buffalo	12	61	66
Sheep and goats	4	9	10
Other domestic	1	4	3
Total, Tg/yr	17	74	79

[a]Extrapolated from Crutzen et al. (1986).
[b]Condensed from Gibbs and Johnson (1993).

To produce more precise estimates for the developing countries requires subdividing gross livestock numbers into age groups and classes (i.e., work, milk production, etc.) similar to the breakdowns shown for the U.S. (Table 3). An example of this approach has been presented for China (Ward and Johnson, 1990).

To estimate methane emissions by ruminant animals, it is necessary to have an estimate of the daily dry matter intake (DMI) and dry matter or energy digestibility as these are the principle parameters that define $\%CH_4$.

Many research projects have evaluated a variety of factors that affect dry matter intake mostly as related to management systems found in the developed countries. Many factors, such as temperature, photoperiod, frequency of watering, pregnancy, body condition, behavioral characteristics, and, of course, availability all affect DMI. Increasing forage digestibility has been suggested to increase forage intake because ruminal fill is reduced and/or ruminal output is increased, thus creating more physical space in the rumen for increased DMI (Ellis et al., 1989). The dynamics of particle mass action in the rumen often are associated with regulation of voluntary intake. Forages high in cell wall contents generally have slower rates of ruminal digestion, thus slower rates of ruminal outflow (Ellis et al., 1989). Sutherland (1988) proposed the concept of buoyancy to determine rates of ruminal outflow. Sutherland suggested that buoyant particles, either buoyant due to naturally low density or reduced "affective" density due to microbial gases entrapped within the ruminal particles' architecture, will have slower ruminal outflow rates than less buoyant particles because of physical placement within the rumen. More buoyant particles are located in the more dorsal regions of the rumen away from the outflow site in the ventral rumen.

Lower quality forages, low in protein, often can have low dry matter intake due to protein or nitrogen deficiencies at the level of the rumen or level of tissue. Ruminal deficiencies of nitrogen result in less than optimum microbial digestion of organic matter in the rumen. Supplementing ruminal nitrogen deficiencies theoretically results in increased ruminal digestion and thus, greater ruminal capacity for intake. Tissue deficiencies of protein equivalents result in some proposed metabolic limitations of intake. Providing more protein at the level of the tissues can enhance or be associated with a greater dry matter intake (Egan, 1977).

Faichney and Boston (1983) and Owens et al. (1991) suggested restricted ruminal outflow does not limit intake of low quality forages; rather fecal output may limit intake. This differs from data of Grovum and Phillips (1978), Ellis et al. (1982), and Ellis et al. (1989), which suggests ruminal outflow of particle residues from medium to high quality forages serves to limit intake rather than post-ruminal fractional flow rate or fecal output.

A rather common assumption has been that diet digestibility is a major regulator of dry matter intake, and regression equations to describe this relation have been proposed. The NRC (1987) points out the limitations of several systems proposed to predict voluntary intake. Their information depicts the differences in predicted vs. observed DMI of yearling steers that were pen-fed with all predictions underestimating intake at body weight of less than 430 kg of a range from approximately 330 to 480 kg (up to 84 days on feed). Under poor quality grazing conditions these equations overestimated intake. Much of this variation could possibly be attributed to inaccurate net energy or metabolizable energy values to enter into the equations to predict intake.

Table 6. Comparison of utilization of leaf and stem material

	Leaf	Stem
DM digestibility %	52.8	55.8
Intake g/d/kg	57.8	39.6
Retention time (hrs.)	23.5	31.8

A study of six tropical grasses in Australia (Minson and McLeod, 1970) indicates a negative relation between dry matter digestibility and intake in a study with leaf and stem material (Table 6). The rate of passage of leaves was higher, which caused lower digestibility but increased intake. This illustration is important in conditions of poor quality forage because animals tend to select leaves rather than stems to the extent possible. The significance of selection is well-known but perhaps best documented by data from Zemmelink (1980) and the two papers by Heaney (1973) and Heaney et al. (1968). Zemmelink reviewed 203 studies on 23 different tropical forages and found that the extent to which surplus feed was offered had a marked effect on intake. The dry matter digestibility of 73 of Zemmelink's forages was below 60%. A study with cattle fed a tropical grass found daily intakes of 48, 59, and 68 when they were offered respectively 69, 93, and 116 g/kg $BW^{.75}$. The range in daily intake (by sheep) was 24 to 128 g/kg $BW^{.75}$ with most of the values between 40 and 80. Heaney (1973) reported that when a 25% excess was fed, daily intake of red clover increased from 65 to 97 g/kg $BW^{.75}$ and alfalfa from 55 to 80 g/kg $BW^{.75}$. Heaney et al. (1968) conclude that animal variation makes intake research comparatively difficult. Dry matter digestibility differences of 4% units can be reliably detected with 3 to 4 sheep while with 8 to 10 sheep, 2% differences can be detected. By comparison, to detect differences of 10 units (g/kg $BW^{.75}$) of daily intake requires 10-15 sheep and differences of 5 units are probably impossible to detect. Heaney included data from 441 trials and the range of intakes was from 12 to 149 for forage crops in use in western Canada. Cocks and Thomson (1988) reported a range of 34 to 71 g/d/kg$W^{.75}$ of daily intake for sheep in Syria with weight changes of -143 to 18 g/d.

The conclusion with respect to the data needed to estimate methane emissions would seem to be that one is forced to make estimates of intake adequate to support assumed functions of maintenance, growth, pregnancy, lactation and work. Estimated dry matter intake can then be multiplied by selected factors for gross energy and methane as a percent of gross energy. Default values to use at present would be those presented above for U.S. cattle.

The principal uncertainty factors and our subjective estimate of the degree of uncertainty are number of animals in each country ($\pm$ 10%), their distribution into age, weight, and production class, the amount of feedstuffs consumed daily ($\pm$ 25%), and the methane yield per unit of diet for each class ($\pm$ 20%). Many errors may cancel, such as the beef-dairy tradeoff mentioned previously. Internationally, two areas particularly warrant attention. One is the estimate of 6.3 Tg emissions by water buffalo with essentially no methane measurements to support the estimate. Also, the methane yield for developing country cattle (Crutzen et al., 1986) currently is estimated at nine percent but may be approximately 30% high judging from citations in our data base, particularly for emissions from low quality diets.

8. Methane emissions from livestock manure

Animal excreta may be further fermented to produce methane as a predominant end product. This fermentation may be extensive, as is desired in anaerobic fermentors designed to trap methane for energy, or it may be inconsequential, as when feces are quickly dried and/or burned. The potential contribution of animal excreta to the global methane pool has only recently been examined, and this examination is far from complete.

Intensive livestock production, such as is practiced in the United States' finishing cattle industry and in all phases of U.S. dairy, swine, and poultry production, results in the accumulation of manure. This mass of manure, whether stockpiled or stored as a slurry in a lagoon, may itself create an anaerobic environment for methane production.

Considerable work has been done to determine the potential methane production from OM in different manures used as feedstocks in anaerobic digesters to produce methane as a power source. For beef cattle, Hashimoto et al. (1981) showed a potential m^3CH_4/kg OM of 0.33 and 0.17, respectively, for cattle on feed and grazing cattle. Likewise, Morris (1976) showed methane production from dairy cattle manure to be 0.24 m^3 CH$_4$/kg OM.

The majority of the world's livestock are free-ranging. Manure deposited in these locations does not undergo the same degree of fermentation as is exhibited in the confined situations. Recent research by Lodman et al. (1993) indicated that manure deposited in this manner was likely to produce less than one percent of the potential methane that would result from anaerobic lagoon disposal. Recent enclosure measurements of methane losses from fecal deposits from dairy cows on pasture (Williams, 1993) also found less than one percent of potential. Maximum loss rate was three percent initially and decreased rapidly over the first and second day of drying and/or aeration.

An accurate estimation of methane produced from manure throughout the world is not possible at this time. Uncertainty about manure handling practices precludes an accurate accounting. Methane production from manure of non-confined animals, less than one percent of potential as indicated above, suggests minor methane emissions from animal manure handled in this manner. The majority of the world's ruminants fall into this management category (i.e., developing countries). Manure handling practices designed to collect methane for fuel energy likely contribute only small amounts through leakage to the global methane pool, as indicated by research in China (Khalil et al., 1990). The principal problem is anaerobic manure storage or anaerobic lagoon disposal of animal manure (including nonruminant species, i.e., poultry and swine). Anaerobic lagoons may ferment almost all of the non-lignin organic matter, thus releasing close to the maximum potential methane (Safley, 1989). Other important factors in anaerobic fermentation are temperature and moisture content of the manure (Safley and Westerman, 1987).

An estimate of 28 Tg/yr of methane emissions from livestock manure has been produced recently (Safley et al., 1992). It is the most extensive qualitative synthesis to date of manure production and disposal globally. The magnitude of total emissions depends heavily on a factor used to estimate methane loss from range cattle manure. This factor is at least ten times higher than the measured rates cited above.

Given the above assumptions along with others concerning numbers and routes of manure disposal, U.S. beef cattle manure is estimated to produce approximately 0.2 million tons of methane and U.S. dairy cattle manure 0.4

million tons of methane annually (Table 7). This is approximately 2% of potential from manure for beef cattle and 10% of potential for dairy cattle. These represent ballpark estimates, and there is a distinct need for better data on numbers and types of manure handling facilities and exactly how they effect fermentation to more accurately predict manure methane production. Uncertainties in manure methane are largely associated with U.S. and European manure disposal methods and are not likely to be of major consequence globally.

Table 7. Estimated potential and current emissions of methane from U.S. cattle manure

Animal	Manure[a] OM/yr (kg) 10^{10}	Estimated potential[b]			Estimated current methane from manure	
		m^3CH_4 kg OM	m^3CH_4 $\times 10^7$	CH_4F^c	m^3CH_4 $\times 10^7$	CH_4 tons $\times 10^6$
Beef						
cows	4.00	0.17	680	0.01	6.8	0.045
heifers	0.54	0.17	92	0.01	0.9	0.006
bulls	0.36	0.17	61	0.01	0.6	0.004
calves	0.83	0.17	141	0.01	1.4	0.009
stockers	1.04	0.17	177	0.01	1.8	0.012
feedlot	1.00	0.33	330	0.05	16.5	0.109
Total	7.77		1481		28.0	0.184
Dairy						
cows	1.80	0.24	432	0.13	56.1	0.37
heifers	0.51	0.24	123	0.04	4.9	0.03
stockers	0.11	0.24	26	0.04	1.0	0.01
calves	0.11	0.24	27	0.04	1.1	0.01
Total	2.53		608		63.1	0.42
Overall	10.30		2089		91.1	0.60

[a] Numbers of animals (Johnson et al., 1990) times organic matter (OM/animal) (Taiganides and Stroshine, 1971).

[b] Beef cattle estimates/kg OM (Hashimoto et al., 1981) and dairy cattle estimates by Morris (1976).

[c] CH_4F = fraction of potential from manure for each class. Grazing or nonconfined animals, .01 (Lodman et al., 1993); feedlot animals, 0.1 times 40% of lots with runoff control (USDA, 1979), others times 0.01 (Lodman et al., 1993); dairy cow estimate of 20% with holding ponds (USDA, 1979) times 0.9 (Overcash et al., 1983) x 0.66 to allow reoxidation at surface, other dairy similar to feedlots.

9. Amelioration potential

The prior literature review and knowledge of management systems suggest several ways to reduce methane emissions. Many could be initiated or more widely implemented with existing technologies. These are listed without consideration of the economic or social factors involved in their adoption. When cereal grain production exceeds human food needs, several improvements result from increased starch feeding. High starch diets shift microbial populations to hydrogen utilizing propionate production, decreasing methane yield. Additionally, rate of gain efficiency and productivity is maximized, decreasing total feed consumed and fermented, which all minimize methane. This effect can be enhanced by feeding starches that are more slowly fermented. More starch bypasses to the small intestine of the ruminant where digestion of the starch proceeds without methane production.

Any feeding system that increases animal productivity results in lower methane per unit of product (i.e., milk, meat, work, or wool). Methane per product will decrease and, in economies saturated with livestock products, will lower total livestock methane. Several feeding and management strategies have been effective, such as providing high quality forages. Harvesting, preserving, and feeding or grazing of forage with high solubles content shifts fermentation toward propionate, away from acetate and the obligatory transfer of excess hydrogen to methane. Additionally, productivity will be enhanced, diluting maintenance requirements and decreasing total feed fed/fermented and corresponding methane.

Eliminating nutrient deficiencies by adding a protein/mineral and/or vitamin supplement to an imbalanced forage diet can dramatically improve productivity (product per unit of time and per animal). This decreases the fraction of feed and corresponding methane wasted in the nonproductive maintenance function. This is of particular importance to developing countries.

Additionally, many performance enhancers have been developed. Implants, growth promotants, additives, and vaccines result in improved rate of gain, composition of gain (low fat), health, and reproduction. Many improvements stem from the application of the science of genetics. Crossbreeding, matching genotype to environment, selecting for improved composition, and efficiency of feed to product conversion are all effective available tools.

Eliminating anaerobic pit or lagoon storage of manure or their conversion to collectors as biogas energy sources would be doubly effective. It would reduce emissions and save the burning of other energy sources.

There are several technological possibilities for future reduction of methane losses from livestock. Persistent ionophores or antibiotics that inhibited methane over the entire feeding period or production cycle could reduce ruminant methane by 20% to 25%. Protozoal inhibitors appear promising. Defaunation reduced methane nearly 50% in cattle limit fed a barley diet and may be effective under many other circumstances.

Enhancing acetogens is another intriguing possibility. This group of microbes produces acetic acid from excess hydrogen and carbon dioxide instead of methane production by methanogens. They exist in the rumen as a minor species, predominate in the gut of some termites, and may be important in the lower gut of several animal species. Developing ways to make them competitive in the rumen or transferring the acetogenesis genes to already successful organisms could be very helpful to animal efficiency and the environment.

10. Conclusions

Livestock are significant contributors to global methane, producing about 79 Tg in 1990. Most of this methane originates from domestic ruminants of which cattle and buffalo are the most important. Methane generated in the rumen as a percentage of feed energy consumed varies from 2% to 12%. On average, highly digestible diets containing large amounts (80-90%) of grain are at the low end of the range (3-4%), while the average for all other forage and mixed diets is similar (6-6.5%) when fed to cattle and sheep ad libitum. Restricted feeding of highly digestible diets can result in the very high 10-12% methane yield, but this management practice almost never occurs in practice.

Livestock-productivity-enhancing additives, supplements, or management practices can markedly decrease the methane per unit of livestock product produced. They are the major mitigation opportunity, particularly in economies saturated with animal products. Growth promotants, nutrient supplements, ionophores, vaccines, and improved reproductive rates are all included in this category. The ionophore feed additives additionally decrease methane per unit of feed for two to three weeks after beginning use. Only the more modest methane decrease through lower total feed requirements (5% to 6%) appears to persist over long-term feeding. Methane estimates from U.S. cattle are 5.9 Tg/yr, of which beef cows produce 40%; cattle in feedlots, 7%; dairy cattle, 31%; and the remainder is from replacements, bulls, and calves.

An extensive literature on methane emissions as affected by diets exists for the developed countries, but little or no data are available for cattle or buffalo or other ruminants fed diets typical for developing countries. The value of 6% of feed energy is probably the best default value for those countries.

More precise estimates for the large populations of ruminants in developing countries requires more analysis of existing data and use of expert opinion to estimate feed types, intakes, and animal productivity. Actual data on CH_4 emission by these animals are sorely needed.

Methane emissions from manure produced by U.S. cattle is estimated to be 0.6 Tg/yr or 11% of that produced directly by U.S. cattle. Methane from manure of livestock in developing countries is estimated to be negligible considering probable manure handling systems.

Acknowledgement. This work was supported in part by NASA Interdisciplinary Research Program in Earth Sciences.

References

Abo-Omar, J.M. 1989. Methane losses by steers fed ionophores singly or alternatively, Ph.D. Dissertation, Colorado State University, Fort Collins.
Armstrong, D.G. 1964. Evaluation of artificially dried grass as a source of energy for sheep, II. The energy value of cocksfoot, timothy, and two strains of rye-grass at varying stages of maturity. *J. Agric. Sci. (Camb.), 62*:399.
Bauchop, T. 1977. Foregut fermentation. In: *Microbial Ecology of the Gut* (R.T.J. Clarke and T. Bauchop, eds.), Academic Press.
Benz, D.A., D.E. Johnson. 1982. The effect of monensin on energy partitioning by forage fed steers. *Proc. West Sec. Amer. Soc. Anim. Sci. 33*:60.
Birkelo, C.P., D.E. Johnson, G.M. Ward. 1986. Net energy value of ammoniated wheat straw. *J. Anim. Sci., 63*:2,044.
Blake, D.R. 1984. Increasing concentrations of atmospheric methane. Ph.D. dissertation, University of California at Irvine. 213 pp.

Blaxter, K.L., N. McGraham. 1956. The effect of the grinding and cubing process on the utilization of the energy of dried grass. *J. Agric. Sci. (Camb.), 47*:207.

Blaxter, K.L., F.W. Wainman. 1964. The utilization of the energy of different rations by sheep and cattle for maintenance and fattening. *J. Agric. Sci. (Camb.), 63*:113.

Blaxter, K.L., J.L. Clapperton. 1965. Prediction of the amount of methane produced by ruminants. *Brit. J. Nutr., 19*:511.

Branine, M.E., C.P. Lofgreen, M.L. Galyean, M.E. Hubbert, A.S. Freeman, D.R. Garcia. 1989. Comparison of continuous with daily and weekly alternate feeding of lasalocid and monensin plus lylosin on performance of growing-finishing steers. *Proc. West Sect. Amer. Soc. Anim. Sci., 40*:353.

Buchner, P. 1965. *Endosymbiosis of Animals With Plant Microorganisms*, Revised English version. Interscience, New York.

Byers, F. 1974. The importance of associative effects of feeds on corn silage and corn grain net energy values. Ph.D. Dissertation, Colorado State University, Fort Collins.

Byers, F.M. 1990. Methane from beef production. *Proc. West Sect. Amer. Soc. Anim. Sci., 41*:144.

Carmean, B.R. 1991. Persistence of monensin effects on nutrient flux in steers. M.S. Thesis, Colorado State University, Ft. Collins.

Carmean, B.R., D.E. Johnson. 1990. Persistence of monensin-induced changes in methane emissions and ruminal protozoa numbers in cattle. *J. Anim. Sci., 65*:(Supp. 1, Abstr.).

Carmean, B.R., K.A. Johnson, D.E. Johnson. 1991. Maintenance energy requirements of the llama. *Am. J. Vet. Res., 53*:1,696.

Cocks, P.S., E.F. Thompson. 1988. Increasing feed resources for small ruminants in the Mediterranean Basin in increasing small ruminant productivity in semi-arid areas. *ICARDA* (E.F. Thomson, F.S. Thomson, eds.).

Corbett, J.L., J.P. Langlands, I. McDonald, J.D. Pullar. 1966. Comparison by direct animal calorimetry of the net energy values of an early and a late season of growth of herbage. *Anim. Prod., 8*:13.

Crutzen, P.J. 1983. Atmospheric Interactions. In: *The Major Biogeochemical Cycles and Their Interactions* (B. Bolin and R. Brook, eds.), John Wiley, p 69-112..

Crutzen, P.J., I. Aselmann, W. Seiler. 1986. Methane production by domestic animals, wild ruminants, and other herbivorous fauna and humans. *TELLUS, 38B*:271.

Czerkawski, J.W. 1972. Fate of metabolic hydrogen in the rumen. *Proc. Nutr. Soc., 31*:141.

Czerkawski, J.W. 1988. *An Introduction to Rumen Studies*, Pergamon Press, Oxford.

Czerkawski, J.W., G. Breckenridge. 1975a. New inhibitors of methane production by rumen micro-organisms. Development and testing of inhibitors in vitro. *Brit. J. Nutr., 34*:429.

Czerkawski, J.W., G. Breckenridge. 1975b. New inhibitors of methane production by rumen micro-organisms. Experiments with animals and other practical possibilities. *Brit. J. Nutr., 34*:447.

Czerkawski, J.W., K.L. Blaxter, F.W. Wainman. 1966. The metabolism of oleic, linoleic, and linolenic acids by sheep with reference to their effects on methane production. *Brit. J. Nutr., 20*:349.

Delfino, J., G.W. Mathison, M.W. Smith. 1988. Effect of lasalocid on feedlot performance and energy partitioning in cattle. *J. Anim. Sci., 66*:136.

Egan, A.R. 1977. Nutritional status and intake regulation in sheep. VIII. Relationships between the voluntary intake of herbage by sheep and the protein/energy ratio in the digestion products. *Aust. J. Agric. Res., 28*:907-915.

Ehhalt, D.H. 1974. The atmospheric cycle of methane. *TELLUS, 26*:58-69.

Ehhalt, D.H., U. Schmidt. 1978. Sources and sinks of atmospheric methane. *PAGEOPH, 116*:452-464.

Ellis, W.C., C. Lascano, R.G. Teeter, F.N. Owens. 1982. Solute and particulate flow markers. In: *Protein Requirements for Cattle: Symposium* (F.N. Owen, ed.), Oklahoma Agriculture Experiment Station Misc. Publ. MP-109, pp 37-52.

Ellis, W.C., M.J. Wylie, J.H. Matis. 1989. Dietary-digestive interactions determining the feeding value of forages and roughages. In: *World Animal Science Subseries B, Vol. 17* (E.R. Orskov, ed.), Elsevier, Amsterdam, pp 177-229.

Faichney, G.J., R.C. Boston. 1983. Interpretation of the fecal excretion patterns of solute and particulate markers introduced into the rumen of sheep. *J. Agric. Sci. (Camb)., 101*:575.

Gibbs, M.J., D.E. Johnson. 1993. Livestock Emissions. In: *International Anthropogenic Emissions of Methane,* U.S. EPA, Washington, D.C. (in review).

Goodrich, R.D., J.E. Garrett, D.R. Gast, M.A. Kirick, D.A. Larson, J.C. Meiske. 1984. Influence of monensin on the performance of cattle. *J. Anim. Sci., 58*:1,484.

Grovum, W.L., G.C. Phillips. 1978. Factors affecting the voluntary intake of food by sheep. The role of distension, flow-rate of digesta and propulsive motility in the intestine. *Br. J. Nutr., 40*:323.

Haaland, G.L. 1978. Protected fat in bovine rations. Ph.D. Dissertation, Colorado State University, Fort Collins.

Hashimoto, A.G., V.H. Varel, Y.R. Chen. 1981. Ultimate methane yield from beef cattle manure: Effect of temperature, ration instruments, antibiotics and manure age. *Agricultural Waste, 3*:241.

Hashizume, T., H. Morimoto, T. Hayer, M. Itch, S. Tanabe. 1967. Utilization of the energy of fattening rations containing ground or steam-rolled barley by Japanese Black Breed Cattle. *4th Energy Symp., EAAP, 12*:261.

Heaney, D.P. 1973. Effects of the degree of selective feeding allowed on forage voluntary intake and digestibility assay results using sheep. *Can. J. Anim. Sci., 53*:431.

Heaney, D.P., G.I. Pritchard, W.J. Pigden. 1968. Variability in ad libitum forage intakes by sheep. *J. Anim. Sci., 27*:159.

Hintz, H.F., D.E. Hogue, E.F. Walker, J.E. Lowe, H.F. Schryver. 1971. *J. Anim. Sci., 32*:245-248.

Hubbert, M.E., M.E. Branine, M.L. Galyean, G.P. Lofgreen, D.R. Garcia. 1987. Influence of alternate feeding of monensin and lasalocid on performance of feedlot heifers -preliminary data. *New Mexico Agric. Exp. Sta., Clayton Livestock Research Center Prog. Rep. No 47.*

Johnson, A.B. 1987. Ionophore rotation research results - pooled five trial summary. *Ionophore Rotation Program Handbook.* Hoffman-LaRoche, Inc., Nutley, N.J.

Johnson, D.E. 1966. Utilization of flaked corn by steers. Ph.D. Dissertation, Colorado State University, Fort Collins.

Johnson, D.E. 1972. Effects of a hemiacetal of chloral and starch on methane production and energy balance of sheep fed a pelleted diet. *J. Anim. Sci., 35*:1,064.

Johnson, D.E. 1974. Adaptational responses in nitrogen and energy balance of lambs fed a methane inhibitor. *J. Anim. Sci., 38*:154.

Johnson, D.E., A.S. Wood, J.B. Stone, E.T. Moran, Jr. 1972. Some effects of methane inhibition in ruminants. *Canadian Journal Anim. Sci., 52*:703.

Johnson, D.E., M. Branine, G.M. Ward. 1990. Methane emissions from livestock. *Proc. Nutr. Symp. Amer. Feed Ind. Assoc.*, Arlington, VA.

Joyner, A.E., L.J. Brown, T.J. Fogg, R.T. Rossi. 1979. Effect of monensin on growth, feed efficiency, and energy metabolism of lambs. *J. Anim. Sci., 48*:1,065.

Khalil, M.A.K., R.A. Rasmussen. 1983. Sources, sinks, and seasonal cycles of atmospheric methane. *J. Geophys. Res., 88*:5,131-5,144.

Khalil, M.A.K., R.A. Rasmussen, M.-X. Wang, L. Ren. 1990. Emissions of trace gases from Chinese rice fields and biogas generators. *Chemosphere, 20*:207.

Krumholtz, L.R., C.W. Forsberg, D.M. Veira. 1983. Association of methanogenic bacteria with rumen protozoa. *Can. J. Microbiol., 29*:676.

Lodman, D.W., M.E. Branine, B.R. Carmean, P. Zimmerman, G.M. Ward, D.E. Johnson. 1993. Estimates of methane emissions from manure of U.S. cattle. *Chemosphere, 26* (1-4):189-200.

MAFF. 1976. Energy Allowances and Feeding Systems for Ruminants. *Technical Bulletin 33*, Ministry of Agriculture, Fisheries and Food, London.

McBee, R.H. 1977. Hindgut fermentation. In: *Microbial Ecology of the Gut* (R.T.J. Clark and T. Bauchop, eds.), Academic Press.
Milchunas, D.G., M.I. Dyer, O.C. Wallmo, D.E. Johnson. 1978. In vivo/in vitro relationships of Colo. mule deer forages. *Spec. Report #43, Colo. Div. of Wildlife*
Minson, D.J., M.N. McLeod. 1970. The digestibility of temperate and tropical grasses. In: *Proceedings 11th International Grassl. Congr.*, p. 719-722.
Moe, P.W., H.F. Tyrrell. 1977. Effects of feed intake and physical form on energy value of corn in timothy hay diets for lactating cows. *J. Dairy Sci.*, 60:752.
Moe, P.W., J.F. Tyrrell. 1980. Methane Production in dairy cows. *8th Energy Symp., EAAP,* 26:12.
Morris, G.R. 1976. Anaerobic fermentation of animal wastes: A kinetic and empirical design fermentation. M.S. Thesis, Cornell University.
NRC. 1984. *Nutrient Requirements of Beef, 6th Ed.*, National Academy Press, Washington, D.C.
NRC. 1985. *Nutrient Requirements of Sheep, 6th Ed.*, National Academy Press, Washington, D.C.
NRC. 1987. *Predicting Feed Intake of Food-producing Animals.* National Academy Press, Washington, D.C.
Overcash, M.R., F.J. Humenik, J.R. Miner. 1983. *Livestock Waste Management,* vol. 1, CRC Press, 2000 Corporate Blvd, NW, Boca Raton, FL, p 37.
Owens, F.N., J. Garza, P. Dubeski. 1991. Advances in amino acid and N nutrition in grazing ruminants. In: *Proceedings of Grazing Livestock Nutrition Conference,* Oklahoma State University Press, pp 109-137.
Robb, J., P.J. Evans, C. Fisher. 1979. A study of the nutritional energetics of sodium hydroxide-heated straw pellets in rations fed to growing lambs. *8th Energy Symp., EAAP,* 26:13.
Robbins, C.T. 1973. The biological basis for the determination of carrying capacity. Ph.D. Thesis, Cornell Univ, Ithaca, NY.
Rumpler, W.V., D.E. Johnson, D.B. Bates. 1986. The effect of a dietary cation concentration on methanogenesis by steers fed diets with and without ionophores. *J. Anim. Sci.* 62:1,737.
Safley, L.M. 1989. Methane productions from animal wastes management systems. *Methane Emissions from Ruminants,* ICF/USEPA Workshop, Palm Springs, CA.
Safley, L.M., P.W. Westerman. 1987. Biogas production from anaerobic lagoons. *Biological Wastes* 23:181.
Safley, L.M., Jr, M.E. Casada, J.W. Woodbury, K.F. Roos. 1992. *Global Methane Emissions from Livestock and Poultry Manure.* U.S. EPA/400/1-91/048.
Sawyer, M.S., W.H. Hoover, C.J. Sniffen. 1974. Effects of a ruminal methane inhibitor on growth and energy metabolism in the ovine. *J. Anim. Sci.* 38:908.

Sheppard, J.C., H. Westberg, J.F. Hopper, K. Genesan, P. Zimmerman. 1982. Inventory of global methane sources and their production rates. *J. Geophys. Res., 87*:1,982.
Sundstol, F. 1982. Energy utilization in sheep fed untreated straw, ammonia treated straw or sodium hydroxide heated straw. *9th Energy Symp., EAAP, 29*:120.
Sundstol, F., A. Ekern. 1976. The nutritive value of frozen, dried, and ensiled grass cut at three different stages of growth. *7th Energy Symposium, EAAP, 19*:241.
Sundstol, F., A. Ekern, P. Lingvall, E. Lindgren, J. Bertilsson. 1979. Energy utilization in sheep fed grass silage and hay. *8th Energy Symp., EAAP, 26*:17.
Sutherland, T. 1988. Particle separation in the forestomachs of sheep. In: *Comparative Aspects of Physiology of Digestion in Ruminants* (A. Dobson and M. Dobson, eds.), Cornell University Press, New York, pp 43-73.
Swift, R.W., J.W. Bratzler, W.H. James, A.D. Tillman, D.C. Meek. 1948. The effect of dietary fat on utilization of the energy and protein of rations by sheep. *J. Anim. Sci., 7*:475.
Taiganides, E.P., R.C. Stroshine. 1971. Impact of farm animal production and processing on total environment. In: *Livestock Waste Management and Pollution Abatement, Proc. Int. Sym. Livestock Wastes,* Am. Soc. Ag. Eng., St. Joseph, MI, p 95.
Theurer, C.B. 1987. Grain processing effects on starch utilization by ruminants. *J. Anim. Sci., 63*:1,649.
Thorton, J.H., F.N. Owens. 1981. Monensin supplementation and in vivo methane production by steers. *J. Anim. Sci., 52*:628.
Tyrrell, H.F., G.A. Varga. 1985. Energy value for lactation of rations containing ground whole ear maize or maize meal both conserved dry or ensiled at high moisture. *10th Energy Symp., EAAP, 32*:306.
Trei, J.E., R.C. Parrish, Y.K. Singh, G.C. Scott. 1971. Effect of methane inhibitors on rumen metabolism and feedlot performance of sheep. *J. Dairy Sci., 54*:536.
USDA Research Report No. 6. 1979. Animal Waste Utilization on Cropland and Pastureland. EPA-600/2-79-059.
Van Soest, P.J. 1982. *Nutritional Ecology of the Ruminant,* O. and B. Books, Corvallis, OR.
Van Nevel, C.J., D.I. Demeyer. 1977. Effect of monensin on rumen metabolism in vitro. *Appl. Environ. Microbiol., 34*:251.
Van der Honing, Y., B.J. Wieman, A. Steg, B. van Donselaar. 1981. The effect of fat supplementation of concentrates on digestion and utilization of energy by productive dairy cows. *Neth. J. Agric. Sci., 29*:79.

Varga, G.A., H.F. Tyrrell, D.R. Waldo, G.B. Huntington, B.P. Glenn. 1985. Effect of alfalfa or orchardgrass silages on energy and nitrogen utilization for growth by Holstein steers. *10th Energy Symp., EAAP, 32*:86.

Ward, G.M., D.E. Johnson. 1990. Methane emission estimates from world's livestock and preliminary calculations for China. *World Resource Review,* 2:238.

Wedegaertner, T.C,. D.E. Johnson. 1983. Monensin effects on digestibility, methanogenesis and heat increment of a cracked corn-silage diet fed to steers. *J. Anim. Sci., 57*:168.

Whitelaw, F.G., J.M. Eadie, L.A. Bruce, W.J. Shand. 1984. Methane formation in faunated and ciliate-free cattle and its relationship with rumen volatile fatty acid production. *Brit. J. Nutr., 52*:261.

Wieser, M.F., C. Wenk. 1970. Effect of plane of nutrition and physical form of ration on energy utilization and rumen fermentation in sheep. *5th Energy Symp., EAAP, 17*:53.

Williams, D.J. 1993. Methane emissions from the manure of free-range cows. *Chemosphere,* 26 (1-4):179-188.

Yen, J.T., J. Nienaber. 1991. Absorption of volatile fatty acids from the gastrointestinal tract of swine. *J. Anim. Sci., 69*:2,001.

Zemmelink, G. 1980. Effect of selective consumption on voluntary intake and digestibility of tropical forages. *Ag. Res. Reports,* Waginingen, p 896.

Chapter 12

Rice Agriculture: Emissions

M.J. SHEARER AND M.A.K. KHALIL

Global Change Research Center
Department of Environmental Science and Engineering
Oregon Graduate Institute, Portland, Oregon 97219 U.S.A.

1. Introduction

Rice agriculture has long been recognized as a major source of methane (CH_4). Global budgets of methane have generally included emissions of about 100 Tg/yr (Tg = Teragrams, 10^{12} grams) from rice agriculture (with a range of 50-300 Tg/yr), constituting about 20% of emissions from all sources (range 14%-40%) (Ehhalt and Schmidt, 1978; Donahue, 1979; Khalil and Rasmussen, 1983; Blake, 1984; Bolle et al., 1986; Bingemer and Crutzen, 1987; Cicerone and Oremland, 1988; Warneck, 1988; Khalil and Rasmussen, 1990). The Intergovernmental Panel on Climate Change (IPCC, 1990) estimates a source of 60 Tg/yr (range 20-150 Tg/yr).

We estimate that rice agriculture contributes some 230 ppbv to the present atmospheric burden of CH_4 in the atmosphere, and may be responsible for some 20-30% of the increase of methane during the last century (Khalil and Rasmussen, 1991).

Estimating the flux of methane from rice fields in various parts of the world requires knowledge of two factors: the emission rates and the regional or global extrapolant. The emission rate or flux depends on different internal and external variables (x_i). Internal variables include: soil characteristics; rice variety; and soil microbiology. External factors include: soil temperature driven by solar radiation; meteorological conditions; water level, which is affected by rainfall and availability of irrigation; and treatments such as the type and amount of fertilizers applied. The global flux, F_G, is usually calculated by equation (1):

NATO ASI Series, Vol. I 13
Atmospheric Methane: Sources, Sinks, and Role in Global Change
Edited by M. A. K. Khalil

$$F_G = \sum_R \bar{\phi}_R T_R A_R \qquad (1)$$

where $\bar{\phi}_R$ (g/m^2/day) is the measured, seasonally averaged emission rate from a region R; A_R (m^2) is the area of the region, presumably with similar characteristics, so that $\bar{\phi}_R$ is an accurate representation of flux for the entire area; and T_R is the growing season (days/year). The global extrapolant is $A_R T_R$. Any smaller region, such as a country, may also be subdivided into similar regions, and the flux of CH_4 from rice agriculture from the country can be calculated based on an equation analogous to equation (1). The nature of the extrapolation process is discussed in more detail by Khalil and Shearer (this volume).

2. Methane emission rates from rice fields

Direct flux measurements over the entire growing period provide precise and accurate values for $\bar{\phi}_R$. Usually, A_R and T_R are determined from agricultural statistics and are also well known. Problems with the extrapolation arise in associating a measured flux with an appropriate area and growing season and in the assumption that the measured flux is representative of the associated area and season.

This empirical approach requires whole season measurements of CH_4 emission rates from as many regions as possible so that the large variations of emissions from one region to another can be properly included in the estimate of global or regional emissions.

Alternative approaches to global extrapolation also exist and are based on the knowledge of the processes or factors that control emission rates. At present there are insufficient data for such approaches to produce reliable estimates of global or regional emissions.

2.1 Flux measurements. During the last decade a number of systematic experiments have been reported on methane emissions from rice fields. All are based on static chamber methods. While there are many variants, the method consists of enclosing a small part (0.1 - 100 m^2) of the rice fields within a chamber and taking periodic samples. The samples are analyzed for methane content usually by gas chromatography using flame ionization detectors

(GC/FID). Methane, emitted from the enclosed area of the rice field, builds up in the chamber. The rate of accumulation is directly proportional to the flux or the rate of emission from the area covered by the chamber. The relationship is:

$$\phi = \frac{\rho\, M\, V}{N_A\, A} \times 10^{-6}\, \frac{dC}{dt} \qquad (2)$$

where ϕ is the flux, ρ is the density of air (molecules/m^3), M is the molecular weight of CH_4, V is the volume of the chamber (m^3), A is the area covered by the chamber (m^2), N_A is Avogadro's number, C is the concentration of methane (ppbv) and dC/dt is in (ppbv/hr). The most common units for reporting fluxes are mg/m^2/hr.

The advantages of chamber methods are that they are inexpensive, easy to use in remote locations and are coupled to a highly sensitive and precise measurement method using GC/FID. Since the fluxes of methane are quite large, the plants need not be exposed to the unnatural conditions of the chamber environment for very long; often 10 minute exposures are sufficient. This fact also makes chamber methods suitable for methane measurements even though they may not be convenient for other gases.

The disadvantages are that placing the chamber can disturb the soil and release abnormal amounts of methane. Several methods have been devised to reduce, if not overcome this problem. In the studies of Khalil et al. (1991) a permanent aluminum base is installed in the soil at the time rice is planted and chambers fit into grooves in this base. In studies reported by Schütz et al. (1989a, 1990) a large permanent chamber is used that has a lid that opens and closes, but the chamber itself is not removed until the rice is harvested. Both these methods create some feedbacks that may affect flux estimates. The chambers also affect the immediate environment of the rice plant by causing heating and a buildup of CO_2, which may affect emissions of methane. Finally, since most chambers are small, the extrapolation of direct flux measurements to large regions may be unreliable because of heterogeneities within fields, within local regions, and within different parts of the same country.

In recent years experiments have been designed to measure the flux throughout the growing season. The earliest experiments of this type made it clear that there are large systematic changes in methane emissions during the growing cycle. Such changes are driven by several factors including the growth of root mass, maturation process for the plants, availability of nutrients and fertilization, and the seasonal change of temperature and length of day (Schütz et al., 1989a; Yagi and Minami, 1990; Khalil et al., 1991).

Direct flux measurements are summarized in Table 1 with a description of the nature of the experiments. These studies were originally designed to quantify the methane produced by rice paddies; later, to establish which factors affect methane emissions; and most recently, to investigate treatments which may reduce methane emissions.

2.2 Factors affecting methane emissions. Methane emissions from rice cultivation result from a combination of three processes: methane production in the paddy soil, methane oxidation at the soil surface and in the root zone, and methane transport from the soil to the atmosphere by diffusion through the floodwater, ebullition, and plant-mediated transport.

Methane is produced in saturated soils by anaerobic bacteria (methanogens), which use chemicals from the decay of organic matter as their food source and produce methane as a by-product. This process takes place only where oxygen is not available, such as in flooded paddy fields and wetlands (Boone, this volume; Neue and Roger, this volume). Temperature, water regime, nutrient availability, and soil factors such as pH and redox potential may affect the growth of the methanogen populations and ultimately the methane flux from the paddy fields.

Experiments to measure the production and emission of methane from paddy soils showed that not all methane produced in the soil is emitted to the atmosphere (Holzapfel-Pschorn et al., 1985; Schütz et al., 1989b, Sass et al., 1992). Laboratory studies of paddy soil have shown that 50% to 90% of the methane produced in the soil can be destroyed before it reaches the surface (summarized in Neue and Roger, this volume). Strains of bacteria that consume methane (methanotrophs) are believed to live in both soil and water. They oxidize the carbon in CH_4 and produce CO_2 as a by-product. This reaction can take place even in the low oxygen environment of a paddy soil. Air transported to the roots of the rice plant creates an oxygenated zone around the roots. There the methanotrophs can break down methane before it is pumped out of the soil by the air circulation system of the plant.

Table 1. Methane flux measurements from rice paddies.

Study	Location	Rice Cultivar	Soil Type	Treatment	Flux mg/m²/hr	‡Emission Season (days)
Cicerone et al., 1983 1982 growing season	California, USA	M101	Vertisol, Capay Clay	Am. phosphate-Am. sulfate + Urea: 113 kg N/ha	10.4	100
Cicerone et al., 1992 1983 growing season	California, USA	Not given	Capay Silty Clay	preplant: 36 kg N/ha + topdressing after planting: 78 kg N/ha	4.0	111
1985 growing season (Only data from plots with rice plants given here)		Not given	Sacramento Clay	No treatment	0.5	128 (approximate)
				2.5 t/ha ground rice straw	6.4	
				5 t/ha ground rice straw	18.9	
				78 kg N/ha (Urea, preplant fertilization)	0.9	
				78 kg N/ha + 2.5 t/ha rice straw	3.0	
				78 kg N/ha + 5 t/ha rice straw	13.9	
Seiler et al., 1984	Andalusia, Spain	Bahia	Not given	Urea (before seeding): 160 kgN/ha; Am. nitrate (at tillering): 40 kg N/ha	4.0	125
Schütz et al., 1989a	Vercelli, Italy	Roma	Sandy Loam	Unfertilized	11.7	114
Data reported from 1984 to 1986; includes data from Holzapfel-Pschorn and Seiler, 1986.			(60% sand, 25% silt, 12% clay, 2.5% organic C)	Rice straw: average	17.6	118
				5 t/ha	24.2	105
				3 t/ha	9.6	109
				6 t/ha	13.5	110
				12 t/ha	22.1	113
				24 t/ha	18.8	
				Compost: 60 t/ha	27.5	113
				CaCN₂: 200 kg N/ha	12.5	118
				Urea: average	10.7	
				200 kg N/ha, raked	9.6	118
				100 kg N/ha, raked	9.6	118
				200 kg N/ha, incorporated	7.9	113
				200 kg N/ha, surface	15.8	113

Table 1. Methane flux measurements from rice paddies (continued).

Study	Location	Rice Cultivar	Soil Type	Treatment	Flux mg/m^2/hr	‡Emission Season (days)
				Straw + Urea: average	22.5	
				6 t/ha + 200 kg N/ha	20.0	113
				12 t/ha + 200 kg N/ha	25.0	113
Schütz et al., 1989a (continued)	Vercelli, Italy	Roma	Sandy Loam	Ammonium sulfate: average	8.2	118
				200 kg N/ha, raked	6.7	109
				200 kg N/ha, incorporated	5.6	105
				100 kg N/ha, incorporated	8.8	105
				50 kg N/ha, incorporated	7.5	113
				200 kg N/ha, surface	12.5	
				Straw + CaCN$_2$: average	16.8	118
				2.5 t/ha + 37.5 kg N/ha	11.7	118
				5 t/ha + 75 kg N/ha	17.9	109
				12 t/ha + 200 kg N/ha	20.6	
Yagi and Minami, 1990	Ibaraki Prefecture, Japan	Koshihikari	Ryugasaki: Gley soil (Eutric Gleysols)	Non-nitrogen (0:60:30)§	2.8	119
				Mineral (60+30:60:30)	2.9	118
				Compost: 12 t/ha + Mineral (above)	3.8	112
				Rice straw: 6 t/ha + Mineral (above)	9.6	117
			Kawachi: peat soil (Dystric Histosols)	Rice straw: 6 t/ha + Mineral (60+25:60:60+25)	16.3	115
			Mito: humic Andosol	Non-nitrogen (0:120:80+30)	1.4	122
				Mineral (50+30:120:80+30)	1.2	125
				Compost: 12 t/ha	1.9	129
				Rice straw: 6 t/ha	3.2	128
				9 t/ha	4.1	128
			Tsukuba: light colored Andosol (volcanic ash soil)	Mineral (70+30:100:70+30)	0.2	125
				Rice straw: 6 t/ha + Mineral (above)	0.4	115

Table 1. Methane flux measurements from rice paddies (continued).

Study	Location	Rice Cultivar	Soil Type	Treatment	Flux mg/m²/hr	‡Emission Season (days)
Dai, 1988; and Schütz et al., 1990	Hangzhou, China	Not given	Not given	Late rice, average Early rice, average	26.4* 6.6*	62 70
Khalil et al., 1991	Sichuan, China	local and hybrid	"purple soil"	Organic	36.6	120
Sass et al., 1990, 1991a 1989 and 1990 data averaged for Lake Charles and Beaumont fertilized sites.	Texas, USA	Jasmine 85	Lk. Charles clay (Typic Pelludert)	Urea: 149 kg N/ha Rice straw: 12 t/ha + Urea: 102 kg N/ha	8.7 15.2	85 86
			Beaumont clay (Entic Pelludert)	Urea: 190 kg N/ha Rice straw: 8 t/ha + Urea: 102 kg N/ha	2.5 5.6	85 86

Sass et al., 1991b	Texas, USA	Jasmine 85	Verland silty clay loam (Vertic Ochraqualf)	All sites, Urea: 190 kg N/ha
				Planted April 9
				  Straw: 6 t/ha
				  w/o straw
				Planted May 18
				  Straw: 6 t/ha
				  w/o straw
				Planted June 15
				  Straw: 6 t/ha
				  w/o straw
					
					23.5
					18.3
					
					18.3
					11.8
					
					12.7
					12.0
						
						85
						
						81
						
						76

Sass et al., 1992	Texas, USA	Jasmine 85	Verland silty clay loam	All sites, Urea: 165 kg N/ha total Normal flood irrigation (control) Mid-season aeration Multiple aeration (3 drained periods) Late season flood irrigation	 4.4 2.3 0.6 6.3	 87 87 87 99

Table 1. Methane flux measurements from rice paddies (continued).

Study	Location	Rice Cultivar	Soil Type	Treatment	Flux mg/m^2/hr	‡Emission Season (days)
Chen et al, 1993	Beijing, China	Huang jinguang	Sandy Loam 1.33% organic matter	610 kg/ha NH$_4$HCO$_3$ + 15 t/ha horse manure: II†	14.6	79
				900 kg/ha NH$_4$HCO$_3$: F	17.5	
				610 kg/ha NH$_4$HCO$_3$ + 15 t/ha horse manure: F	35.9	
				710 kg/ha NH$_4$HCO$_3$ + 30 t/ha horse manure: F	48.9	
				610 kg/ha NH$_4$HCO$_3$ + 15 t/ha horse manure: D	-0.0	
Chen et al, 1993 (continued)	Nanjing, China	Shanyou 63	"yellow-brown earth" 2.29% organic matter	190 kg/ha Urea + 15 t/ha barnyard manure: L	10.8	101
				45 t/ha barnyard manure: II	9.5	
				600 kg/ha Ammonium Sulfate: F	2.6	
				190 kg/ha Urea + 3 t/ha rapestraw: M	14.3	
				190 kg/ha Urea + 15 t/ha barnyard manure: Semi-arid	6.6	

Partial Studies: studies where measurements were not taken for a full season or are not yet reported in full (Wassman et al.), and are not comparable to the preceding studies. Ranges of emissions are shown.

Cicerone and Shetter (1981)	California, USA				1.3 - 7.8	
Khalil et al. (1990)	Sichuan, China				1 - 50	
Parashar et al. (1991)	New Delhi, Karnal, Dehradun, and Hyderabad, India			no fertilizer	0.07 - 80.0	
				694 kg/ha K$_2$SO$_4$/KCl	7.4 - 47.0	
				694 kg/ha K$_2$SO$_4$/KCl + 1042 kg/ha rapeseed cake	7.7 - 35.4	
Wassman et al. (1993)	Zhejian, China			rapeseed cake	7.9 - 44.0	
				1042 kg/ha rapeseed cake	6.0 - 50.6	

‡ Number of days that methane was actually emitted from the rice paddies.
* Values for Schütz et al. (1990) were digitized from the figures.
§ (N:P$_2$O$_5$:K$_2$O) in kg/ha; rates of mineral fertilizer are expressed as basal + supplementary applications.
† F: flood irrigation; II: intermittent irrigation; D: dry culture.

In the same experiments, the pathways of methane emission to the atmosphere throughout the growing season were also measured. Most of the methane (up to 90%) was found to be transported through the inter-cellular spaces in the plant. About 10% of the methane was transported in bubbles from the soil, while less than 1% of the methane diffused through the soil and water to the atmosphere. At the beginning of the growing season all the methane was emitted through ebullition; as the rice plants developed, an increasing amount was transported through the plant.

2.3 Internal factors. The "internal factors" of soil microbiology, soil properties, and different rice cultivars are at present the most difficult to incorporate into the global estimate of methane flux from rice paddies. While the life cycle of methanogens is understood (see Boone, this volume, for a summary), predicting the population fluctuations in rice fields is not yet possible. Methanogens do not directly break down organic matter but require other bacteria to produce the substrates they need for food. The population dynamics of the various interdependent strains of bacteria may account for year to year variation found in field studies where all other variables are nearly the same. Chemicals exuded from the root system of the plants may also influence the growth of methanogens or competing bacteria populations (Neue and Roger, this volume).

Neue et al. (1990) identified soil texture, mineralogy, and Eh/pH buffer systems as the soil properties most important in affecting methane production. Experiments by Yagi and Minami (1990) in Japan, and Sass et al. (1990, 1991a) in the U.S., have directly measured the effect of different soil textures during the same growing season. The differences in fluxes between the soils studied in Japan may reflect the ability of the soils to sustain a reducing environment. The study in Texas found lower methane fluxes from the soil with a higher clay content, which perhaps favored a buildup of soil toxins. Chen et al. (1993), also found differences in methane flux between Nanjing and Beijing, P.R.C., which may be related to different soil types.

Most of the methane emitted from rice paddies passes through the stems of the rice plant (Schütz et al., 1989b). Different rice cultivars may have different capacity for gas transport, though this has not been tested. The only published study using two different cultivars in the same environment for flux measurements (Khalil et al., 1991) found no significant differences between the two rice varieties. Different rice cultivars also have different growing season lengths, requiring different periods of standing water in the paddies. While most studies report the cultivar used, there is not enough information at present to relate cultivar type to variations in methane flux.

Whether the crop is directly seeded or is transplanted may affect the length of time the crop is kept flooded and thus affect the total CH_4 emissions over the growing season. The Khalil et al. (1991) data were obtained from a region where the rice was transplanted to the fields, then standing water was maintained in the fields until harvest about 120 days later. In the study by Sass et al. (1991a) the crop was seeded, and permanently flooded after the young plants were established. Out of a growing season of 140 days, the crop was flooded for only 85 days.

2.4 External factors. External factors affecting the flux of methane from rice may be the result of the weather, including solar radiation which influences soil temperature, and rainfall; or agricultural practices, such as irrigation and fertilization. The former are mainly factors which change with time, or season of growth; the latter vary with area, depending on availability of water and mineral fertilizers, traditional agricultural methods, and local economics.

Schütz et al. (1989a) found a positive relationship between soil temperature and daily CH_4 fluxes, which changed slightly throughout the growing season. Khalil et al. (1991) found a good correlation between soil temperature and methane flux, with a Q_{10} of about 2 (see Figure 1). Temperature relationships may change from season to season or from site to site within a season. Khalil et al. (1991) found a different relationship for measurements during 1988 and 1989 at the same site.

Changes in flux over the growing season have been found in every experiment (Schütz et al., 1989a; Yagi and Minami, 1990; Khalil et al., 1991). These researchers described the flux in terms of the stage of growth of the plant. Planting date affects the stage of the crop during the longest and warmest days. Schütz et al. (1990) found significant differences between average fluxes for early and late rice crops in Hangzhou, China. Sass et al. (1991b) compared the differences in flux over a planting season by planting a crop at one-month intervals. The later plantings ripened more quickly, were irrigated for a shorter period, and differences in CH_4 fluxes between the fertilizer treatments decreased.

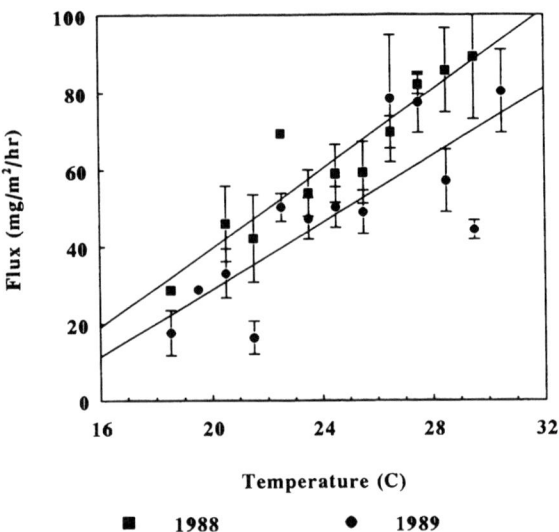

Figure 1. Relationship between flux and soil temperature in a paddy field in Sichuan, China (Khalil et al., 1991).

Rice agriculture may be divided into wetland and dryland (or upland) culture (Grist, 1986; Neue et al., 1990). In wetland culture, the soil is prepared (by puddling) to reduce water loss, and dikes or levees are built to contain the water. Wetland culture may be separated into irrigated and rainfed rice. While rainfed rice is planted during the wet season, all water is provided by rainfall, and dry spells decrease methane fluxes. Dryland rice culture is usually a low-yield, subsistence agriculture, highly susceptible to drought, where no special preparations are made to retain water in the fields. Dryland rice may not be a source of methane at all, as the soil is not saturated long enough for methanogen populations to build up.

Most of the studies in Table 1 were carried out where flood irrigation is used. However, Sass et al. (1992) compared normal flooding to three other irrigation regimes: one that aerated the soil at mid-season; one that drained the field at three periods during the season; the third a delayed flooding scheme. They found that the multiple aeration regime produced the least methane but used the most water, while the late flood lengthened the total time the crop needed to mature, leading to the highest emissions. Chen et al. (1993) used a similar "intermittent" irrigation regime where water was allowed to evaporate until there was no standing water; then the fields were flooded again. This procedure also reduced fluxes from their experimental plot. Field measurements on wetlands and on tropical soils (Harriss et al., 1982; Keller et al., 1986) found that methane flux stops and methane oxidation begins once the soil dries out. Khalil et al. (1990) found low CH_4 emissions from fallow rice fields where the soils were still saturated, but no emissions from paddy soils allowed to dry after harvest.

Fertilizer treatments have also been found to affect methane flux. The addition of rice straw, manure, compost, and other organic fertilizers appears to enhance methane production, probably by providing a food supply for the methanogens (Schütz et al., 1989a, 1990; Yagi and Minami, 1990; Sass et al., 1991a; Cicerone et al., 1992; Chen et al., 1993). Nitrogen fertilizers, especially ammonium sulfate, may inhibit methane production, possibly by changing the soil pH (Schütz et al., 1989a; Chen et al., 1993).

Use of mineral fertilizers in rice fields has been increasing globally (Neue et al., 1990); however, there are still large regions where they are unavailable or not economic. In particular, organic manures are still widely used in Asia and may result in higher fluxes for the largest rice growing areas.

3. Methodology for estimating emissions

3.1 Regional and global extrapolations. Every methane budget includes an estimate of emissions from rice paddies; however, only three studies have included details of regional emissions. Two of these studies were done to provide data for global transport-chemistry models for methane; they concentrated on different variables, and the authors made different types of information available (Aselmann and Crutzen, 1989; Matthews et al., 1991). The third study is by Khalil and Shearer (1992).

Aselmann and Crutzen (1989) provided detailed tables of the percent of the area in 2.5° latitude by 5° longitude boxes. The percentages reflect the area in rainfed and irrigated rice only for Asia (no upland or dryland rice was included) but show total area for Africa, Central and South America. Matthews et al. (1991) published detailed rice crop calendars, indicating the months of possible cultivation of rice by country, each Indian state, and each province of China. Area was allocated by 1° latitude by 1° longitude cells. For our (Khalil and Shearer) estimates we developed an inventory of direct flux measurements (see Table 1) and modified the information from Matthews et al. on growing seasons to estimate global and regional annual emission rates.

Estimates of global emissions of rice agriculture usually concentrate on regions in south and east Asia, where 90% of the area planted to rice is located (United Nations [U.N.], 1991), and where agricultural practices are most likely to favor methane production in rice paddies. Nine of the top ten rice growing countries are in Asia (by 1990 area: India, China, Bangladesh, Indonesia, Thailand, Vietnam, Myanmar, Brazil, Philippines, and Pakistan). Of the four non-Asian countries in the top twenty rice planting countries, three (Brazil, Madagascar, and Nigeria) have 60% to 80% of their total area planted in dryland rice (Grist, 1986); the fourth country (USA), uses different cultivation practices than most other large rice growing countries.

The flux of CH_4 from rice fields is usually calculated by estimating the length of the season of methane emission, the area in rice (irrigated and rainfed), and the seasonal average flux of methane ($mg/m^2/hr$) (see Equation 1). Each of these main variables is discussed separately.

3.2 Season length. The estimate of the season length from the literature may be confused by whether "season" refers to the season of methane emission, the rice growing season (planting or transplanting to harvest), the total growing season (the frost free period), or the total growing period, which in the case of transplanted rice includes the time in the seedling beds. For example, in the study of Sass et al. (1991a) the total growing season is about 245 days, the rice growing season was 140 days from planting to harvest, while methane was emitted during only 85 days (flood irrigated period). Because the number of areas where the methane emission season has actually been measured is quite limited, some other variable must be used to approximate it.

Matthews et al. (1991) used crop calendars to estimate the months of the rice growing season by country; autumn, winter and summer crops by Indian state; and early rice, double/late crop rice, and single (mid-season) crop rice by Chinese province. In Table 2 we compare the growing season and season of methane emission estimated from the literature (see Table 1) with the growing season estimated from Matthews et al. With the exception of Italy, the difference between the growing season estimated using crop calendars and season of methane emission is about 40 to 50 days.

In the Khalil and Shearer estimate, Matthews et al.'s growing seasons were modified as follows: growing seasons of 140 to 170 days were reduced by 30 days; growing seasons of 110 to 140 days were reduced by 20 days; and growing seasons of fewer than 110 days were reduced by 10 days. This adjustment was made to better reflect results of the field measurements for the period of methane emissions.

Table 2. Seasonal factors for methane emissions from rice fields

References	Planting Season Days	CH_4 Emission Period Days	Growing Season days[†]
Cicerone et al. (1983): USA	145	100	152
Holzapfel-Pschorn and Seiler (1986): Italy	147	126	122
Yagi and Minami (1990): Japan	140	115	152
Sass et al. (1991a): USA	140	85	152
Khalil et al. (1991): Sichuan, PRC	120	120	168
Chen et al. (1993): Beijing, PRC	100	87	137

[†] Estimated from Matthews et al. (1991). See text.

3.3 Methane flux factors. Ideally, each rice growing region should have an average methane flux factor associated with it which uniquely reflects the soil, climate and cultivation practices of the area. Practically, estimators must rely on the information that is available and their best judgement to apply measured fluxes from one region to another where data is not available.

Aselmann and Crutzen (1989) calculated fluxes from the work of Holzapfel-Pschorn and Seiler (1986) in Italy, with a base rate of 300 mg/m^2/day (12.5 mg/m^2/hr) for average soil temperatures of 20°C and below, and assuming a linear relationship between flux and temperature up to 1000 mg/m^2/day (42 mg/m^2/hr) for soil temperatures of 30°C. Matthews et al. (1991) used a flux rate of 0.5 g/m^2/day (21 mg/m^2/hr) for all areas.

Khalil and Shearer used averaged fluxes from the studies listed in Table 1 for the major seasonal divisions of Chinese and Indian rice, for other countries where fluxes have been measured, and to estimate fluxes for other rice growing countries. Flux rates were reduced by 40% for areas of rainfed rice, approximately the reduction in flux found by Chen et al. (1993) for their intermittent irrigation regime.

3.4 Area planted to rice. Estimates of rice growing area are usually taken from published agricultural statistics supplied by the United Nations, the International Rice Research Institute, or the agricultural agencies of individual countries (see, for example: U.N. Food and Agriculture Organization [FAO] Yearbook; Bachelet and Neue, 1993; China Agricultural Yearbook). For convenience, agricultural statistics are reported by political or administrative subdivisions. Aselmann and Crutzen (1989) and Matthews et al. (1991) did careful global rice area allocations, described earlier.

Khalil and Shearer calculated the fraction of the total areas in each province in China and state in India by taking the average of the areas reported in Matthews et al. (1991), the China Agriculture Yearbooks (all years), and in Bansil (1984). The total areas for each country were reduced by the percentage of dryland (upland) rice grown. Estimates of the percentages were taken from tables in De Datta (1975), Huke (1982), Morris et al. (1984), and Grist (1986). Where there was more than one estimate per country, an average was taken. If there was no estimate for a particular country but there was reason to suppose a significant percentage of the rice area was in upland rice, as for example in Africa or Central America, an average was taken of the published estimates for neighboring countries, or the average for the continent was used (usually for African nations). Areas of rainfed rice were estimated in the same manner.

4. Regional and global emissions and role of rice agriculture in causing the increase of methane

Recent estimates of the global source range from 50 to 100 Tg (10^{12} grams) per year. Source estimates by country vary greatly with the assumptions made on the importance of different factors affecting the methane flux, and the information on the factors currently available. Analysis of the role of rice paddies in the increase of atmospheric methane suggests that it was an important factor, particularly from the middle of this century to the present. However, the rate of increase has slowed in the last decade. Predicting the rice agriculture source of methane using either the trend of past emissions or the population will probably lead to large over estimates of the future source (Khalil and Rasmussen, 1990; Khalil et al., 1993).

4.1 Global emissions. Three recent estimates of the global source of CH_4 from rice agriculture are shown in Table 3. (Matthews et al. (1991) assumed a global source of 100 Tg.)

The major difference between these estimates is probably in the way a weighted average flux rate is calculated. The Khalil and Shearer estimate takes a seasonally averaged flux, based on the field studies shown in Table 1, and applies it to the seasonal area in wetland rice culture. The Aselmann and Crutzen estimate used temperature adjusted fluxes as described earlier. The IPCC/OECD estimate uses the early and late growing season fluxes measured by Schütz et al. (1990) in Hangzhou, P.R.C., to calculate a range of global emissions, and may therefore be unreliable since only a few data are used.

Table 3. Estimates of global methane emissions from rice agriculture.

	Khalil and Shearer (1992)	Aselmann and Crutzen (1989)	IPCC (1992)
Tg/year	66	92 (53†)	60
Year of Estimate	1990	1985	Not given

† Number in parentheses assumes a constant flux of 13 $mg/m^2/hr$

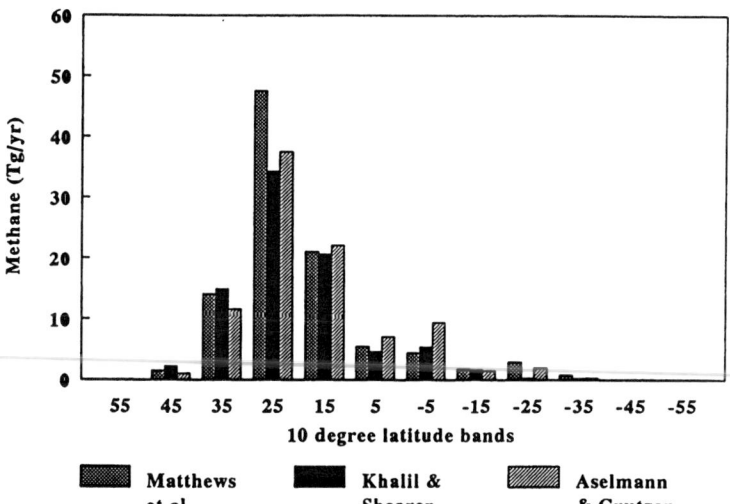

Figure 2. A comparison of latitudinal allocation of methane emissions from rice agriculture for three global studies.

Table 4. Estimates of methane emission by country for 1990 using five sets of assumptions.

Country	Khalil and Shearer	Matthews[†] et al.	Bachelet and Neue[‡]	Bachelet and Neue/Soil[§]
Bangladesh	4.0	6.7	4.1	4.0
China	23.0	21.6	21.3	14.7
India	15.3	27.6	18.5	17.5
Indonesia	6.2	3.7	4.5	3.5
Japan	0.2	1.1	1.0	0.8
Cambodia	0.9	0.9	0.4	0.4
N. Korea	0.3	0.4	0.5	0.4
S. Korea	1.1	0.6	0.8	0.6
Laos	0.5	0.3	0.1	0.1
Malaysia	0.3	0.3	0.3	0.2
Myanmar*	1.1	3.7	0.2	1.4
Nepal	0.2	0.3	0.3	0.2
Pakistan	0.7	1.1	0.8	0.4
Philippines	1.2	1.4	1.2	0.8
Sri Lanka	0.4	0.5	0.4	0.3
Taiwan	0.4	0.5	0.4	0.4
Thailand	4.7	6.9	3.8	2.2
Vietnam	2.6	4.3	2.7	1.8
TOTALS	63	82	61	50

*   Formerly Burma
†   Estimates made by Bachelet and Neue (1993) using the Matthews et al. (1991) database and a constant methane flux of 0.5 g/m^2-day.
‡   Estimates made by Bachelet and Neue using International Rice Research Institute (IRRI) rice yield data, and regional estimates of incorporated organic matter.
§   The Bachelet and Neue estimate modified using the FAO soil map, and authors' best judgement on the methane potential of different soil types.

4.2 Regional emissions. Regional emissions are compared by latitudinal bands, and by countries. The methane source, apportioned to 10 degree latitude bands, is shown in Figure 2. The overall similarity between the estimates is the result of the allocation of rice area; all three studies use the FAO Agricultural Yearbook data. The Khalil and Shearer estimate and Aselmann and Crutzen (1989) differ from Matthews et al. (1991) in the 20° to 30° N latitude band because the former studies reduce the total area by the estimated area in dryland cultivation. The Khalil and Shearer estimate also reduces rice area in Africa and Central and South America by the percent in dryland cultivation.

Estimates of methane emission by country are shown in Table 4, comparing the countries of South and East Asia by 5 estimation techniques. These countries produce about 90% of the estimated CH_4 from rice agriculture. The countries with the largest methane source from wetland rice paddies are all in Asia and are shown in Table 4. The estimates by Khalil and Shearer, OECD/IPCC, and Matthews et al. have been described earlier. Bachelet and Neue (1993) calculated a range of 40 to 80 Tg/yr as the rice source for South and East Asia, making different assumptions about the effects of soil type, rice yield, temperature, and organic matter incorporated in the soil.

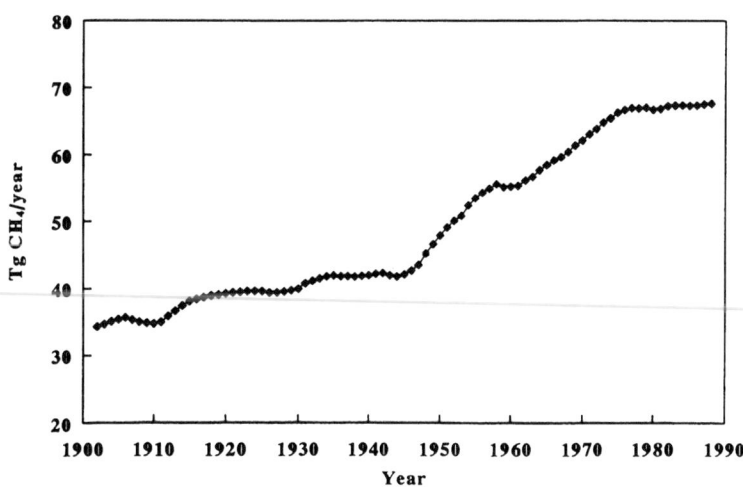

Figure 3a. The time series of global CH_4 emissions from rice agriculture.

5. Trends: The role of rice agriculture in the budget of atmospheric methane.

The increase in rice agriculture was likely one of the main contributors to the increase of methane during the last century. We estimate that rice agriculture contributes some 300 ppbv of methane to the present atmosphere and may be responsible for some 20% of the increase of methane during the last century (Khalil and Rasmussen, 1991). However, factors that were important causes of the increase in the past are changing and probably will not be as important in the future (see Figure 3). The time series of global methane emissions from rice estimated by Khalil and Shearer is largely influenced by the tremendous increase in area planted to rice in the last four decades. By the early 1980s this growth had slowed significantly.

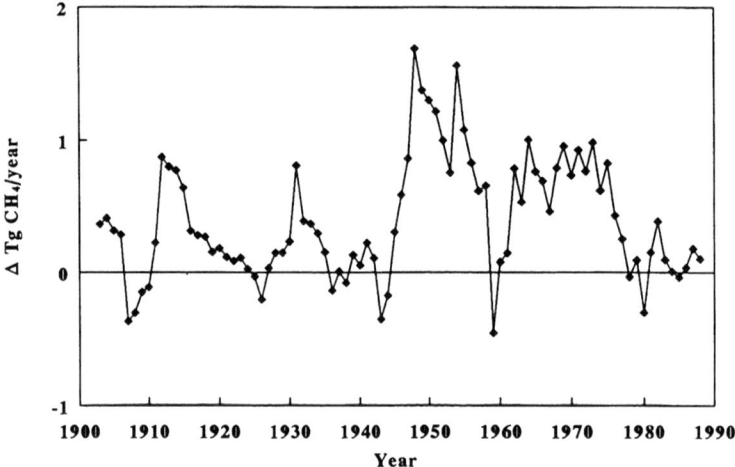

Figure 3b. The difference in global emissions from rice paddies between consecutive years (Khalil and Shearer, 1992).

We constructed time series of harvested rice area for all rice growing countries listed in the 1988 FAO Production Yearbook (U.N., 1989). Historical statistics compiled by Mitchell (1980, 1982 and 1983), and more recent data available from the United Nations (FAO Yearbook, various years between 1956 and 1988) were the primary sources of information. Additional statistics for China came from the China Agriculture Yearbook (various years) and USDA (1984).

Figure 4 shows the time series of per capita rice field methane emissions for China (Khalil et al., 1993). Except for a brief period in the 1950s, per capita emissions have steadily declined, despite the increase in area planted (nearly 10 million hectares). We believe this makes the future trends of population of little use in predicting methane emissions from rice growing countries. The efforts to increase rice yield without increasing the area planted, such as irrigation, fertilizers, and high yield rice cultivars, will affect the future methane emissions from rice paddies.

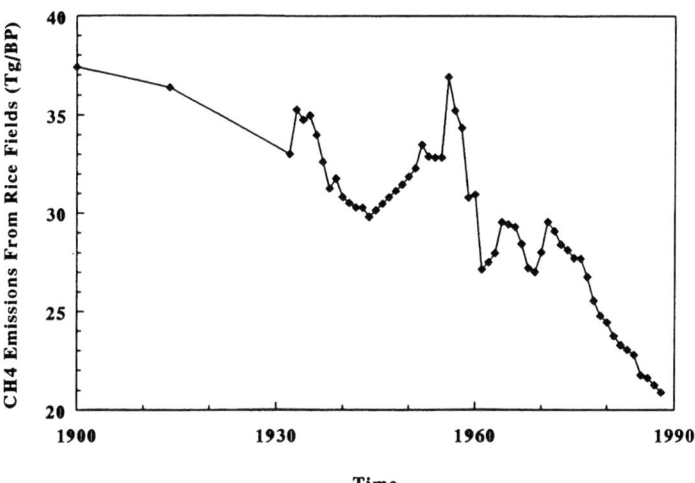

Figure 4. The per capita emissions of methane from rice fields in China during the last century. The per capita emissions have declined by 35% from the 1930s to the 1980s (Khalil et al., 1993).

6. Summary

Global estimates of methane emission from rice agriculture are derived by summing up regional estimates, which are calculated by multiplying flux, area and seasonal factors (Eq. 1). Any part of the equation may introduce uncertainty. There is no way to assess whether a seasonal average flux measurement is correctly associated with area and growing season extrapolants, since seasonal fluxes can vary widely even in nearby areas.

Table 1 lists the areas where methane fluxes from rice agriculture have been measured. The most comprehensive controlled studies have been made in the United States and Europe, which do not necessarily represent the most important rice planting areas. Systematic seasonal measurements have been made in three locations in China but none have yet been published for India, with the largest rice growing area in the world. No seasonal measurements have been made for the tropical rice growing areas of Southeast Asia or for the continents of Africa and South America.

The assumption that dryland rice is not a significant source of methane uses the best information we have available, but it has never been tested. The difficulties in estimating the season of methane emission from the growing season for rice was demonstrated in Table 2. However, except for the limited number of areas where sampling has been done, the season of methane emission must be estimated from growing season. Using the growing season without adjustments will probably overestimate the methane emission season.

Global estimates from three different sources have good general agreement on the latitudinal emissions from rice agriculture. When the regions of interest are individual countries, different assumptions lead to a much larger disagreements.

Rice agriculture is an important component of the methane budget, and in the past century was important in the increase of atmospheric methane. However, in the 1980s, emissions from rice agriculture appear to have nearly stabilized. Future emissions from rice agriculture will depend on efforts to increase production without increasing the area planted, particularly through fertilization, irrigation, and planting high productivity hybrid rice cultivars.

Acknowledgements. We thank Dr. R.A. Rasmussen, R. Dalluge, D. Stearns, F. Moraes, Y. Lu, and R. MacKay for their contributions. Financial support for this project was provided in part by the U.S. Department of Energy (grant number DE-FG06-85ER60313). Additional support was provided by the Andarz Co.

References

Aselmann, I., P.J. Crutzen. 1989. Global distribution of natural freshwater wetlands and rice paddies, their net primary productivity, seasonality and possible wetland emissions. *J. of Atmos. Chem.,* 8:307-358.

Bachelet, D., H.U. Neue. 1993. Methane emissions from wetland rice areas of Asia. *Chemosphere, 26* (1-4):219-238.
Bansil, P.C. 1984. *Agricultural Statistics in India, A Guide.* Third Rev. Edition. Oxford & IBH Publishing Co., New Delhi, India.
Bingemer, H.G., P.J. Crutzen. 1987. The production of methane from solid wastes. *J. Geophys. Res., 92*:2,181-2,187.
Blake, D.R. 1984. Increasing concentrations of atmospheric methane. Ph.D. dissertation, Univ. of California at Irvine.
Bolle, H.-J., W. Seiler, B. Bolin. 1986. Other greenhouse gases and aerosols. In: *The Greenhouse Effect, Climate Change and Ecosystems (SCOPE 29)*, Chapter 4 (J. Wiley and Sons, N.Y., 1986), 157-198.
Chen, Z., L. Debo, K. Shao, B. Wang. 1993. Features of CH_4 emission from rice paddy fields in Beijing and Nanjing, China. *Chemosphere, 26* (1-4):239-246.
China Agriculture Yearbook 1985, 1986, 1987, 1989. Agribookstore, Hampton, VA.
Cicerone, R.J., J.D. Shetter. 1981. Sources of atmospheric methane: measurements in rice paddies and a discussion. *J. Geophys. Res., 86* (C8):7,203-7,209.
Cicerone, R.J., R.S. Oremland. 1988. Biogeochemical aspects of atmospheric methane. *Global Biogeochemical Cycles, 2*:299-327.
Cicerone, R.J., J.D. Shetter, C.C. Delwiche. 1983. Seasonal variation of methane flux from a California rice paddy. *J. Geophys. Res., 88* (C15):11,022-11,024.
Cicerone, R.J., C.C. Delwiche, S.C. Tyler, P.R. Zimmerman. 1992. Methane emissions from California rice paddies with varied treatments. *Global Biogeochemical Cycles, 6(3)*:233-248.
Dai, Aiguo. 1988. CH_4 Emission rates from rice paddies in HangZhou China during fall of 1987. M.S. Thesis, Institute of Atmospheric Physics, Academia Sinica, Beijing, P.R.C.
De Datta, Surajit K. 1975. Upland rice around the world. In: *Major Research in Upland Rice.* International Rice Research Institute, Los Baños, Philippines.
Donahue, T.M. 1979. The atmospheric methane budget. In: *Proceedings of the NATO Advanced Study Institute on Atmospheric Ozone: Its Variation and Human Influences* (A.C. Aikin, ed.), U.S. Department of Transportation, Washington D.C.
Ehhalt, D., U. Schmidt. 1978. Sources and sinks of atmospheric methane. *PAGEOPH, 116*:452-464.
Grist, D.H. 1986. *Rice.* Longman, Inc., New York, U.S.A. 6th Edition.
Harris, R.C., D.I. Sebacher, F.P. Day, Jr. 1982. Methane flux in the Great Dismal Swamp. *Nature, 297*:673-674.
Huke, R.E. 1982. *Rice Area by Type of Culture: South, Southeast, and East Asia.* International Rice Research Institute (IRRI), Los Baños, Philippines.
Holzapfel-Pschorn, A., R. Conrad, W. Seiler. 1985. Production, oxidation and

emission of methane in rice paddies. *FEMS Micro. Ecol., 31*:343-351.

Holzapfel-Pschorn, A., W. Seiler. 1986. Methane emission during a cultivation period from an Italian rice paddy. *J. Geophys. Res., 91* (D11):11,803-11,814.

IPCC. 1990. *Climate Change, The IPCC Scientific Assessment* (J.T. Houghton, B.A. Callander, and S.K. Varney, eds.), Cambridge University Press, G.B.

Keller, M., W.A. Kaplan, S.C. Wofsy. 1986. Emissions of N_2O, CH_4 and CO_2 from tropical forest soils. *J. Geophys. Res., 91* (D11):11,791-11,802.

Khalil, M.A.K., R.A. Rasmussen. 1983. Sources, sinks and seasonal cycles of atmospheric methane. *J. Geophys. Res., 88*:5,131-5,144.

Khalil, M.A.K., R.A. Rasmussen. 1990. Constraints on the global sources of methane and an analysis of recent budgets. *Tellus, 42B*:229-236.

Khalil, M.A.K., R.A. Rasmussen. 1991. The global methane cycle. In: *Proceedings of the Indo-US Workshop on Impact of Global Climatic Changes on Photosynthesis and Plant Productivity*, New Delhi, India, Jan. 8-12, 1991.

Khalil, M.A.K., M.J. Shearer. 1993. Methane emissions from rice agriculture. *Chemosphere* (in press).

Khalil, M.A.K., R.A. Rasmussen, M.-X. Wang, L. Ren. 1990. Emissions of trace gases from Chinese rice fields and biogas generators: CH_4, N_2O, CO, CO_2, chlorocarbons, and hydrocarbons. *Chemosphere, 20* (1-2):207-206.

Khalil, M.A.K., R.A. Rasmussen, M.-X. Wang, L. Ren. 1991. Methane emissions from rice fields in China. *Environ. Sci. Tech., 25*:979-981.

Khalil, M.A.K., M.J. Shearer, R.A. Rasmussen. 1993. Methane sources in China: Historical and current emissions. *Chemosphere, 26* (1-4):127-142.

Matthews, E., I. Fung, J. Lerner. 1991. Methane emission from rice cultivation: geographic and seasonal distribution of cultivated areas and emissions. *Global Biogeochemical Cycles, 5* (1):3-24.

Mitchell, B.R. 1980. *European Historical Statistics*, 2nd Rev. Ed. Facts on File, New York, U.S.A.

Mitchell, B.R. 1982. *International Historical Statistics, Africa and Asia*. New York University Press.

Mitchell, B.R. 1983. *International Historical Statistics, The Americas and Australasia*. Gale Research Co., Detroit, Michigan.

Morris, R.A., D.J. Greenland, R.E. Huke. 1984. Remote sensing and rice research. In: *Applications of Remote Sensing for Rice Production*, pp. 33-54 (A. Deepak and K.R. Rao, eds.), A. Deepak Publishing, Hampton, VA, U.S.A.

Neue, H.-U., P. Becker-Heidmann, H.W. Scharpenseel. 1990. Organic matter dynamics, soil properties, and cultural practices in rice lands and their relationship to methane production. In: *Soils and the Greenhouse Effect* (A.F. Bouwman, ed.), J. Wiley and Sons, N.Y.

Parashar, D.C., J. Rai, P.K. Gupta, N. Singh. 1991. Parameters affecting methane emission from paddy fields. *Indian Journal of Radio & Space Physics, 20*:12-17.

Sass, R.L., F.M. Fisher, P.A. Harcombe, F.T. Turner. 1990. Methane production and emission in a Texas rice field. *Global Biogeochemical Cycles, 4* (1):47-68.
Sass, R.L., F.M. Fisher, P.A. Harcombe, F.T. Turner. 1991a. Mitigation of methane emissions from rice fields: possible adverse effects of incorporated rice straw. *Global Biogeochemical Cycles, 5* (3):275-287.
Sass, R.L., F.M. Fisher, F.T. Turner, M.F. Jund. 1991b. Methane emission from rice fields as influenced by solar radiation, temperature, and straw incorporation. *Global Biogeochemical Cycles, 5* (4):335-350.
Sass, R.L., F.M. Fisher, Y.B. Wang, F.T. Turner, M.F. Jund. 1992. Methane emission from rice fields: the effect of floodwater management. *Global Biogeochemical Cycles 6* (3):249-262.
Schütz, H., A. Holzapfel-Pschorn, R. Conrad, H. Rennenberg, W. Seiler. 1989a. A 3-year continuous record on the influence of daytime, season, and fertilizer treatment on methane emission rates from an Italian rice paddy. *J. Geophys. Res., 94* (D13):16,405-16,416.
Schütz, H., W. Seiler, R. Conrad. 1989b. Processes involved in formation and emission of methane in rice paddies. *Biogeochemistry, 7*:33-53.
Schütz, H., W. Seiler, H. Rennenberg. 1990. Soil and land use related sources and sinks of methane (CH_4) in the context of the global methane budget. In: *Soils and the Greenhouse Effect* (A.F. Bouwman, ed.), J. Wiley and Sons, N.Y.
Seiler, W., A. Holzapfel-Pschorn, R. Conrad, D. Scharffe. 1984. Methane emission from rice paddies. *J. of Atmos. Chem., 1*:241-268.
United Nations. 1977, 1978, 1980, 1982, 1984, 1986, 1989, 1990, 1991. *FAO Production Yearbook, vols. 30, 31, 33, 35, 37, 39, 42, 43, 44.* Food and Agriculture Organization of the United Nations, Rome.
U.S. Dept. of Agriculture. 1984. *Agricultural Statistics of People's Republic of China, 1949-1982.* International Economics Division, Economic Research Service, Statistical Bulletin No. 714. U.S. Govt. Printing Office, Washington, D.C.
Warneck, P. 1988. *Chemistry of the Natural Atmosphere* (Academic Press, N.Y.)
Wassmann, R., H. Papen, H. Rennenberg. 1993. Methane emission from rice paddies and possible mitigation strategies. *Chemosphere, 26* (1-4):201-218.
Winchester, J.W., Song-Miao Fan, Shao-Meng Li. 1988. Methane and nitrogen gases from rice fields of China - possible effects of microbiology, benthic fauna, fertilizer, and agricultural practice. *Water, Air and Soil Pollution, 37*:149-155.
Yagi, K., K. Minami. 1990. Effect of organic matter application on methane emission from some Japanese paddy fields. *Soil Sci. Plant Nutr., 36* (4):599-610.

Chapter 13

Rice Agriculture: Factors Controlling Emissions

H.-U. NEUE[1] AND P.A. ROGER[2]

[1] The International Rice Research Institute, P.O. Box 933, Manila, Philippines
[2] Laboratoire de Microbiologie ORSTOM, Université de Provence, 3 Place Victor Hugo
13331 Marseille Cedex 3, France

1. Introduction

Recent atmospheric measurements indicate that concentrations of greenhouse gases are increasing. Atmospheric methane concentration has increased at about 1% annually to 1.7 ppmV during the last decades (Khalil and Rasmussen, 1987). The resulting effect on global temperature is highly significant because the warming efficiency of methane is up to 30 times that of carbon dioxide (Dickinson and Cicerone, 1986). Data from polar ice cores indicate that tropospheric methane concentrations have increased by a factor of 2-3 over the past 200-300 years (Khalil and Rasmussen, 1989). The increase of methane concentrations in the troposphere correlate closely with global population growth and increased rice production (Figure 1), suggesting a strong link to anthropogenic activities. The total annual global emission of methane is estimated to be 420-620 Tg/yr (Khalil and Rasmussen, 1990), 70-80% of which is of biogenic origin (Bouwman, 1990). Methane emissions from wetland rice agriculture have been estimated up to 170 Tg/yr, which account for approximately 26% of the global anthropogenic methane budget. Flooded ricefields are probably the largest agricultural source of methane, followed by ruminant enteric digestion, biomass burning, and animal wastes (summarized by Bouwman, 1990).

Projected global population levels indicate that the demand for rice will increase by 65% over the next 30 years, from 460 million t/yr today to 760 million t/yr in the year 2020 (IRRI, 1989). The growing demand is most likely to be met by the existing cultivated wetland rice area through intensifying rice production in all rice ecologies, mainly in irrigated and rainfed rice. Coupled with existing rice production technologies, global methane emissions from wetland rice agriculture are likely to increase. Mitigation of methane emissions is needed to stabilize or even lower atmospheric methane concentrations.

This paper discusses principles and prospects of rice cultivation in view of methane formation, methane fluxes, and mitigation options.

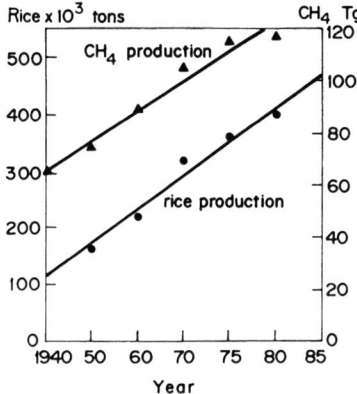

Figure 1. Rice production and methane emission rates.

2. Rice environments

Rice is cultivated under a wider variety of climatic, soil, and hydrological conditions than any other crop. It is grown from the equator to as far as 50°N and 40°S, and from sea level to altitudes of more than 2500 m. The temperature may be as low as 4°C during the seedling stage and as high as 40°C at flowering. Rice is irrigated in arid areas and is grown in rainfed areas with only 500 mm rain/yr. Rice is cultivated as an upland crop and in soils that are submerged more than 1 m. Rice is the only major crop grown on flooded soils.

Rice cultural systems have developed to suit the physical, biological, and socioeconomic conditions of different regions. Because the water regime during the growing season is the most discriminating physical factor, ricelands can be grouped into two main systems: wetlands and uplands. Terms used to differentiate rice cultures are, for example, lowland rice, irrigated rice, rainfed rice, deepwater rice, swamp rice, upland rice, hill rice, dryland rice, and pluvial rice. Many other terms have been evolved in different regions reflecting specific characteristics of and constraints to rice cultivation in these areas. These terms reflect the wide range of agroecologies in which rice is grown and are very reasonable in the context they evolved. But the general use of these terms, although understandable, is often semantically and technically incorrect (Moormann and Van Breemen, 1978). A comprehensive classification of rice ecologies has been outlined by Neue (1989). He defined three major rice ecologies with a total of seven subecologies by hierarchically applying floodwater source and floodwater depth as diagnostic criteria (Table 1). Further differentiation is done by modifiers related to climate, landform, floodwater regime, soil, and cropping system.

Table 1. Classification of rice ecologies.

Floodwater source	Irrigation		Pluvial, phreatic, surface flow or tidal				
Floodwater depth (cm)	1-5	5-25	0-25	25-50	50-100	>100	<0
Rice ecologies	-- Irrigated rice --		-- Rainfed rice --				Upland rice
Subecologies	Shallow	Medium	Shallow	Medium	Deep	Very deep	Upland rice
Land ecosystem			-- Wetland --				Upland

Rice ecologies are major discriminators for the potential of methane production in ricefields because of their distinct floodwater regimes. The potential of upland rice for methane production is not significant since upland rice is never flooded for a significant period of time. Aerobic soils, including upland rice soils, seem to be important sites for deposition and microbial oxidation of atmospheric CH_4 (Seiler and Conrad, 1987; Cicerone and Oremland, 1988). Irrigated rice has the highest potential to produce CH_4 because flooding and, consequently, anoxic conditions are assured and controlled. The potential for methane production in rainfed rice should vary widely in time and space since floodwater regimes are primarily controlled by rainfall within the watershed. Periods of severe droughts or floods during the growing season are characteristic for rainfed rice. Subecologies are determined by floodwater depth, which likely affects methane fluxes. Emission rates and harvested area of each rice ecology determine the global methane emission. Rice areas harvested in different regions of the world are given in Table 2.

Table 2. Distribution of harvested ricelands (million ha) by rice ecologies (FAO, 1988).

Region	Irrigated	Rain-fed	Deep water	Upland	Total area	Yield (t/ha)	Rough rice production (10^6 ton)
East Asia[a]	34.0	2.8	-	-	36.8	5.4	200.0
Southeast Asia[b]	13.9	13.7	3.75	4.65	36.0	2.9	102.5
South Asia[c]	19.4	20.0	7.3	6.7	53.4	2.0	105.5
Near East[d]	1.25	-	-	-	1.25	3.3	4.1
South/Central Am. Caribbean and USA	2.5	0.5	0.4	5.65	9.05	2.9	26.5
Africa	0.9	1.95	-	2.70	5.5	1.8	9.9
USSR	0.66	-	-	-	0.66	4.1	2.7
Europe	0.42	-	-	-	0.42	5.4	2.3
Oceania	0.12	-	-	-	0.12	6.6	0.79
Australia	0.11	-	-	-	0.11	7.1	0.76
World	73.26	38.95	11.45	19.70	143.4	3.2	455.05

[a]China Taiwan Korea DPR Korea RP Japan; [b]Burma Cambodia Indonesia Laos Malaysia Philippines Thailand Vietnam; [c]Bangladesh Bhutan India Nepal Pakistan Sri Lanka; [d]Afghanistan Iran Iraq.

Especially since the 1960s, rice production dramatically increased because of high-yielding rice cultivars, large investments in irrigation schemes, and improved soil, water, and crop management. The developed irrigation schemes and the shorter growth duration of modern cultivars increased the harvested area by allowing 2 to 3 crops per year. However, expansion of residential and industrial areas as well as diversification of crops resulted in only a slight increase in the total harvested area of rice (Figure 2). Though many factors determine the relative contribution of each rice ecology to rice supplies in the future, irrigated areas will continue to dominate rice production. At present, about 50% of the harvested area is in irrigated rice but it contributes about 70% of total production.

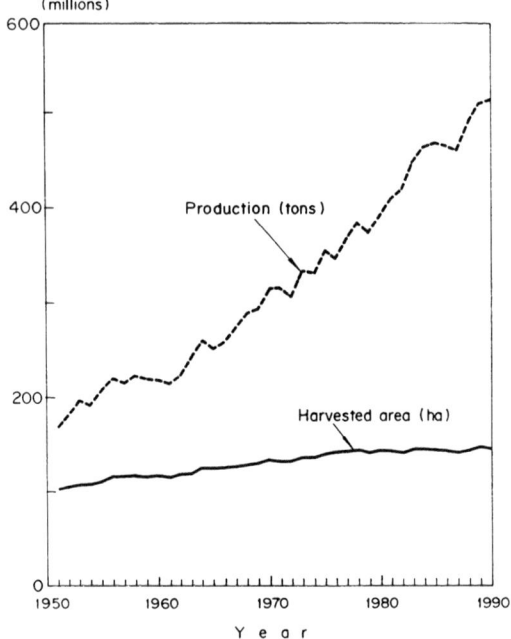

Figure 2. Rough rice production and harvested area (IRRI, 1991)

3. Microbiology of methane emission

3.1 Methanogens. Biogenic methane production is exclusively accomplished by methanogenic bacteria that can metabolize only in strict absence of oxygen and at redox potentials of less than −200 mV. Oxygen causes an irreversible disassociation of the F_{420}-hydrogenase enzyme complex probably due to the lack of protective superoxide dismutase (Schönheit et al., 1981). Methanogens are found in strictly anaerobic environments of freshwater, brackish and marine sediments, hot springs, mid-ocean ridges, decomposing algal mats, heart wood of living trees, intestinal tracts of man and animals (especially the rumen of herbivores), and sewage digesters. In ricefields, methanogenesis occurs in the reduced soil of wetland rice and, possibly, in anoxic water of deepwater rice.

Recent reviews on methanogenic bacteria deal with their biogeochemistry (Oremland and Capone, 1988; Boone, this volume), taxonomy, and ecology (Garcia, 1990).

Table 3. Simplified classification of methanogenic bacteria (adapted from Garcia 1990) and their habitat (Garcia, ORSTOM, personal communication).

Methanobacteriales	
Methanobacteriaceae	
Methanobacterium	Various freshwater habitats. Half of the species are thermophilic; few are alkaliphilic.
Methanobrevibacter	Specialized habitats such as trees (Zeikus and Henning, 1975), rumen (Smith and Hungate, 1958), sewage sludge, intestinal tracts of animals (Miller and Wolin, 1985).
Methanosphaera	Feces or digestive tracts of animals (Biavati et al., 1988).
Methanothermaceae	
Methanothermus	Extreme thermophile from volcanic springs
Methanococcales	
Methanococcaceae	
Methanococcus	Isolated mostly from marine or coastal environments
Methanomicrobiales	
Methanomicrobiaceae	
Methanolacinia	Marine sediments.
Methanospirillum	Mesophilic strains from various habitats.
Methanoculleus	
Methanocorpusculum	Sewage sludge lacustrine sediments (Zhao et al., 1989).
Methanomicrobium	
Methanogenium	
Methanoplanaceae	Symbiont of marine ciliate.
Methanoplanus	
Methanosarcinaceae	Freshwater and marine sediments, ricefields, lagoons, anaerobic sewage-sludge digestors, and rumen (Raimbault, 1981).
Methanosarcina	

Genera not ascribed to a family (mostly isolated from salty biotopes):
Methanolobus Methanococcoides Methanohalophilus Halomethanococcus Methanohalobium Methanothrix and *Methanopyrus*

Twenty genera of methane-producing bacteria have been described (Table 3) but only a few, including *Methanobacterium* and *Methanosarcina*, have been isolated from rice soils (Rajagopal et al., 1988). *Methanospirillum* and *Methanocorpusculum*, which were isolated from freshwater sediments, as well as methanogens found as endosymbionts in sapropelic amoeba should also be present in wetland ricefields (Garcia, ORSTOM, personal communication).

The distribution of methanogens in natural environments depends on their adaptation to temperature, pH, and salinity ranges. Most methanogens are mesophilic with temperature optima of 30-40°C. Thermophilic (40-70°C) species account for 20% of the strains (Garcia, 1990) and some extreme thermophilic (up to 97°C) are also known. Most methanogens are neutrophilic with a relatively narrow pH range of 6-8. A few alkaliphilic isolates with optimum growth at pH 8-9 have been reported in the genera *Methanosarcina*, *Methanobacterium* (Blotevogel et al., 1985; Worakit et al., 1986), and *Methanohalophilus* (Mathrani et al., 1988). No acidophilic strains have been reported. A strain isolated from peat tolerated a pH of 3, but its optimum was 6-7 (Williams and Crawford, 1984, 1985).

Methanogens can only metabolize a limited number of simple carbon compounds and hydrogen availability is a key factor for methanogenesis. As summarized by Garcia (1990):

Hydrogenotrophic methanogens (77% of the 68 described species) oxidize H_2 and reduce CO_2 to form methane. According to Conrad et al. (1985), H_2-dependent methanogenesis in sediments results mostly from H_2 transfer between microbial associations within flocks or consortia.

Methylotrophic methanogens (28% of the species) can use methyl compounds as methanol, methylamines, or dimethylsulfide; 10 species have been identified as obligate methylotrophs.

Acetotrophic methanogens (14% of the species) utilize acetate. The growth of virtually all methanogens is stimulated by acetate and its importance as a methane precursor in sediments has been documented (Cappenberg, 1974;, Cappenberg and Prins, 1974; Winfrey and Zeikus, 1977). Sixty percent of the hydrogenotrophic species also use formate. A few species use H_2 to reduce methanol to methane (hydrogeno-methylotrophic methanogens); others form methane in the presence of CO_2 and alcohols as hydrogen donors (alcoholotrophic methanogens). The importance of methanol and methylated amines as methane precursor in sediments varies with the abundance of decomposing plant materials such as algal mats (King, 1988). Methanol and methylated amines might be abundant in wetland ricefields after fertilizer application has induced the formation of large algal mats.

Methanogenesis in sediments is characterized by a complete degradation of organic matter while in the rumen of ruminants and the intestine of most animals, mineralization is incomplete since intermediate products are absorbed as food. The anaerobic degradation of organic matter to methane in sediments requires the cooperation of several types of bacteria within a substrate chain to provide the simple carbon compounds needed by methanogens. According to Conrad (1989), four types of bacteria are needed: a) hydrolytic and fermenting bacteria, b) H^+-reducing bacteria, c) homoacetogenic bacteria, and d) methanogenic bacteria. The first group hydrolyzes polymers and ferments the resulting monomers to smaller molecules such as alcohols, short chain fatty acids, H_2, and CO_2. Methanogens can immediately convert H_2/CO_2, formate, acetate, and a few other simple compounds including methanol, methylamines, and dimethylsulfide to CH_4 and CO_2. Fermentation products such as fatty acids, alcohols, aromates, and others cannot directly be utilized. They are oxidized by obligate H^+-producing bacteria to acetate and CO_2. Homoacetogens are very versatile bacteria that can use sugars, alcohols, fatty acids, purines, and aromatic compounds as well as methanol, formate, H_2, and CO_2 to produce acetate as the sole fermentation product (Dolfing, 1988).

All methanogens use NH_4^+ as a nitrogen source, and a few species are known to fix molecular nitrogen (Belay et al., 1984; Murray and Zinder, 1984).

3.2 Inhibitors of methane formation. Mineral terminal electron acceptors like nitrate or sulfate inhibit methanogenesis in sediments by channeling the electron flow to thermodynamically more efficient bacteria like denitrifiers or sulfate reducers (Balderston and Payne, 1976; Ward and Winfrey, 1985). Manganese and iron oxides should have the same effect. Methanogens, sulfate-reducers and homoacetogenic bacteria compete for H_2 produced by fermentative bacteria. Hydrogenotrophic homoacetogens do not significantly compete with methanogens for H_2 in sediments (Lovley and Klug, 1983). Since H_2 concentration is usually very low in such environments (Strayer and Tiedje, 1978), sulfate reducers are able to out-compete hydrogenotrophic methanogens in the presence of sulfate because of their higher affinity for H_2 and faster growth

(Winfrey and Zeikus, 1977; Abram and Nedwell, 1978).

NaCl inhibits pure cultures of methanogens, though high concentrations (about 0.2 M) are required for several strains (Patel and Roth, 1977). In general, adding NaCl to a nonsaline soil inhibits methanogenesis (Koyama et al., 1970). Methanogenesis is inhibited by brackish water (Garcia et al., 1974, De Laune et al., 1983, Holzapfel-Pschorn et al., 1985, Bartlett et al., 1987). Inhibitory effects and interactions with sulfate-reducing bacteria are given as possible reasons (Mitsch and Gosselink, 1986). Competition for H_2 and toxicity of sulfide are the likely mechanisms. However, methanogenesis and sulfate reduction are not mutually exclusive when methane is produced from methanol or methylated amines for which sulfate reducers show little affinity (Oremland et al., 1982; Oremland and Polcin, 1982; Kiene and Visscher, 1987). Methanol is formed during anaerobic decomposition of plant pectins (Schink and Zeikus, 1980). In saline environments, degradation of osmoregulatory compounds such as glycinebetaine produces methylamines (King, 1984). Obligately methylotrophic methanogens constitute about half of the methanogenic population present in salt marsh sediments (Franklin et al., 1988), and methylotrophs are found in rice soils (Rajagopal et al., 1988).

Chemical substances inhibiting methanogenesis have been reviewed by Oremland and Capone (1988). The 2-bromoethane-sulfonic acid (BES), an analog to Coenzyme M, is a specific inhibitor of methanogenesis. Several chlorinated CH_4 analogues such as chloroform and methyl chloride have been identified to inhibit methanogenesis. Chloroform completely suppressed methane production in a paddy soil but did not hamper the turnover of glucose (Krumböck and Conrad, 1991) although evidence for glucose-utilizing H_2-syntrophic methanogenic bacterial associations has been found in glucose amended paddy soil (Conrad et al., 1989). Substances encountered in ricefields that inhibit methanogenesis and methane oxidation include DDT (McBride and Wolfe, 1971), acetylene (Raimbault, 1975), and nitrapyrin, an inhibitor of nitrification (Salvas and Taylor, 1980). Slow release of acetylene from calcium carbide, encapsulated in fertilizer granuals highly reduced methane emission (Bronson and Mosier, 1991).

3.3 Methane-oxidizing bacteria. Methane oxidation may greatly limit the flux of CH_4 to the atmosphere (Bont et al., 1978). Holzapfel-Pschorn et al. (1986) reported that 67% of the CH_4 produced during a rice growing season in an Italian ricefield was oxidized. Sass et al. (1991) found that 58% was oxidized in a Texas ricefield. Schütz et al. (1989) reported that up to 90% of CH_4 generated at the late growth stage was oxidized.

Methane can be oxidized by aerobic and anaerobic bacteria. Several reviews have been published on methane-oxidizing bacteria and aerobic methane oxidation (Whittenbury et al., 1970 a,b; Higgins et al., 1981; Anthony, 1982; Crawford and Hanson, 1984). Aerobic methane-oxidizing (methanotrophic) bacteria constitute a group of eubacteria that grow only on methane or carbon compounds lacking carbon-carbon bonds such as methanol, formate, and methylated amines. One species (*Methylobacterium organophilum*) can also grow on more complex organic compounds in combination with methane (Patt et al., 1974). All aerobic methanotrophs sequentially oxidize CH_4 to CO_2 via methanol, formaldehyde, and formate. Oxygen is essential for the growth of methane-oxidizing bacteria, but the required partial pressure may be low (Cicerone and Oremland, 1988), especially when methanotrophs fix nitrogen (Murrell and Dalton, 1983) or grow with nitrate as a nitrogen source (Toukdarian and Lidstrom, 1984). Aerobic methanotrophs, which require both methane and oxygen, are most active in ricefields at the interface of aerobic and anaerobic environments (floodwater-soil interface, rice rhizosphere).

Anaerobic oxidation of methane is poorly understood but appears to be an important methane sink in sulfate-containing environments, such as marine sediments or anoxic water (Alperin and Reeburgh, 1984; Iversen et al., 1987). The process has also been reported to occur in freshwater systems (Panganiban et al., 1979).

4. Flooded rice soils as site for methane emission

In general, flooded rice soils provide an optimum environment for methane production and emission, especially in the tropics. Flooding a soil causes the essential low redox potential and anaerobic decomposition of organic matter and stabilizes the soil pH near neutral. In the tropics, the temperature of the reduced puddled layer becomes optimal for methanogenesis. Rice plants highly enhance the emission of methane. Variations in methane fluxes from rice paddies are caused by variations in soil properties, crop management, and related growth of rice.

4.1 Rice soils. Neue (1989) characterized a typical soil profile of a flooded rice soil during the middle of a growing season as follows:

Horizon   Description

Ofw A layer of standing water that becomes the habitat of bacteria, phytoplankton, macrophytes (submerged and floating weeds), zooplankton, and aquatic invertebrates and vertebrates. The chemical status of the floodwater depends on the water source, soil, nature, and biomass of aquatic fauna and flora, cultural practices, and rice growth. The pH of the standing water is determined by the alkalinity of the water source, soil pH, algal activity, and fertilization. Because of the growth of algae and aquatic weeds, the pH and oxygen content undergo marked diel fluctuations. During daytime, the pH may increase up to 11 and the standing water becomes oversaturated with O_2 due to photosynthesis of the aquatic biomass. Standing water stabilizes the soil water regime, moderates the soil temperature regime, prevents soil erosion, and enhances C and N supply.

Apox The floodwater-soil interface that receives sufficient O_2 from the floodwater to maintain a pE + pH above the range where NH_4^+ becomes the most stable form of N. The thickness of the layer may range from several mm to several cm depending on pedoturbation by soil fauna and the percolation rate of water.

Apg The reduced puddled layer is characterized by the absence of free O_2 in the soil solution and a pE + pH low enough to reduce iron oxides.

Apx This layer has increased bulk density, high mechanical strength, and low permeability. It is frequently referred to as plow pan or traffic pan.

B The characteristics of the B horizon depend highly on water regime. In epiaquic moisture regimes the horizon generally remains oxidized, and mottling occurs along cracks and in wide pores. In aquic moisture regimes, the whole horizon or at least the interior of soil peds remain reduced during most years.

The chemistry and biology of rice soils have frequently been reviewed (Ponnamperuma, 1972, 1981, 1984a, 1985; Patrick and Reddy, 1978; De Datta, 1981; Watanabe and Roger, 1985; Yu, 1985; Patrick et al., 1985; Roger et al., 1987; Neue, 1988).

The duration and pattern of flooding and saturation are important criteria for methane formation. Saturation can be caused by groundwater (aquic moisture regime) or surface water (epiaquic moisture regime). Flooding an air-dried cultivated soil drastically changes the hydrosphere, atmosphere, and biosphere of that soil. Flooding highly limits diffusion of air into the soil. The O_2 supply cannot meet the demand of aerobic organisms, and facultative and anaerobic organisms proliferate using oxidized soil substrates as electron acceptors in their respiration. Consequently, the redox potential falls sharply according to a sequence predicted by thermodynamics and CO_2 and HCO_3^- concentrations increase to very high levels. As a result, the soil pH of acid soils increases while that of sodic and calcareous soils decreases, stabilizing between 6.5 and 7.2. Flooding and puddling render most soils an ideal growth medium for rice by supplying abundant water, buffering soil pH near neutral, enhancing N_2 fixation, and increasing diffusion rates, mass flow, and availability of most nutrients. In less favorable soils, flooding may result in toxicities of Fe, H_2S, or organic acids, or deficiencies of Zn or S.

The anaerobic fermentation produces an array of organic substances, many of them transitory and not found in aerobic soils. The major gaseous end products are CO_2, H_2S, and CH_4. The description of the paddy soil profile clearly indicates that methane formation mainly takes place in the reduced Apg horizon. In aquic moisture regimes, the B horizon may also become a source of methane. But in general carbon contents of B horizons are low and their organic matter is less degradable. In epiaquic moisture regimes, methane oxidation may predominate in the B horizon. The same holds true for the Apox layer. Harrison and Aiyer (1913) established early on that all methane diffusing into the aerobic surface layer is oxidized. This was reconfirmed by Bont et al. (1978). They found that 10 ml of a suspension of rice soil oxidized 2 ml of methane within 24 hours when incubated aerobically. Methane may also be oxidized in shallow floodwater since it is often oversaturated with O_2 due to assimilation of the aquatic flora.

In deepwater ricefields, the deeper layers of the floodwater may also become anoxic during the crop cycle (Whitton and Rother, 1988), permitting methanogenesis from the large quantity of organic material available from rice culms, nodal roots, and dead aquatic biomass.

4.2 Temperature regimes of rice soils. Rice is grown under widely differing temperature regimes. The temperature of flooded soils at planting may range from 15°C in northern latitudes to 40°C in equatorial wetlands. Rice physiologists have studied extensively the effects of air and water temperature on rice growth characteristics (Matsushima et al., 1964 a,b; Yoshida, 1981), but there is only little information on the temperature regimes of flooded rice soils and their effects on the chemistry of the soils (Kondo, 1952; Cho and Ponnamperuma, 1971; Gupta, 1974; Sharma and De Datta, 1985). Seasonal and diel temperature changes likely influence methane formation and emission. Holzapfel-Pschorn and Seiler (1986) reported a marked influence of soil temperature on the methane flux with doubling of emission rates when temperature increased from 20 to 25°C. Diel variation of methane emission is correlated with temperature fluctuation (Schütz et al., 1989).

Most isolates of methanogenic bacteria are mesophilic with temperature optima of 30 to 40°C (Acharya, 1935; Vogels et al., 1988). Psychrophilic acetate-utilizing methanogens with a temperature optimum below 20°C seem to occur in acidic peat, which generally shows substantial rates of methane production (Svensson, 1984). The temperature optimum for the production of methanogenic substrates by fermenting bacteria may not concur with the optimum for methanogenesis. In subtropical regions or at high altitudes, the accumulation of intermediate metabolites may reach toxic levels, especially early in the rice-growing season, because of low temperatures. Specific drainage techniques with increased percolation rates and/or intermittent aeration periods are practiced to remedy such accumulations. In tropical lowlands, high temperature throughout the growing seasons stimulates degradation and methane production.

In flooded conditions, soil temperature varies in response to the meteorological regime acting upon the atmosphere-floodwater and floodwater-soil interfaces. The changing properties of soil and floodwater (i.e., temporal changes in reflectivity, heat capacity, thermal conductivity, incoming water temperature, and water flow) as well as vegetation interact with these external influences. Hackman (1979) reported that floodwater temperatures are above minimum air temperature but below maximum air temperature if daily amplitudes of air temperature are high, while water temperatures are above maximum air temperatures if daily fluctuations are low. Neue (1988) reported that floodwater temperature in Philippine ricefields always exceeded ambient air temperature and showed lower daily fluctuations. The temperature of the puddled layer closely followed the temperature of the floodwater and decreased with depth. The annual mean soil temperature at 2:00 p.m. was 33°C at 7 cm depth, its daily maximum equaled or exceeded the maximum air temperature on most days.

Floodwater transmits short-wave radiation to the soil while reducing the upward escape of emitted long-wave radiation. Thus, a "greenhouse effect" is produced, heating floodwater and soil. Diel temperature amplitudes of the floodwater are highly moderated because of the high heat capacity of water and because evaporation of water consumes energy from the floodwater but not directly from the soil. The high thermal conductivity of flooded and puddled soils, in which the bulk densities may be reduced to only 0.2-0.5 g/cm^3, enhances the downward conduction to the dense layer. Dissolved and suspended particles and aquatic biomass in the floodwater change the absorption of radiation, and depth of floodwater changes the heat capacity. The temperature of both floodwater and soils may rise above 40°C in unplanted soils with muddy floodwater of shallow depth. Floodwater temperature is lowered by canopy shading, flow of water, and through rainfall. In ricefields where floodwater has been drained for transplanting or seeding, soil temperatures may reach 50°C in the top centimeter because of increased heat absorption and reduced heat capacity, thermal conductivity, evaporation, and ventilation.

Aselmann and Crutzen (1990) computed monthly distributions of global methane emissions from linearly temperature-dependent methane fluxes in the range from 300 to 1000 mg/m^2 per day for temperatures from 20 to 30°C and constant emission of 300 mg/m^2 per day for temperatures below 20°C. Emissions of methane in the northern hemisphere reveal low monthly values (1.5 - 3 Tg) in December to April and a bell-shaped distribution between May and November with a clear peak of about 16 Tg in August. The southern hemisphere reveals highest emission rates (up to 2 Tg) in the months of February and March. The largest sources where computed between 20 and 30°N (South China, North India, Pakistan, Bangladesh, North Myanmar) with 37.6 Tg/yr, followed by 10 to 20°N (South India, South Myanmar, Thailand, Cambodia, Laos, Vietnam, North and Central Philippines, Brunei, Kalimantan) with 22 Tg/yr, 30 to 40°N (Central China, Japan, Korea) with 8 Tg/yr, 0 to 10°S with 6.5 Tg/yr (most of Indonesia) and 0-10°N (Sri Lanka, Malaysia, South Philippines, Brunei, Kalimantan) with 5 Tg/yr. Though rice ecologies have not been discriminated explicitly, the computed distribution of emission rates clearly reflects the importance of irrigated rice ecologies since double- and triple-cropped areas account for a major part of irrigated rice.

4.3 Organic matter accumulation and decomposition. Readily mineralizable organic matter derived from primary production or organic amendments are the main source for methane formation in wetland rice soils. The net primary production of wetland rice soils (Table 4) can be deduced from yield statistics and estimates of aquatic biomass and weed growth during fallow periods.

In 1988, worldwide wetland rice production was 477 million tons (t), of which upland rice contributed about 28 million t. Based on a shoot/grain ratio of 3/2 (Ponnamperuma, 1984b) and a root/shoot ratio of 0.17 (Yoshida, 1981; Watanabe and Roger, 1985), the total dry matter production of wetland rice amounts to 1123 million t. Adding 74 million t dry matter of aquatic biomass [600 kg/ha season (Roger and Watanabe, 1984; Watanabe and Roger, 1985)] and 200 million t of weed dry matter [2 t/ha during fallow periods (Buresh and De Datta, 1991)] amounts to a total dry matter production of 1512 million t or 1500 g/m^2 per yr.

Table 4. Annual net primary production in wetland ricefields.

Source	Dry Weight	Returned to soil	
		%	(million t/yr)
Wetland rough rice	449	--	--
Wetland rice straw[a]	674	15	101
Wetland rice roots[b]	115	100	115
Aquatic biomass (algaeweeds)	74	100	74
Fallow weeds	200	50	100
TOTAL	1512	26	390

[a] Shoot:grain ratio = 1.50; [b] Root:shoot ratio = 0.17

It is assumed that, on an average, 15% of the straw, 50% of the weeds, and all roots and aquatic biomass amounting to 390 million t dry matter or 156 million t carbon are returned to the soil. If a maximum of 30% of the returned carbon is transformed to methane as found by Neue (1985) in studies with ^{14}C-labeled straw in soils prone to methane formation, 60 Tg of methane would be globally produced in wetland ricelands annually. The input of degradable organic carbon is likely higher due to organic amendments. Reliable data on amounts of organic manures added are lacking. Based on long-term yield trials in the Philippines, a relation among soil C content, N fertilizer rates, and rice grain yields was established. The optimum C content in puddled and flooded soils was found to be 2 - 2.5%, corresponding to 0.20 - 0.25% total nitrogen (Neue, 1985; Smith et al., 1987). Since almost 90% of the tropical soils studied by Kawaguchi and Kyuma (1977) had less than the optimum total nitrogen content, moderate organic amendments seem to be essential to sustain or increase soil fertility and rice yields. In some instances, the returned net primary production of organic matter seems to be sufficient.

The rate and pattern of organic matter addition and decomposition control the rate and pattern of methane formation. Anaerobic fermentation produces an array of organic substances, many of them transitory and not found in well-aerated soils. Ponnamperuma (1984a) listed various gases, hydrocarbons, alcohols, carbonyls, volatile fatty acids, nonvolatile fatty acids, phenolic acids, and volatile S compounds. Methanogens constitute the last step in the electron transfer chain generated by the anaerobic degradation of organic matter. Submergence of soils retards initial decomposition of rice straw in the field only slightly compared with upland soils (Neue and Scharpenseel, 1987; Neue, 1988). The rate of decomposition decreases with soil depth (Neue, 1985). Decomposition of the remaining, more resistant metabolites and residues is similar, with half-lives of about 2 years in all soils and water regimes if the following conditions for flooded soils are met:

- soil is intensively puddled each cropping season;
- soil temperature of the puddled layer is 30-35°C;
- neutral pH;
- low soil bulk density and wide soil/water ratio;
- shallow floodwater;
- high and balanced nutrient supply;
- no long-lasting accumulation of organic acids;
- permanent supply of energy-rich photosynthetic aquatic and benthic biomass;

- high diversity of micro- and macroorganisms that provide successive fermentation down to CO_2, CH_4, H_2, and NH_3;
- supply of O_2 into the reduced layer by rice root excretion and oligochaete population; and
- diel oversaturation of the floodwater with O_2 due to photosynthetic aquatic biomass enhancing the aeration function of oligochaetes.

Decomposition and, accordingly, methane production are retarded in wetland rice soils with low and imbalanced nutrient supply, high bulk density, and low biological diversity and activity, as demonstrated in the Aeric Paleaquult of Northeast Thailand (Snitwongse et al., 1988). If the biological activity is restricted to bacteria, as in laboratory experiments, the decomposition of rice straw in flooded soils is highly retarded (Capistrano, 1988). Only 7-18% of the incorporated straw was decomposed after 100 days following the order San Manuel clay loam (pH 6.6) > Maahas clay loam (pH 5.5) > Louisiana clay (pH 4.9). These results clearly demonstrate the high limitations of laboratory incubation studies.

In calcareous and alkaline soils, methane production may occur within hours after flooding an air-dried soil, while in acid soils it may take weeks before methane is formed (Figure 3). In very acid soils, methane may not be formed at any time. The formation of methane is preceded by the production of volatile acids. Short-term H_2 evolution immediately follows the disappearance of O_2 after flooding. Thereafter, CO_2 production increases, and finally, with decreasing CO_2, CH_4 formation increases (Takai et al., 1956; Neue and Scharpenseel, 1984). The addition of organic substrates enhances the fermentation process. With increasing temperature up to 35°C, decomposition starts earlier and is more vigorous in every case. At high temperatures, the formation of CO_2 and CH_4 occurs earlier and is stronger (Yamane and Sato, 1961). The period of occurrence and the amount of the gaseous products and volatile acids depend largely on temperature and reducing conditions.

The ratio of CO_2 to CH_4 formation is controlled by the fermentation chain and the ratio of the oxidizing capacity (amount of reducible O_2, NO_3^-, $Mn(IV)$, and $Fe(III)$) to the reducing capacity (Takai, 1961). The actual capacity is highly influenced by O_2 diffusion from the atmosphere, floodwater, and plant roots; the soil bulk density (soil-water ratio); and fertilization. Less CH_4 and higher accumulation of volatile acids are found in soils with higher bulk density (lower soil:water ratio). Digging tubificidae (earthworms) in the top soil decrease methane formation but increase methane emission by enhancing fluxes of gases.

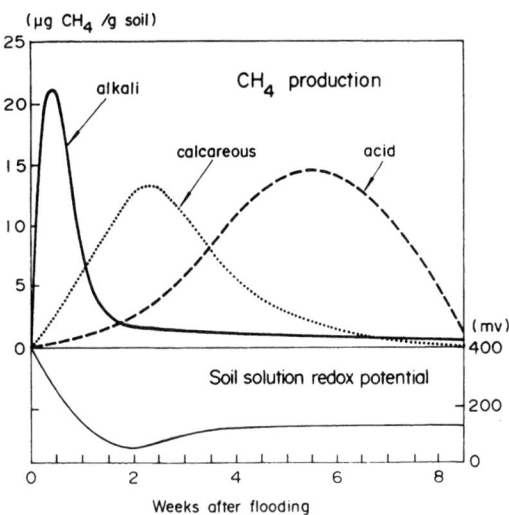

Figure 3. Methane formation in alkaline, calcareous, and acid soils.

Consecutive addition of organic substrates through plant growth and photosynthetic biomass production in the floodwater maintain the fermentation chain. The low specific activity of CH_4 produced after adding ^{14}C-labeled rice straw in field experiments (IRRI, 1981) was caused by degradation of newly produced photosynthetic biomass and root exudates. In permanently flooded soils, methane is only produced in significant amounts after soil-borne production or addition of readily mineralizable organic substrates. Humification of organic matter in wetland rice soils is less than that in aerobic soils. Humus of seasonal flooded soils has lower H_2 and N contents, its degree of unsaturation and its content of carboxyl and phenolic groups is lower, but its alcoholic and methoxyl groups are higher (Kuwatsuka et al., 1978; Tsutsuki and Kuwatsuka, 1978; Tsutsuki and Kumada, 1980). Humification indices in flooded soils, given as the ratio of nonhumified and humified materials (Sequi et al., 1986), are high (low humification) in topsoils and decrease with depth. Very acid rice soils have more humified materials.

Submergence is often equated with retarded decomposition and accumulation of organic matter. But wetland rice soils in the tropics fall into wet soils with high temperature in all seasons that show rapid mineralization and weak humification (Bonneau, 1982; Neue and Scharpenseel, 1987), both of which favor methane formation.

4.4 Redox potential. The supply of biodegradable carbon and the activity of the edaphon are the key to most of the characteristic biochemical and chemical processes in flooded soils (Neue, 1988). These processes include soil reduction and associated electrochemical changes; N immobilization and fixation; production of an array of organic compounds, especially organic acids; and release of NH_4^+, CO_2, H_2S, and CH_4. Since methane is produced only by strictly anaerobic bacteria (methanogens), a sufficient low redox potential is required.

The magnitude of reduction is determined by the amount of easily degradable organic substrates; their rate of decomposition; and the amounts and kinds of reducible nitrates, iron and manganese oxides, sulfates, and organic compounds. A rapid initial decrease of Eh after flooding in most soils is caused by high decomposition rates of organic substrates and a low buffer of nitrates and Mn oxides. The most important redox buffer systems in rice soils are Fe(III) oxyhydroxides/Fe(II) and organic compounds stabilizing the Eh somewhat between +100 and −100 mV in most soil solutions. Measurements in the bulk soil may reveal Eh values as low as −300 mV because of direct contact with reduced surfaces of soil particles. The most important interacting chemical changes after flooding an air-dried acid soil are shown in Figure 4a,b.

Although the reduction of flooded soils proceeds stepwise in a thermodynamic sequence (Ponnamperuma, 1972; Patrick and Reddy, 1978), the given oxidation-reduction systems are only partially applicable to field conditions. The mineral phases present in soils are not pure and often unknown, and a large portion of reduced Fe^{2+} and Mn^{2+} ions are held at ion exchange sites (Tsuchiya et al., 1986). Changes in pH and activities of reactants and resultants can also alter the order of redox reactions. As a consequence, reduction potentials of a given redox reaction span a fairly wide range not only because of variations at microsites. Nevertheless, redox potentials of the bulk soil (corrected to pH 7) of at least −50 mV are needed for the formation of CH_4.

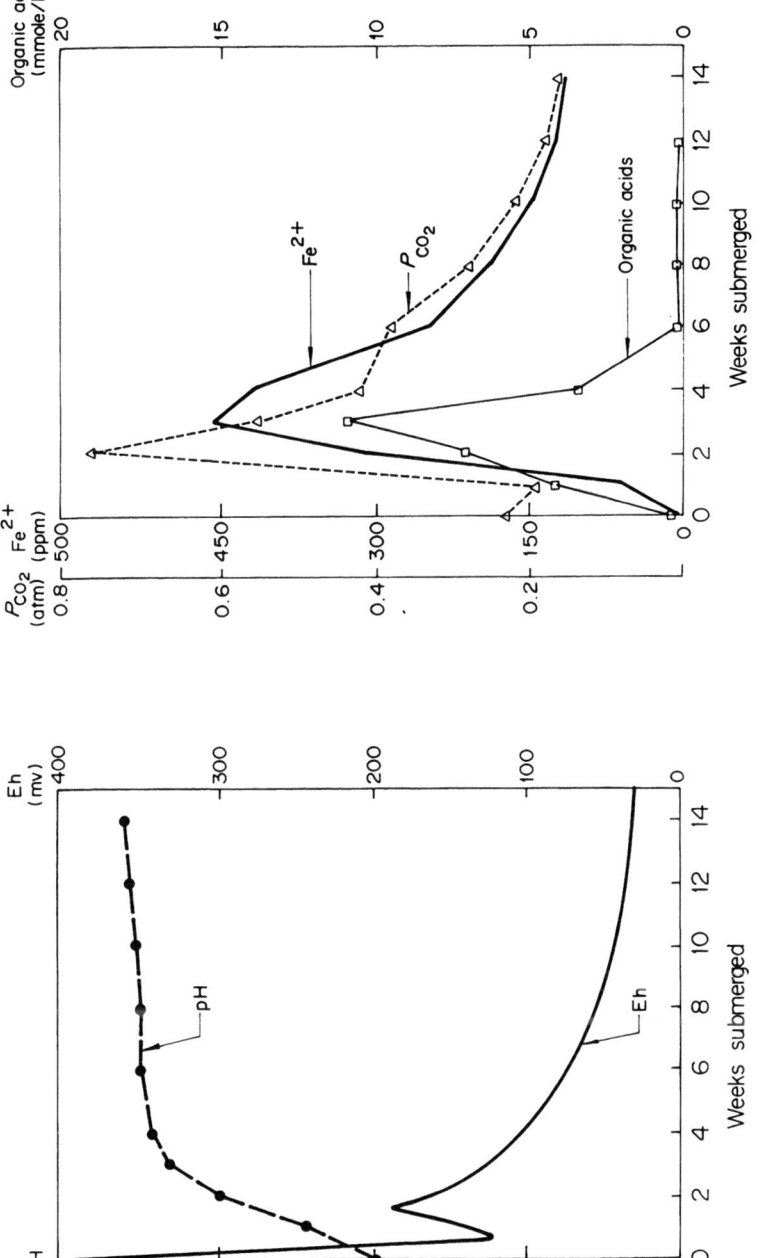

Figure 4a. Kinetics of pH and Eh in the soil solution of a flooded Ultisol at 30°C (adapted from Ponnamperuma, 1985).

Figure 4b. Kinetics of P_{CO_2}, water-soluble Fe^{2+}, and organic acids in the soil solution of a flooded Ultisol at 30°C (adapted from Ponnamperuma, 1985).

Chemical reactions that are favored thermodynamically are not necessarily favored kinetically. The lack of effective coupling and the slowness of redox reactions mean that catalysis is required if equilibrium is to be attained. In soils, the catalysis of redox reactions is mediated by microbial organisms. Equilibrium is dependent entirely on the growth and ecological behavior of the soil microbial population and the degree to which the reagents and products can diffuse and mix. Soil organisms are important with regard to kinetic aspects of redox by affecting the rate of a redox reaction but not its standard free energy change (Sposito, 1981).

Soils low in active iron with high organic matter may attain Eh values of −200 to −300 mV within 2 weeks after submergence (Ponnamperuma, 1972). In soils high in both iron and organic matter, the Eh may rapidly fall to −50 mV and then slowly decline over weeks and level off. Soils where the redox potential is controlled by a ferritic, ferruginous, or oxidic mineralogy and/or the soil reaction is strong acidic or allic are less prone to methane formation (Neue et al., 1990).

4.5 Partial pressure of CO_2. The partial pressure of CO_2 directly influences CH_4 production since CO_2 is a carbon source for methane. It also affects CH_4 production indirectly because the accumulation of CO_2 coupled with the formation of HCO_3^- buffers the pH near neutral in all flooded soils.

The increase in pH of acid soils is initially brought about by soil reduction of Fe-oxyhydroxides. The pH decrease of sodic and calcareous soil and the final regulation of the pH rise in acid soils are the result of CO_2 accumulation. The pH values at steady state of flooded alkaline, calcareous, and acid soils are highly sensitive to the partial pressure of CO_2. Carbon dioxide that accumulates in large amounts profoundly influences the chemical equilibria of almost all divalent cations (Ca^{2+}, Mg^{2+}, Fe^{2+}, Mn^{2+}, Zn^{2+}) in flooded soils as well as methane formation. Parashar et al. (1990) found the highest emission rates of CH_4 at a pH of 8.2. Acharya (1935) reported that the preliminary stage of acid formation is more tolerant to pH reactions, but gas formation is greatly impeded outside the range of pH 7.5 - 8.

Table 5. Coefficient of correlation between soil characteristics and the number of methanogenic bacteria/methane production potential (calculated from data from 29 soils given by Garcia et al., 1974).

	Range	Mean	log no. of methanogens per g of soil	log methane production potential[a]
Clay (%)	2.8-66	28	− 0.488**	− 0.524 **
Silt (%)	7.9-58	23	− 0.227	− 0.145
Sand (%)	1.7-82	37	+ 0.491**	+ 0.486 **
ECs (dS/m)	0.04-5.3	0.95	− 0.543**	− 0.384 *
pH (7 d after flooding[DAF])	3.4-6.8	5.4	+ 0.589**	+ 0.522 **
Eh [7 (DAF) (mV)	+400-135	+116	− 0.661**	− 0.646 **
Carbon content (%)	0.4-9.0	2.1	− 0.067	− 0.071
Total nitrogen (%)	0.04-0.31	0.12	+ 0.044	+ 0.192
C/N	9.9-29	17	− 0.299	− 0.400 *
$S-SO_4^{2-}$ [zero day] (mg/kg)	53-1690	380	− 0.437*	− 0.265
$N-NO_3^-$ [zero day] (mg/kg)	0-41.8	2.7	− 0.235	+ 0.005
Cl^- [zero day] meq/100g	0-42.9	7.1	− 0.496**	− 0.358
No. of denitrifiers (log 10 No./g)	1.7-5.4	3.9	+ 0.305	+ 0.153
Denitrification potential[b]	40->1500	> 370	− 0.637**	− 0.509 **
No. of sulfate-reducing bacteria (log 10 No./g)	1.8-5.9	3.4	− 0.144	− 0.071
Sulfate-reducing index[c]	0-5	2.2	− 0.257	− 0.227
Rice growth index[d]	0-7.8	3.3	+ 0.321	+ 0.473 **

[a] Methane produced during anaerobic incubation of soil at 37°C during 8 to 12 days after flooding;
[b] Denitrification of 100 mg/kg $N-NO_3$ (KNO_3) at 30°C;
[c] Percentage of dead rice plants in standardized conditions of growth favoring sulphatoreduction;
[d] Weight of grain produced in pot experiment.
* and ** = significant at 1% and 5% levels, respectively.

Up to 2.6 t CO_2/ha is produced in the puddled layer during the first few weeks of flooding (IRRI, 1964). After the addition of organic substrates, the partial pressure of CO_2 in a flooded soil may reach a peak of almost 100 kPa (Ponnamperuma, 1985; Neue and Bloom, 1989). Typical values in flooded soils range from 5 to 20 kPa (Kundu, 1987; Patra, 1987). Carbon dioxide concentrations greater than 15 kPa retard root development, leading to wilting and reduced nutrient uptake (Dent, 1986).

At soil temperatures found in flooded tropical soils, CO_2 and CH_4 formation occur sooner and in larger amounts than in cooler climates (Tsutsuki and Ponnamperuma, 1987). The amount of CH_4 found in the soil solution and in gas bubbles of flooded soils may be up to 3 times higher than that of CO_2 after the initial stage of flooding (Martin et al., 1983). The change in favor of CH_4 is likely caused by assimilation of CO_2 and precipitation of carbonates rather than reduction of CO_2 to CH_4, but the controlling processes still need elucidation. According to Takai (1970), the bulk of CH_4 is formed through decarboxylation of acetic acid, which would result in a 1:1 ratio of CO_2 and CH_4 formation.

4.6 Correlations between soil properties and methane formation. Neue et al. (1990) identified four crucial parameters for high methane production in wetland rice soils into four crucial parameters aside from carbon supply and water regime: temperature, texture and mineralogy, Eh/pH buffer, and salinity. He suggested that soils are not prone to high methane production if one or more of the following soil characteristics following limits of Soil Taxonomy (USDA, 1975) are met:

- EC > 4 dS/m while flooded,
- acidic or allic reaction,
- ferritic, gibbsitic, ferruginous, or oxidic mineralogy,
- > 40% of kaolinitic or halloysitic clays,
- < 18% clay in the fine earth fraction if the water regime is epiaquic, and
- drought-prone during cultivation period.

Soils comprising these features are Oxisols, most of the Ultisols, and some of the Aridisols, Entisols, and Inceptisols. Rice soils that are prone to methane production mainly belong to the orders of Entisols, Inceptisols, Alfisols, Vertisols, and Mollisols. Correlations between methane formation and physicochemical features (Tables 5 and 6), calculated from data of 29 soils given by Garcia et al. (1974), support this concept.

Table 6. Correlation between the number of methanogenic bacteria/methane production potential and pH/Eh at different days after flooding (DAF)(calculated from data of 29 soils given by Garcia et al., 1974).

	DAF methanogens	Log no. of production potential	Log methane
pH	0	+ 0.112	+ 0.030
	7	+ 0.589**	+ 0.522**
	14	+ 0.411*	+ 0.293
	21	+ 0.541**	+ 0.385*
	28	+ 0.539**	+ 0.268
Eh	0	− 0.107	− 0.197
	7	− 0.661**	− 0.646**
	14	− 0.438*	− 0.361*
	21	− 0.443**	− 0.303
	28	− 0.380*	− 0.097

* and ** = significant at 1% and 5% levels respectively.

Methane production potential is negatively correlated with Eh, ECs, chloride content, sulfate content, and C-N ratio but positively correlated with pH. The pH/Eh values 1 week after flooding, the denitrification potential, and the rice growth index show higher correlations. The high negative correlation with the clay content and positive correlation with the sand content indicate dominance of kaolinitic clays, and/or effect of the large number of salt affected mangrove soils in the sample. Considering only the 11 non-saline soils results in a highly significant correlation between soil carbon as well as soil nitrogen content and methane production potential but not with the number of methanogenic bacteria. This shows that the carbon content influences the activity but not necessarily the density of methanogenic bacteria. In the sample of non-saline soils there is no correlation between methane production potential and clay as well as sand. The positive correlation between methane production and the rice growth index, measured as rice grain yield, clearly indicates that increasing rice production enhances methane formation. Since the rice growth index is also an index for soil fertility of rice soils, it is evident that improving wetland soil fertility combined with current cropping technologies increases methane production and likely methane emission.

5. Rice cultivars and methane emission

Rice plants play an important role in the flux of methane. Up to 90% of the methane released from the rice soil to the atmosphere is emitted via the rice plant (Bont et al., 1978; Seiler, 1984; Holzapfel-Pschorn et al., 1986). The aerenchyma and intracellular space of rice plants mediate the transport of CH_4 from the reduced soil to the atmosphere (Raimbault et al., 1977). However, up to 80% of the methane produced is apparently oxidized in the rhizosphere (Holzapfel-Pschorn et al., 1985) and the oxidized soil floodwater interface (Figure 5).

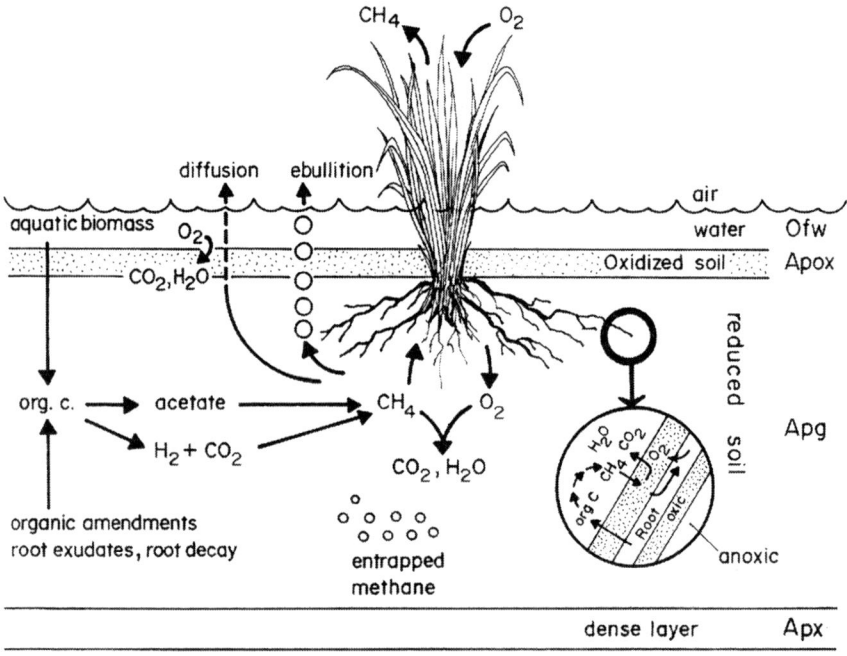

Figure 5. Schematic of production, reoxidation, and emission of CH_4 in a paddy field. (Modified from Schütz et al., 1989.)

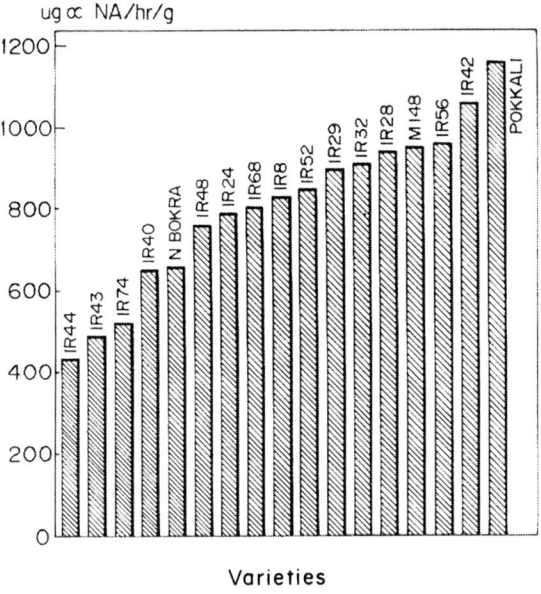

Figure 6. Root oxidizing power of selected rices.

The aerenchyma of rice plants acts as a chimney but the transport mechanisms still have to be elucidated. The well-developed air spaces in leaf blades, leaf sheath, culm and roots provide an efficient gas exchange between the atmosphere and the anaerobic soil. Atmospheric O_2 is supplied via the aerenchyma to the roots for respiration. Oxygen diffusion from rice roots seems to constitute an important part of the root-oxidizing power aside from enzymatic oxidation due to hydrogen peroxide production. Because of the abundance of methane-oxidizing bacteria present in the rhizosphere, its potential for methane oxidation is very high. At tillering, Bont et al. (1978) counted in the rhizosphere 10 times more methane-oxidizing bacteria than in the bulk anaerobic soil and 1/3 more than in the oxidized soil-water interface. They found significant increases in CH_4 emission from cultivar IR36 when suppressing CH_4 oxidation with acetylene at the soil-water interface. However, acetylene had only a small effect on emission rates when applied to the rhizosphere. Bont et al. (1978) concluded that the utilization of O_2 by reduced substances and microbial activity other than methanotrophs in the root-soil interfacial region exceeds the supply of O_2 by the root. Consequently, the aerobic zone surrounding the root is too thin to get the diffusing CH_4 oxidized or the rhizosphere is, for the most part, anaerobic. Rice plants may not only mediate the flux of CH_4, they enhance biological activities in soils and their root exudates and degrading roots may be an important source of CH_4 formation. Sass et al. (1991) found that spatial variability of methane production coincided with spatial distribution of roots in wetland ricefields.

Large cultivar differences in root oxidation power (Figure 6) and in emission rates (Parashar et al., 1990) open up the possibility of breeding rices that emit less methane. The inheritance of underlying traits has still to be elucidated.

6. Agronomic practices affecting CH_4 production.

Various rice culture systems have been developed to suit the physical, biological, and socioeconomic conditions of different regions and environments. Little is known about the effect of agronomic practices on methane fluxes in wetland ricefields.

6.1 Water control. Water control (irrigation and drainage) is one of the most important factors in rice production. In many ricefields, crops suffer from either too much or too little water because of rainfall pattern and topography.

During the monsoonal rainy season in tropical Asia, ricelands are naturally flooded. Excess water is a serious constraint in river, lacustrine, and coastal floodplains. Bunding (raising levees around the field) of ricefields and terracing and levelling of sloping land considerably change the water regime of that land. The overall effect is that uncontrolled runoff of water is minimized and more water, whether from natural sources or from irrigation is retained on or in the soil. Bunding and leveling of ricefields is a perfect measure of erosion control and allows efficient water harvesting and water conservation.

The following parameters of floodwater regimes are important for rice growth and should affect methane fluxes as well:

- duration and depth of flooding,
- regularity of flooding as determined by climatic relief and regime of rivers, and
- degree and pattern to which flooding is controlled by irrigation, flood protection, and drainage.

Single ricefields may have two distinct flooding regimes in a given year, especially in pronounced wet-dry monsoonal climates. Ricefields may be naturally shallow to moderately deep flooded during the rainy season and dry out or become shallow flooded by irrigation water during the dry season.

Floodwater control is the prerequisite for most technology changes in rice cultivation. Floodwater and soil water regime primarily determine methanogenesis and will likely be key issues for reducing methane fluxes. Aerating wetland soils to reduce methane fluxes without hampering rice production is a tempting mitigation technology.

But water stress at any growth stage may reduce yield. Moisture stress of 50 kPa (slightly above field capacity) may reduce grain yield to 20-25% of the yield of continually flooded treatments (De Datta, 1981). The rice plant is most sensitive to water deficit during the reproductive stage causing a high percentage of sterility (Yoshida, 1981). Water deficits during the vegetative stage may reduce plant height, tiller number, and leaf area that may also highly reduce yields if plants do not recover before flowering. The duration of a moisture stress is more important than the growth stage at which the stress occurred.

Short aeration periods at the end of the tillering stage and just before heading may improve wetland rice yields (Wang Zhaoqian, 1986) but only if it is followed by flooding. Intermittent irrigation or keeping soils only saturated considerably lowers rice yields (Borrell et al., 1991). Intermittent drying periods or percolation rates of up to 35 mm/day are associated with maximum rice yields in subtropical China and Japan (Wang Zhaoqian, 1986; Iwata et al., 1986). Percolation rates significantly increase yields only above production levels of 6 t/ha (Honya, 1966). Permanent year-round flooding or saturation, which may favor methane production, increases gleying and reduces soil fertility (Li Shi-jun and Li Xue-yuan, 1981), except in acid sulfate soils and some iron toxic soils. Although yields in a triple rice cropping system at IRRI have slightly declined over the past 22 years (Greenland, 1985), there is little evidence that high percolation rates, intermittent drying, or dry fallow periods are needed for high rice yields in most tropical wetland soils. The magnitude of aeration (oxidation) needed is likely dependent and interlinked with decomposition pattern and accumulation of organic and inorganic toxins, and nutrient imbalances involving Fe, Mn, Zn, S, and P (Neue, 1988, 1989). Much more information is needed to understand these interrelationships.

The high water demand for wetland rice becomes a constraint in areas and seasons of limited water resources. Wetland rice requires, on an average, 1240 mm of water (Yoshida, 1981), while upland crops may need less than half (Maesschalck et al., 1985). No methane is produced when upland crops are cultivated on riceland in seasons in which the land is naturally not flooded. This is commonly practiced in areas of limited water resources, especially since modern rice cultivars with short growth duration are available leaving sufficient water in the soil for a following upland crop. If water supply is assured, shifts to upland crops, which would reduce methane emission, are highly dependent on socioeconomic conditions.

6.2 Land preparation. Tillage operations vary according to water availability, soil texture, topography, rice culture, and resources available. Kawaguchi and Kyuma (1977) found that 40% of the tropical rice soils they studied had at least 45% clay. Soils with such high clay content have a poor structure and are hard when dry. Since hand- and animal- powered tillage are still common in most Asian countries and the principal form of mechanization is only the 10-15 hp tiller (hand tractor), wet tillage is the preferred land preparation. Wet tillage comprises land soaking until the soil is saturated, then plowing, puddling and harrowing. One third of the total water required for a rice crop is needed for the wet field preparation. Two weeks are required to prepare the field for transplanting.

According to De Datta (1981), the advantages of wet tillage are:

- improved weed control;
- ease of transplanting;
- establishment of reduced soil condition which improves soil fertility and fertilizer management;
- reduced draft requirement;
- reduced water percolation;
- reliability of monsoon rains by the time land preparation is completed; and
- higher fertilizer efficiency, especially for N fertilizer.

Flooding and puddling a soil provide an ideal growth medium for rice by supplying abundant water, buffering soil pH near neutral, enhancing N fixation and carbon supply, and increasing diffusion rates, mass flow, and availability of most nutrients. Standing water stabilizes the soil moisture regime, moderates soil temperature, and prevents soil erosion.

When initial crop growth at early monsoonal rainfall becomes essential because of subsequent floods, as in deepwater rice ecosystems, tillage and seeding are done in dry soils. Less than 200 mm of rainfall for the planting month leads to dryland preparation and seeding. Labor constraints associated with seedbed preparation, land preparation, and transplanting may also force farmers in other rainfed rice ecosystems to dryland tillage and seeding to ensure timely crop establishment. In most rice-growing countries where large power units can be employed because of available capital (as in the United States, Australia, most of Latin America, and Europe), dryland tillage is commonly practiced. In the United States and Australia rice is mostly also sown in dry soil, which is flooded after crop establishment. Upland ricelands are never flooded and tillage is, of course, the same as for other upland crops.

Information on the effect of land preparation on methane fluxes is lacking. Dryland tillage and dry seeding shorten the anaerobic phase and may slow down the decrease of the redox potential resulting in delayed and likely lower methane production. Minimum or zero tillage should have similar effects. However, these land preparation and seeding techniques require likely new rice cultural types, higher fertilizer rates, higher powered tillage implements, and alternative seeding and weeding techniques.

6.3 Seeding and transplanting. Transplanted rice is the major practice of rice culture in most of tropical Asia. Direct seeding of pregerminated rice in wet prepared soils becomes popular in areas with good water control and if manpower is lacking or becoming expensive. Pregerminated seeds are mostly broadcast onto puddled fields without standing water (saturated soil moisture) and the field is flooded after crop establishment. The crop duration (vegetative phase) is shorter in direct seeded rice, avoiding delay due to seedbed preparation, transplanting, and reduced initial growth because of the transplanting shock. Weed control in broadcast seeded rice and possible moisture stresses because of insufficient water control are the main obstacles in direct wet seeded rice. The yield potential for direct wet seeded rice is similar to that of transplanted rice (De Datta, 1981).

The advantages of transplanting rice seedlings are:

- lower seed requirement (increased tillering),
- save seed establishment (control of pests and fertilization),
- variable schedule of field establishment is possible without risk,
- less requirements for floodwater control, and
- tolerance to biotic and abiotic stresses increases with seedling age.

Methane fluxes should vary between direct seeded and transplanted rice because

- crop duration is shorter in direct seeded rice,
- soil surface is aerated for 7-14 days after land preparation in direct seeded rice,
- growth pattern and canopy development differ, and
- transplanting causes additional soil disturbances.

6.4 Fertilization. The most deficient nutrient for high wetland rice yields is nitrogen, followed by P, K, and Zn. The choice of nitrogen source depends on the method and time of application. Most farmers apply nitrogen fertilizer in two or three parts. The first part is applied during final land preparation or shortly after planting and the remainder as topdressing at later growth stages, especially at the early panicle stage. The most common source of N-fertilizer in wetland rice is urea followed by ammonium-containing fertilizers like ammonium sulfate. The source of nitrogen used as topdressing at later growth stages is less critical because of rapid uptake. In general, K and P are basically applied during the final land preparation. Potassium chloride is the principal fertilizer source of K and superphosphate is the primary source of P fertilizer. On acid rice soils, phosphate rock may be applied. Zn may be added by seed treatments, dipping seedling roots in ZnO solution, or broadcasting Zn salts at the time Zn deficiency symptoms occur.

Studies on fertilizer use and crop management to minimize nitrogen losses (up to 60% due to volatilization of NH_3, nitrification denitrification) and to increase the efficiency of fertilizer have recently been reviewed (De Datta, 1981; De Datta and Patrick, 1986; De Datta, 1987). For basal application, ammonium-containing or ammonium producing (urea) N fertilizer are recommended (De Datta, 1981) to minimize denitrification losses. To reduce volatilization losses incorporation or deep placement of N fertilizer has to be done without standing water at final harrowing. Broadcasting basal N-fertilizer into floodwater results in extensive N losses (as ammonia) to the atmosphere due to high pH values as a result of the algal assimilation or alkaline irrigation water (Fillery and Vlek, 1986).

Reports on the influence of the mineral fertilizer application (source, mode, and rate) on CH_4 production and emission are inconsistent. The complex interrelationships of fertilization on the biochemistry (CH_4 production and oxidation) of flooded soils and on plant growth (plant-mediated emission) still have to be elucidated. Increasing the number of tiller/m^2 and enhancing root growth in methane enriched soil layers through fertilization will obviously increase methane emission. As discussed above, encapsulating methane inhibitors in fertilizer seems to be very promising.

It is evident that organic amendments of flooded soils increase CH_4 production and emission (Schütz et al., 1989) by lowering the Eh and providing carbon sources. Addition of plant residues accelerate and intensify Eh and pH changes (Katyal, 1977). The effect of vetch, which has a narrow C-N ratio, is greater than that of rice straw (Yu, 1985). Changes are more pronounced when organic substrates are added to soils low in organic matter (Nagarajah et al., 1989). Increasing the soil bulk density of flooded soils retards organic matter decomposition, increases the concentration and residue time of organic acids, and reduces the speed of Eh and pH changes as well as methane formation.

Though organic amendments are propagated to sustain soil resources, actual application of organic substrates into wetland ricefields seems on the decline (see Figure 7). In China the production of green manure increased sharply after 1960 and peaked sometime in the 1970s (13.2 million ha), followed by a steep decline to only 6.6 million ha in 1987 (Stone, 1990). In Japan, the decline of green manure cultivation started already in the 1950s. According to Kanazawa (1984), the total addition of organic substrates to ricefields in Japan decreased from 6 t/ha in 1965 to 2.7 t/ha in 1980.

Based on the content of readily mineralizable carbon, humified substrates like compost should produce less methane per unit carbon while rice straw or green manures likely produce more. Application of compost did not remarkably enhance methane emission while application of rice straw significantly increased methane emission irrespective of soil type (Yagi and Minami, 1990). Sound technologies have to consider both maintaining or increasing soil fertility and reducing methane emission. In a sustainable wetland rice system, it may be advisable to minimize rather than to maximize organic soil amendments.

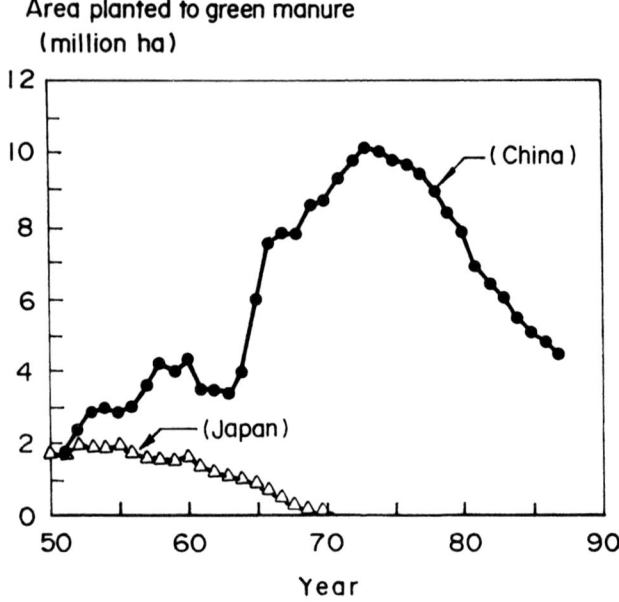

Figure 7. Area planted to green manure in China (1952-1987) (Stone, 1990) and in Japan (Watanabe, 1984).

6.5 Pest control. Control of pests in wetland rice ranges from varietal resistance through cultural control, biological control, and chemical control. Application of pesticides to the floodwater, soil surface, or into the soil may have significant effects on methane fluxes, especially if it adversely affects the aquatic and soil flora and fauna. Many cultural control measures, such as cropping pattern, crop residue management, tillage, water, fertility, weeding, or plant spacing and population, affect pests in rice but should affect methane fluxes at the same time. For example, the mechanical disturbance of the soil during weeding (2-3 times per season by hand or small implements) increases the release of gases trapped in the soil. Weeds become an additional source of methane since they are commonly returned to the reduced soil. On the other hand, aquatic weeds may provide a more efficient pathway for CH_4 to the atmosphere than rice plants, as indicated by Holzapfel-Pschorn et al. (1986).

7. Summary

Ricefields provide ideal environments for methanogenesis, especially irrigated ricefields in the tropics, because of anaerobic conditions at neutral pH, optimum temperature, and high easily degradable C inputs. Rice plants favor methane fluxes by supplying carbon and acting as chimney. The high C input in rice soils due to a high primary production both by the crop and the photosynthetic aquatic biomass and organic amendments favor methane emission from ricefields.

Irrigated rice ecologies seem to be the major source for increased global methane emissions from ricefields. The assured water supply and control, the intensive soil preparation, and the resultant improved growth of rice, mediating the methane flux to the atmosphere, favor methane production and emission. Methane emissions should be much lower and highly variable in space and time in rainfed rice because of drought periods during the growing season and the poorer growth of rice. In deepwater rice, methane production may be high but related emission rates are unknown. Upland rice is believed not to be a source of methane emission because upland rice is never flooded for a significant period of time.

Global extrapolations of emission rates are still highly uncertain and tentative. Accounting for variations of emission rates due to climate, soil properties, duration and pattern of flooding, organic amendments, fertilization and rice growth reveals that most published extrapolations are too high. Adjusting a basic emission rate of 0.5 g m^{-2} day^{-1} according to rice ecologies or soil types results in global emission rates of about 40 Tg year^{-1} (Neue, 1992; Bachelet and Neue, 1993). But reported measurements of CH_4 fluxes in rice fields do not account for ebullition induced by soil disturbance due to wet tillage, transplanting, fertilization, weeding, pest control and harvest. A large proportion of soil entrapped methane that may account for up to 90% of the methane generated and is oxidized in undisturbed rice fields likely escapes to the atmosphere during these cultural practices. Therefore, the global CH_4 emission rate from rice fields is likely higher than 40 Tg year^{-1}.

Methane from ricefields may contribute up to one third of the global anthropogenic methane emission, and mitigation technologies are required to stabilize atmospheric methane concentration in the long term. Possible mitigation technologies include reducing C inputs of easily degradable carbon, increasing soil and plant-mediated methane oxidation, reducing emission pathways through the selection and breeding of rice cultivars, and preventing or reducing methane formation through intermittent aeration, sources and mode of fertilizer, and application of chemical inhibitors. The most effective mitigation technology would be to shift from wetland rice to upland cropping whenever possible.

However, technologies that will be accepted by farmers have to be not only environmentally but also socio-economically sound. Because of the limited existing knowledge and the complexity of any methane-mitigating technology, a comprehensive interdisciplinary research approach of various biological and social sciences and common sense are needed to develop and implement feasible technologies.

Acknowledgement. We thank Dr. J. L. Garcia (ORSTOM, France) for providing us unpublished data and for his useful comments.

References

Abram, J.W., D.B. Nedwell. 1978. Inhibition of methanogenesis by sulfate reducing bacteria competing for transferred hydrogen. *Arch. Microbiol., 117*:89-92.

Acharya, C.N. 1935. Studies on the anaerobic decomposition of plant materials. II. Some factors influencing the anaerobic decomposition. *Biochem. J., 29*:953-960.

Alperin, M.J., W.S. Reeburgh. 1984. Geochemical observations supporting anaerobic methane oxidation. In: *Microbial Growth on C-1 Compounds* (R.L. Crawford and R.S. Hanson, eds.), American Society of Microbiology, Washington D.C. p. 282-289.

Anthony, C. 1982. *The Biochemistry of Methylotrophs.* Academic Press, San Diego California.

Aselmann, I., Crutzen, P.J. 1990. Global inventory of wetland distribution and seasonality net primary production and estimated methane emission. In: *Soils and the Greenhouse Effect* (A.F. Bouwman, ed.), John Wiley & Sons, Chichester, England, p 441-450.

Bachelet, D., H.U. Neue. 1993. Methane emissions from wetland rice areas of Asia. *Chemosphere, 26* (1-4):219-246.

Balderston, W.L., W.J. Payne. 1976. Inhibition of methanogenesis in salt marsh sediments and whole-cell suspensions of methanogenic bacteria by nitrogen oxides. *Appl. Environ. Microbiol., 32*:264-269.

Bartlett, K.B., D.S. Bartlett, R.C. Harriss, D.I. Sebacher. 1987. Methane emissions along a salt marsh salinity gradient. *Biogeochemistry,* 4:183-202.
Belay, N.R., R. Sparling, L. Daniels. 1984. Dinitrogen fixation by a thermophilic methanogenic bacterium. *Nature,* 312:286-288.
Biavati, B., M. Vasta, J.G. Ferry. 1988. Isolation and characterization of *Methanosphaera cuniculi* sp.nov. *Appl. Environ. Microbiol.,* 54:786-771.
Blotevogel, K.H., U. Fischer, M. Mocha, S. Jannsen. 1985. *Methanobacterium thermoalcapiphilum* sp. nov. a new moderately alkaliphilic and thermophilic autotrophic methanogen. *Arch. Microbiol.,* 142:211-217.
Bolle, H.J., W. Seiler, B. Bolin. 1986. Other greenhouse gases and aerosols, assessing their role for atmospheric radiative transfer. In: *The Greenhouse Effect, Climatic Change, and Ecosystems* (B. Bolin, B.R. Döös, J. Jäger, and R.A. Warrick, eds.), Chichester, New York, Brisbane, Toronto, Singapore, Wiley and Sons; p 157-203.
Bonneau, M. 1982. Soil temperature. In: *Constituents and Properties of Soils* (M. Bonneau and B. Souchier, eds.), Academic Press, London, England, p 366-371.
Bont, J.A.M. de, K.K. Lee, D.F. Bouldin. 1978. Bacterial oxidation of methane in rice paddy. *Ecol. Bull.,* 26:91-96.
Borrell, A.K., S. Fukai, A.L. Garside. 1991. Irrigation methods for rice in tropical Australia. *Int. Rice Res. Newsl.,* 16 (3):28.
Bouwman, A.F. 1990. *Soils and the Greenhouse Effect.* (A.F. Bouwman, ed.), John Wiley.
Bronson, K.F., A.R. Mosier. 1991. Effect of encapsulated calcium carbide on dinitrogen, nitrous oxide, methane and carbon dioxide emissions in flooded rice. *Biology and Fertility of Soils,* 3:116-120.
Buresh, R.J., S.K. De Datta. 1991. Nitrogen dynamics and management in rice-legume cropping systems. *Adv. Agron.,* 45:1-59.
Capistrano, R.F. 1988. Decomposition of ^{14}C-labelled rice straw in 3 submerged soils under controlled laboratory conditions. M.S. thesis, University of the Philippines at Los Baños Laguna, Philippines.
Cappenberg, T.E. 1974. Interrelations between sulfate-reducing and methane-producing bacteria in bottom deposits of a fresh-water lake. I. Field observation. *Anton. Leeuwenhoek J. Microbiol. Serol.,* 40:285-295.
Cappenberg, T.E., R.A. Prins. 1974. Interrelations between sulfate-reducing and methane-producing bacteria in bottom deposits of a fresh-water lake. III. Experiments with ^{14}C-labelled substrates. *Anton. Leeuwenhoek J. Microbiol. Serol.,* 40:457-469.
Cho, D.Y., F.N. Ponnamperuma. 1971. Influence of soil temperature on the chemical kinetics of flooded soils and the growth of rice. *Soil Sci.,* 112:184-194.
Cicerone, R.J., R.S. Oremland. 1988. Biogeochemical aspects of atmospheric methane. *Global Biogeochem. Cycles,* 2:299-327.

Conrad, R. 1989. Control of methane production in terrestrial ecosystems. In: *Exchange of Trace Gases Between Terrestrial Ecosystems and the Atmosphere* (M.O. Andreae and D.S. Schimel, eds.), S. Bernhard Dahlem Konferenzen. Wiley, New York, p 39-58.

Conrad, R., R. Bonjour, M. Aragno. 1985. Aerobic and anaerobic microbial consumption of hydrogen in geothermal spring water. *FEMS Microbiol. Lett.*, 29:201-206.

Conrad, R., H.P. Mayer, M. Wüst. 1989. Temporal change of gas metabolism by hydrogen-syntrophic methanogenic bacterial association in anoxic paddy soil. *FEMS Microbiol. Ecol.*, 62:265-274.

Crawford, R.L., R.S. Hanson (eds.) 1984. Microbial growth on C1 compounds. *Proceedings of the 4th International Symposium American Society for Microbiology*, Washington D.C.

De Datta, S.K. 1981. *Principles and Practices of Rice Production.* John Wiley and Sons New York USA.

De Datta, S.K. 1987. Advances in soil fertility research and nitrogen fertilizer management for lowland rice. In: *Efficiency of Nitrogen Fertilizer for Rice.* International Rice Research Institute, P.O. Box 933, Manila, Philippines, p 27-41.

De Datta, S.K., W.H. Patrick (eds.). 1986. Nitrogen economy of flooded rice soils. *Development in Plant and Soil Sciences.* Martin Nijhoff Publication, Dodrecht, The Netherlands.

De Laune, R.D., E.J. Smith, W.H. Patrick. 1983. Methane release from Gulf Coast wetlands. *Tellus, 35B*:8-15.

Dent, D. 1986. Acid sulfate soils: a baseline for research and development. *ILRI Publication 39.* Wageningen, The Netherlands.

Dickinson, R.E., R.J. Cicerone. 1986. Future global warming from atmospheric trace gases. *Nature, 319*:109-115.

Dolfing, J. 1988. Acetogenesis. In: *Biology of Anaerobic Microorganisms* (A.J.B. Zehnder, ed.) Wiley, New York, p 417-468.

FAO - Food and Agriculture Organization. 1988. *Quarterly Bulletin of Statistics. Vol. 1 No. 4.* FAO, Rome, Italy.

Fillery, I.R.P., P.L.G. Vlek. 1986. Reappraisal of the significance of ammonia volatilization as a N loss mechanism in flooded ricefields. In: *Development in Plant and Soil Sciences* (S.K. De Datta and W.H. Patrick, eds.), Martin Nijhoff Publ., Dodrecht, The Netherlands, p. 79-98.

Franklin, N.J., W.J. Wiebe, W.B. Whitman. 1988. Populations of methanogenic bacteria in Georgia salt marsh. *Appl. Environ. Microbiol.*, 54:1,151-1,157.

Garcia, J.L. 1990. Taxonomy and ecology of methanogens. *FEMS Microbiol. Rev.*, 87:297-308.

Garcia, J.L., M. Raimbault, V. Jacq, G. Rinaudo, P. Roger. 1974. Activities microbiennes dans les sols de rizieres du senegal: relations avec les proprietes physicochimiques et influence de la rhizosphere. *Rev. Ecol.*

Biol., 11 (2):169-185.
Greenland, D.J. 1985. Physical aspects of soil management for rice-based cropping systems. In: *Soil Physics and Rice*. International Rice Research Institute, P.O. Box 933, Manila, Philippines, p. 1-16.
Gupta, G.P. 1974. The influence of temperature on the chemical kinetics of submerged soils. Ph.D. thesis, Indian Agricultural Research Institute, New Delhi, India.
Hackman, Ch.W. 1979. Rice field ecology in Northeastern Thailand. The effect of wet and dry season on a cultivated aquatic ecosystem. In: *Monogr. Biol., 34* (J. Illies, ed.), W. Junk Publisher, 22 p.
Harrison, W.H., P.A.S. Aiyer. 1913. The gases of swamp rice soil. I. Their composition and relationship to the crop. Memoires, Department of Agriculture, India. *Chem. Ser., 5*(3): 65-104.
Higgins, I.J., D.J. Best, R.C. Hammond, D.C. Scott. 1981. Methane-oxidizing microorganisms. *Microbiol. Rev., 45*:556-590.
Holzapfel-Pschorn, A., W. Seiler. 1986. Methane emission during a cultivation period from an Italian rice paddy. *J. Geophys. Res., 91*:11,803-11,814.
Holzapfel-Pschorn, A., R. Conrad, W.W. Seiler. 1985. Production oxidation and emission of methane in rice paddies. *FEMS Microbiol. Ecol., 31*:343-351.
Holzapfel-Pschorn, A., R. Conrad R, W. Seiler. 1986. Effects of vegetation on the emission of methane from submerged paddy soil. *Plant Soil, 92*:223-233.
Honya, K. 1966. *Fundamental conditions for high yields of rice.* [in Japanese]. Nobunkyo Publishing, Tokyo.
IRRI - International Rice Research Institute. 1964. Annual report for 1963. P.O. Box 933, Manila, Philippines, 201 p.
IRRI - International Rice Research Institute. 1981. Annual report for 1980. P.O. Box 933, Manila, Philippines, 306 p.
IRRI - International Rice Research Institute. 1989. IRRI toward 2000 and beyond. P. O. Box 933, Manila, Philippines.
IRRI - International Rice Research Institute. 1991. World rice statistics 1990. P. O. Box 933, Manila, Philippines.
Iversen, N., R.S. Oremland, M.J. Klug. 1987. Big Soda Lake (Nevada) 3 pelagic methanogenesis and anaerobic methane oxidation. *Limnol. Oceanogr., 32*:804-814.
Iwata, S., S. Hasegawa, K. Adachi. 1986. Water flow balance and control in rice cultivation. In: *Wetlands and Rice in Subsaharan Africa* (A.S.R. Juo and J.A. Lowe, eds.), IITA Ibadan Nigeria, p. 69-86.
Kanazawa, N. 1984. Trends and economic factors affecting organic manures in Japan. In: *Organic Matter and Rice*. International Rice Research Institute, P.O. Box 933, Manila Philippines, p 557-568.
Katyal, J.C. 1977. Influence of organic matter on chemical and electrochemical properties of some flooded soils. *Soil Biol., 9*:259-266.

Kawaguchi, K., K. Kyuma. 1977. *Paddy Soils in Tropical Asia: Their Material Nature and Fertility.* The University Press of Hawaii, Honolulu, Hawaii, USA.
Khalil, M.A.K., R.A. Rasmussen. 1987. Atmospheric methane: trends over the last 10000 years. *Atmos. Environ., 21* (11):2,445-2,452.
Khalil, M.A.K., R.A. Rasmussen. 1989. Climate induced feedback for the global cycles of methane and nitrous oxide. *Tellus, 41B*:554-559.
Khalil, M.A.K., R.A. Rasmussen. 1990. Constraints on the global sources of methane and an analyses of recent budgets. *Tellus, 428*:229-236.
Kiene, R.P., P.T. Visscher. 1987. Production and fate of methylated sulfur compounds from methionine and dimethylsulfoniopropionate in anoxic marine sediments. *Appl. Environ. Microbiol., 53*:2,426-2,434.
King, G.M. 1984. Metabolism of trimethylamine choline and glycine betaine by sulfate-reducing and methanogenic bacteria in marine sediments. *Appl. Environ. Microbiol., 48*:719-725.
King, G.M. 1988. Methanogenesis from methylated amines in a hypersaline algal mat. *Appl. Environ. Microbiol., 54*:130-136.
Kondo, Y. 1952. Physiological studies on cool-weather resistance of rice varieties. Nogyo Gijutsi Kenkyusho Hokodu Di seiri Inde. Sakrimotsu Ippan (*National Institute of Agriculture Science Bulletin Japan Series*) D 3:113-228.
Koyama, T., M. Hishida, T. Tomino. 1970. Influence of sea salts on the soil metabolism. II. On the gaseous metabolism. *Soil Sci. Plant Nutr., 16*:81-86.
Krumböck, M., R. Conrad. 1991. Metabolism of position-labelled glucose in anoxic methanogenic paddy soil and lake sediment. *FEMS Microbiol. Ecol., 85*:247-256.
Kundu, D.K. 1987. Chemical kinetics of aerobic soils and rice growth. Ph.D. thesis, Indian Agricultural Research Institute, New Delhi, India.
Kuwatsuka, S., K. Tsutsuki, K. Kumada. 1978. Chemical studies on humic acids. I. Elementary composition of humic acid. *Soil Sci. Plant Nutr., 23*:337-347.
Li Shi-jun, Li Xue-yuan. 1981. Stagnancy of water in paddy soils under the triple cropping system and its improvement. In: *Proceedings of Symposium on Paddy Soil.* Institute of Soil Science. Academica Sinica, ed. Science Press, Beijing, and Springer-Verlag, Berlin, p. 509-516.
Lovley, D.R., M.J. Klug. 1983. Methanogenesis from methanol and from hydrogen and carbon dioxide in the sediments of a eutrophic lake. *Appl. Environ. Microbiol., 45*:1,310-1,315.
Maesschalck, G.H., M. Verplancke, M. De Boodt. 1985. Water use and wateruse efficiency under different management systems for upland crops. In: *Soil Physics and Rice.* International Rice Research Institute, P.O. Box 933, Manila, Philippines, p. 397-408.
Martin, U., H.U. Neue, H.W. Scharpenseel, P.M. Becker. 1983. Anaerobe Zersetzung von Reisstroh in einem gefluteten Reisboden auf den

Philippinen. *Mitt. Dtsch. Bodenkdl. Gesellsch., 38*:245-250.
Mathrani, I.M., D.R. Boone, R.A. Mah, G.E. Fox, P.P. Lau. 1988. *Methanohalophilus zhilinae* sp.nov., an alkaliphilic halophilic methylotrophic methanogen. *Int. J. Sys. Bacteriol., 38*:139-142.
Matsushima, S., T. Tanaka, T. Hoshino. 1964a. Analysis of yield- determining process and its application to yield prediction and culture improvement of lowland rice. LXX combined effect of air temperature and water temperature at different stages of growth on the grain yield and its components of rice plants. *Proc. Crop Sci. Soc. Jpn., 33*:53-58.
Matsushima, S., T. Tanaka, T. Hoshino. 1964b. Analysis of yield- determining process and its application to yield prediction and culture improvement of lowland rice. LXX combined effect of air temperature and water temperature at different stages of growth on the growth and morphological characteristics of rice plants. *Proc. Crop Sci. Soc. Jpn., 33*:135-140.
McBride, B.C., R.S. Wolfe. 1971. Inhibition of methanogenesis by DDT. *Nature, 234*:551.
Miller, T.L., M.J. Wolin. 1985. *Methanosphaera stadtmaniae* gen.nov.sp.nov.: a species that forms methane by reducing methanol with hydrogen. *Arch. Microbiol., 141*:116-122.
Mitsch, W.J., J.G. Gosselink. 1986. *Wetlands*. Van Nostrand Reinhold Company New York, USA.
Moormann, F.R., N. Van Breemen. 1978. *Rice: Soil Water Land*. International Rice Research Institute, P.O. Box 933, Manila, Philippines.
Murray, P.A., S.H. Zinder. 1984. Nitrogen fixation by a methanogenic bacterium. *Nature, 312*:284-286.
Murrell, J.C., H. Dalton. 1983. Nitrogen fixation in obligate methanotrophs. *J. Gen. Microbiol., 129*:3,481-3,486.
Nagarajah, S., H.U. Neue, M.C.R. Alberto. 1989. Effect of Sesbania Azolla and rice straw incorporation on the kinetics of NH_4, K, Fe, Mn, Zn, and P in some flooded rice soils. *Plant Soil, 116*:37-48.
Neue, H.U. 1985. Organic matter dynamics in wetland soils. In: *Wetland Soils: Characterization, Classification and Utilization*. International Rice Research Institute, P.O. Box 933, Manila, Philippines, p. 109-122.
Neue, H.U. 1988. Holistic view of chemistry of flooded soil. In: *Proceedings of the First International Symposium on Paddy Soil Fertility, 6-13 December 1988*. International Board for Soil Research and Management, Bangkok, p. 21-56.
Neue, H.U. 1989. Rice growing soils: Constraints utilization and research needs. Pages 1-14 *in* Classification and management of rice growing soils. *FFFTC Book Series No. 39*. Food and Fertilizer Technology Center for the ASPAC Region, Taiwan, R.O.C.
Neue, H.U. 1992. Agronomic practices affecting methane fluxes from rice cultivation. In: *Trace Gas Exchange in a Global Perspective, Ecol. Bull.*

(Copenhagen), 42:174-182 (D.S. Ojima and B.H. Svensson, eds.).
Neue, H.U., H.W. Scharpenseel. 1984. Gaseous products of the decomposition of organic matter in submerged soils. In: *Organic Matter and Rice.* International Rice Research Institute, P.O. Box 933, Manila, Philippines, p. 311-328.
Neue, H.U., H.W. Scharpenseel. 1987. Decomposition pattern of ^{14}C-labelled rice straw in aerobic and submerged rice soils of the Philippines. *Science Total Environ.*, 62:431-434.
Neue, H.U., P.R. Bloom. 1989. Nutrient kinetics and availability in flooded soils. In: *Progress in Irrigated Rice Research.* International Rice Research Institute, P.O. Box 933, Manila, Philippines, p 173-190.
Neue, H.U., P. Becker-Heidmann, H.W. Scharpenseel. 1990. Organic matter dynamics soil properties and cultural practices in ricelands and their relationship to methane production. In: *Soils and the Greenhouse Effect* (A.F. Bouwman, ed.), John Wiley & Sons, Chichester, England, p. 457-466.
Oremland, R.S., S. Polcin. 1982. Methanogenesis and sulfate-reduction:competitive and noncompetitive substrate in estuarine sediments. *Appl. Environ. Microbiol.*, 44:1,270-1,276.
Oremland, R.S., D.G. Capone. 1988. Use of "specific" inhibitors in biogeochemistry and microbial ecology. *Adv. Microbiol. Ecol.*, 10:285-383.
Oremland, R.S., L.M. Marsh, S. Polcin. 1982. Methane production and simultaneous sulfate reduction in anoxic salt marsh sediments. *Nature (London)*, 296:143-145.
Panganiban, A.T., T.E. Patt, W. Hart, R.S. Hanson. 1979. Oxidation of methane in the absence of oxygen in lake water samples. *Appl. Environ. Microbiol.*, 37:303-309.
Parashar, D., C.J. Rai, P.K. Gupta, N. Singh. 1990. Parameters affecting methane emission from paddy fields. *Indian J. Radio Space Physics*, 20:12-17.
Patel, G.B., L.A. Roth. 1977. Effect of sodium chloride on growth and methane production of methanogens. *Can. J. Microbiol.*, 6:893.
Patra, P.K. 1987. Influence of water regime on the chemical kinetics of soils and rice growth. Ph.D. thesis, Indian Agricultural Research Institute, New Delhi, India.
Patrick, W.H., Jr., C.N. Reddy. 1978. Chemical changes in rice soils. In: *Soils and Rice.* International Rice Research Institute, P.O. Box 933, Manila, Philippines, p 361-380.
Patrick, W.H., Jr., D.S. Mikkelsen, B.R. Wells. 1985. Plant nutrient behavior in flooded soil. In: *Fertilizer Technology and Use, 3d Ed.* Soil Science Society of America Madison Wisconsin.
Patt, T.E., G.C. Cole, J. Bland, R.S. Hanson. 1974. Isolation and characterisation of bacteria that grow on methane and organic compounds as sole source of carbon and energy. *J. Bacteriol.*, 120:955-964.

Ponnamperuma, F.N. 1972. The chemistry of submerged soils. *Adv. Agron.,* 24:29-96.
Ponnamperuma, F.N. 1981. Some aspects of the physical chemistry of paddy soils. In: *Proceedings of the Symposium on Paddy Soils.* Science Press, Beijing People's Republic of China, p 59-94.
Ponnamperuma, F.N. 1984a. Effects of flooding on soils. In: *Flooding and Plant Growth* (T.T. Kozlowski, ed.), Academic Press, New York, USA, p 9-45.
Ponnamperuma, F.N. 1984b. Straw as a source of nutrients for wetland rice. In: *Organic Matter and Rice.* International Rice Research Institute, P.O. Box 933, Manila, Philippines, p 117-136.
Ponnamperuma, F.N. 1985. Chemical kinetics of wetland rice soils relative to soil fertility. In: *Wetland Soils: Characterization Classification and Utilization.* International Rice Research Institute, P.O. Box 933, Manila, Philippines, p 71-89.
Raimbault, M. 1975. Étude de l'influence inhibitrice de l'acétylene sur la formation biologique du méthane dans un sol de riziére. *Ann. Microbiol. (Inst. Pasteur), 126a*:217-258.
Raimbault, M. 1981. Inhibition de la formation de methane par l'acétylène chez Methananosarcina bakerii. *Cah. ORSTOM, Ser. Biol.,* 43:45-51.
Raimbault, M., G. Rinaudo, J.L. Garcia, M. Boureau. 1977. A device to study metabolic gases in the rice rhizosphere. *Biol. Biochem.,* 9:193-196.
Rajagopal, B.S., N. Belay, L. Daniels. 1988. Isolation and characterization of methanogenic bacteria from rice paddies. *FEMS Microbiol. Ecol.,* 53:153-158.
Roger, P.A., I. Watanabe. 1984. Algae and aquatic weeds as source of organic matter and plant nutrients for wetland rice. In: *Organic Matter and Rice.* International Rice Research Institute, P.O. Box 933, Manila, Philippines, p. 147-168.
Roger, P.A., I.F. Grant, P.N. Reddy, I. Watanabe. 1987. The photosynthetic aquatic biomass in wetland ricefields and its effect on nitrogen dynamics. In: *Efficiency of N Fertilizers for Rice.* International Rice Research Institute, P.O. Box 933, Manila, Philippines, p 43-68.
Salvas, P.L., B.F. Taylor. 1980. Blockage of methanogenesis in marine sediments by the nitrification inhibitor 2-chloro-6-(trichloromethyl) pyridine (Nitrapin or N-serve) *Curr. Microbiol.,* 4:305.
Sass, R.L., F.M. Fischer, P.A. Harcombe, F.T. Turner. 1991. Methane production and emission in a Texas rice field. *Global Biogeochem. Cycles,* 4:47-68.
Schink, B., J.G. Zeikus. 1980. Microbial methanol formation: a major end product of protein metabolism. *Curr. Microbiol.,* 4:387-389.
Schönheit, P., H. Keweloh, R.K. Thauer. 1981. Factor F_{420} degradation in *Methanobacterium thermoautotrophicum* during exposure to oxygen. *FEMS Microbiol. Lett.,* 12:347-349.

Schütz, H., A. Holzapfel-Pschorn, R. Conrad, H. Rennenberg, W. Seiler. 1989. A three-year continuous record on the influence of daytime season and fertilizer treatment on methane emission rates from an Italian rice paddy field. *J. Geophys. Res., 94*:16,405-16,416.

Seiler, W. 1984. Contribution of biological processes to the global budget of CH_4 in the atmosphere. In: *Current Perspectives in Microbial Ecology* (M.J. Kleig and C.A. Reddy, eds.), American Society of Microbiology, Washington D.C., p 468-477.

Seiler, W., R. Conrad. 1987. Contribution of tropical ecosystems to the global budget of trace gases especially CH_4 H_2 CO and N_2O. In: *The Geography of Amazonia: Vegetation and Climate Interactions* (R.E. Dickinson, ed.), Wiley, N.Y., p 133-162.

Sequi, P., M. De Nobili, L. Leita L, G. Cerciguani. 1986. A new index of humification. *Agrochemical, 30*:175-179.

Sharma, P.K., S.K. De Datta. 1985. Effects of puddling on soil physical properties and processes. In: *Soil Physics and Rice*. International Rice Research Institute, P.O. Box 933, Manila, Philippines, p 217-234.

Smith, J., H.U. Neue, G. Umali. 1987. Soil nitrogen and fertilizer recommendations for irrigated rice in the Philippines. *Agric. Sys., 24*:165-181.

Smith, P.H., R.E. Hungate. 1958. Isolation and characterization of *Methanobacterium ruminantium* n.sp. *J. Bacteriol., 75*:713-718.

Snitwongse, P., S. Pongpan, H.U. Neue. 1988. Decomposition of ^{14}C-labelled rice straw in a submerged and aerated rice soil in Northeastern Thailand. In: *Proceedings of the First International Symposium on Paddy Soil Fertility*, 6-13 December 1988. International Board for Soil Research and Management, Bangkok, p 461-480.

Sposito, G. 1981. *The Thermodynamics of Soil Solutions*. Clarendon Press, Oxford.

Stone, B. 1990. Evolution and diffusion of agricultural technology in China. In: *Sharing Innovation Global Perspectives on Food Agriculture and Rural Development* (N.G. Kotler, (ed.), International Rice Research Institute, P.O. Box 933, Manila, Philippines, p 35-93.

Strayer, R.F., J.M. Tiedje. 1978. Kinetic parameters of the conversion of methane precursors to methane in hypereutrophic lake sediment. *Appl. Environ. Microbiol., 36*:330-340.

Svensson, B.H. 1984. Different temperature optima for methane formation when enrichments from acid peat are supplemented with acetate or hydrogen. *Appl. Environ. Microbiol., 48*:389-394.

Takai, Y. 1961. Reduction and microbial metabolism in paddy soils (3) [in Japanese English summary]. *Nogyo Gijitsu (Agro. Technol.), 19*:122-126.

Takai, Y. 1970. The mechanism of methane fermentation in flooded soils. *Soil Sci. Plant Nutr., 16*:238.

Takai, Y., T. Koyama, T. Kamura. 1956. Microbial metabolism in reduction process of paddy soil. Part I. *Soil Plant Food, 2*:63-66.

Toukdarian, A.E., M.E. Lidstrom. 1984. Nitrogen metabolism in a new obligate methanotroph *Methylosinus* strain 6. *J. Gen. Microbiol., 130*:1,827-1,837.

Tsuchiya, K., H. Wada, Y. Takai. 1986. Leaching of substances from paddy soils. 4. Water solubilization of inorganic components in submerged soils. *Jpn. J. Soil Sci. Plant Nutr., 57* (6):593-597.

Tsutsuki, K., S. Kuwatsuka. 1978. Chemical studies on soil humic acids. II. Composition of oxygen-containing functional groups of humic acids. *Soil Sci. Plant Nutr., 24*:547-560.

Tsutsuki, K., K. Kumada. 1980. Chemistry of humic acids [in Japanese; English summary]. *Fert. Sci., 3*:93-171.

Tsutsuki, K., F.N. Ponnamperuma. 1987. Behavior of anaerobic decomposition products in submerged soils. Effects of organic material amendment soil properties and temperature. *Soil Sci. Plant Nutr., 33* (1):13-33.

USDA -- United States Department of Agriculture, Soil Conservation Service Soil Survey Staff (1975) Soil taxonomy: a basic system of soil classification for making and interpreting soil surveys. *USDA Agric. Handb. 436.* U.S. Government Printing Office, Washington, D.C.

Vogels, G.D., J.T. Keltjens, C. Van der Drift. 1988. Biochemistry of methane production. In: *Biology of Anaerobic Microorganisms* (A.J.B. Zehnder, ed.), Wiley New York, p 707-770.

Wang, Zhaoqian. 1986. Rice-based systems in subtropical China. In: *Wetlands and Rice in Subsaharan Africa* (A.S.R. Juo and J.A. Lowe, eds.), IITA Ibadan Nigeria, p 195-206.

Ward, D.M., M.R. Winfrey. 1985. Interactions between methanogenic and sulfate-reducing bacteria in sediments. *Adv. Aquatic Microbiol., 3*:141-179.

Watanabe, I. 1984. Use of green manures in Northeast Asia. In: *Organic Matter and Rice.* International Rice Research Institute, P.O. Box 933, Manila, Philippines, p 229-233.

Watanabe, I., P.A. Roger. 1985. Ecology of flooded ricefields. In: *Wetland Soils: Characterization Classification and Utilization.* International Rice Research Institute, P.O. Box 933, Manila, Philippines, p 229-246.

Whittenbury, R., K.A. Phillips, J.K. Wilkinson. 1970a. Enrichment isolation and some properties of methane-utilizing bacteria. *J. Gen. Microbiol., 61*:205-218.

Whittenbury, R., S.L. Davies, J.F. Davey. 1970b. Exospores and cysts formed by methane-utilizing bacteria. *J. Gen. Microbiol., 61*:219-226.

Whitton, B.A., J.A. Rother. 1988. Environmental features of deepwater ricefields in Bangladesh during the flood season. In: *1987 International Deepwater Rice Workshop.* International Rice Research Institute, P.O. Box 933, Manila, Philippines, p 47-54.

Williams, R.T, R.L. Crawford. 1984. Methane production in Minnesota

peatlands. *Appl. Environ. Microbiol., 47*:1,266-1,271.
Williams, R.T., R.L. Crawford. 1985. Methanogenic bacteria including an acid tolerant strain from peatlands. *Appl. Environ. Microbiol., 50*:1,542-1,544.
Winfrey, M.R, J.G. Zeikus. 1977. Effect of sulfate on carbon and electron flow during microbial methanogenesis in freshwater sediments. *Appl. Environ. Microbiol., 33*:275-281.
Worakit, S., D.R. Boone, R.A. Mah, M.E. Abdel-Samie, M.M. El-Halwagi. 1986. *Methanobacterium alcaliphilum* sp. nov. an H_2-utilizing methanogen that grows at high pH values. *Int. J. Syst. Bacteriol., 36*:380-382.
Yagi, K., K. Minami. 1990. Effects of organic matter application on methane emission from Japanese paddy fields. In: *Soil and the Greenhouse Effects* (A.F. Bouwman, ed.), John Wiley, p 467-473.
Yamane, I., S. Sato. 1961. Effect of temperature on the formation of gases and ammonium nitrogen in the waterlogged soils. *Rep. Inst. Agric. Res. Tokoku Univ., 12*:1-10.
Yoshida, S. 1981. *Fundamentals of Rice Crop Science.* International Rice Research Institute, P.O. Box 933, Manila, Philippines. 269 p.
Yu, T. 1985. *Physical Chemistry of Paddy Soils.* Springer-Verlag, Berlin.
Zeikus, J.G., D.L. Henning. 1975. *Methanobacterium arboriphilus* sp.nov. an obligate anaerobe isolated from wetwood of living trees. Antonie van Leeuwenhoek. *J. Microbiol. Serol., 41*:543-552.
Zhao, Y., D.R. Boone, R.A. Mah, J.E. Boone, L. Xun. 1989. Isolation and characterization of *Methanocorpusculum labreanum* sp.nov. from the LaBrea Tar Pits. *Int. J. Syst. Bacteriol., 39*:10-13.

Chapter 14

Biomass Burning

JOEL S. LEVINE,[1] WESLEY R. COFER III[2], AND JOSEPH P. PINTO[1]

[1]*Atmospheric Sciences Division, NASA Langley Research Center*
Hampton, Virginia 23665-5225

[2]*Atmospheric Research and Exposure Assessment Laboratory*
U.S. Environmental Protection Agency, Research Triangle Park, North Carolina 27711

1. Introduction

Our planet is a unique object in the solar system due to the presence of a biosphere with its accompanying biomass and the occurrence of fire (Levine, 1991a). The burning of living and dead biomass is a very significant global source of atmospheric gases and particulates. Crutzen and colleagues were the first to assess biomass burning as a source of gases and particulates to the atmosphere (Crutzen et al., 1979; and Seiler and Crutzen, 1980). However, in a recent paper, Crutzen and Andreae (1990) point out that "Studies on the environmental effects of biomass burning have been much neglected until rather recently but are now attracting increased attention." The "increased attention" includes the Chapman Conference on Global Biomass Burning. Much of the information presented here is based on material from this conference (Levine, 1991b). Biomass burning and its environmental implications have also become important research elements of the International Geosphere-Biosphere Program (IGBP) and the International Global Atmospheric Chemistry (IGAC) Project (Prinn, 1991).

The production of atmospheric methane (CH_4) by biomass burning will be assessed. Field measurements and laboratory studies to quantify the emission ratio of methane and other carbon species will be reviewed. The historic database suggests that global biomass burning is increasing with time and is controlled by human activities. Present estimates indicate that biomass burning contributes between about 27 and 80 Teragrams per year (Tg/yr; Tg = 10^{12} grams) of methane to the atmosphere. This represents 5 to 15% of the global annual emissions of methane. Measurements do indicate that biomass burning is the overwhelming source of CH_4 in tropical Africa. However, if the rate of global biomass burning increases at the rate that it has been over the last few decades, then the production of methane from biomass burning may become much more important on a global scale.

2. Gaseous emissions due to biomass burning

Biomass burning includes the combustion of living and dead material in forests, savannas, and agricultural wastes, and the burning of fuel wood. Under the ideal conditions of complete combustion, the burning of biomass material produces carbon dioxide (CO_2) and water vapor (H_2O), according to the reaction:

$$CH_2O + O_2 \rightarrow CO_2 + H_2O \qquad (1)$$

where CH_2O represents the average composition of biomass material. Since complete combustion is not achieved under any conditions of biomass burning, other carbon species, including carbon monoxide (CO), methane (CH_4), nonmethane hydrocarbons (NMHCs), and particulate carbon, result by the incomplete combustion of biomass material. In addition, nitrogen and sulfur species are produced from the combustion of nitrogen and sulfur in the biomass material.

While CO_2 is the carbon species overwhelmingly produced by biomass burning, its emissions into the atmosphere resulting from the burning of savannas and agricultural wastes are largely balanced by its reincorporation back into biomass via photosynthetic activity within weeks to years after burning. However, CO_2 emissions resulting from the clearing of forests and other carbon combustion products from all biomass sources including CH_4, CO, NMHCs, and particulate carbon, are largely "net" fluxes into the atmosphere since these products are not reincorporated into the biosphere when the land is converted to another use.

Biomass material contains about 40% carbon by weight, with the remainder hydrogen (6.7%) and oxygen (53.3%) (Bowen, 1979). Nitrogen accounts for between 0.3 and 3.8% and sulfur for between 0.1 and 0.9%, depending on the nature of the biomass material (Bowen, 1979). The nature and amount of the combustion products depend on the characteristics of both the fire and the biomass material burned. Hot, dry fires with a good supply of oxygen produce mostly carbon dioxide with little CO, CH_4, and NMHCs. The flaming phase of the fire approximates complete combustion, while the smoldering phase approximates incomplete combustion, resulting in greater production of CO, CH_4, and NMHCs. The percentage production of CO_2, CO, CH_4, NMHCs, and carbon ash during the flaming and smoldering phases of burning based on laboratory studies is summarized in Table 1 (Lobert et al., 1991). Typically for forest fires, the flaming phase lasts on the order of an hour or less, while the smoldering phase may last up to a day or more, depending on the type of fuel, the fuel moisture content, wind velocity, topography, etc. For savanna grassland and agricultural waste fires, the flaming phase lasts a few minutes and the smoldering phase lasts up to an hour.

Table 1. Percentage of production of CO_2, CO, CH_4, and NMHCs during flaming and smoldering phases of burning based on laboratory experiments (Lobert et al., 1991).

	Percentage in burning stage	
	Flaming	Smoldering
CO_2	63	37
CO	16	84
CH_4	27	73
NMHCs	33	67

3. Emission ratios

The total mass of the carbon species (CO_2 + CO + CH_4 + NMHCs + particulate carbon) M_C is related to the mass of the burned biomass (m) by M_C = f × m, where f = mass fraction of carbon in the biomass material, i.e., 40-45%. To quantify the production of gases other than CO_2, we must determine the emission ratio (ER) for each species. The emission ratio for each species is defined as:

$$ER = \frac{\Delta X}{\Delta CO_2} \qquad (2)$$

where ΔX is the concentration of the species X produced by biomass burning, and ΔCO_2 is the concentration of CO_2 produced by biomass burning. $\Delta X = X^*$

$- \overline{X}$ where X^* is the measured concentration of X in the biomass burn smoke plume, and $\overline{X}$ is the background (out of plume) atmospheric concentration of the species. Similarly, $\Delta CO_2 = CO_2^* - \overline{CO_2}$, where CO_2^* is the measured concentration in the biomass burn plume, and $\overline{CO_2}$ is the background (out of plume) atmospheric concentration of CO_2.

In general, all species emission factors are normalized with respect to CO_2, as the concentration of CO_2 produced by biomass burning may be directly related to the amount of biomass material burned by simple stoichiometric considerations as discussed earlier. Furthermore, the measurement of CO_2 in the background atmosphere and in the smoke plume is relatively simple.

For the reasons outlined above, it is most convenient to quantify the combustion products of biomass burning in terms of the species emission ratio (ER). Measurements of the emission ratio for CH_4 and CO normalized with respect to CO_2 for diverse ecosystems (for example, wetlands, chaparral, and boreal; for different phases of burning, flaming, smoldering phases and combined flaming and smoldering phases, called "mixed") are summarized in Table 2. Measurements of the emission ratio for CH_4 normalized with respect to CO_2 for various burning sources in tropical Africa are summarized in Table 3.

Table 2. Emission ratios for CO, CH_4, and NMHCs for diverse ecosystems (in units of $\Delta X/\Delta CO_2$, in percent; ± = standard deviation) (Cofer et al., 1991).

		CO	CH_4	NMHCs
Wetlands	Flaming	4.7 ± 0.8	0.27 ± 0.11	0.39 ± 0.17
	Mixed	5.0 ± 1.1	0.28 ± 0.13	0.45 ± 0.16
	Smoldering	5.4 ± 1.0	0.34 ± 0.12	0.40 ± 0.15
Chaparral	Flaming	5.7 ± 11.6	0.55 ± 0.23	0.52 ± 0.21
	Mixed	5.8 ± 2.4	0.47 ± 0.24	0.46 ± 0.15
	Smoldering	8.2 ± 1.4	0.87 ± 0.23	1.17 ± 0.33
Boreal	Flaming	6.7 ± 1.2	0.64 ± 0.20	0.66 ± 0.26
	Mixed	11.5 ± 2.1	1.12 ± 0.31	1.14 ± 0.27
	Smoldering	12.1 ± 1.9	1.21 ± 0.32	1.08 ± 0.18

Table 3. Emission ratio for CH_4 for different fires in tropical Africa (in units of $\Delta CH_4/\Delta CO_2$ in percent; ± = standard deviation) (Delmas et al., 1991).

Type of Combustion	Emission Ratio = $\Delta CH_4/\Delta CO_2$	
	Mean	Range
Natural savanna bushfire	0.28 ± 0.04	0.23 – 0.34
Forest fire	1.23 ± 0.60	0.56 – 2.22
Emissions from traditional charcoal oven	2.06 ± 2.86	6.7 – 14.2
Firewood	1.79 ± 0.81	1.04 – 3.2
Charcoal	0.14	

Table 4. Average emission factors for CO_2, CO, and CH_4 for diverse ecosystems (in units of grams of combustion product carbon to kilograms of fuel carbon; ± = standard deviation) (Radke et al., 1991).

	CO_2	CO	CH_4
Chaparral-1	1644 ± 44	74 ± 16	2.4 ± 0.15
Chaparral-2	1650 ± 31	75 ± 14	3.6 ± 0.25
Pine, Douglas fir and brush	1626 ± 39	106 ± 20	3.0 ± 0.8
Douglas fir, true fir and hemlock	1637 ± 103	89 ± 50	2.6 ± 1.6
Aspen, paper birch and debris from jack pine	1664 ± 62	82 ± 36	1.9 ± 0.5
Black sage, sumac, and chamise	1748 ± 11	34 ± 6	0.9 ± 0.2
Jack pine, white and black spruce	1508 ± 16	175 ± 91	5.6 ± 1.7
"Chained" and herbicidal paper birch and poplar	1646 ± 50	90 ± 21	4.2 ± 1.3
"Chained" and herbicidal birch, polar and mixed hardwoods	1700 ± 82	55 ± 41	3.8 ± 2.8
Debris from hemlock, deciduous and Douglas fir	1600 ± 70	83 ± 37	3.5 ± 1.9
Overall average	1650 ± 29	83 ± 16	3.2 ± 0.5

Table 5. Emission factors for CO_2, CO, CH_4, and NMHC and ash based on laboratory experiments (in % of fuel carbon; ± = standard deviation) (Lobert et al., 1991).

	Mean	Range
CO_2	82.58	49.17 – 98.95
CO	5.73	2.83 – 11.19
CH_4	0.42	0.14 – 0.94
NMHC (as C)	1.18	0.14 – 3.19
Ash (as C)	5.00	0.66 – 22.28
Total sum C	94.91	

Some researchers present their biomass burn emission measurement in the ratio of grams of carbon in the gaseous and particle combustion products to the mass of the carbon in the biomass fuel in kilograms. Average emission factors for CO_2, CO, and CH_4 in these units for diverse ecosystems are summarized in Table 4 and emission factors for CO_2, CO, CH_4, NMHCs, and carbon ash in terms of percentage of fuel carbon based on laboratory experiments are summarized in Table 5. Inspection of Tables 2-5 indicates that there is considerable variability in both the emission ratio and the emission factor for carbon species as a function of ecosystem burning and the phase of burning (flaming or smoldering).

A recent compilation of CO_2-normalized emission ratios for carbon species is listed in Table 6. This table gives the range for both field measurements and laboratory studies and provides a "best guess."

Table 6. CO_2-Normalized emission ratios for carbon species: summary of field measurements and laboratory studies (in units of grams C in each species per kilograms of C in CO_2) (Andreae, 1991).

	Field Measurements	Laboratory Studies	"Best Guess"
CO	6.5 – 140	59 – 105	100
CH_4	6.2 – 16	11 – 16	11
NMHCs	6.6 – 11.0	3.4 – 6.8	7
Particulate organic carbon (including elemental carbon)	7.9 – 54		20
Element carbon (black soot)	2.2 – 16		5.4

4. Emission of methane

Once the mass of the burned biomass (M) and the species emission ratios (ER) are known, the gaseous and particulate species produced by biomass combustion may be calculated. The mass of the burned biomass (M) is related to the area (A) burned in a particular ecosystem by the following relationship (Seiler and Crutzen, 1980):

$$M = A \times B \times \alpha \times \beta \qquad (3)$$

where B is the average biomass material per unit area in the particular ecosystem (g/m^2), α is the fraction of the average above-ground biomass relative to the total average biomass B, and β is the burning efficiency of the above-ground biomass. Parameters B, α, and β vary with the particular ecosystem under study and are determined by assessing the total biomass before and after burning.

The total area burned during a fire may be assessed using satellite data. Recent reviews have considered the extent and geographical distribution of biomass burning from a variety of space platforms: astronaut photography (Wood and Nelson, 1991), the NOAA polar orbiting Advanced Very High Resolution Radiometer (AVHRR) (Brustet et al., 1991a; Cahoon et al., 1991; Robinson, 1991a, 1991b), the Geostationary Operational Environmental Satellite (GOES) Visible Infrared Spin Scan Radiometer Atmospheric Sounder (VAS) (Menzel et al., 1991); and the Landsat Thematic Mapper (TM) (Brustet et al., 1991b).

Hence, the contribution of biomass burning to the total global budget of methane or any other species depends on a variety of ecosystem and fire parameters, including the particular ecosystem that is burning (which determines the parameters B, α and β), the mass consumed during burning, the nature of combustion (complete vs. incomplete), the phase of combustion (flaming vs. smoldering), and knowledge of how the species emission factors (EF) vary with changing fire conditions in various ecosystems. The contribution of biomass burning to the global budgets of any particular species depends on precise knowledge of all these parameters. While all these parameters are known imprecisely, the largest uncertainty is probably associated with the total mass (M) consumed during biomass burning on an annual basis (and there are large year-to-year variations in this parameter!). The total biomass burned annually according to source of burning is summarized in Table 7 (Seiler and Crutzen, 1980; Hao et al., 1990; Crutzen and Andreae, 1990; Andreae, 1991). The estimate for carbon released of 3940 Tg/yr includes all carbon species produced by biomass combustion (CO_2 + CO + CH_4 + NMHCs + particulate carbon). About 90% of the released carbon is in the form of CO_2 (about 3550 Tg/yr).

Table 7. Global estimates of annual amount of biomass burning and the resulting release of carbon to the atmosphere (Seiler and Crutzen, 1980; Crutzen and Andreae, 1990; Hao et al., 1990; and Andreae, 1991).

Source of burning	Biomass burned (Tg/yr)[1]	Carbon released (Tg C/yr)[2]	CH_4 released (Tg CH_4/yr)[3]
Savanna	3690	1660	21.9
Agricultural waste	2020	910	12.0
Fuel wood	1430	640	8.4
Tropical forests	1260	570	7.5
Temperature and boreal forests	280	130	1.7
Charcoal	21	30	0.4
World total	8700	3940	51.9

[1] 1 Tg (teragram) = 10^6 metric tons = 10^{12} grams.
[2] Based on a carbon content of 45% in the biomass material. In the case of charcoal, the rate of burning has been multiplied by 1.4.
[3] Assuming that 90% of the carbon released is in the form of CO_2 and that the "best guess" emission ratio of C:CH_4 to C:CO_2 is 1.1% (see Table 5), and CH_4 (Tg) = C:CH_4 (Tg) * 16/12.

Knowledge of the CO_2-normalized emission ratio for CH_4 coupled with information on the total production of CO_2 due to biomass burning allows us to estimate the total annual global production of CH_4 due to biomass burning. Field measurements and laboratory studies indicate that the emission ratio for CH_4 is in the range of 6.2 to 16 grams of carbon in the form of CH_4 (C:CH_4) per kilogram of carbon in the form of CO_2 (C:CO_2) (see Table 6), which corresponds to a C:CH_4 to C:CO_2 emission ratio in the range of 0.62 to 1.6%. Using a "best guess" of 1.1% and assuming that biomass burning produces about 3550 Tg/yr of C:CO_2, then the global annual production of C:CH_4 due to biomass burning is in the range of 21.7 to 56 Tg/yr, which converts to 29 to 75 Tg/yr of CH_4, with a "best guess" of 52 Tg/yr of CH_4. The production of CH_4 by different burning sources on a global scale is summarized in Table 7. A detailed study using a chemical transport model with a 1° x 1° spatial grid yielded an annual average CH_4 production due to biomass burning of 63.4 Tg (Taylor and Zimmerman, 1991), which is somewhat smaller than the maximum CH_4 production value calculated here of 74.7 Tg CH_4/yr. Assuming that the total annual global production of CH_4 from all sources is about 500 Tg CH_4 (Cicerone and Oremland, 1988), then the range of CH_4 we calculate corresponds to between 6% and 15% of the global emissions of CH_4, while the calculations of Taylor and Zimmerman (1991) suggest that biomass burning produces about 14% of the global emissions of CH_4. Considering all of the uncertainties in these calculations, there is very good agreement between these two estimates. While an upper limit range of about 15% for the production of methane due to biomass burning may not seem very significant, the importance of this source is enhanced when we consider that the largest single global sources of methane do not produce much more than about 20% of the total.

Delmas et al. (1991) have studied the CH_4 budget of tropical Africa. They considered the emission of CH_4 from biogenic processes in the soil and from biomass burning. They found that the dry African savanna soil is always a net sink for CH_4. They measured an average soil uptake rate for atmospheric CH_4 of 2×10^{10} CH_4 molecules cm^{-2} s^{-1}. They calculated the production of CH_4 (and CO_2) due to biomass burning and found that biomass burning supplies about 9.22 Tg CH_4 yr (and 3750 Tg CO_2 yr) (see Table 8). Hence, in tropical Africa, biomass burning, not biogenic emissions from the soil, controls the CH_4 budget.

In addition to the direct production of CH_4 by the combustion of biomass material, there is recent evidence to suggest that burning stimulates biogenic emissions of CH_4 from wetlands. Flux chamber measurements indicate higher fluxes of CH_4 from wetlands following burning. It has been suggested that combustion products, carbon dioxide, carbon monoxide, acetate, and formate entering the wetlands following burning are used by methanogenic bacteria in the metabolic production of CH_4 (Levine et al., 1990).

At present, biomass burning is a significant global source of several important radiatively and chemically active species. Biomass burning may supply 40% of the world's annual gross production of CO_2 or 26% of the world's annual net production of CO_2 (due to the burning of the world's forests) (Seiler and

Crutzen, 1980; Crutzen and Andreae, 1990; Hao et al., 1990; Levine, 1990; Andreae, 1991; Houghton, 1991). Biomass burning supplies 10% of the world's annual production of CH_4, 32% of the CO; 24% of the NMHCs, excluding isoprene and terpenes; 21% of the oxides of nitrogen (nitric oxide and nitrogen dioxide); 25% of the molecular hydrogen (H_2); 22% of the methyl chloride (CH_3Cl); 38% of the precursors that lead to the photochemical production of tropospheric ozone; 39% of the particulate organic carbon (including elemental carbon); and more than 86% of the elemental carbon (Levine, 1990; Andreae, 1991).

Table 8. Total emissions of CO_2 and CH_4 from the burning of biomass in tropical Africa.

Source	Biomass burned[1]	CO_2 Emission Factor[2]	CH_4 Emission Factor[2]	CO_2 Emissions[3]	CH_4 Emissions[4]
Savanna bushfires	2.52	1370	1.65	3.45	4.14
Forest fires	0.13	957	6.94	0.12	0.90
Firewood burning	0.12	957	5.42	0.11	0.65
Charcoal production	0.11	641	21.0	0.07	2.31
Total	2.88			3.75	9.22

[1] Biomass burned in units of Gigatons dry matter = 10^9 metric tons = 10^3 Tg = 10^{15} grams
[2] Emission factors units of g gas/kg dry matter
[3] CO_2 emissions in units of Gigatons/yr
[4] CH_4 emissions in units of Teragram/yr

5. Historic changes in biomass burning

It is generally accepted that the emissions from biomass burning have increased in recent decades, largely as a result of increasing rates of deforestation in the tropics. Houghton (1991) estimates that gaseous and particulate emissions to the atmosphere due to deforestation have increased by a factor of 3 to 6 over the last 135 years. He also believes that the burning of grasslands, savannas, and agricultural lands has increased over the last century because rarely burned ecosystems, such as forests, have been converted to frequently burned ecosystems, such as grasslands, savannas, and agricultural lands. In Latin America, the area of grasslands, pastures, and agricultural lands increased by about 50% between 1850 and 1985. The same trend is true for South and Southeast Asia. In summary, Houghton (1991) estimates that total biomass burning may have increased by about 50% since 1850. Most of the increase results from the ever-increasing rates of forest burning, with other contributions of burning (grasslands, savannas, and agricultural lands) having increased by 15% to 40%. The increase in biomass burning is not limited to the tropics. In analyzing 50 years of fire data from the boreal forests of Canada, the U.S.S.R., the Scandinavian countries, and Alaska, Stocks (1991) has reported a dramatic increase in area burned in the 1980s. The largest fire in the recent past destroyed more than 12 million acres of boreal forest in the People's Republic of China and Russia in a period of less than a month in May 1987 (Cahoon et al., 1991).

The historic data indicate that biomass burning has increased with time and that the production of greenhouse gases from biomass burning has increased with time. Furthermore, the bulk of biomass burning is human-initiated. As greenhouse gases build up in the atmosphere and the Earth becomes warmer, there may be an enhanced frequency of fires. The enhanced frequency of fires may prove to be an important positive feedback in a warming Earth. However, it has been suggested that the bulk of biomass burning worldwide may be significantly reduced (Andrasko et al., 1991). Policy options for mitigating biomass burning have been developed by Andrasko et al. (1991). For mitigating burning in the tropical forests, where much of the burning is aimed at land clearing and conversion to agricultural lands, policy options include the marketing of timber as a resource and improved productivity of existing agricultural lands to reduce the need for conversions of forests to agricultural lands. Improved productivity will result from the application of new agricultural technology, e.g., fertilizers. For mitigating burning in tropical savanna grasslands, animal grazing could be replaced by stall feeding since savanna burning results from the need to replace nutrient-poor tall grass with nutrient-rich short grass. For mitigating burning on agricultural lands and croplands, incorporate crop wastes into the soil, instead of burning, as is the present practice throughout the world. The crop wastes could also be used as fuel for household heating and cooking rather than cutting down and destroying forests for fuel as is presently done.

6. Uncertainties and future research

The construction of a global emissions inventory for methane from biomass burning must account for the high degree of variability of these emissions in both space and time. Biomass burning exhibits strong seasonal and geographic variations. As shown earlier, methane emissions from biomass burning are highly dependent on the type of ecosystem being burned, which determines the total amount of biomass consumed and the extent of flaming and smoldering phases during combustion. The calculations by Taylor and Zimmerman (1991) go a long way towards deriving a global inventory in that they have simulated the variability of biomass burning. They scaled the burning rate inversely with precipitation as global data sets are currently not available. Satellite techniques, when they are developed, offer a promising way to obtain global coverage.

Taylor and Zimmerman also used a constant emission ratio in their calculations since measurements of the emission ratio for methane are lacking for many different ecosystems. While some data exist for mid-latitude ecosystems, measurements are needed to better define the contributions from burning tropical forests and savannas. In addition, airborne measurements are limited to the outer edges of biomass burn plumes so little is known about variability across the plume. The use of long path remote measurements across plumes is also planned for the future.

7. Summary

Biomass burning may be the overwhelming regional or continental-scale source of CH_4 as in tropical Africa and a significant global source of CH_4. Our best estimate of present methane emissions from biomass burning is about 51.9 Tg/yr, or 10% of the annual methane emissions to the atmosphere. Increased frequency of fires that may result as the Earth warms up may result in increases in this source of atmospheric methane.

It is appropriate to conclude this chapter with an observation of fire historian, Stephen Pyne (1991):

"We are uniquely fire creatures on a uniquely fire planet, and through fire the destiny of humans has bound itself to the destiny of the planet."

References

Andrasko, K.J., D.R. Ahuja, S.M. Winnett, D.A. Tirpak. 1991. Policy options for managing biomass burning to mitigate global climate change. In: *Global Biomass Burning: Atmospheric, Climatic, and Biospheric Implications* (J.S. Levine, ed.), The MIT Press, Cambridge, Massachusetts, 445-456.

Andreae, M.O. 1991. Biomass burning: Its history, use, and distribution and its impact on environmental quality and global climate. In: *Global Biomass Burning: Atmospheric, Climatic, and Biospheric Implications* (J.S. Levine, ed.), The MIT Press, Cambridge, Massachusetts, 3-21.

Bowen, H.J.M. 1979. *Environmental Chemistry of the Elements.* Academic Press: London, England.

Brustet, J.M., J.B. Vickos, J. Fontan, K. Manissadjan, A. Podaire, F. Lavenue. 1991a. Remote sensing of biomass burning in West Africa with NOAA-AVHRR. In: *Global Biomass Burning: Atmospheric, Climatic, and Biospheric Implications* (J.S. Levine, ed.), The MIT Press, Cambridge, Massachusetts, 47-52.

Brustet, J.M., J.B. Vickos, J. Fontan, A. Podaire, F. Lavenue. 1991b. Characterization of active fires in West African savannas by analysis of satellite data: Landsat thematic mapper. In: *Global Biomass Burning: Atmospheric, Climatic, and Biospheric Implications* (J.S. Levine, ed.), The MIT Press, Cambridge, Massachusetts, 53-60.

Cahoon, D.R., Jr., J.S. Levine, W.R. Cofer, III, J.E. Miller, P. Minnis, G.M. Tennille, T.W. Yip, B.J. Stocks, P.W. Heck. 1991. The great Chinese fire of 1987: A view from space. In: *Global Biomass Burning: Atmospheric, Climatic, and Biospheric Implications* (J.S. Levine, ed.), The MIT Press, Cambridge, Massachusetts, 61-66.

Cicerone, R.J., R.S. Oremland. 1988. Biogeochemical aspects of atmospheric methane. *Global Biogeochem. Cycles,* 2:299-327.

Cofer, W.R. III, J.S. Levine, E.L. Winstead, B.J. Stocks. 1991. Trace gas and particulate emissions from biomass burning in temperate ecosystems. In: *Global Biomass Burning: Atmospheric, Climatic, and Biospheric Implications* (J.S. Levine, ed.), The MIT Press, Cambridge, Massachusetts, 203-208.

Crutzen, P.J., L.E. Heidt, J.P. Krasnec, W.H. Pollock, W. Seiler. 1979. Biomass burning as a source of atmospheric gases CO, H_2, N_2O, NO, CH_3Cl, and COS. *Nature,* 282:253-256.

Crutzen, P.J., M.O. Andreae. 1990. Biomass burning in the tropics: Impact on atmospheric chemistry and biogeochemical cycles. *Science,* 250:1,669-1,678.

Delmas, R.A., A. Marenco, J.P. Tathy, B. Cros, J.G.R. Baudet. 1991. Sources and sinks of methane in the African savanna. CH_4 emissions from biomass burning. *J. Geophys. Res.,* 96:7,287-7,299.

Hao, W.M., M.H. Liu, P.J. Crutzen. 1990. Estimates of annual and regional release of CO_2 and other trace gases to the atmosphere from fires in the tropics, based on the FAO statistics for the period 1975-1980. In: *Fire in the Tropical Biota: Ecosystem Processes and Global Challenges* (J.G. Goldammer, ed.), Springer-Verlag, Berlin-Heidelberg, Germany, 440-462.

Houghton, R.A. 1991. Biomass burning from the perspective of the global carbon cycle. In: *Global Biomass Burning: Atmospheric, Climatic, and Biospheric Implications* (J.S. Levine, ed.), The MIT Press, Cambridge, Massachusetts, 321-325.

Levine, J.S. 1990. Global biomass burning: Atmospheric, climatic, and biospheric implications. *EOS Transactions (AGU), 71:*1,075-1,077.

Levine, J.S. 1991a. The biosphere as a driver for global atmospheric change. In: *Scientists on Gaia* (S.H. Schneider and P.J. Boston, eds.), The MIT Press, Cambridge, Massachusetts, 353-361.

Levine J.S. (ed.). 1991b. *Global Biomass Burning: Atmospheric, Climatic, and Biospheric Implications*, The MIT Press, Cambridge, Massachusetts, 569 pages.

Levine, J.S., W.R. Cofer, III, D.I. Sebacher, R.P. Rhinehart, E.L. Winstead, S. Sebacher, C.R. Hinkle, P.A. Schmalzer, A.J. Koller, Jr. 1990. The effects of fire on biogenic emissions of methane and nitric oxide from wetlands. *J. Geophys. Res., 95:*1,853-1,864.

Lobert, J.M., D.H. Scharffe, W.-M. Hao, T.A. Kuhlbusch, R. Seuwen, P. Warneck, P.J. Crutzen. 1991. Experimental evaluation of biomass burning emissions: Nitrogen and carbon containing compounds. In: *Global Biomass Burning: Atmospheric, Climatic, and Biospheric Implications* (J.S. Levine, ed.), The MIT Press, Cambridge, Massachusetts, 289-304.

Menzel, W.P., E.C. Cutrim, E.M. Prins. 1991. Geostationary satellite estimation of biomass burning in Amazonia during BASE-A. In: *Global Biomass Burning: Atmospheric, Climatic, and Biospheric Implications* (J.S. Levine, ed.), The MIT Press, Cambridge, Massachusetts, 41-46.

Prinn, R.G. 1991. Biomass burning studies and the International Global Atmospheric Chemistry (IGAC) Project. In: *Global Biomass Burning: Atmospheric, Climatic, and Biospheric Implications* (J.S. Levine, ed.), The MIT Press, Cambridge, Massachusetts, 22--28.

Pyne, S.J. 1991. Sky of ash, earth of ash: A brief history of fire in the United States. In: *Global Biomass Burning: Atmospheric, Climatic, and Biospheric Implications* (J.S. Levine, ed.), The MIT Press, Cambridge, Massachusetts, 504-511.

Radke, L.F.; D.A. Hegg, P.V. Hobbs, J.D. Nance, J.H. Lyons, K.K. Laursen, R.E. Weiss, P.J. Riggan, D.E. Ward. 1991. Particulate and trace gas emissions from large biomass fires in North America. In: *Global Biomass Burning: Atmospheric, Climatic, and Biospheric Implications* (J.S. Levine, ed.), The MIT Press, Cambridge, Massachusetts, 209-224.

Robinson, J.M. 1991a. Fire from space: Global fire evaluation using IR remote sensing. *Intern. Journ. of Remote Sensing, 12*:3-24.
Robinson, J.M. 1991b. Problems in global fire evaluation: Is remote sensing the solution? In: *Global Biomass Burning: Atmospheric, Climatic, and Biospheric Implications* (J.S. Levine, ed.), The MIT Press, Cambridge, Massachusetts, 67-73.
Seiler, W., P.J. Crutzen. 1980. Estimates of grass and net fluxes of carbon between the biosphere and the atmosphere from biomass burning. *Climatic Change, 2*:207-247.
Stocks, B.J. 1991. The extent and impact of forest fires in north circumpolar countries. In: *Global Biomass Burning: Atmospheric, Climatic, and Biospheric Implications* (J.S. Levine, ed.), The MIT Press, Cambridge, Massachusetts, 197-202.
Taylor, J.A., P.R. Zimmerman. 1991. Modeling trace gas emissions from biomass burning. In: *Global Biomass Burning: Atmospheric, Climatic, and Biospheric Implications* (J.S. Levine, ed.), The MIT Press, Cambridge, Massachusetts, 345-350.
Wood, C.A., R. Nelson. 1991. Astronaut observations of global biomass burning. In: *Global Biomass Burning: Atmospheric, Climatic, and Biospheric Implications* (J.S. Levine, ed.), The MIT Press, Cambridge, Massachusetts, 29-40.

Chapter 15

Wetlands

ELAINE MATTHEWS

Hughes STX Corporation, NASA Goddard Space Flight Center
Institute for Space Studies, 2880 Broadway, New York, New York 10025, U.S.A.

1. Introduction

Wetlands are most likely the largest natural source of methane to the atmosphere (Khalil and Rasmussen, 1983; Cicerone and Oremland, 1988; Fung et al., 1991), accounting for ~20% of the current global annual emission of ~450-550 Tg (10^{12} g). Measurements of methane from Greenland and Antarctic ice cores indicate atmospheric concentrations of ~350 ppbv during the Last Glacial Maximum rising to 650 ppbv during the pre-industrial Holocene (Stauffer et al., 1988; Chappellaz et al., 1990).

Pre-industrial source strengths of methane, consistent with historical concentrations and estimates based on isotopes, have been estimated at ~180-380 Tg methane (Stauffer et al., 1985; Khalil and Rasmussen, 1987; Chappellaz et al., 1993). Wetlands are understood to have been the dominant source with smaller contributions from wild fires, animals and oceans. During the last two hundred years, atmospheric methane concentrations have more than doubled to ~1700 ppbv and are increasing at 0.8-1.0%/year (Rasmussen and Khalil, 1981; Craig and Chou, 1982; Khalil and Rasmussen, 1982, 1983, 1985; Ehhalt et al., 1983; Blake, 1984; Rasmussen and Khalil, 1984; Stauffer et al., 1985; Blake and Rowland, 1986, 1988; Pearman et al., 1986; Steele et al., 1987; Khalil et al., 1989; Lang et al., 1990a,b). Currently, the total annual emission of methane is about twice that estimated for the pre-industrial period, but both the relative and absolute contribution of wetlands is smaller than in the past due to reductions in wetland areas. Climate and related biological interactions that presently control the spatial extent of wetlands and their methane emissions are expected to change during the next 50 to 100 years.

The role of wetlands in the cycle of methane has been studied for several decades. Early estimates of methane emissions from this source were based on very few measurements and highly uncertain information about the areal extent of wetlands (e.g., Koyama, 1964; Ehhalt, 1974; Ehhalt and Schmidt, 1978; Blake, 1984; Seiler, 1984; Holzapfel-Pschorn and Seiler, 1986; Seiler and Conrad, 1987). While all these studies relied on the same wetland area of $2.6 \times 10^{12} m^2$ (from Twenhofel, 1926, 1951), the estimated global methane emission ranged from 11 to 300 Tg/yr, although the more recent estimates lie in the lower half of the range. This wide range of emission figures reflects differing assumptions about the magnitude and annual duration of methane fluxes. Sebacher et al. (1986) suggested a potential wetland area of $4.5-9.0 \times 10^{12} m^2$ for northern peatlands and estimated an annual methane emission from them of 45-106 Tg.

Using newer information about wetland distributions and environmental characteristics, several recent estimates have converged around 100 Tg (Matthews and Fung, 1987; Aselmann and Crutzen, 1989; Bartlett et al., 1990) using global wetland areas of $5-6 \times 10^{12} m^2$. However, the similarities mask some major uncertainties about seasonal methane-production periods as well as important differences in the relative importance of the role of climatically and ecologically distinct wetland ecosystems, i.e., tropical/subtropical wetlands whose methane emissions are governed generally by large scale precipitation and flood cycles, and high-latitude wetlands whose highly seasonal emissions are controlled via interactions between temperature and water cycles. A complex suite of environmental parameters including soil chemistry, substrate quality, and soil water status influence emissions in all these environments, further complicating the task of evaluating emissions.

The last decade has seen major developments in understanding the role of wetlands in the global methane cycle. The total data set of measured methane fluxes and concentrations in various components of wetland ecosystems has expanded significantly. Measurements of the isotopic composition of methane in the atmosphere and from a variety of sources are increasing (see Stevens, this volume). A globally distributed network of stations measuring atmospheric constituents has yielded important information on seasonal, temporal, and geographic variations of atmospheric methane. Modeling and synthesis techniques for analyzing the information have improved. In regard to wetland ecosystems in particular, data expansion has occurred in several domains. (1) Measurement studies now cover many ecosystems characteristic of global wetland areas not studied a decade ago (e.g., Africa, South America). (2) Periodic or single-date warm-season measurements characteristic of early field studies have been augmented (a) with measurements that span full growing seasons or full years in high-latitude environments (Alaska, Minnesota, Canada), (b) with measurements conducted during wet and dry seasons in tropical ecosystems (South America, Africa), and (c) in the case of Alaskan and Canadian ecosystems, with time series of several years. (3) Large-scale interdisciplinary field campaigns, designed to characterize chemistry and dynamics of the regional troposphere, have been conducted in Alaska, Canada, Central Africa, and the Amazon Basin, integrating ground-based measurements (chambers, balloons, towers, floating platforms) with aircraft flights, and satellite overpasses.

This paper provides an overview of the role of natural wetlands in the global cycle of methane by addressing the following questions: What is the current status of information on the extent and environmental characteristics of wetlands on a global scale? How well represented are major wetland ecosystems in the flux measurements and what are the large-scale patterns of methane emission? Where do the largest uncertainties remain and how can they be reduced? Section 2 discusses wetland distributions, characteristics and seasonality on a global scale. Flux measurements at local sites and emission estimates at regional and global scales are discussed in Section 3. Section 4 focuses on wetlands and climate change, reporting on a series of studies evaluating the potential response of wetlands to predicted future changes in temperature and water status. The final section provides a summary of recent progress in understanding the role of natural wetlands in the methane budget, and outlines some remaining research questions.

2. Distribution and characteristics of natural wetlands

2.1 Areal distributions and seasonality. Early global estimates of methane emission from wetlands relied on very general information about wetland areas. As noted, most of these estimates were based on a global wetland area of $2.6 \times 10^{12} m^2$ put forward in the 1920s (Twenhofel, 1926). More recently, several groups have compiled data sets specifically designed to evaluate methane emissions from natural wetlands. Matthews and Fung (1987) derived a data set by integrating three global digital data bases on vegetation, ponded soils and fractional inundation at 1° latitude by 1° longitude resolution, while that of Aselmann and Crutzen (1989) was compiled at 2.5° resolution from regional and local wetland reports. Global wetland areas derived from these studies are 5.3 and 5.7 x $10^{12} m^2$, respectively. Relative regional distributions of areas from the two works are very similar (Figure 1). About one-half of the total area lies between 50° and 70°N. This high-latitude region is characterized by peat-rich ecosystems (bogs and fens) and a temperature-restricted thaw season resulting in highly seasonal emissions of methane. Approximately 35% of the global wetland area is broadly distributed in the latitude zone extending from 20°N to 30°S. This region is co-dominated by forested and nonforested swamps and marshes, with a smaller contribution from alluvial or floodplain formations. Some of these tropical wetlands, particularly floodplain habitats, undergo large seasonal expansion and contraction in response to precipitation cycles. Since many of them lie along river courses with exceptionally level topography over large distances, rapid and significant changes in inundation during the year are common.

Differences between the two studies are discussed in Aselmann and Crutzen (1989). In brief, Aselmann and Crutzen's areas are slightly lower in the northern subtropics (10°-30°N) and in the southern zone between 20°S and 40°S, which accounts for 9% of the Matthews and Fung total and 2% of the Aselmann and Crutzen total. The southern subtropical differences come primarily from the inclusion by Matthews and Fung of what are probably infrequently-flooded ephemeral wetlands in arid regions of Australia. The two studies exhibit a larger areal discrepancy in the southern tropics from the equator to 10°S. Aselmann and Crutzen indicate that ~20% of the global total occurs in this narrow tropical zone, whereas Matthews and Fung arrived at a value equal to only ~10% of the global total. Causes for these tropical discrepancies are not clear although the studies relied on different sources and compilation methodologies. A sizeable portion of the locations that disagree coincide with river systems that are associated with the largest areal uncertainties.

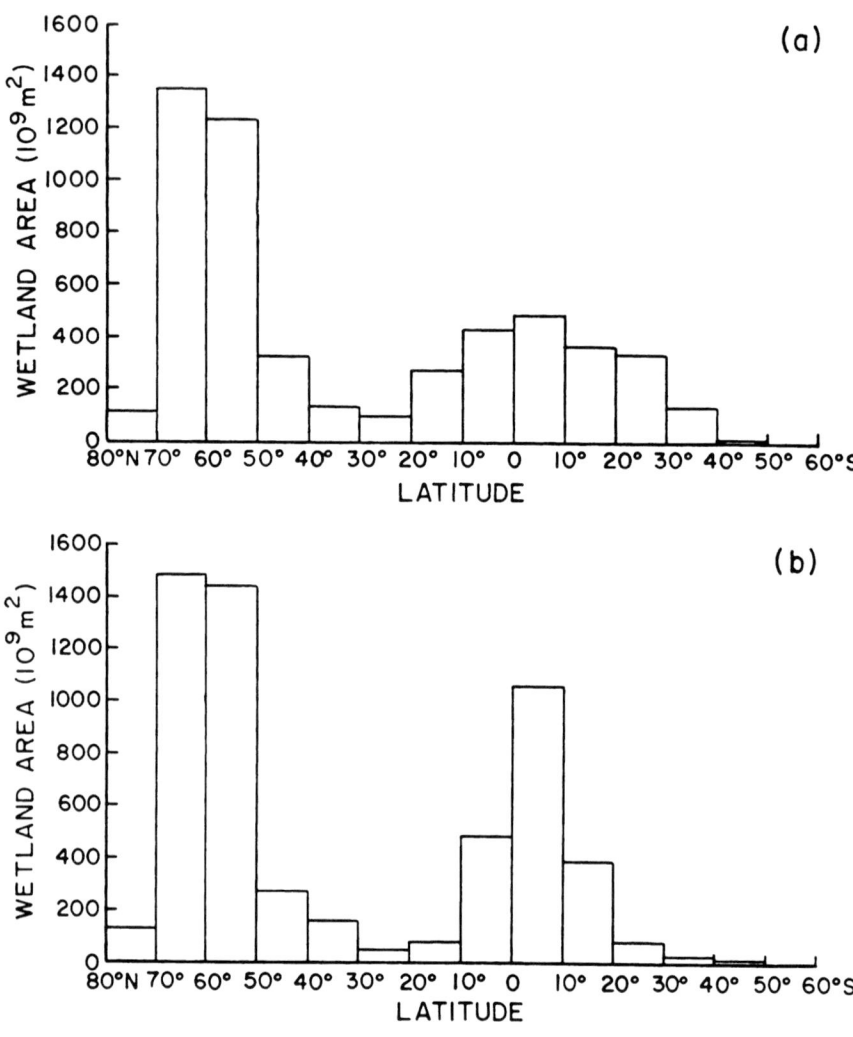

Figure 1. Latitudinal distribution of wetland areas from the data base of (a) Matthews and Fung (1987) and (b) Aselmann and Crutzen (1989). Total wetland areas are 5.3 and 5.7 x $10^{12} m^2$, respectively.

2.2 Improving data on wetland areas and seasonality. Uncertainties in estimating methane emissions from natural wetlands result from difficulties in determining areal coverage of various wetland habitats, which may vary during the year, and seasonality of inundation periods. Aselmann and Crutzen (1989) attempted a global estimate of maximum and minimum extent of wetland areas and months of methane production using hydrological and meteorological observations and wetland descriptions. They and others (see regional case studies included in Gore, 1983b) confirm that considerable uncertainties remain with respect to the seasonality of both areas and inundation. Seasonal expansion and contraction of tropical/subtropical swamps in response to precipitation and flood cycles is particularly difficult to estimate. Bartlett et al. (1990) note that estimates of total Amazon floodplain area range over an order of magnitude and areas covered by the three main categories of floodplain habitats within the Amazon are poorly known. In contrast, the methane production season for high latitude ecosystems is regulated largely by temperature and habitats undergo minor changes in area over seasonal timescales. Satellite-based remote sensing techniques (microwave, visible and near-infrared channels, radar) may offer some opportunities for improving areal estimates of wetland habitats on a seasonal basis (Rose and Rosendahl, 1983; Ormsby et al., 1985; Bartlett et al., 1989). However, these techniques may be useful primarily in herbaceous environments characterized by standing water or for determining extent of open water in pond-dominated northern wetlands such as the Hudson Bay Lowlands and some parts of Alaska. Seasonal series of Landsat and TM imagery, which could improve areal estimates of habitats and inundated conditions are difficult to obtain and analyze for tree-dominated or cloudy environments of tropical regions. Satellite remote sensing appears promising for evaluating distributions of northern wetland habitats (Morrissey and Ennis, 1981; Walker et al., 1982; Bartlett et al., 1992; Roulet et al., 1992a) despite the difficulty of distinguishing variations in water status of soils in boreal and arctic environments.

2.3 Wetland classification. Both data bases discussed above used simplified groupings of detailed wetland information primarily because these generalizations matched the ecosystem classes represented in the methane-flux measurements available in the late 1980s. Matthews and Fung (1987) classified 28 wetland vegetation types as described in the UNESCO (1973) vegetation classification system, in addition to ~100 other vegetation types occupying locations identified as wetlands using information on ponded soils and inundation, into five major groups. The simplification was guided by methane-important ecological and environmental criteria such as seasonality, soil organics, and vegetative cover. The resulting groups are forested and nonforested bog, forested and nonforested swamp, and alluvial formations.

Following regional wetland classifications of the local sources used in data collection, Aselmann and Crutzen (1989) identified 45 freshwater wetland types globally, which were grouped into six broad categories following Level I of the hierarchical wetland classification devised for Canada by Zoltai and Pollett (1983). To accommodate tropical environments, they added floodplains to Zoltai and Pollett's (1983) system, which classes wetlands according to physiognomic features such as peat structure, vegetation cover, and inundation depth and seasonality. The resulting groups are bog, fen, swamp, marsh, floodplain, and shallow lakes. With the exception of shallow lakes, the generalized classes of Aselmann and Crutzen (1989) and Matthews and Fung (1987) are comparable.

In the few years since publication of these data bases, field measurements characterizing methane production and emission from varied wetland ecosystems over extended time periods have increased significantly. This broadening of habitat coverage might now justify using more of the information contained in the wetland data sets to either reclassify or expand the classification of methane-significant ecosystems (Gore, 1983a,b; Glaser, 1987; National Wetlands Working Group, 1988). Both data bases are flexibly designed to allow for alternative horizontal or vertical groupings. Furthermore, the geographic structure of the data bases allows integration of information from compatible data sets of, e.g., temperature, precipitation, soil type, soil-water status, etc.

3. Distribution and characteristics of methane emission from wetlands
   3.1 Flux Measurements. Several extensive syntheses of methane measurements and of techniques for estimating methane emission from wetlands have been provided recently by Bartlett et al. (1990) for temperate, subtropical,

and tropical wetlands and by Harriss et al. (1993) for northern high-latitude wetlands. These works have been integrated and expanded (Bartlett and Harriss, 1993). The reader is referred to those publications for a critical discussion of the measurements.

Methane emissions from natural wetlands show large variability resulting from the complex suite of environmental factors that affect the production of methane via anaerobic decomposition of organic material and the factors that affect consumption and release of methane to the atmosphere. Fluxes measured at field sites and from soil samples have been independently correlated with local environmental and ecological factors. These factors include temperature (Koyama, 1963; Baker-Blocker et al., 1977; King and Wiebe, 1978; Harriss et al., 1983; Mayer, 1982; Moore and Knowles, 1987, 1990; Crill et al., 1988a; Wilson et al., 1989), water status (Svensson, 1976, 1980; Harriss et al., 1982; Svensson and Rosswall, 1984; Sebacher et al., 1986; Harriss et al., 1988b; Whalen and Reeburgh, 1988; Moore and Knowles, 1989; Morrissey and Livingston, 1992; Roulet et al., 1992a), nutrient input and organic accumulation (Harriss and Sebacher, 1981; Svensson and Rosswall, 1984; Wilson et al., 1989; Morrissey and Livingston, 1992), substrate characteristics (Sebacher et al., 1986), vegetation characteristics and phenology (Dacey and Klug, 1979; Cicerone and Shetter, 1981; Sebacher et al., 1985; Wilson et al., 1989), redox potential (Svensson and Rosswall, 1984), and biomass and carbon exchange (Aselmann and Crutzen, 1989; Whiting et al., 1991). Although fluxes have been correlated with a variety of individual environmental variables by various researchers, no general quantitative relationships have been found or are expected to be found that apply to all wetland environments. Studies of several years' duration in high latitude fens and tundras (Moore et al., 1990; Whalen and Reeburgh, 1992) lead these researchers to conclude that factors influencing fluxes are not entirely independent and that variables that serve as environmental integrators of methane production and consumption processes may be more successful predictors of fluxes.

A major development in understanding the role of tropical wetland environments in the global methane budget was provided by a series of large-scale inter-disciplinary field campaigns designed to characterize the chemistry and dynamics of the regional troposphere. The ABLE 2 and CAMREX missions (Bartlett et al., 1988, 1990; Crill et al., 1988b; Devol et al., 1988, 1990; Harriss et al., 1988a, 1990) were carried out in Amazonian Brazil in the 1985 dry season (ABLE 2A) and the 1987 wet season (ABLE 2B). The TROPOZ (Tropospheric Ozone) and DECAFE (Dynamique et Chimie de l'Atmosphere en Foret Equatoriale) experiments were carried out in wet and dry seasons in Central Africa during several field seasons during 1987-1989 (Delmas et al., 1992; Fontan et al., 1992; Tathy et al., 1992;). Until these campaigns, low latitude wetlands had received little attention. In addition, early studies in low latitudes had concentrated in subtropical environments of the southeastern U.S. characterized by very low emissions (Harriss and Sebacher, 1981; Harriss et al., 1982; Bartlett et al., 1985; Barber et al., 1988). The overall results from these later missions confirm that tropical wetlands are larger sources of methane than early studies suggested. This larger role is due primarily to higher emission rates, measured at a series of representative wetland sites, than the rates of more subtropical/temperate environments measured earlier in the 1980s and used as tropical proxies in early estimates. Preliminary results from the Arctic Boundary Layer Experiments (ABLE 3A in Alaska in the summer of 1988 and ABLE 3B in Canada in the summer of 1990), as well as additional measurements in Alaska and Canada (Whalen and Reeburgh, 1988, 1990, 1992; Moore et al., 1990; Morrissey and Livingston, 1992; Ritter et al., 1992; Roulet et al., 1992a,b), show a pattern of fluxes from northern wetlands lower than those found earlier by Sebacher et al. (1986) and Crill et al. (1988a). The low values are consistent with some of those measured by Svensson (1980) and Svensson and Rosswall (1984). These patterns partially reflect full-season field measurements that capture variations around the peak fluxes measured earlier as well as the inclusion of less productive sites that may occupy significant areas in boreal and arctic regions (Ritter et al., 1992).

Using the compilation of Bartlett and Harriss (1993) as a base, Table 1 shows continental/regional wetland areas along with a summary of emission measurements in wetland environments of these regions. (Note that a "measurement" here is a single measurement or suite of measurements in one wetland habitat for a time period lasting from minutes to an entire season and therefore can encompass hundreds of chamber emplacements.) Despite the increase in field studies during the last few years, some gaps remain. Northern high latitude systems, which account for about half the world's wetland area, are more completely covered than are tropical and subtropical habitats. In North America, this comprehensive coverage includes a breadth of wetland habitats measured over complete seasonal cycles and over several years. In contrast, although measurements in South America have been conducted in both wet and dry seasons, they are exclusively in environments closely associated with the Amazon River and do not include nonriverine seasonally flooded grasslands such as the Pantanal. The largest wetlands in the former Soviet Union, a country which accounts for ~30% of the world's wetlands, are the Siberian Lowlands in which flux measurements have not yet been made. Observational data from African forested wetlands is still scanty and is completely lacking from flooded herbaceous wetlands in Africa. Methane flux measurements in Asian wetlands are absent from the published literature although some measurements have been attempted in the bog forests of Malaysia (Paul Glaser, personal communication).

Table 1. Regional/continental distribution of wetland areas and emission measurements. (Note: "measurement" here is a single measurement or suite of measurements in a wetland habitat for a time period lasting from minutes to an entire season.) Areas from Aselmann and Crutzen (1989); measurements derived from compilations of Bartlett and Harriss (1993), Harriss et al. (1993), Bartlett et al. (1990) and others. Numbers in parentheses are Alaskan subset of USA data.

Region	Area		# Measurements
	$10^9 m^2$	% Total	
Central/South America	1542	27	34
USSR	1512	27	?
Canada	1268	22	48
USA (Alaska)	553 (325)	10 (6)	80 (50)
Africa	355	6	14
Asia/Oceania	307	5	?
Europe	154	3	6
Total	5691	100	182

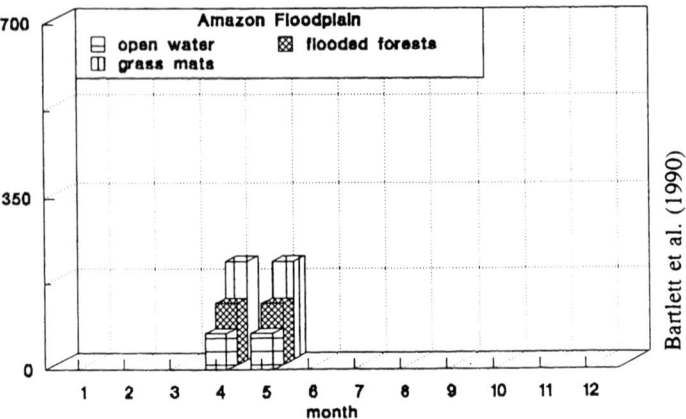

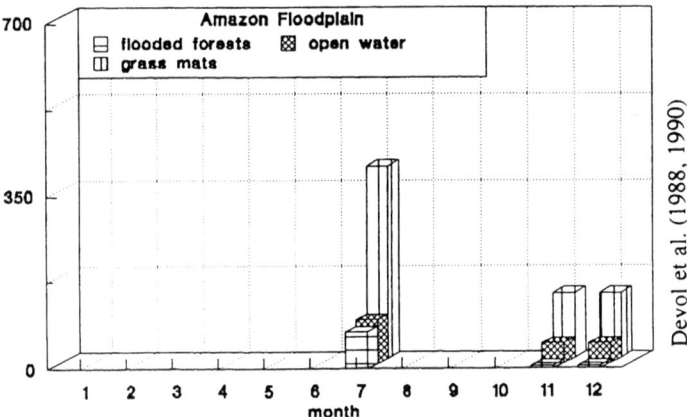

Figure 2. Sample of measurements of seasonal methane flux from (a) tropical, (b) subtropical, (c) temperate/low boreal, and (d) high boreal/arctic wetland ecosystems. (Compiled from Bartlett et al., 1990; Harriss et al., 1993; Bartlett and Harriss, 1992; and others). Unit for methane flux (Y-axis) for all panels is mg/m^2/d.

Figure 2a Tropical (continued)

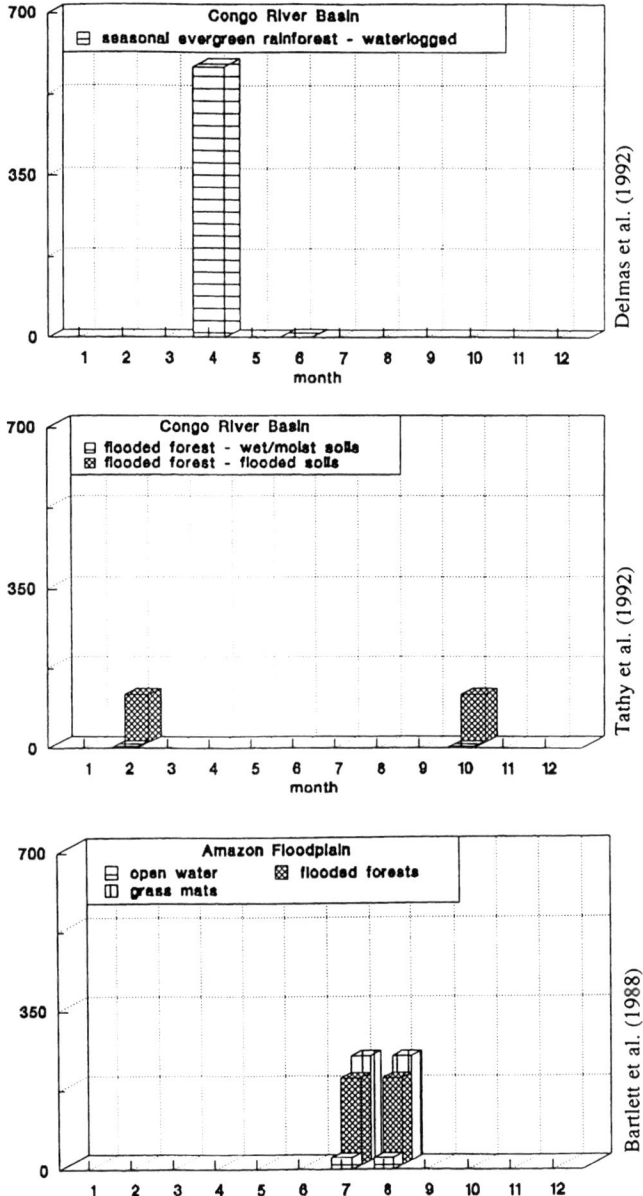

Figure 2a Tropical (continued)

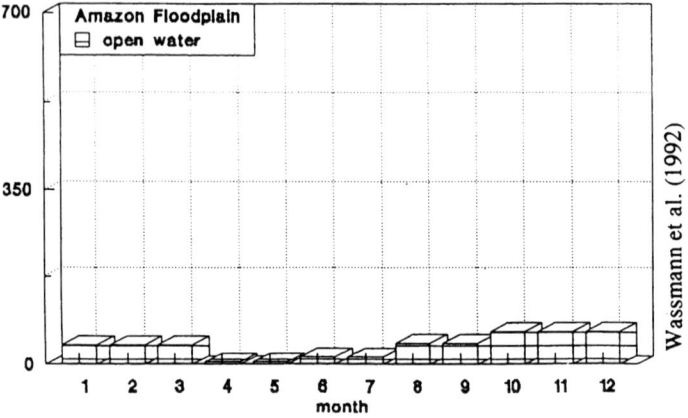

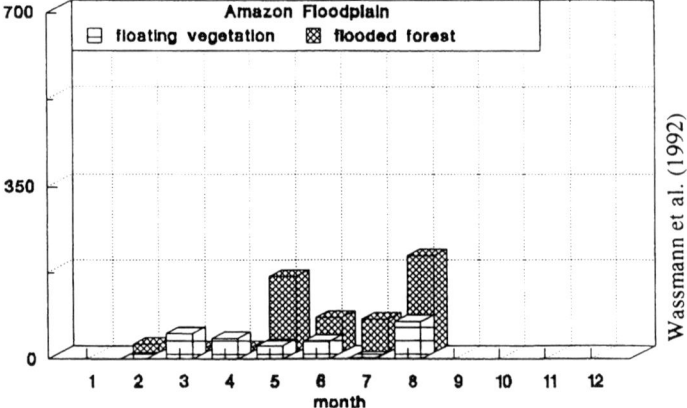

Figure 2b Subtropical

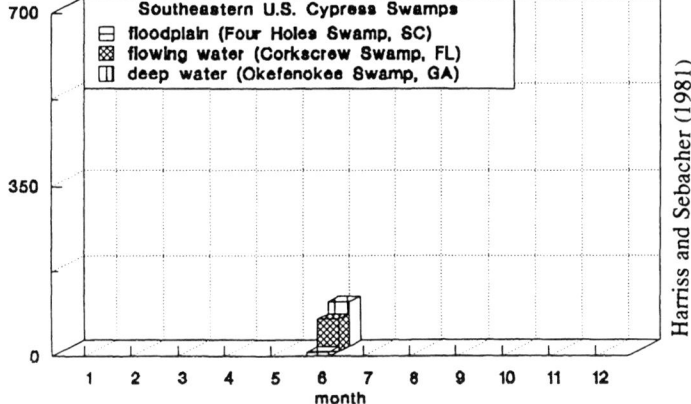

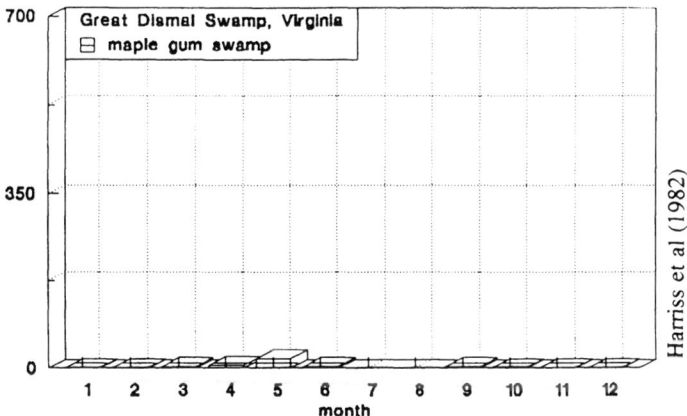

Figure 2b Subtropical (continued)

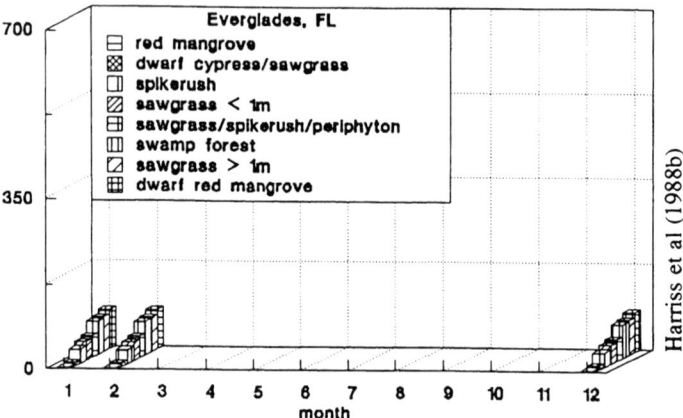

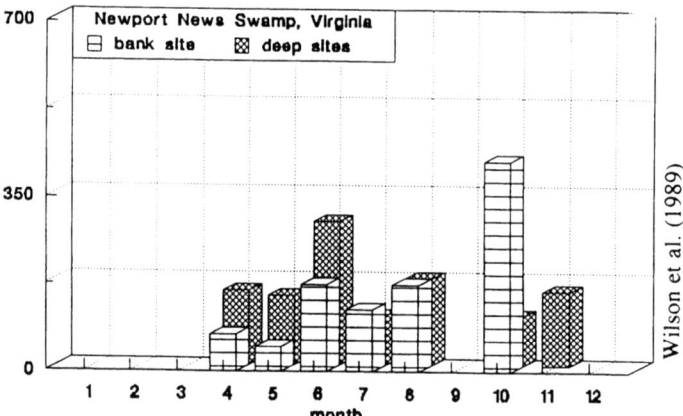

Figure 2c Temperate/Low Boreal

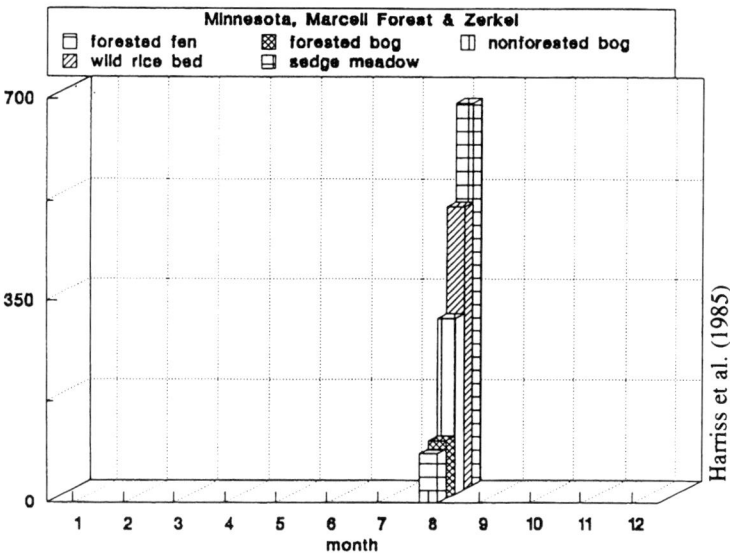

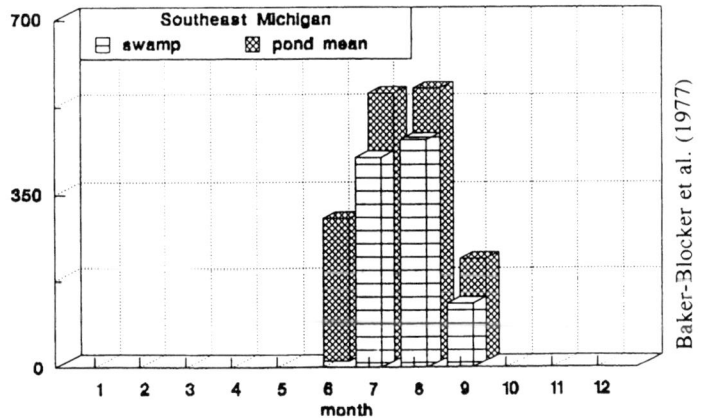

Figure 2c Temperate/Low Boreal (continued)

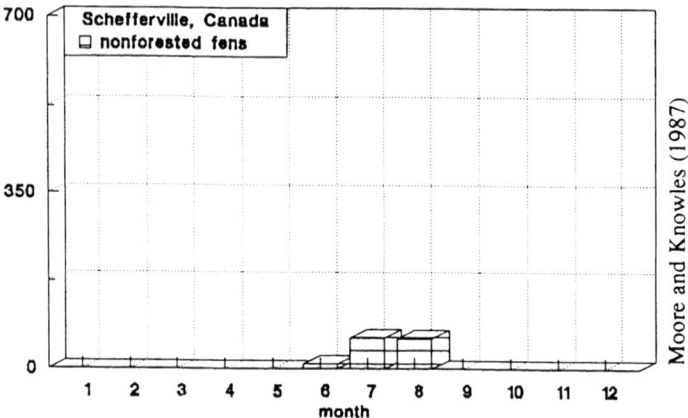

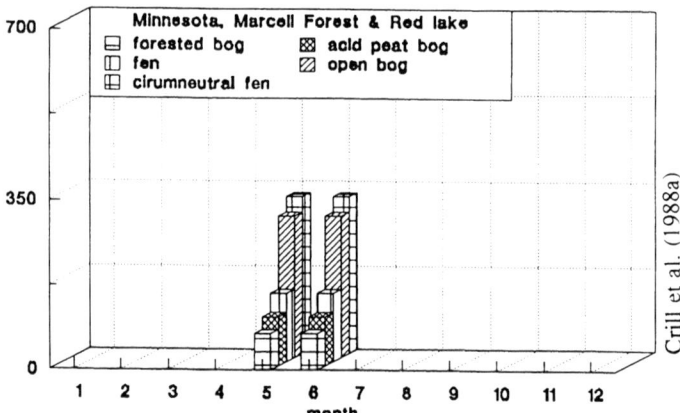

Figure 2c Temperate/Low Boreal (continued)

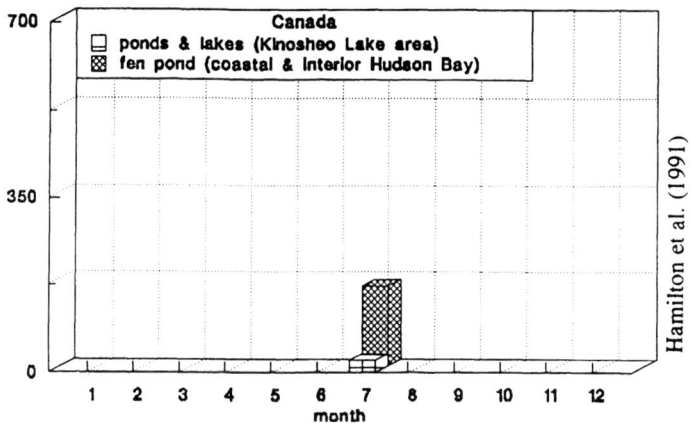

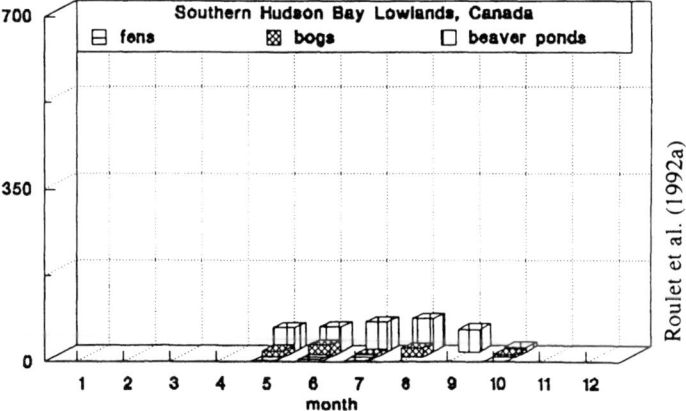

Figure 2d High Boreal/Arctic

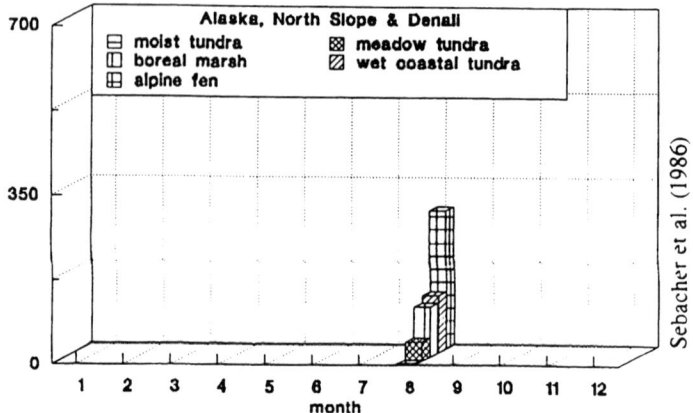

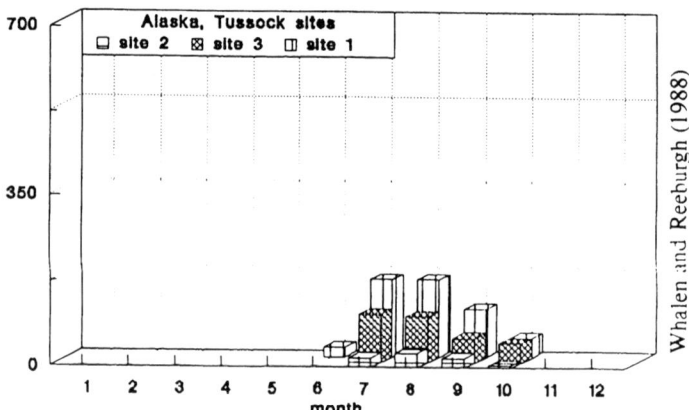

Figure 2d High Boreal/Arctic (continued)

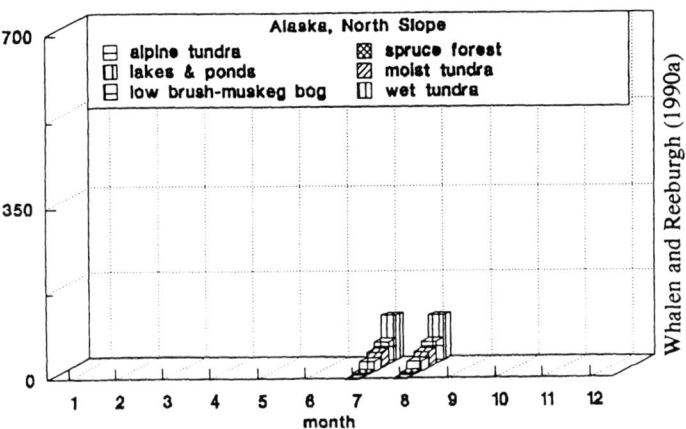

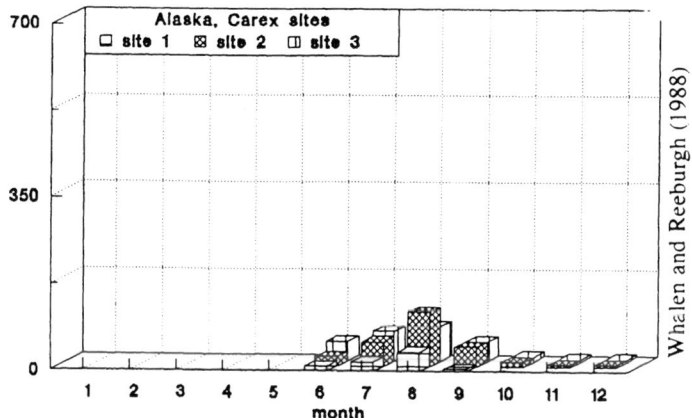

Figure 2d High Boreal/Arctic (continued)

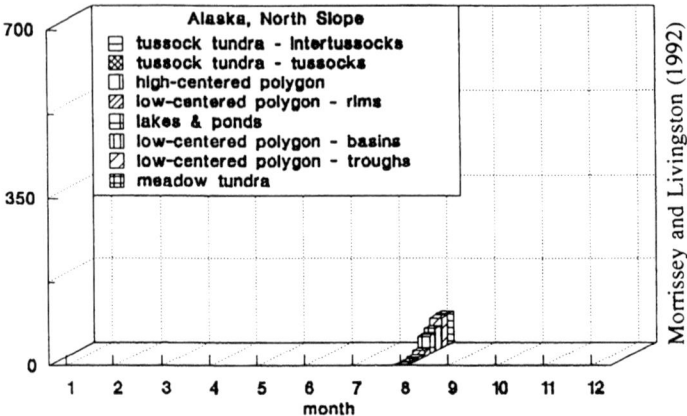

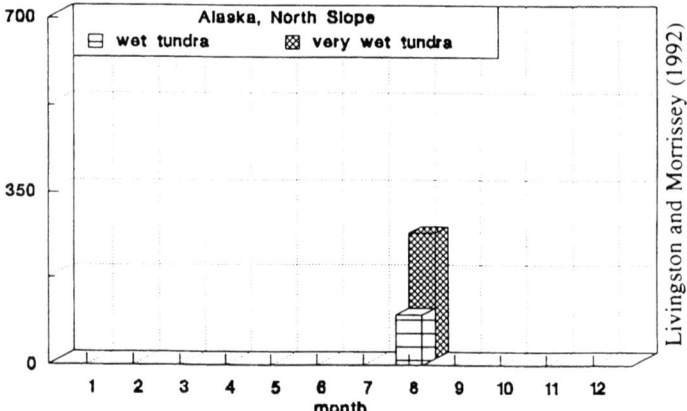

Figure 2d High Boreal/Arctic (continued)

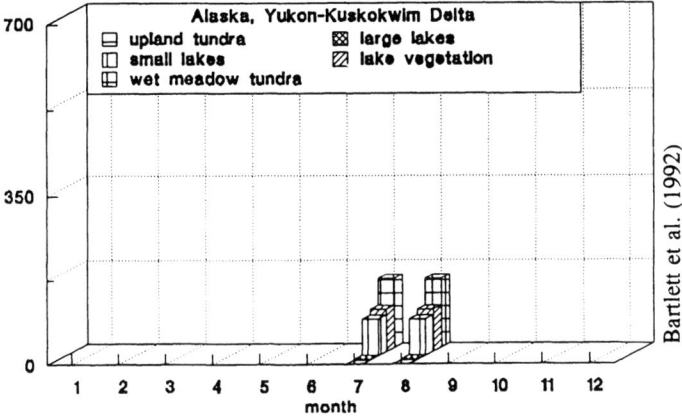

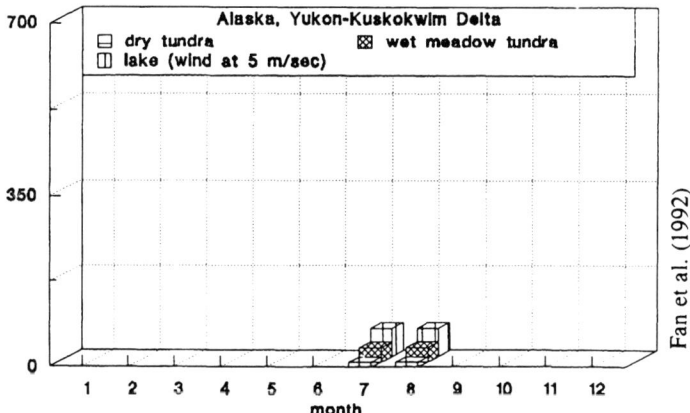

Figure 2 provides a sample of methane flux measurements from wetland habitats grouped into (a) tropical, (b) subtropical, (c) temperate/low boreal, and (d) high boreal/arctic regions. The presentation here is designed to illustrate general patterns in seasonal variations of fluxes, and in flux magnitudes and variability among ecosystems for a range of climatic zones. Note that some studies of several months' duration report mean values for individual ecosystems; in Figure 2, these means are shown for each month of the experiment.

Most tropical studies have been carried out in Amazonian riverine habitats - flooded forests, floating grass mats, and lakes and channels (Figure 2a). Bartlett et al. (1990) conclude that wet- and dry-season methane fluxes from Amazon floodplain environments may be relatively constant adding the cautionary note that emissions during transition periods are not yet measured and may be higher. In general, methane fluxes from open water are lower and less variable than those from flooded forests and floating mats. The work of Wassmann et al. (1992) suggests differences in the seasonal pattern of peak fluxes from open water and several vegetated environments. Fluxes are similar during wet (Bartlett et al., 1990; Devol et al., 1990) and dry seasons (Bartlett et al., 1988; Devol et al., 1988) partially due to very high variability. Although there are significant seasonal changes in inundated areas of tropical riverine systems associated with the Amazon, these habitats are not characterized by the large seasonal pulses of organic inputs from litterfall or temperature-regulated pulses of microbial activity found in higher latitudes. However, episodic events can play a significant role in seasonal emissions in the tropics and elsewhere (Table 2). Field techniques designed to measure separately the direct and ebullitive contributions to methane fluxes in tropical riverine and lake environments, as well as calculations of the role of bubbling in overall fluxes, confirm that episodic ebullition may commonly account for 20-75% or more of the total seasonal emission of methane in these environments (Bartlett et al., 1988, 1990; Crill et al., 1988b; Devol et al., 1988, 1990; Wilson et al., 1989; Keller, 1990; Wassmann et al., 1992) further complicating measurement and modeling of fluxes from these dynamic environments. The relative role of bubbling apparently varies with ecosystem; ebullient fractions increase from open water to grass mats and flooded forest. Furthermore, bubbling events appear to be more pronounced during periods of falling and low water and in shallow lake waters.

Fewer measurements are available for subtropical wetlands, and are confined to the southeastern U.S. (Virginia, South Carolina, Georgia, and Florida) (Figure 2b). Wilson et al. (1989) demonstrate that seasonal trends in methane flux from the Newport News Swamp in Virginia are strongly correlated with temperature best represented by a step function. They suggest pulses of organic substrate inputs as the likely driver for the series of emission peaks observed during the year. The spring peak reflects mineralization of labile organic matter accumulated during the winter followed by temperature-triggered decomposition to substrates for methanogenesis while summer and autumn peaks are related to root exudates and litter input. The remaining subtropical measurements show considerably smaller methane fluxes from low latitude swamps. Until a few years ago, the studies of Harriss and Sebacher (1981) and Harriss et al. (1982) were the only published low latitude measurements to serve as proxies for tropical wetlands.

Early methane flux measurements in temperate and low boreal regions (Figure 2c) suggested these wetlands as extremely productive environments. For example, Harriss et al. (1985) and Baker-Blocker et al. (1977) showed methane fluxes in Minnesota and Michigan in the range of 200 to ~600 mg $CH_4/m^2/d$ in summer and fall. High fluxes measured in diverse wetlands in Minnesota (increasing from forested fens and bogs, to nonforested bogs and sedge meadows) were the ecosystem fluxes used to represent boreal wetlands by Matthews and Fung (1987). Later studies at Minnesota sites (Crill et al., 1988a) confirm high fluxes from open bogs and circumneutral fens similar to results obtained in Alaskan alpine fens (Sebacher et al., 1986). More recent reports from the Hudson Bay Lowlands (HBL) (Hamilton et al., 1991; Roulet et al., 1992a) reveal overall lower flux rates for these low boreal wetland habitats; ponds (fen and beaver) exhibit fluxes in the upper range of HBL wetlands while bogs and fens average fluxes of 13 and 3 mg $CH_4/m^2/d$, respectively, over the season from May to October. Individual sites within these wetland types periodically exhibit larger rates of emission such as a thicket swamp with emissions of 40-60 $mg/m^2/d$ in May and June, rising to 120-160 $mg/m^2/d$ in July and remaining between 60-120 $mg/m^2/d$ through August (Roulet et al., 1992a). Moore et al.

(1990) found that episodic degassing pulses associated with lowered water tables following several weeks of reduced precipitation can account for 18-65% of the seasonal emission of methane from subarctic boreal fens in Canada (Table 2).

Table 2. Contribution of episodic events to seasonal methane emission from a series of wetland habitats.

Wetland Habitat	Mechanism/Magnitude
Subarctic boreal fens (Moore et al., 1990)	degassing pulse with lowered water table after 3 weeks of low rainfall; accounted for 18-65% of seasonal emission, depending on habitat
Temperate freshwater swamp (Wilson et al., 1989)	(direct measurement of ebullient flux) ebullition: observed in 19% of measurements, accounted for 34% of the seasonal emission
Amazon floodplain (Devol et al., 1988, 1990)	ebullition loss: 73% of emission in rising-water season, 59% of emission in low-water season
Amazon lake (Lago Calado) (Crill et al., 1988a)	(direct measurement of diffusive and ebullitive flux); ebullition loss: 70% of flux from open-water lake areas
Amazon floodplain (Bartlett et al., 1988)	(direct measurement of diffusive and ebullient flux); ebullition loss: 49% of open-water flux, 54% of flooded-forest flux, 64% of grass-mat flux
Amazon floodplain (Wassmann et al., 1992)	ebullition loss: 80% of open-water flux, 91% of flooded-forest flux, 67% of floating grass-mats flux, 80% of total flux from Varzea

Methane flux measurements in high boreal/arctic wetlands are shown in Figure 2d. Whalen and Reeburgh (1988) measured year-round methane fluxes at a series of permanent sites in a subarctic muskeg and along a pond margin. These tussock and carex sites were chosen as representative of arctic tussock tundra and wet meadow tundra. Although situated in similar climatic regimes, tussock sites show positive fluxes from July through October with highest values in July-August while the carex sites show positive fluxes from June through December peaking in August. Dise (1993) reports that winter methane fluxes from Minnesota peatlands may account for up to 20% of the annual emission from these environments. The dry upland tundra and large lakes measured by Bartlett et al. (1992) during the July-early August period of the ABLE 3A experiment show extremely low summer fluxes (mean 2-4 mg $CH_4/m^2/d$), while the smaller lake, lake vegetation and wet meadow tundra exhibit mean values of 77, 89, and 144 mg $CH_4/m^2/d$. The chamber measurements of Bartlett et al. (1992) are generally comparable to the fluxes of Fan et al. (1992) and Ritter at al. (1992) derived from tower and airborne eddy-correlation measurements at the same time.

3.2 Regional and global emission estimates

Although there is general agreement concerning the global area and distribution of wetlands, uncertainties remain as to the dynamics of methane producing areas and periods in addition to overall magnitudes of characteristic ecosystem fluxes and factors that control production and emission. The total methane emission from natural wetlands from Aselmann and Crutzen (1989), estimated to be 80 Tg with a range of 40-160 Tg, is similar to the 110 Tg estimate of Matthews and Fung (1987). However, the relative contribution of high- and low-latitude ecosystems to the total emission is reversed in the two studies (Figure 3a,b). The high daily fluxes (Table 3) and emission periods of 100 or 150 days applied to boreal wetlands by Matthews and Fung (1987) result in about 60% of the total emission confined to the region from 50°-70°N; tropical/subtropical emission periods of 180 days and lower flux rates applied to these low-latitude wetlands (area-weighted mean of swamps = ~100 mg/m^2/d) result in about 30% of the total wetland area contributing about 25% of the total annual emission. Aselmann and Crutzen (1989) assumed slightly higher daily flux rates for the swamps and marshes predominating in the low latitudes (area-weighted mean = ~120 mg/m^2/d) along with emission periods of more than 250 days, resulting in a total emission largely concentrated in the tropics. Boreal and polar wetlands, with production periods averaging almost six months and area-weighted daily flux rates about one-quarter those assumed by Matthews and Fung

(1987), play a smaller role in the emission of methane, contributing about one-third of the total annual emission. The large sensitivity of emission estimates to the assumed length of methane-production seasons highlights the crucial importance of improving information on seasonal changes in areas and methane-production conditions for wetland ecosystems.

Table 3. Comparison of methane fluxes assumed for wetland emission estimates of Matthews and Fung (1987) and of Bartlett et al. (1990). Areas and length of methane-production seasons from Matthews and Fung (1987) were used in both estimates.

Wetland Type	Flux (mg/m^2/d)		Area 10^{12}m^2
	Matthews & Fung (1987)	Bartlett et al. (1990)	Matthews & Fung (1987)
Forested bog	200	100	2.1
Nonforested bog	200	160	0.9
Forested swamp	70	120	1.1
Nonforested swamp	120	200	1.0
Alluvial formations	30	160	0.2
Total Annual Emission, Tg	111	110	5.3

Bartlett et al. (1990) estimated the global methane emission from natural wetlands using areas, ecosystem classes and inundation periods of Matthews and Fung (1987) combined with newly-derived fluxes for major ecosystems based on measurements from the Amazon Boundary Layer Experiment 2 (ABLE 2) campaigns in 1985 (dry season) and 1987 (wet season) and new measurements from northern wetlands. The re-evaluated fluxes (Table 3) are higher for tropical/subtropical swamps; alluvial formations are also assumed to be higher methane emitters with a flux of 160 mg/m^2/d in contrast to 30 mg/m^2/d of Matthews and Fung (1987). Fluxes for boreal habitats are lower, equal to about 50-75% of those used by Matthews and Fung (1987). In the estimate of Bartlett et al. (1990), tropical/subtropical ecosystems contribute about two-thirds to the total emission of methane from wetlands, about twice that estimated by Matthews and Fung (1987) and similar to the pattern indicated by Khalil and Rasmussen (1983), Aselmann and Crutzen (1989) and Fung et al., (1991) (Table 4).

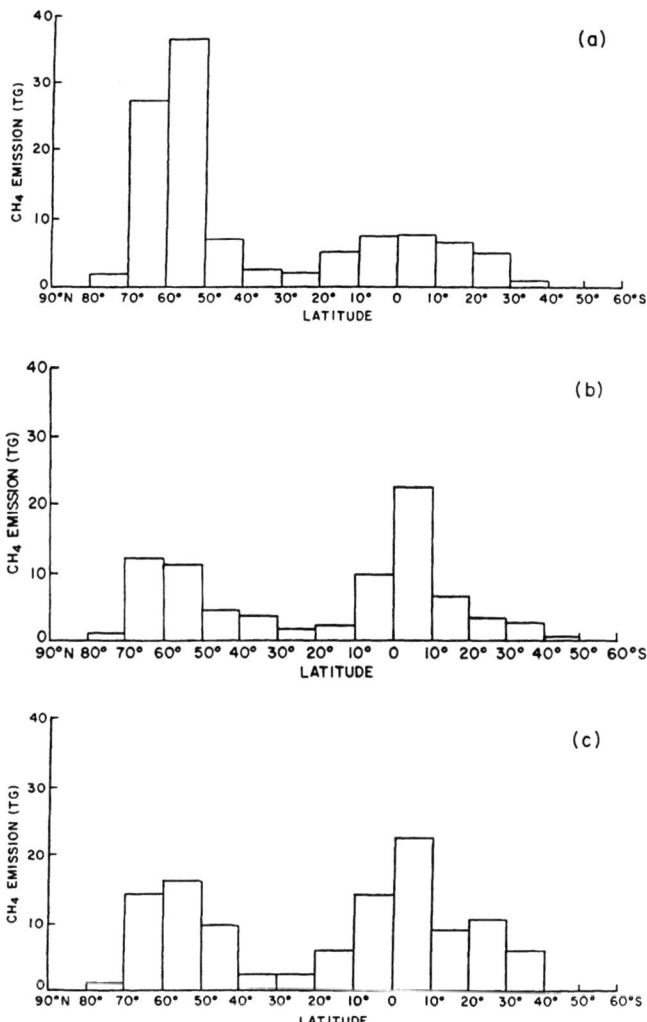

Figure 3. Latitudinal distribution of methane emission estimated by (a) Matthews and Fung (1987) using areas shown in Figure 1a; (b) Aselmann and Crutzen (1989) using areas shown in Figure 1b; and (c) Fung et al. (1991) using areas shown in Figure 1a in conjunction with an expanded suite of flux measurements, modeled inundation seasons, and temperature-dependent high-latitude fluxes.

Table 4. Regional wetland areas and associated methane emissions from studies published between 1983 and 1992.

Tropical		Temperate		Boreal-Arctic		Global		Reference & Comments
Area 10^{12} m^2	Emissions Tg	Area 10^{12} m^2	Emissions Tg	Area 10^{12} m^2	Emissions Tg	Area 10^{12} m^2	Emissions Tg	
--	90	included in boreal	--	--	66	--	156	Khalil & Rasmussen, 1983
--	--	--	--	4.5-9.0	45-106	--	--	Sebacher et al., 1986: peatlands
--	38±17	--	--	--	--	--	47±22	Seiler & Conrad, 1987
2.0	34	0.6	12	2.7	65	5.3	111	Matthews & Fung, 1987
0.1-0.5	3-17	--	--	--	--	--	--	Bartlett et al., 1988: Amazon floodplain
--	--	--	--	--	72	--	--	Crill et al., 1988a
--	8-13	--	--	--	--	--	--	Devol et al., 1988: Amazon floodplain
2.1	45	1.1	11	2.4	25	5.7	80	Aselmann & Crutzen, 1989
2.0	55	0.6	17	2.7	39	5.3	111	Bartlett et al., 1990
0.1	5	--	--	--	--	--	--	Devol et al., 1990: Amazon floodplain
--	--	--	--	1.5	14-19	--	--	Moore et al., 1990: fens
2.0	71	0.6	12	2.7	32	5.3	115	Fung et al., 1991

Table 4. Regional wetland areas and associated methane emissions from studies published between 1983 and 1992.

Tropical		Temperate		Boreal-Arctic		Global		Reference & Comments
Area 10^{12} m^2	Emissions Tg	Area 10^{12} m^2	Emissions Tg	Area 10^{12} m^2	Emissions Tg	Area 10^{12} m^2	Emissions Tg	
0.1	2-3	--	--	--	--	--	--	Tathy et al., 1992: Congo Basin
--	--	--	--	7.3	44	--	--	Ritter et al., 1992: tundra
--	--	--	--	7.3	14-42 (1987)	--	--	Whalen & Reeburgh, 1992: tundra
				7.3	26-78 (1988)			
				7.3	24-67 (1989)			
				7.3	69-135 (1990)			
2.0	66	0.6	5	2.7	34	5.3	105	Bartlett & Harriss, 1993

However, dominance of the low-latitude source in the work of Bartlett et al. (1990) and Fung et al. (1991) (Figure 3c) is a function of higher daily flux rates for tropical wetland ecosystems based on tropical measurements whereas the relative dominance of the tropics/subtropics in the analysis of Aselmann and Crutzen (1989) results from the combined effects of larger tropical wetland areas, moderately higher daily flux rates and substantially longer production seasons for these low-latitude wetlands.

Wetland areas and their geographic distribution contribute little to variations among the global-scale estimates shown in Table 4 since the two data bases commonly used in these estimates, discussed above, are very similar. Differences among the studies reflect differing assumptions about the duration of methane productive seasons as well as incorporation of methane fluxes from the expanding suite of new measurements. While estimates of annual methane emission from wetlands may be converging around 100 Tg, with about two-thirds emanating from the low latitudes, the agreement stems in part from reliance on the same inundation periods in several of the studies.

Whalen and Reeburgh (1992) estimated annual tundra emissions for 1987-1990 by extrapolating mean fluxes measured in a series of habitats for each of the four years. This range of tundra estimates encompasses all the boreal/arctic values in Table 4, which represent the full complement of northern wetland ecosystems. The total tundra area of $7.3 \times 10^{12} m^2$ used in this study is ~3 times the boreal/arctic wetland area suggested by others. However, the measurements of Whalen and Reeburgh (1992) and others indicate that significant portions of drier tundra may emit methane at low rates. Several contemporaneous Alaskan studies using enclosures (Bartlett et al., 1992), tower-based eddy correlations (Fan et al., 1992) and aircraft-derived eddy correlations (Ritter et al., 1992), to develop habitat weighted fluxes for all tundras, suggest emissions of 10, 11, and 44 Tg/yr, respectively, from tundra environments north of 60°N. The latter is suggested as an upper limit partially because the sampled areas may be biased toward particularly productive habitats (Ritter et al., 1992). Considering uncertainties in extrapolations and ecosystem classification, these tundra estimates derived using different methods show very good agreement.

The length of productive seasons as well as seasonal variations in extent of inundation remain major uncertainties. Furthermore, most emission estimates assume constant daily fluxes that vary only with ecosystem type. Workers have attempted to introduce the seasonally varying flux rates observed in the field (Figure 2) using temperature-flux relationships for northern ecosystems (Fung et al., 1991). While this initial approach may provide reasonable emission patterns on a seasonal basis, other processes such as seasonal water-table variations or substrate input are likely candidates for inclusion in such models. While uncertainties in global methane estimates resulting from spatial and ecosystem variability can be reduced further by expanding the measurement base, uncertainties in the seasonality of wetland areas and methane-producing inundated conditions will be reduced through integrated approaches combining remote-sensing and modeling techniques.

4. Wetlands and climate change.

About half of the global wetland area occurs in latitudes from 50° to 70°N, regions expected to undergo temperature increases on the order of several degrees C during the next 100 years (Hansen et al., 1988). These changes could lead to (1) a lengthened thaw season and associated expansion of biological activity, (2) increased net primary productivity due either to direct fertilization from increases in CO_2 concentration or to indirect temperature effects, and/or (3) larger areas subjected to thaw and anaerobic conditions. Based on temperature increases alone, methane emissions would be expected to increase in high latitudes providing a positive feedback on the climate system. However, other factors may moderate this response. For example, nutrient limitation may limit productivity increases and microbial adaptation to the current thermal regime may be inelastic. Available water supply in high latitudes may decline in response to increased evaporation under warmer conditions. These changes in water status may produce lowered water tables and dry soil conditions in previously waterlogged or inundated environments, thereby increasing methane oxidation, reducing methane emissions and perhaps causing former wetlands to act as methane sinks. Seasonal water supply is the major controller of tropical and subtropical methane emissions, affecting both area and length of inundation periods in the low latitudes. While General Circulation Models generally predict greater precipitation for high-latitude wetland areas, low-latitude regions may be subject to reduced precipitation. However, predictions of hydrologic perturbations over the next 50-100 years are highly uncertain, leaving open the question of whether current wetlands will become larger or smaller methane sources or perhaps sinks in the future. Furthermore, difficulties in establishing the geologic, edaphic, topographic and climatic underpinnings of wetland creation on short time scales suggest that very local conditions may determine whether current wetland sites expand or contract and whether per-unit-area fluxes increase or decline (Chappellaz et al., 1993).

Several investigators have evaluated potential changes in methane emissions from high latitude wetland ecosystems in response to predicted climate alterations using current relationships between methane flux and climate variables (Table 5). Frolking (1993) and Harriss and Frolking (1993) evaluated possible inter-annual oscillations in high latitude summer methane emission from wetlands as a function of 20th century temperature variations. Compared to baseline long-term summer averages, temperature anomalies range from about -2°C for the coolest years to +2°C for the warmest. Based on reconstructed summer temperature anomalies in high latitude wetland regions and correlations between temperature and methane flux measurements in Alaska and Minnesota, results indicate that summer methane emissions from boreal wetlands during the 20th century have varied approximately 5 Tg around an estimated mean of 32 Tg. This result suggests that wetland emissions are moderately sensitive to the magnitude of temperature variations that might be expected in the early stages of warming predicted for the next century. Frolking (1993) further suggests that a 3-5°C temperature rise, with no change in water status, might increase boreal emissions to more than twice their current estimated value. Livingston and Morrissey (1991) evaluated the sensitivity of regional North Slope (Alaska) methane emissions to potential changes in both temperature and water status. In situ flux measurements show a three-fold increase in response to the ~5°C mid-summer soil temperature elevation observed in the study area between the 1987 and 1989 field seasons. Extrapolation to the North Slope region suggests a potential four-to-five-fold increase in emission under conditions wetter than

Table 5. Potential wetland methane emission response to climate change.

Description	Methane-Emission Response
Livingston and Morrissey (1991) Relationship of regional North Slope (Alaska) tundra emissions to temperature; simulated changes in water status via changes in vegetation distributions; extrapolation using measurements from 1987-1989.	local: 4-fold increase with 4°C summer soil temperature rise; regional: 4-5-fold increase under 4°C warmer and wetter conditions; 2-fold increase under 4°C warmer and drier conditions; *large sensitivity of emission to warming*
Frolking (1993), Harriss and Frolking (1993) Sensitivity of high latitude emissions to historical summer temperature variations; no evaluation of water status; temperature-flux correlations and regional 20th century temperature anomalies used to model historical flux anomalies	± 2°C anomalies gave ±5 Tg variation around mean annual emission of ~30 Tg; suggestion: 3-5°C rise could double emissions; *moderate sensitivity of emission to early warming*
Whalen and Reeburgh (1992) Four-year time series of flux measurements at fixed Alaskan tundra sites extrapolated to estimate interannual variations in high-latitude tundra emissions.	4-fold emission variation over four years: 14-42 Tg (1987, driest), 26-78 Tg (1988), 24-67 Tg (1989), 69-135 Tg (1990, anomalously wet) *large sensitivity of emission to combined temperature and water effects*
Roulet et al. (1992c) Measurements of methane emission and related environmental factors in low boreal Canadian fens; water-table position predicted with hydrologic/thermal model for peatlands	15% emission increase with 2°C elevation in soil temperature; 75-80% emission decline with 14-cm drop in water table; *moderate sensitivity of emission to warming, large sensitivity of emission to water status*

those of 1987 and a doubling of methane emission with a 4°C elevation even under conditions moderately drier than 1987. Based on a 4-year time series of methane measurements at a broad suite of tundra sites in Alaska, Whalen and Reeburgh (1992) estimated annual tundra emissions for 1987-1990 based on mean emission values for tundra habitats measured during the four-year period. Mean fluxes varied considerably during the period which encompassed consistent summer temperature anomalies of +2°C and precipitation variations ranging from ~60% (1987-1989) to 190% (1990) of long-term averages. Extrapolated tundra emissions vary by a factor of four, from 14-42 Tg CH_4 in the driest year (1987) to 69-135 Tg in 1990, the anomalously wet year. These latter two analyses based on field measurements indicate potential large sensitivity of high latitude methane emission to the combined effects of temperature and precipitation changes. The analysis of Roulet et al. (1992c) on the sensitivity of methane emissions to temperature and to water-table variations in Canadian fens indicates a moderate dependence of the emissions on temperature and a very strong dependence of emission on water table depth. Their modeling study suggests an emission increase of ~15% with a 2°C elevation in temperature at 10-cm soil depth but ~75-80% declines in emissions from floating and non-floating fens following a 14-cm drop in water table.

A primary determinant of high latitude wetland responses to climate change lies in the extent to which anaerobic conditions are maintained under circumstances of reduced moisture availability. Seasonal, areal, and vertical components of methane production and consumption activities will likely adjust differently to moisture changes.

5. Summary.

Table 6 provides a summary of progress in understanding the role of natural wetlands in the global methane cycle during the last decade; Table 7 outlines some remaining questions and uncertainties.

Table 6. Summary of progress and understanding on the role of natural wetlands in the global methane cycle.

Measurements of methane characteristics in most major wetland environments.

Characterization of seasonal and inter-annual flux variability via full-season and multi-year measurements.

Development and improvement of spatial averaging techniques:
eddy correlation;
integrated series data bases for extrapolations (vegetation, soils, drainage, temperature, precipitation, permafrost).

Identification of general patterns in flux behavior and controls for some major wetland ecosystems.

Development of quantitative relationships between flux and controlling variables/environmental integrators (temperature, water status) for various ecosystems.

Similarity of wet- and dry-season fluxes in some Amazon floodplain habitats (flooded forest, open water); increases utility of periodic opportunistic measurements.

Interdisciplinary campaigns to characterize tropospheric chemistry in a variety of regimes
    Tropical: ABLE 2A and 2B, CAMREX, DECAFE, TROPOZ
    Boreal/Arctic: ABLE 3A and 3B.

Increasing suite of isotopic measurements of atmosphere and sources.

Availability of analysis techniques - photochemical, transport models - to synthesize and constrain sources/sinks.

Confirmation, through flux measurements and models, of current importance of low-latitude wetlands and smaller role of northern ecosystems in methane emissions.

Achievement of comparable emissions results across spatial scales via different flux measurement techniques:
    chambers, towers, aircraft.

Improved remote sensing of wetland habitats and their seasonality for regional extrapolations.

Presently-available information on the distribution of wetland methane sources exceeds information on production, oxidation and emission characteristics and the factors that control them. Independent data sources generally agree on latitudinal and environmental profiles of wetlands. The expanding suite of ground-based methane flux measurements conducted in a broad array of geographically and ecologically diverse ecosystems has characterized large-scale features of the role of natural wetlands in the global methane cycle. Model studies including photochemical and transport processes (Fung et al., 1991), as well as regional emissions estimates derived from eddy-correlation techniques (Fan et al., 1992; Ritter et al., 1992), suggest patterns consistent with the ground-based measurements. However, uncertainties with respect to inundation periods, seasonal changes in wetland habitat areas, and inter-annual variations in inundation extent remain large.

Tropical and subtropical wetlands, which account for about one-third of the global wetland area, likely contribute ~50-75% to the annual emission of methane from natural wetlands. Precipitation-driven inundation periods may last from a few to 12 months a year in these environments. More than half the world's wetlands occur in boreal and arctic habitats north of 50°N. As in the tropical riverine wetlands, many of these environments are characterized by complex landscapes composed of herbaceous and open-water features with distinctive methane emissions. At present boreal methane emissions, telescoped into a few months of the year during thaw seasons, probably contribute about one-third of the world's total wetland emissions. However, their response to climate changes predicted for the next century is highly uncertain. Depending on local interactions among temperature, water status, nutrients, microbial populations etc., boreal/arctic ecosystems may become larger or smaller methane sources, or methane sinks.

Measurements of methane fluxes in natural wetlands continue to confirm that methane fluxes vary by orders of magnitude among ecosystems at local scales, and by factors of three to four inter-annually in response to temperature and water-status variations. Although wetland fluxes have been correlated with environmental parameters, these quantitative relationships do not appear to be constant over the broad range of wetland ecosystems on the earth. Variables that serve as environmental integrators of production and consumption processes are the most promising predictors of methane fluxes. Episodic events have been shown to contribute substantially to total seasonal emissions in a broad array of wetlands. Ebullition may account for 20-75% of seasonal methane emissions in tropical environments; similar fractions due to degassing pulses associated with lowered water tables have been observed in Canadian fens. Wind-triggered methane releases have been observed in lakes in Alaska (Fan et al., 1992) and Panama (Keller, 1990).

Table 7. Continuing research questions on the role of natural wetlands in the global methane cycle.

Expanded measurements confirm large spatial, seasonal and inter-annual variability.

Measurements confirm importance of highly local controls (topography, water table, organic content) on fluxes.

Measurements confirm differential relationships between controls and fluxes among wetland habitats.

Siberian Lowlands and non-riverine tropical grasslands not yet measured.

Similarities (isotopes, seasonality, magnitude) between methane emitted from wetlands and from rice in tropics.

Large uncertainty about wetland response to changed climate:
    change in wetland habitat distribution and/or seasonality
    integrated effects of temperature, water status, etc.

Importance of episodic events: ebullition, degassing, hydrostatic pressure changes, wind.

Spatial averaging vis-a-vis representativeness of ecosystem measurements.

Area and seasonal variation of wetland environments: floodplains, non-riverine flooded grasslands.

Remote sensing of wetland habitat distribution and seasonality:
    tropics: cloud problems
    boreal/arctic: difficulty in estimating water status of wet/moist soils with no standing water.

High latitude wetlands are a major contributor to annual oscillations of atmospheric methane concentration and to maintenance of its characteristic north-south gradient (Khalil and Rasmussen, 1983; Quay et al., 1988; Fung et al., 1991). High frequency measurements of atmospheric methane concentrations can be used to constrain hypotheses about magnitudes and locations of methane sources. While locations of measurement stations were initially designed to provide information on clean-air background concentrations (Steele et al., 1987, Lang et al., 1990a,b), information from stations that encompass source regions can be used to diagnose sources at local and regional scales. In addition, monitoring patterns and trends in intra- and inter-annual variations of high latitude methane concentrations may signal perturbations in the release of methane from climatically-sensitive boreal wetlands.

Increasing information on isotopic signatures and variations of atmospheric methane and methane emitted from terrestrial sources provides additional quantitative constraints on the budget of methane (Stevens, this volume; Whiticar, this volume). Measurements of ^{13}C and ^{14}C of atmospheric and terrestrial methane have been carried out by several groups (see Games and Hayes, 1976; Deines, 1980; Schoell, 1980; Rust, 1981; Stevens and Engelkemeir, 1982; Tyler, 1986; Burke et al., 1988; Tyler et al., 1988; Quay et al.,1988, 1991; Stevens, 1988; Wahlen et al., 1988, 1989; King et al., 1989). The $\delta^{13}C$ of atmospheric methane has been measured at approximately -46 to -47 ‰ in the late 1980s. Despite variations, methane emitted from wetlands (both natural and rice paddies) shows a distinctive isotopic signature more depleted in ^{13}C than any other source ($\delta^{13}C$ of -60 to -70 ‰) while landfill sources are less depleted at about -50 ‰. By contrast, abiogenic sources such as natural gas (-35 to -45 ‰) and biomass burning (-32 to -24 ‰) are heavier than atmospheric methane.

Models to test the consistency between atmospheric methane variations and hypotheses about sources and sinks include globally-averaged height-dependent photochemical models to analyze the balance between sources and sinks (e.g., Chameides et al., 1977; Thompson and Cicerone, 1986; Quay et al., 1988) and photochemical-transport models in one, two or three dimensions (e.g., Mayer et al., 1982; Crutzen and Gidel, 1983; Khalil and Rasmussen, 1983; Blake, 1984; Isaksen and Hov, 1987; Wahlen et al., 1989; Fung et al., 1991; Quay et al., 1991; Taylor et al., 1991; Conrad and Rasmussen, this volume). Hypotheses about source/sink scenarios can be evaluated for their ability to simulate observed variations in methane and its isotopic composition in the atmosphere.

Progress in understanding the role of natural wetlands in the global methane cycle will be achieved primarily through coherent programs of long-term atmospheric, isotopic, and field measurements covering representative sources/sinks, environments and seasons. Photochemical-transport models that synthesize and evaluate suites of source/sink distributions provide a crucial integrative pathway to narrowing current uncertainties in the role of natural wetlands in the global methane budget.

Acknowledgments. I thank S. Frolking, K. Bartlett and B. Harriss for providing preprints. In addition to providing preprints and copies of difficult-to-acquire papers, N. Roulet and P. Glaser shared their expertise in discussions about wetlands and their classification as did K. Bartlett, P. Crill, J. Chanton, and B. Reeburgh. Comments on early versions of the manuscript by A. Khalil, M. Shearer and B. Reeburgh are gratefully acknowledged. I thank J. John and L. DelValle for graphics and drafting support. Support from the Office of Strategic Planning of the U.S. Environmental Protection Agency, NASA Interdisciplinary Research Program, and the Goddard Director's Discretionary Fund is gratefully acknowledged.

References

Aselmann, I., P. Crutzen. 1989. Global distribution of natural freshwater wetlands and rice paddies: their net primary productivity, seasonality and possible methane emissions. *J. Atmos. Chem.,* 8:307-358.

Baker-Blocker, A., T.M. Donohue, K.H. Mancy. 1977. Methane flux from wetland areas. *Tellus,* 29:245-250.

Barber, T.R., R.A. Burke, Jr., W.M. Sackett. 1988. Diffusive flux of methane from warm wetlands. *Global Biogeochem. Cycles,* 2:411-425.

Bartlett, K.B., R.C. Harriss. 1993. Review and assessment of methane emissions from wetlands. *Chemosphere,* 26 (1-4):261-320.

Bartlett, K.B., R.C. Harriss, D.I. Sebacher. 1985. Methane flux from coastal salt marshes. *J. Geophys. Res.,* 90:5,710-5,720.

Bartlett, K.B., P.M. Crill, D.I. Sebacher, R.C. Harriss, J.O. Wilson, J.M. Melack. 1988. Methane flux from the central Amazonian floodplain. *J. Geophys. Res.,* 93:1,571-1,582.

Bartlett, D.S., K.B. Bartlett, J.M. Hartman, R.C. Harriss, D.I. Sebacher, R. Pelletier-Travis, D.D. Dow, D.P. Brannon. 1989. Methane emissions from the Florida Everglades: Patterns of variability in a regional wetland ecosystem. *Global Biogeochem. Cycles,* 3:363-374.

Bartlett, K.B., P.M. Crill, J.A. Bonassi, J.E. Richey, R.C. Harriss. 1990. Methane flux from the Amazon River floodplain: Emissions during rising water. *J. Geophys. Res.,* 95:16,773-16,788.

Bartlett, K.B., P.M. Crill, R.L. Sass, R.C. Harriss, N.B. Dise. 1992. Methane emissions from tundra environments in the Yukon-Kuskokwim Delta, Alaska. *J. Geophys. Res.,* 97:16,645-16,660.

Blake, D.S. 1984. Increasing concentrations of atmospheric methane. Ph.D. thesis, Univ. Cal. at Irvine, 213p.

Blake, D.R., F.S. Rowland. 1986. Worldwide increase in tropospheric methane, 1978 to 1983. *J. Atmos. Chem.,* 4:43-62.

Blake, D.R., F.S. Rowland. 1988. Continuing worldwide increase in tropospheric methane, 1978 to 1987. *Science,* 239:1,129-1,131.

Blake, D.R., E.W. Mayer, S.C. Tyler, Y. Makide, D.C. Montague, F.S. Rowland. 1988. Global increase in atmospheric methane concentrations between 1978 and 1980. *Geophys. Res. Lett.,* 9:477-480.

Burke, R.A., Jr., T.R. Barber, W.M. Sackett. 1988. Methane flux and stable hydrogen and carbon isotope composition of sedimentary methane from the Florida Everglades. *Global Biogeochem. Cycles,* 2:329-340.

Chameides, W.L., C.S. Liu, R.J. Cicerone. 1977. Possible variations in atmospheric methane. *J. Geophys. Res.,* 82:1,795-1,798.

Chappellaz, J.A., J.M. Barnola, D. Raynaud, Y.S. Korotkevich, C. Lorius. 1990. Ice-core record of atmospheric methane over the past 160,000 years. *Nature,* 345:127-131.

Chappellaz, J.A., I.Y. Fung, A.M. Thompson. 1993. The atmospheric CH_4 increase since the Last Glacial Maximum, 1. Source estimates. *Tellus, 45B* (in press).

Cicerone, R.J., J.D. Shetter. 1981. Sources of atmospheric methane: Measurements in rice paddies and a discussion. *J. Geophys. Res.,* 86:7,203-7,209.

Cicerone, R.J., R.S. Oremland. 1988. Biogeochemical aspects of atmospheric methane. *Global Biogeochem. Cycles,* 2:299-327.

Craig, H., C.C. Chou. 1982. Methane: The record in polar ice cores. *Geophys. Res. Lett.,* 9:1,221-1,224.

Crill, P.M., K.B. Bartlett, R.C. Harriss, E. Gorham, E.S. Verry, D.I. Sebacher, L. Madzar, W. Sanner. 1988a. Methane flux from Minnesota peatlands. *Global Biogeochem. Cycles,* 2:371-384.

Crill, P.M., K.B. Bartlett, J.O. Wilson, D.I.Sebacher, R.C. Harriss. 1988b. Tropospheric methane from an Amazonian floodplain lake. *J. Geophys. Res.,* 93:1,564-1,570.

Crutzen, P.J., L.T. Gidel. 1983. A two-dimensional photo-chemical model of the atmosphere 2: The tropospheric budgets of the anthropogenic chlorocarbons CO, CH_4, CH_3, Cl and the effect of various NOx sources on tropospheric ozone. *J. Geophys. Res., 88*:6,641-6,661.

Dacey, J.W.H., M. Klug. 1979. Methane flux from lake sediments through water lilies. *Science, 203*:1,253-1,255.

Deines, P. 1980. The isotopic composition of reduced organic carbon. In: *Handbook of Environmental Isotope Chemistry, Vol. 1* (P. Fritz and J. Fontes, eds.) Elsevier, New York, p. 329

Delmas, R.A., J. Servant, J.-P. Tathy, B. Cros, M. Labat. 1992. Sources and sinks of methane and carbon dioxide exchanges in mountain forest in Equatorial Africa. *J. Geophys. Res., 97*:6,169-6,179.

Devol, A.H., J.E. Richey, W.A. Clark, S.L. King. 1988. Methane emissions to the troposphere from the Amazon Floodplain. *J. Geophys. Res., 93*:1,583-1,592.

Devol, A.H., J.E. Richey, B.R. Forsberg, L.A. Martinelli. 1990. Seasonal dynamics of methane emissions from the Amazon River floodplain to the troposphere. *J. Geophys. Res., 95*:16,417-16,426.

Dise, N.B. 1993. Factors affecting methane production under rice. *Global Biogeochemical Cycles, 7* (1):123-142.

Ehhalt, D.H. 1974. The atmospheric cycle of methane. *Tellus, 26*:58-70.

Ehhalt, D.H., U. Schmidt. 1978. Sources and sinks of atmospheric methane. *Pageoph., 116*:452-464.

Ehhalt, D.H., R.J. Zander, R.A. Lamontagne. 1983. On the temporal increase of tropospheric CH_4. *J. Geophys. Res., 88*:8,442-8,446.

Fan, S.-M., S.C. Wofsy, P.S. Bakwin, D.J. Jacob, S.M. Anderson, P.L. Kebabian, J.B. McManus, C.E. Kolb, D.R. Fitzjarrald. 1992. Micrometeorological measurements of CH_4 and CO_2 exchange between the atmosphere and the Subarctic tundra. *J. Geophys. Res., 97*:16,627-16,643.

Fontan, J., A. Druilhet, B. Benech, R. Lyra, B. Cros. 1992. The DECAFE experiments: Overview and meteorology. *J. Geophys. Res., 97*:6,123-6,136.

Frolking, S. 1993. Methane from northern peatlands and climate change. In: *Proceedings of Carbon Cycling in Boreal Forest and Sub-Arctic Ecosystems,* 9-12 September 1991, Corvallis, Oregon (in press).

Fung, I., J. John, J. Lerner, E. Matthews, M. Prather, L.P. Steele, P.J. Fraser. 1991. Three-dimensional model synthesis of the global methane cycle. *J. Geophys. Res., 96*:13,033-13,065.

Games, L.M., J.M. Hayes. 1976. On the mechanisms of CH_4 and CO_2 production in natural anaerobic environments. In: *Proceedings of the 2nd International Symposium on Environmental Biogeochemistry* (J.O. Nraigue, ed.), Butterworth, Stoneham, MA, p 51.

Glaser, P.H. 1987. The Ecology of Patterned Boreal Peatlands of Northern Minnesota: A community Profile, *U.S. Fish and Wildlife Service Biological Report, 85(7.14)*. U.S. Department of the Interior, Washington D.C., 98p.
Gore, A.J.P. (ed.) 1983a. *Ecosystems of the World, Mires: Swamp, Bog, Fen and Moor, General Studies, 4A*. Elsevier, New York, 440p.
Gore, A.J.P. (ed.) 1983b. *Ecosystems of the World, Mires: Swamp, Bog, Fen and Moor, Case Studies, 4B*. Elsevier, New York, 479p.
Hamilton, J.D., C.A. Kelley, J.W.M. Rudd. 1991. Methane and carbon dioxide flux from ponds and lakes of the Hudson Bay Lowlands. *EOS, 72*:84.
Hansen, J., I. Fung, A. Lacis, D. Rind, S. Lebedeff, R. Ruedy, G. Russell, P. Stone. 1988. Global climate changes as forecast by Goddard Institute for Space Studies three-dimensional model. *J. Geophys. Res., 93*:9,341-9,364.
Harriss, R.C., D.I. Sebacher. 1981. Methane flux in forested freshwater swamps of the southeastern United States. *Geophys. Res. Lett., 8*:1,002-1,004.
Harriss, R.C., S. Frolking. 1993. The sensitivity of methane emissions from northern freshwater wetlands to global warming. In: *Climate Change and Freshwater Ecosystems* (P. Firth, S. Fisher, eds.), Springer-Verlag, New York, p. 48.
Harriss, R.C., D.I. Sebacher, F.P. Day, Jr. 1982. Methane flux in the Great Dismal Swamp. *Nature, 297*:673-674.
Harriss, R.C., E. Gorham, D.I. Sebacher, K.B. Bartlett, P.A. Flebbe. 1985. Methane flux from northern peatlands. *Nature, 315*:652-653.
Harriss, R.C., S.C. Wofsy, M. Garstang, E.V. Browell, L.C.B. Molion, R.J. McNeal, J.M. Hoell, Jr, R.J. Bendura, S.M. Beck, R.L. Navarro, J.T. Riley, R.L. Snell. 1988a. The Amazon Boundary Layer Experiment (ABLE 2A): Dry season 1985. *J. Geophys. Res., 93*:1,351-1,360.
Harriss, R.C., D.I. Sebacher, K.B. Bartlett, D.S. Bartlett, P.M. Crill. 1988b. Sources of atmospheric methane in the south Florida environment. *Global Biogeochem. Cycles, 2*:231-243.
Harriss, R.C., M. Garstang, S.C. Wofsy, S.M. Beck, R.J. Bendura, J.R.B. Coelho, J.W. Drewry, J.M. Hoell, Jr, P.A. Matson, R.J. McNeal, L.C.B. Molion, R.L. Navarro, V. Rabine, R.L. Snell. 1990. The Amazon Boundary Layer Experiment (ABLE 2B): Wet season 1987. *J. Geophys. Res., 95*:16,721-16,736.
Harriss, R.C., K. Bartlett, S. Frolking, P. Crill. 1993. Methane emissions from northern peatlands: a review and assessment. In: *Biogeochemistry of Global Change: Radiatively Active Trace Gases* (R.S. Oremland, ed.), Chapman and Hall, New York (in press).
Holzapfel-Pschorn, A., W. Seiler. 1986. Methane emission during a cultivation period from an Italian rice paddy. *J. Geophys. Res., 91*:11,803-11,814.
Isaksen, I.S.A., O. Hov. 1987. Calculation of trends in the tropospheric concentration of O_3, OH, CO, CH4 and NOx. *Tellus B, 39*:122-139.

Keller, M.M. 1990. Biological sources and sinks of methane in tropical habitats and tropical atmospheric chemistry. Ph.D. thesis. Princeton University and National Center for Atmospheric Research, 216p.

Khalil, M.A.K., R.A. Rasmussen. 1982. Secular trends of atmospheric methane (CH_4). *Chemosphere, 11*:877-883.

Khalil, M.A.K., R.A. Rasmussen. 1983. Sources, sinks and seasonal cycles of atmospheric methane. *J. Geophys. Res., 88*:5,131-5,144.

Khalil, M.A.K., R.A. Rasmussen. 1985. Causes of increasing methane: Depletion of hydroxyl radicals and the rise of emissions. *Atmos. Environ., 19*:397-407.

Khalil, M.A.K., R.A. Rasmussen. 1987. Atmospheric methane: Trends over the last 10,000 years. *Atmos. Environ., 21*:2,445-2,452.

Khalil, M.A.K., R.A. Rasmussen, M.J. Shearer. 1989. Trends of atmospheric methane in the 1960s and 1970s. *J. Geophys. Res., 94*:18,279-18,288.

King, G.M., W.J. Wiebe. 1978. Methane release from soils of a Georgia salt marsh. *Geochim. Cosmochim. Acta, 42*:343-348.

King, S.L., P.D. Quay, J.M. Lansdown. 1989. The $^{13}C/^{12}C$ kinetic isotope effect for soil oxidation of methane at ambient atmospheric concentrations. *J. Geophys. Res., 94*:18,273-18,277.

Koyama, T. 1963. Gaseous metabolism in lake sediments and paddy soils and the production of atmospheric methane and hydrogen. *J. Geophys. Res., 68*:3,971-3,973.

Koyama, T. 1964. Biogeochemical studies on lake sediments and paddy soils in the production of atmospheric methane and hydrogen. In: *Recent Researches in the Fields of Hydrosphere, Atmosphere and Nuclear Geochemistry* (Y. Miyake and T. Koyama, eds.), Muruzen Co. Ltd., Tokyo, p 143.

Lang, P.M., L.P. Steele, R.C. Martin, K.A. Masarie. 1990a. Atmospheric methane data for the period 1983-1985 from the NOAA/CMDL global cooperative flask sampling network. *Tech. Mem. ERL CMDL-1*. Natl. Oceanic Atmos. Admin., Boulder, CO.

Lang, P.M., L.P. Steele, R.C. Martin. 1990b. Atmospheric methane data for the period 1986-1988 from the NOAA/CMDL global cooperative flask sampling network. *Tech. Mem. ERL CMDL-2*. Natl Oceanic Atmos Admin, Boulder, CO.

Livingston, G.P., L.A. Morrissey. 1991. Methane emissions from Alaskan arctic tundra in response to climatic change. In: *International Conference on the Role of Polar Regions in Global Change* (G. Weller, C.L. Wilson, and B.A.B. Severin, eds.), Proceedings of a Conference held June 11-15, 1990, at the University of Alaska Fairbanks. Geophysical Institute and Center for Global Change and Arctic System Research, University of Alaska Fairbanks, Fairbanks, Alaska, p 372.

Matthews, E., I. Fung. 1987. Methane emission from natural wetlands: global distribution, area, and environmental characteristics of sources. *Global Biogeochem. Cycles, 1*:61-86.

Mayer, E.W., D.R. Blake, S.C. Tyler, Y. Makide, D.C. Montague, F.S. Rowland. 1982. Methane: Interhemispheric concentration gradient and atmospheric residence time. *Proc. Natl. Acad. Sci. U.S.A., 79*:1,366-1,370.

Moore, T.R., R. Knowles. 1987. Methane and carbon dioxide evolution from subarctic fens. *Can. J. Soil Sci., 67*:77-81.

Moore, T.R., R. Knowles. 1989. The influence of water table levels on methane and carbon dioxide emissions from peatland soils. *Can. J. Soil Sci., 69*:33-38.

Moore, T.R., R. Knowles. 1990. Methane emissions from fen, bog and swamp peatlands in Quebec. *Biogeochem., 11*:45-61.

Moore, T., N. Roulet, R. Knowles. 1990. Spatial and temporal variations of methane flux from subarctic/northern boreal fens. *Global Biogeochem. Cycles, 4*:29-46.

Morrissey, L.A., R.A. Ennis. 1981. Vegetation mapping of the National Petroleum Reserve in Alaska using Landsat digital data. *U.S. Geol. Surv. Open File Report 81-315*, U.S. Geol. Surv., Reston, VA, 25p.

Morrissey, L.A., G.P. Livingston. 1992. Methane emissions from Alaska Arctic tundra: An assessment of local spatial variability. *J. Geophys. Res., 97*:16,661-16,670.

National Wetlands Working Group. 1988. *Wetlands of Canada*. Ecological Land Classification Series No. 24. Sustainable Development Branch, Environment Canada, Ottawa and Polyscience Publications, Montreal, 452p.

Ormsby, J.P., B.J. Blanchard, A.J. Blanchard. 1985. Detection of lowland flooding using active microwave systems. *Photogram. Eng. Rem. Sens., 51*:317-328.

Pearman, G.I., D. Etheridge, F. De Silva, P.J. Fraser. 1986. Evidence of changing concentrations of atmospheric CO_2, N_2O and CH_4 from air bubbles in Antarctic ice. *Nature, 320*:248-250.

Quay, P., S.K. King, J.M. Lansdown, D.O. Wilbur. 1988. Isotopic composition of methane released from wetlands: Implications for the increase in atmospheric methane. *Global Biogeochem. Cycles, 2*:385-397.

Quay, P., S.K. King, J. Stutsman, D.O. Wilbur, L.P. Steele, I. Fung, R.H. Gammon, T.A. Brown, G.W. Farwell, P.M. Grootes, F.H. Schmidt. 1991. Carbon isotopic composition of atmospheric CH_4: Fossil and biomass burning source strengths. *Global Biogeochem. Cycles, 5*:25-47.

Rasmussen, R.A., M.A.K. Khalil. 1981. Atmospheric methane (CH_4): Trends and seasonal cycles. *J. Geophys. Res., 86*:9,826-9,832.

Rasmussen, R.A., M.A.K. Khalil. 1984. Atmospheric methane in the recent and ancient atmospheres: Concentrations, trends and interhemispheric gradient. *J. Geophys. Res., 89*:11,599-11,605.

Ritter, J.A., J.D.W. Barrick, G.W. Sachse, G.L. Gregory, M.A. Woerner, C.E. Watson, G.F. Hill, J.E. Collins. 1992. Airborne flux measurements of trace species in an arctic boundary layer. *J. Geophys. Res., 97*:16,601-16,625.

Rose, P.W., P.C. Rosendahl. 1983. Classification of Landsat data for hydrologic application, Everglades National Park. *Photogram. Eng. Rem. Sens., 49*:505-511.

Roulet, N.T., R. Ash, T.R. Moore. 1992a. Low boreal wetlands as a source of atmospheric methane. *J. Geophys. Res., 97*:3,739-3,749.

Roulet, N.T., J. Ritter, A. Jano, C. Kelly, L. Klinger, T.R. Moore, R. Protz, W.R. Rouse. (1992b). The Hudson Bay Lowland as a source of atmospheric methane, *J. Geophys. Res.* (in press).

Roulet, N.T., T. Moore, J. Bubier, P. LaFleur. 1992c. Northern fens: Methane flux and climatic change. *Tellus, 44B*:100-105.

Rust, F.E. 1981. Ruminant methane d($^{13}C/^{12}C$) values: Relationship to atmospheric methane. *Science, 211*:1,044-1,046.

Schoell, M. 1980. The hydrogen and carbon isotopic composition of methane from natural gases of various origins. *Geochem. Cosmochim. Acta, 44*:649-661.

Sebacher, D.I., R.C. Harriss, K.B. Bartlett. 1985. Methane emissions to the atmosphere through aquatic plants. *J. Environ. Qual., 14*:40-46.

Sebacher, D.I., R.C. Harriss, K.B. Bartlett, S.M. Sebacher, S.S. Grice. 1986. Atmospheric methane sources: Alaskan tundra bogs, an alpine fen, and a subarctic boreal marsh. *Tellus, 38B*:1-10.

Seiler, W. 1984. Contribution of biological processes to the global budget of CH4 in the atmosphere. In: *Current Perspectives in Microbial Ecology* (M. Klug and C. Reddy, eds.), Amer. Soc. Microbiol., Washington, D.C., p 468.

Seiler, W., R. Conrad. 1987. Contribution of tropical ecosystems to the global budgets of trace gases, especially CH_4, H_2, CO and N_2O. In: *The Geophysiology of Amazonia, Vegetation and Climate Interactions* (R.E. Dickinson, ed.), John Wiley, New York, p 133.

Stauffer, B., F. Fischer, A. Neftel, H. Oeschger. 1985. Increase of atmospheric methane recorded in Antarctic ice core. *Science, 229*:1,386-1,388.

Stauffer, B., E. Lochbronner, H. Oeschger, J. Schwander. 1988. Methane concentration in the glacial atmosphere was only half that of the preindustrial Holocene. *Nature, 332*:812-814.

Steele, L.P., P.J. Fraser, R.A. Rasmussen, M.A.K. Khalil, T.J. Conway, A.J. Crawford, R.H. Gammon, K.A. Masarie, K.W. Thoning. 1987. The global distribution of methane in the troposphere. *J. Atmos. Chem., 5*:125-171.

Stevens, C.M. 1988. Atmospheric methane. *Chem. Geol., 71*:11-21.
Stevens, C.M., A. Engelkemeir. 1982. Stable carbon isotopic composition of methane from some natural and anthropogenic sources. *J. Geophys. Res., 87*:4,879:4,882.
Svensson, B.H. 1976. Methane production in tundra peat. In: *Microbial Production and Utilization of Gases (H_2, CH_4, CO)* (H.G. Schlegel, G. Gottschalk, and N. Pfennig, eds.), Gottingen, p 135.
Svensson, B.H. 1980. Carbon dioxide and methane fluxes from ombrotrophic parts of a subarctic mire. *Ecol. Bull. (Stockholm), 30*:235-250.
Svensson, B.H., T. Rosswall. 1984. In situ methane production from acid peat in plant communities with different moisture regimes in a subarctic mire. *Oikos, 43*:341-350.
Tathy, J.-.P, B. Cros, R.A. Delmas, A. Marenco, J. Servant, M. Labat. 1992. Methane emission from flooded forest in Central Africa. *J. Geophys. Res., 97*:6,159-6,168.
Taylor, J.A., G. Brasseur, P. Zimmerman, R. Cicerone. 1991. A study of the sources and sinks of methane and methyl chloroform using a global three-dimensional Lagrangian tropospheric tracer transport model. *J. Geophys. Res., 96*:3,013-3,044.
Thompson, A.M., R.J. Cicerone. 1986. Atmospheric CH_4, CO and OH from 1960 to 1985. *Nature, 321*:148-150.
Twenhofel, W.H. 1926. *Principles of Sedimentation.* McGraw-Hill, New York.
Twenhofel, W.H. 1951. *Principles of Sedimentation.* McGraw-Hill, New York.
Tyler, S.C. 1986. Stable carbon isotope ratios in atmospheric methane and some of its sources *J. Geophys. Res., 91*:13,232-13,238.
Tyler, S.C., P.R. Zimmerman, C. Cumberbatch, J.P. Greenberg, C. Westberg, J.P.E.C. Darlington. 1988. Measurements and interpretation of $d^{13}C$ of methane from termites, rice paddies, and wetlands in Kenya. *Global Biogeochem. Cycles, 2*:341-355.
UNESCO. 1973. *International Classification and Mapping of Vegetation.* UNESCO, Paris, 93p.
Wahlen, M., N. Tanaka, R. Henry, T. Yoshinari, R.G. Fairbanks, A. Shemesh, W.S. Broecker. 1988. ^{13}C, D, and $_{14}C$ in methane. In: *Report to Congress and Environmental Protection Agency on NASA Upper Atmosphere Research Program*, NASA, p 315.
Wahlen, M., N. Tanaka, R. Henry, B. Deck, J. Zeglen, J.S. Vogel, J. Southon, A. Shemesh, R. Fairbanks, W. Broecker. 1989. Carbon-14 in methane sources and in atmospheric methane: the contribution from fossil carbon. *Science, 245*:286-290.
Walker, D.A., W. Acevedo, K.R. Everett, L. Gaydos, J. Brown, P.J. Webber. 1982. Landsat-assisted environmental mapping in the Arctic National Wildlife Refuge, Alaska. U.S. Cold Regions Res Eng Lab, Hanover, NH.

Wassmann, R., U.G. Thein, M.J. Whiticar, H. Rennenberg, W. Seiler, W.J. Junk. 1992. Methane emissions from the Amazon floodplain: Characterization of production and transport. *Global Biogeochem. Cycles, 6*:3-13.

Whalen, S.C., W.S. Reeburgh. 1988. A methane flux time series for tundra environments. *Global Biogeochem. Cycles, 2*:399-409.

Whalen, S.C., W.S. Reeburgh. 1990. A methane flux transect along the trans-Alaska pipeline haul road. *Tellus, 42B*:237-249.

Whalen, S.C., W.S. Reeburgh. 1992. Interannual variations in tundra methane emission: A 4-year time-series at fixed sites. *Global Biogeochem. Cycles, 6*:139-159.

Whiting, G.J., J.P. Chanton, D.S. Bartlett, J.D. Happell. 1991. Relationships between CH_4 emission, biomass and CO_2 exchange in a subtropical grassland. *J. Geophys. Res., 96*:13,067-13,071.

Wilson, J.O., P.M. Crill, K.B. Bartlett, D.I. Sebacher, R.C. Harriss, R.L. Sass. 1989. Seasonal variation of methane emissions from a temperate swamp. *Biogeochemistry, 8*:55-71.

Zoltai, S.C., F.C. Pollett. 1983. Wetlands in Canada: Their classification, distribution and use. In: *Ecosystems of the World, Mires: Swamp, Bog, Fen and Moor, Case Studies, 4B.* (A.J.P. Gore, ed.), Elsevier, New York, p 245.

Chapter 16

Waste Management

SUSAN A. THORNELOE[1], MORTON A. BARLAZ[2], REBECCA PEER[3], L.C. HUFF[3], LEE DAVIS[3], JOE MANGINO[3]

[1]*United States Environmental Protection Agency, Office of Research and Development Air and Energy Engineering Research Laboratory, Research Triangle Park, North Carolina, U.S.A.*

[2]*North Carolina State University, Civil Engineering Department Raleigh, North Carolina, U.S.A.*

[3]*Radian Corporation, Research Triangle Park, North Carolina, U.S.A.*

1. Introduction

Landfills, wastewater treatment lagoons, and livestock waste management are operations representing sources of methane. Estimates of CH_4 emissions from these sources suggest approximately 70 (54-95) Tg/yr globally or 14% of total global CH_4 emissions of 500 Tg/yr (IPCC, 1992). This chapter begins with a brief overview of how CH_4 is generated from the anaerobic decomposition of waste and then discusses generation of CH_4 in detail in landfills, wastewater treatment lagoons, and livestock waste management. Current techniques for estimating CH_4 emissions from waste are summarized, and sources of uncertainty are identified.

The potential control of CH_4 emissions from waste management has been targeted by the United States (U.S.) and other countries as part of greenhouse gas reduction programs designed to meet the goals of treaties signed at the United Nations Conference on Environment and Development (UNCED) held in 1992. Consequently, reducing the uncertainty associated with CH_4 emission estimates is a high priority.

NATO ASI Series, Vol. I 13
Atmospheric Methane: Sources, Sinks, and Role in Global Change
Edited by M. A. K. Khalil

2. Methane production during the anaerobic decomposition of waste

The anaerobic decomposition of organic matter is a complex process that requires that several groups of microorganisms act synergistically under favorable environmental conditions (see Boone, this volume). The pathway described below has been demonstrated to apply to anaerobic decomposition in sludge digesters and in livestock waste management systems. This anaerobic pathway is also expected to occur in landfills and anaerobic wastewater lagoons (Barlaz et al., 1989a).

Three trophic groups of anaerobic bacteria must be present to produce CH_4 from biological polymers such as, cellulose, hemicellulose, and protein: (1) hydrolytic and fermentative microorganisms, (2) obligate proton-reducing acetogens, and (3) methanogens (Wolfe, 1979; Zehnder et al., 1982). The hydrolytic and fermentative group is responsible for the hydrolysis of biological polymers. The initial products of polymer hydrolysis are soluble sugars, amino acids, long-chain carboxylic acids, and glycerol. Following polymer hydrolysis, the hydrolytic and fermentative microorganisms ferment the initial products of decomposition into short-chain carboxylic acids, alcohols, carbon dioxide (CO_2), and hydrogen. Acetate, a direct precursor of CH_4, is also formed.

The second group of bacteria -- obligate proton-reducing acetogens -- convert the fermentation products of the hydrolytic and fermentative microorganisms to CO_2, hydrogen, and acetic acid. The conversion of fermentation intermediates, such as butyrate, propionate, and ethanol, is thermodynamically favorable only at very low hydrogen concentrations. Thus, these substrates are utilized only when the obligate proton-reducing acetogenic bacteria can function in syntrophic association with hydrogen scavengers, such as CH_4-producing or sulfate-reducing organisms.

The third group of bacteria necessary for the production of CH_4 are methanogens. Major substrates utilized by methanogens for the production of CH_4 are acetate, formate, methanol, methylamines, and hydrogen plus CO_2 (Wolin and Miller, 1985).

While CH_4 and CO_2 are the terminal products of anaerobic decomposition, CO_2 and water are the terminal products of aerobic decomposition. Aerobic decomposition occurs in management facilities where waste is exposed to air, such as when compost is turned for aerating, and in uncontrolled dumps, such as when refuse is spread in thin layers or otherwise exposed to oxygen (as by scavenging). When refuse is buried in large piles, whether at an open dump or in a sanitary landfill, the oxygen entrained at burial is consumed rapidly, and the majority of decomposition will occur under anaerobic conditions (Bhide et al., 1990).

3. Methane production from waste burial

The proportion of waste generated in developing countries has been projected to increase over the next several decades, while the proportion of waste generated by developed countries is expected to decline. This trend can be attributed to projections of higher population increases in developing countries and not to increased per capita waste generation. Despite efforts toward source reduction and recycling programs, per capita waste generation is expected to increase in the U.S. (Kaldjian, 1990) and in other industrialized countries. The much lower rates of population growth in industrialized countries are expected to result in slower growth in municipal solid waste (MSW) production, as compared to developing countries. However, this scenario will not be realized if per capita income decreases in developing countries. Recently, declining economic conditions have resulted in reduced MSW generation in Caracas, Venezuela, Mexico City, Mexico, and Buenos Aires, Argentina (Bartone et al., 1991).

Methods of managing MSW vary widely, ranging from open dumps and open burning to sanitary landfills with leachate collection systems and landfill gas control. The majority of the world's MSW is managed using either sanitary landfills or open dumps. In the U.S., recent estimates indicate that 72% of MSW is buried in landfills (Kaldjian, 1990). Anaerobic decomposition prevails in landfills. Both anaerobic and aerobic processes occur at open dumps. The CH_4 potential of other waste management processes such as incineration, recycling, and composting is considered insignificant in comparison to landfills and open dumps.

Landfilled waste contains numerous constituents that have the potential to biodegrade under anaerobic conditions. However, optimal conditions for anaerobic decomposition within a landfill may not exist and may thus result in overestimated emissions. Many methodologies for estimating emission assume that optimal conditions exist. In a recent study field data were gathered to develop an empirical model that is intended to reflect actual emissions to the atmosphere (Peer et al., 1993). This model adjusts for gas recovery efficiency and CH_4 oxidation. Estimates are presented later in this chapter, along with updated estimates using the approach developed by Bingemer and Crutzen (1987).

3.1 Factors affecting CH_4 potential of buried waste. The traditional method of classifying MSW according to sortable categories (such as paper, plastic, food waste, yard waste, glass, metals, rubber, wood, textiles, dirt, and miscellaneous (Kaldjian, 1990)) is appropriate for recycling studies and overall solid waste management planning. However, data specific to the chemical composition of refuse are more applicable to analyses of refuse decomposition. Studies of refuse in Madison, Wisconsin, showed cellulose plus hemicellulose to be about 60% of landfill waste and to account for 91% of the CH_4 potential of refuse (Barlaz, 1985, 1988; Barlaz et al., 1989b). The components of MSW that contain significant biodegradable fractions are food waste, yard waste, and paper, which have a combined cellulose and hemicellulose content of 50 to 100%. Lignin is the other major organic component of refuse; however, lignin does not undergo significant decomposition under anaerobic conditions (Young and Frazer, 1987).

Methane formation does not occur immediately after refuse is placed in a landfill or dump. It can take months or years for the proper environmental conditions and the required microbiological populations to become established. Numerous factors control decomposition, including moisture content, nutrient concentrations, presence and distribution of microorganisms, particle size, water flux, pH level, and temperature. For a review of the factors affecting CH_4 production see Halvadakis et al. (1983), Pohland and Harper (1987), and Barlaz et al. (1990).

The two factors that appear to have the greatest effect on CH_4 production are moisture content and pH. The effect of refuse moisture content has been summarized by Halvadakis et al. (1983), although some of their data relate to manure and not to municipal waste. The broadest data sets are those of Emberton (1986) and Jenkins and Pettus (1985). Emberton measured CH_4

production rates in excavated landfill samples under laboratory conditions. Jenkins and Pettus sampled refuse from landfills and tested how CH_4 production was affected by the moisture content of refuse. In both studies, the CH_4 production rate increased with increasing moisture content, despite differences in refuse density, age, and composition. It is difficult to translate the results of these laboratory studies to actual landfills. An attempt by the U.S. EPA's Air and Energy Engineering Research Laboratory (AEERL) to identify a statistically significant correlation between landfill gas recovery and precipitation (which affects refuse moisture content) found no such correlation (Peer et al., 1992).

A second key factor influencing the rate and onset of CH_4 production is pH. The optimum pH level for activity by methanogenic bacteria is between 6.8 and 7.4. CH_4 production rates decrease sharply with pH values below about 6.5 (Zehnder et al., 1982). When refuse is buried in landfills, there is often a rapid accumulation of carboxylic acids; this results in a pH decrease and a long time lapse between refuse burial and the onset of CH_4 production.

Neutralizing leachate and recycling it back through refuse has been shown to enhance the onset and rate of CH_4 production in laboratory studies (Pohland, 1975; Buivid et al., 1981; Barlaz et al., 1987, International Energy Agency, 1992). Given that moisture and pH have been reported as the two most significant factors limiting CH_4 production, the stimulatory effect of leachate neutralization and recycling is expected. Neutralization of leachate provides a means of externally raising the pH of the refuse ecosystem. Recycling neutralized leachate back through a landfill increases and stabilizes refuse moisture content and substrate availability and provides mixing in what would otherwise be an immobilized batch reactor.

Notably, field experience with leachate recycling systems is limited and more information is needed to fully document their value. In addition, the lapsed time preceding the onset of CH_4 production in landfills is an important aspect when considering the management of individual landfills for biogas recovery or emissions mitigation. The age at which landfills and uncontrolled dumps begin to produce CH_4 is of lesser importance when evaluating global CH_4 emissions from MSW management systems. In this case, the total CH_4 production potential is more critical.

3.2 Determination of the CH_4 potential of landfills and dumps. Knowledge of the chemical composition of refuse buried in a landfill makes it possible to estimate the volume of CH_4 that may be produced. The mass of CH_4 that would be produced if all of a given constituent were converted to CH_4, CO_2, and ammonia may be calculated from Equation 1 (Parkin and Owen, 1986):

$$C_nH_aO_bN_c + [n - 1/4a\ 1/2b + 3/4c]H_2O \rightarrow \quad (1)$$

$$[1/2n - 1/8a + 1/4b + 3/8c]CO_2 + [1/2n + 1/8a - 1/4b - 3/8c]CH_4 + cNH_3$$

Using this stoichiometry, the CH_4 potential of cellulose ($C_6H_{10}O_5$) and hemicellulose ($C_5H_8O_4$) is 415 and 424 liters (l) CH_4 at standard temperature and pressure (0°C, 1 atmosphere) per dry kilogram (kg), respectively (18.5 and 18.9 grams [g] CH_4/dry kg).

These methane potentials represent maximum CH_4 production if 100% of the cellulose and hemicellulose were converted to CH_4. However, decomposition of these constituents in landfills is well below 100%, mainly because (1) some cellulose and hemicellulose is surrounded by lignin or other recalcitrant materials (such as plastic) and, therefore, is not biologically available; and (2) without active intervention, buried refuse is not evenly exposed to moisture, microorganisms, and nutrients. Barlaz et al. (1989b) applied mass balances to shredded refuse incubated in laboratory-scale lysimeters with leachate recycle. Carbon recoveries of 87 to 111% were obtained, where a perfect mass balance would give a carbon recovery of 100%. Mineralization of 71% of the cellulose and 77% of the hemicellulose was measured in a container sampled after 111 days. Mass balances were useful for documenting the decomposition of specific chemical constituents and demonstrating the relationship between cellulose and hemicellulose decomposition and CH_4 production.

Mass balances may be used to estimate the CH_4 potential remaining in a landfill by sampling the refuse, performing the appropriate chemical analyses, and calculating the CH_4 potential. Ideally, the initial chemical composition and CH_4 potential of the refuse would be known, in which case comparing that initial CH_4 potential with the potential at the time of sampling would provide information on the fraction of the refuse that has been degraded. Indisputably, representative sampling of a full-scale sanitary landfill is not realistic. However, it is possible to obtain multiple samples at presumably representative locations within a landfill to get an estimate of the range and extent of decomposition. Samples should be as large as can reasonably be handled and reduced.

Another technique for assessing the CH_4 potential of refuse is the biochemical CH_4 potential (BMP) test (Shelton and Tiedje, 1984; Bogner, 1990). In the BMP test, the anaerobic biodegradability of a small sample of refuse (5 to 10 g) is measured in a small batch reactor (100 to 200 m l). While the BMP represents an upper bound of CH_4 potential from refuse, it will be lower than the stoichiometric estimate described above. BMPs also require representative sampling in landfills.

Comparison of CH_4 production data between field-scale landfills and laboratory experiments is difficult because there are essentially no data in the open literature on CH_4 production rates in field-scale facilities. Data from field-scale landfills are complicated by questions regarding the mass of refuse responsible for production of a measured volume of gas and the efficiency of gas collection. There are more CH_4 production data collected under laboratory conditions than field conditions. However, the laboratory data are not always comparable to experimental conditions. For instance, moisture, particle size, and temperature are not uniform between studies. In addition, most laboratory experiments were conducted to explore techniques for enhancing CH_4 production. The enhanced CH_4 production rates would not normally be expected at field-scale landfills.

CH_4 yields of 1.9 to 5.4 g CH_4/dry kg refuse have been reported in laboratory tests conducted with leachate recycle and neutralization (Buivid et al., 1981; Barlaz et al., 1987; Kinman et al., 1987; Barlaz, 1988). These studies show significant variation in CH_4 production rate and CH_4 yield. Some of the differences can be explained by differences in experimental design. For example, the data reported by Barlaz et al. (1987) and Barlaz (1988) differ in reactor volume (100 vs. 2 ℓ), temperature (25 vs. 41°C), and the rate of leachate recycle. Also, Buivid et al. (1981) used refuse with an abnormally high paper content.

CH$_4$ yields were measured in field-scale test cells as part of the Controlled Landfill Project in Mountain View, California (Pacey, 1989). Yields of 38.6 to 92.2 ℓ (1.7 to 4.1 g) CH$_4$ /dry kg of refuse were measured after 1597 days. However, mass balance data on cellulose losses from individual cells suggested that more CH$_4$ was produced than was measured in certain test cells.

A number often used in engineering practice as an estimate of CH$_4$ production in field scale landfills is 0.1 ft^3 CH$_4$/wet lb-yr, over a 10 to 20 year production period. This value has not been documented in open literature. Assuming refuse is buried at 20% moisture content, this converts to 7.8 ℓ (0.3 g) CH$_4$/dry kg-yr, a number comparable to some of the lower values reported in the literature.

Even in landfills with venting systems, some of the CH$_4$ is likely to escape from the landfill through the final cover. The fraction released through the final cover will be a function of the type of gas venting system in place and the type of cover. Probably not all the CH$_4$ that escapes from landfills is released to the atmosphere as CH$_4$. CH$_4$ that passes through the cover soil may be converted to CO$_2$ in the presence of oxygen by aerobic methanotrophic bacteria. CH$_4$ oxidation has been documented in landfill cover soil studied under laboratory conditions (Whalen et al., 1990). However, there are no data on the quantitative significance of CH$_4$ oxidation above landfills. CH$_4$ escaping through cracks in a landfill cover likely will not reside in the cover for a period sufficient to undergo significant biodegradation.

3.3 Emissions estimate methodology for landfills and open dumps. Two techniques for estimating emissions from landfills are reviewed here: the Organization of Economic Cooperation and Development (OECD) and EPA/AEERL methods. These methods were developed to estimate global CH$_4$ emissions. Models that estimate CH$_4$ production from individual landfills are reviewed by Augenstein and Pacey (1990) and Peer et al., (1993). While global estimates focus on ultimate CH$_4$ release, models of individual landfills emphasize the rate and duration of CH$_4$ production as these factors affect the economics of landfill gas recovery projects.

3.4 OECD methodology. OECD (1991) used the mass balance approach developed by Bingemer and Crutzen (1987), where an instantaneous release of

CH_4 is assumed to enter the atmosphere during the same year that refuse is placed in a landfill. This method also assumes that (1) all of the CH_4 that is produced escapes to the atmosphere (none is oxidized on its way to the atmosphere) and (2) all developing nations generate and dispose of MSW at the same per capita rate.

To calculate the annual emission from MSW, OECD used the following equation from Bingemer and Crutzen: CH_4 Emission = Total MSW Generated (kg/yr) x MSW Landfilled (%) x DOC in MSW (%) x Fraction Dissimilated DOC (%) x 0.5 g CH_4/g Biogas x Conversion Factor (16 g CH_4/12 g C) − Recovered CH_4 (kg/yr) where DOC is degradable organic carbon; Fraction Dissimilated DOC is the portion of carbon in substrates that is converted to landfill gas; and "Recovered CH_4" is the amount of CH_4 that is recovered through gas recovery systems and never emitted to the atmosphere. The uncertainties of this approach are attributed to assumptions regarding anaerobic decomposition. Many factors inhibit this process, and this approach tends to overestimate potential emissions. Moreover, this methodology does not adjust for CH_4 oxidation, which is known to occur.

3.5 EPA/AEERL's regression model methodology. The EPA/AEERL methodology uses an empirical model derived using landfill gas recovery data. The quantity of CH_4 estimated by this model is much less than that predicted by stoichiometric analyses or by laboratory studies (EMCON, 1982; Barlaz et al., 1989b, 1990; Peer et al., 1992). The data gathered from U.S. landfills that were used to develop this model represent a broad range of climate zones and waste composition as described below (Campbell et al., 1991; Peer et al., 1992). This model is intended to reflect the amount of gas that is ultimately released to the atmosphere by adjusting for gas recovery efficiency and CH_4 oxidation. For the estimates presented in this chapter, it was assumed that the recovery efficiency is 80% and that 10% of non-recovered CH_4 is oxidized. Refinements of this methodology include adjustments for recovery efficiency and CH_4 oxidation based on factors derived through an uncertainty analysis. Comparison of the refined estimates with the earlier estimates indicates a slight increase.

Data from 21 landfills were used to determine if there is a correlation between CH_4 recovery and landfill characteristics such as waste quantity, age, depth, and climate. Selection of variables for the regression models was based on the results of the correlation and scatter plots of selected variables (such as climate, landfill depth, refuse mass, and refuse age). The main conclusion of the study was that the annual CH_4 recovery rate was linearly correlated with the mass of refuse in the landfill, and with landfill depth (Peer et al., 1992). The data, regression line, and the 95% confidence limits of the regression coefficient are shown in Figure 1. The regression was significant ($P < 0.01$), but much of the variability in the data is unexplained (adjusted $R^2 = 0.50$). The intercept was not significant, so the final model was forced through the zero. Peer et al. (1992) provide detailed information on the characteristics of each site that may contribute to data variability.

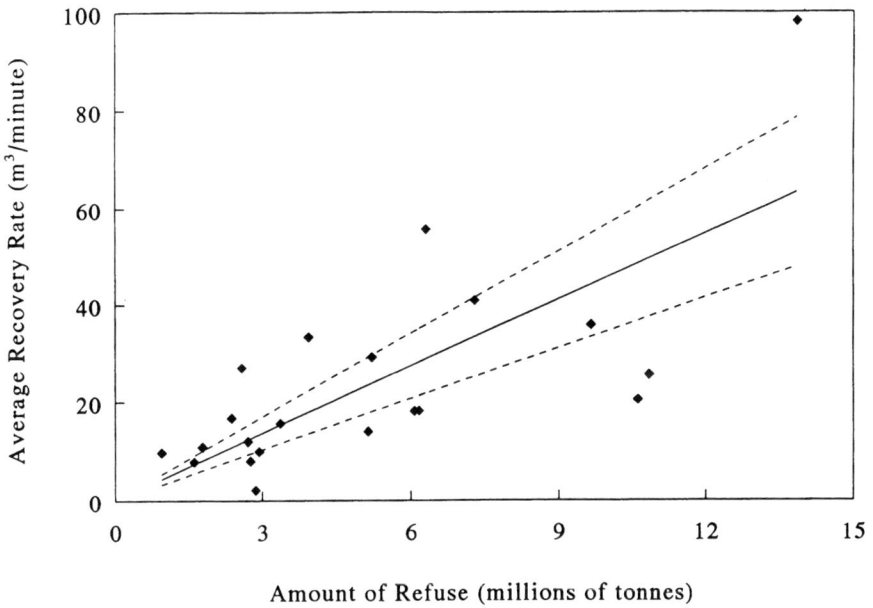

Figure 1. Methane recovery data as a function of buried refuse (Peer et al., 1992).

No statistically significant relationships were identified between annual CH_4 recovery and climate variables such as precipitation, temperature, and dew point. The effect of refuse age on gas production was also analyzed. Gas recovery correlated most strongly with refuse between 10 and 20 years old. Although these results were not conclusive, they suggest that the generation time for gas production is 20 to 30 years (Peer et al., 1992). This generation time is within the range of generation times assumed in many landfill gas recovery models (EMCON, 1982; Augenstein and Pacey, 1990).

One advantage of using the EPA/AEERL model is that refuse mass is the only variable required to estimate CH_4 emissions. Furthermore, it will be relatively easy to update the relationship between CH_4 recovery and refuse mass as more data become available. The confidence limits of the regression coefficient can be used to bound emission estimates:

(1) The upper and lower 95% confidence limits are 6.5 and 2.5 m^3 CH_4/min/10^6 Mg wet refuse (4.6 to 1.8 g CH_4/min/kg of wet refuse), respectively.
(2) Assuming an average generation time of 25 years gives an average CH_4 recovery of 59.4 m^3 CH_4/Mg wet refuse (42 g CH_4/kg of wet refuse).
(3) A range of 33 to 86 m^3 CH_4/10^6 Mg wet refuse (24 to 61 g CH_4/kg wet refuse) results.

These values were derived using data collected at U.S. MSW landfills. The use of this factor assumes that other countries have a waste composition similar to U.S. landfills. However, U.S. MSW generally has a higher organic content than most other countries (Bingemer and Crutzen, 1987). Therefore, the use of this factor may overestimate landfill emissions for other countries. Future refinements of the model will adjust for waste composition using gas potential data for different biodegradable waste streams.

3.6 Assumptions and data used to estimate waste generation rates. To estimate global CH_4 emissions it was necessary to make several assumptions regarding waste generation and disposal. Emission factors in all industrialized countries were assumed to be equal to those in the U.S. The CH_4 emission factors for the less-developed countries (LDCs), where adequate data were not available, were assumed to be 25 to 75% of the average U.S. estimate. Several factors were taken into account to develop this range. The composition of waste is different in third world countries. For example, much of the garbage is scavenged before it is placed in the landfill, especially paper, textiles, and metal products. More putrescibles end up in dumps and probably do not generate as much CH_4 because more oxidation (aerobic process) takes place. In addition, garbage is often burned, which decreases the amount of material available for anaerobic decomposition, but may increase CH_4 emissions from inefficient combustion. Finally, most landfills in the LDCs are open dumps which are scavenged by humans and wild and domestic animals. While anaerobic conditions may form in some of these dumps, the potential for aerobic decomposition is much greater than in sanitary landfills. Based on these assumptions, the 25 to 75% range was chosen as the default value for LDCs, since no data on actual CH_4 emissions are available.

Estimates of waste generation and burial are presented in Table 1. This table was developed using the data presented in the references shown. A recent review of global waste management trends found that information on MSW generation in developing countries is difficult to obtain; in many cases, it is anecdotal (Davis et al., 1992). As shown in the reference list for Table 1, much country-specific data were available to determine MSW generation and land disposal values, especially for developed countries. Where no data were available, data from similar countries were used.

Most of the available data for developing countries are provided on a per capita basis for only the larger cities; this information was combined with population (United Nations, 1990) and percent of the country that is considered urban (Population Reference Bureau, Inc., 1989), to determine the amount of waste generated in urban centers of these countries. An estimate for rural per capita refuse generation rate (Kessler, 1990) was then combined with the rural population value to determine rural waste generation. Once the amount of waste generated for the entire country was estimated, a default value of 50% disposed on land (whether in landfills or open dumps) was used to estimate the amount of waste that may degrade anaerobically. The 50% default was chosen to represent waste that is actually collected in some manner and disposed of in landfills or large enough open dumps for anaerobic conditions to occur. The remaining 50% of waste generated is assumed to be (1) incinerated or combusted, (2) dumped in rivers or other bodies of water, or (3) scattered or buried in small piles that degrade anaerobically. These disposal methods do not produce large amounts of CH_4.

The amount of MSW landfilled in the U.S. is approximately 189 million tonnes (U.S. EPA, 1988). Paper was the largest single component of the DOC fraction in both the U.S. and Canada. Per capita MSW generation was in the range of 1.7 to 1.8 kg/person/day for both the U.S. and Canada (Kaldjian, 1990; El Rayes and Edwards, 1991), and the average DOC content of paper or MSW is 20%. The average MSW generation rate in other OECD countries is 1.1 kg/person/day. MSW in these countries has a DOC content of approximately 15.3%. The value used for the U.S. is for MSW only; an additional 15 Tg/yr of biodegradable industrial solid waste is also landfilled (U.S. EPA, 1987), which is unaccounted for in the present estimates of landfill CH_4. In most cases, country-specific information does not state whether industrial waste is included with MSW.

Information on the amount of MSW generated and landfilled in the European countries that are not OECD members and in the former Soviet Union is limited. Nozhevnikova et al. (1993) used 0.8 kg/person/day for MSW generation in the former USSR. Average MSW generation for Greece, the former Soviet Union, and Eastern Europe is approximately 0.6 kg/person/day (Bingemer and Crutzen, 1987; Frantzis, 1988; Papachristou, 1988; Peterson and Perlmutter, 1989), and the available data indicate that putrescibles make up a large portion of the MSW (estimates range from 32 to 60%). This MSW contains approximately 15% DOC (Bingemer and Crutzen, 1987; Frantzis, 1988; Papachristou, 1988; Peterson and Perlmutter, 1989; Zsuzsa, 1990).

For most Asian countries, estimates of MSW generation were identified for one or two major cities, but not for the entire country. National per capita MSW generation estimates were identified for Indonesia, Sri Lanka, the Philippines, Singapore, Taiwan, and Pakistan. These estimates range from 0.4 kg/person/day for the Philippines to 1.0 kg/person/day for Singapore. The average per capita MSW generation for these countries is estimated to be 0.6 kg/person/day (Davis et al., 1992).

Table 1. Waste totals (Tg) by geographic region used to develop emission estimates

Geographic Region	Waste	Waste Landfilled	References*
Africa	80	38	1-14
Asia and the Middle East	363	175	15-28
Europe	224	165	29-46
North America	301	216	47-49
Oceania	14	12	50-51
South and Central America	64	47	52-57

* Reference Key

1. El Halwagi et al., 1988.
2. El Halwagi et al., 1986.
3. Kaltwasser, 1986.
4. UNDP et al., 1987.
5. Holmes, 1984.
6. Monney, 1986.
7. Cointreau, 1984.
8. Cointreau, 1987.
9. The World Bank, 1985.
10. Mwiraria et al., 1991.
11. United Republic of Tanzania, 1989.
12. Verrier, 1990.
13. World Resources Institute, 1990.
14. Rettenberger and Weiner, 1986.
15. Bhide and Sundaresan, 1983.
16. United Nations, 1989.
17. Maniatis et al., 1987.
18. Lohani and Thanh, 1980.
19. Ahmed, 1986.
20. Pairoj-Boriboon, 1986.
21. Gadi, 1986.
22. Mei-Chan, 1986.
23. U.S. EPA, 1990.
24. Diaz and Goulueke, 1987.
25. Xianwen and Yanhua, 1991.
26. Cossu, 1990a.
27. Hayakawa, 1990.
28. Swartz, 1989.
29. Lechner, 1990.
30. World Resources Institute, 1990.
31. Carra and Cossu, 1990.
32. Ettala, 1990.
33. Stegmann, 1990.
34. Ernst, 1990.
35. Cossu and Urbini, 1990.
36. Beker, 1990.
37. Carra and Cossu, 1990.
38. Gandolla, 1990.
39. Cossu, 1990b.
40. Swartz, 1989.
41. Richards, 1989.
42. U.S. EPA, 1990.
43. Scheepera, 1990.
44. Bartone and Haley, 1990.
45. Bartone, 1990a.
46. Bingemer and Crutzen, 1987.
47. U.S. EPA, 1988.
48. U.S. EPA, 1990.
49. El Rayes and Edwards, 1990.
50. Bateman, 1988.
51. Richards, 1989.
52. Kessler, 1990.
53. U.S. EPA, 1990.
54. World Resources Institute, 1990.
55. Diaz and Golueke, 1987.
56. Bartone et al., 1991.
57. Yepes and Campbell, 1990.

Few data are available on MSW production and management in Central America, South America, the Caribbean Islands, and Mexico. Most of the available information is only for the larger cities. The average per capita MSW generation rate in seven South and Central American countries (Brazil, Colombia, Chile, Paraguay, Peru, Venezuela, Costa Rica), and Mexico is estimated to be 0.8 kg/person/day. The components are mainly vegetable and putrescible waste paper and cardboard. The average DOC for the seven South and Central American countries and Mexico is 17% (Davis et al., 1992).

Information on MSW generation and disposal for African and Middle Eastern countries is also very limited. In Africa, it appears that toxic and hazardous industrial and commercial wastes are purposely or inadvertently disposed of with the MSW stream. Some information pertaining to generation rates for African countries was located; but information for only two Middle Eastern countries, Israel and Yemen, was obtained. Based on the very limited information for these two continents, it is estimated that per capita generation rates range from 0.3 to 1.1 kg/person/day, and the DOC content ranges from 3 to 20%.

3.7 Global and country-specific estimates of CH_4 emissions from landfills and dumps.

3.7.1. OECD Proposed Methodology. The MSW generation and landfill disposal data in Table 1 were used to calculate CH_4 emissions for each country using the OECD methodology discussed earlier.

This methodology is identical to that of Bingemer and Crutzen (1987), but makes use of more recent waste generation data and the results of an exhaustive study by the EPA/AEERL to gather country-specific waste generation and disposal data. As shown in Table 2, the global estimate of landfill CH_4 emissions using this methodology is 57 Tg/yr. The estimate for the U.S. (i.e., 21 Tg/yr) has been adjusted for the amount of CH_4 that is recovered for energy utilization (1.2 Tg/yr, Thorneloe, 1992).

3.7.2 EPA's Regression Model Methodology. Based on the EPA's regression model methodology, global landfill CH_4 emissions are estimated to range from 11 to 32 Tg/yr, with a midpoint of 21 Tg/yr. Additional refinements in the methodology may lend to adjustment of these estimates. Table 2 presents country-specific global estimates using the methodology described in Peer et al. (1992) and provides the lower, most probable, and upper-bound estimates of CH_4

emissions by continent. This estimate for the U.S. (i.e., 4 to 12 Tg/yr) has also been adjusted for the amount of CH_4 that is recovered for utilization based on a recently completed survey (i.e., 1.2 Tg/yr, Thorneloe, 1992). Emission estimates for countries other than the U.S. have not been adjusted for the amount of CH_4 that is recovered for utilization. Richards (1989) and Thorneloe (1992) estimate that worldwide there are 269 sites in 20 countries where landfill gas is recovered, including 114 sites in the U.S. No estimates are available, however, on the amount of CH_4 recovered in other countries.

3.8 Uncertainty associated with estimating landfill CH_4 emissions. There are several sources of uncertainty in estimating emissions of CH_4 from landfills, including:

(1) The quantity of CH_4 that is actually produced from the waste in the landfill;
(2) The quantity of CH_4 that is actually emitted to the atmosphere (the question as to how much is emitted and how much is oxidized as it passes through the landfill); and
(3) The quantity and composition of landfilled waste.

Two issues contribute to the difficulty of estimating CH_4 potential from open dumps: (1) the physical characteristics (size, configuration, temperature, moisture, compaction) of open dumps are unknown, and (2) the quantity and composition of open-dumped waste are also unknown. However, CH_4 is generated from open dumps. Bhide et al. (1990) reported biogas recovery from two uncontrolled landfills in Nagpur, India. Each of these sites was about 8 hectares in surface area and about 3 to 5 m deep. Neither site contained any cover material, and the older of the two landfills accepted waste from 1971 to 1984. Most of the organic matter had decomposed by the time the tests were performed, but biogas was obtained from wells 50 mm in diameter at a rate of 0.240 m^3/hr (CH_4 content not identified). Waste had been deposited in the second site "only recently" and the rate of biogas recovery was from 5 to 9 m^3/hr. The CH_4 content of the biogas from the second site was 30 to 40%. The work of Bhide et al. (1990) suggests that open dumps are a source of CH_4. Therefore, they have been included in the emission estimates.

The CH_4 potential of other types of landfills, such as those containing industrial and hazardous wastes, is not well understood. Industrial waste contains waste streams that will decompose under anaerobic conditions. Certain industrial waste streams, such as those of the food industry, may have high organic content and are, therefore, potentially significant sources of CH_4. However, landfills containing hazardous waste will have a low CH_4 potential because of the low moisture content and the requirement that only solid materials are accepted. In addition, the chemicals in the waste stream may be toxic to the microbes. Therefore, we believe the global emissions are negligible compared to those from MSW landfills or for industrial landfills.

The disposal of industrial and hazardous waste with MSW was common in the U.S. until 1975. Many closed landfill sites in the U.S. and worldwide contain biodegradable mixtures of these waste streams. Some industrialized countries such as the United Kingdom consider landfills as an acceptable treatment option for hazardous and industrial wastes. However, regulations being considered by the European Commission may prevent this practice. Waste streams in developing countries are less controlled and mingling of MSW, industrial wastes, and raw sewage in landfills is common (Cointreau, 1982). Co-disposal sites will generate CH_4 and may have emission potentials similar to MSW landfills having no history of co-disposal. Currently, published estimates of CH_4 emissions specific to open dumps and industrial and hazardous waste landfills are not available. Estimates presented in Table 2 include open dump emissions, but do not specifically consider the CH_4 potential of landfilled industrial and hazardous waste.

3.9 Trends in waste management and their impact on CH_4 emissions. As methods of MSW disposal change, there will be changes in CH_4 emissions. Trends in global waste management and their impact on CH_4 release are discussed here.

3.9.1 U.S. and Canada. Landfilling is the predominant MSW management method in both the U.S. and Canada. However, there is a trend in both countries toward more recycling, more incineration (especially in the U.S.), and less landfilling (U.S. EPA, 1988; Swartz, 1989; Kaljian, 1990; El Rayes and Edwards, 1991; Alter, 1991). Both countries also have a growing number of landfill gas recovery sites. The U.S. generates 12 times as much MSW as Canada.

Although the percentage of MSW to be placed in landfills is predicted to decline, increases in the amount of MSW generated and in the percent DOC will cause the fraction of degradable MSW in landfills to remain close to current levels. For example, if reliance on landfill disposal in the U.S. were to decline to 50% in the year 2010, 114 of the predicted 227 million Mg of MSW generated would be placed in landfills. Assuming an increase in carbon content to 22.2%, it is reasonable to assume that the U.S. will place 25 million Mg of DOC in landfills in 2010. Therefore, it can also be assumed that the current rate of landfill gas recovery will continue for several more decades (Boyner et al., 1988; Willumsen, 1990).

Although landfill gas production in the U.S. and Canada is expected to remain relatively steady over the next two decades, the startup of new landfill gas recovery systems is expected to shift the balance of landfill gas emissions. The amounts of CH_4 emitted to the atmosphere will decrease as more is controlled through flaring or utilization. Concurrently, the amount of CO_2 released will increase (Bonomo and Higginson, 1988; Boyner et al., 1988; El Rayes and Edwards, 1991; Willumsen, 1990; Rathje, 1991; Thorneloe, 1992). U.S. landfills are currently recovering about 1.2 Tg/yr of CH_4 and producing 344 MW_e of power (Thorneloe, 1992). Clean Air Act regulations proposed in May 1991 are expected to have a major impact on reducing landfill CH_4 emissions from both new and existing MSW landfills in the U.S. The proposed air emission regulations are expected to result in an additional emission reduction ranging from 5 to 7 Tg/yr of CH_4 (Federal Register, 1991; U.S. EPA, 1991).

3.9.2 OECD countries. In the future, most OECD countries are considering policies that would increase the amount of MSW handled by recycling and incineration and to decrease the amount placed in landfills. Landfills will be large, regional sites that also will be used to dispose of incinerator ash. The decreases in the amount of MSW landfilled, coupled with increases in landfill gas recovery and incineration, are anticipated to lead to reduced CH_4 emissions and increased emissions of CO_2 and other combustion gases.

3.9.3 European countries and the former Soviet Union. Sanitary landfills or open dumps (e.g., placing refuse in a scattered fashion near residences or along roadsides) are used almost exclusively for MSW management in Greece, Hungary, Portugal, Poland, Romania, Bulgaria, Yugoslavia, and the former Soviet Union (Curi, 1988; Bartone, 1990a; Bartone and Haley, 1990; Mnatsaknian, 1991). In the future, Poland plans to close open dumps and dispose of MSW in larger, regional sanitary landfills. The former Soviet Union hopes to establish an effective recycling program. No landfill gas recovery sites were identified in these countries.

3.9.4 Asian countries. Some Asian countries are upgrading their collection methods by introducing compactor trucks and covered containers. These changes could serve to decrease the amount of scavenging and increase the amount of MSW that is dumped in an uncontrolled fashion. Economic constraints, in tandem with a history of slow MSW management development, indicate that the use of sanitary landfills will not increase markedly in the near future for most of Asia. However, increases in population and total MSW will likely lead to increased CH_4 emissions.

3.9.5 South America, Central America, Mexico, and the Caribbean Islands. In the future, Brazil hopes to build recycling and composting plants, and sanitary landfills. Mexico also hopes to increase its number of sanitary landfills. Few landfill gas recovery sites are currently operating in South America and Mexico; there seems to some interest in increasing the number of landfill gas recovery sites in Brazil (Richards, 1988; Kessler, 1990).

3.9.6 Africa and the Middle East. In Africa the only MSW management methods reported to hold promise for expanded and successful application are recycling, composting, or possibly biogas recovery if markets and appropriate technologies can be developed to support these systems (Conner, 1978; Cointreau, 1982; Betts, 1984; Oluwande, 1984; Vogler, 1984; El-Halwagi et al., 1988). Much of the recyclable material in the MSW stream is currently being recovered (Cointreau, 1982; Wright et al., 1988), at least when comparing the amount of material recycled in low-income countries with that of middle-income and developed nations. However, recyclable materials are still available in the waste streams and, because wages are low in the developing countries, further recycling may be a viable waste management option (Cointreau, 1982).

4. Wastewater treatment and septic sewage systems

Wastewater from domestic, commercial, and industrial facilities is also a source of CH_4. Wastewater treatment lagoons, particularly those that treat wastewater with high biochemical oxygen demand (BOD), are suspected of emitting significant amounts of CH_4. Global CH_4 emissions from wastewater treatment lagoons are estimated to be about 25 Tg/yr (Orlich, 1990). This estimate is based on data from one wastewater lagoon study in Thailand and the assumption that 300 ℓ (13.4 g at standard temperature and pressure) of CH_4 is produced per kg of BOD. This estimate is considered uncertain due to a lack of field data on the CH_4 production potential for different types of lagoons. In addition, uncertainty results from limitations in available data on quantities of wastewater being treated and the characteristics of lagoons worldwide that affect CH_4 production.

Lagoons are commonly used for wastewater treatment and disposal in developing countries, primarily because land is available, operations are relatively simple, minimal energy is needed, and capital and operating expenses are low. Anaerobic conditions are favored because of the limited maintenance and control of these lagoons and limited use of expensive aeration devices. Moreover, the average temperature in many tropical and subtropical developing countries is close to the optimum biological temperature of methanogenic bacteria (i.e., 35°C). This results in greater biological activity and a higher CH_4 production potential, as compared to lagoons found at cooler latitudes (Gloyna, 1971).

Mean daily gas production rates were estimated by Toprak (1993) to be 51,000 m^3/day for wastewater treatment lagoons for a treatment plant being constructed in Izmir to be completed in the year 2000. This city is one of the fastest growing metropolitan areas in Turkey. Plans are being considered to utilize the CH_4 as an alternative source of energy.

The World Bank predicts that, because lagoons are relatively inexpensive and easy to operate, they will continue to be a preferred wastewater treatment method for developing countries (Bartone, 1990c). Furthermore, with increasing population growth and urbanization in developing countries, the number and variety of sources discharging into lagoons will increase, resulting in higher BOD loading rates. Increased CH_4 emissions from lagoons will result as these changes occur.

The U.S. EPA estimates that 130 billion ℓ (34 billion gal.) of domestic, commercial, and industrial wastewater is treated in the U.S. each day (U.S. EPA, 1987). Approximately 433 domestic and industrial lagoon systems receive various types and quantities of industrial wastewater, in addition to domestic wastewater. Furthermore, another estimated 5,000 municipal lagoons contain domestic wastes from residential, commercial, and institutional sources (U.S. EPA, 1987). Insufficient data are available to characterize the CH_4 emission potential of wastewater treatment lagoons both in the U.S. and globally. An empirical model for wastewater treatment lagoons using field data would be of value in relating BOD and any other significant factors to CH_4 emissions.

The CH_4 potential is expected to be minimal for the lagoons that receive pretreated wastewater. However, when no pretreatment occurs or pretreatment is minimal, lagoons may be a significant source of CH_4. CH_4 formation varies depending on temperature, retention time, BOD loading, lagoon depth, oxygen content, and the frequency at which the lagoon is dredged. The majority of the U.S. lagoons are facultative; that is, aerobic decomposition occurs in the upper strata and anaerobic degradation occurs in the lower strata. It is likely that facultative lagoons proceed to a more anaerobic state as BOD loading increases and surface aeration diminishes (e.g., due to low wind speed). In particular, information on lagoons used for wastewater from food-processing industries and other industries that have high BOD wastewater streams would be of value.

Most of the world's population, including about 25% of the U.S. population, relies on individual septic systems. This is the most common treatment and disposal method for domestic wastewater. Septic tanks are anaerobic digesters of the simplest form and release CH_4 to the atmosphere at a rate that is dependent on temperature, retention time, and system configuration. However, a certain portion of the CH_4 will be oxidized as the gas diffuses through the soil. These findings were confirmed in work by Khalil et al. (1990), who showed that very little CH_4 escapes to the atmosphere from underground biogas pits in China.

The process of CH_4 production in septic tanks is similar in principle to small biogas pits that are used extensively in China and India to recover gas for energy. The primary difference between the septic tanks used in the U.S. and the biogas pits of China and India is that septic tanks are not designed to produce gas for energy recovery. Rather, septic tanks typically treat only domestic sewage generated by household residences and commercial establishments; CH_4 emitted from the tanks is not normally collected.

Table 2. Methane emission estimates from buried refuse using the OECD and regression models

Country	OECD Method	Regression Model Method		
		Lower Bound	Midpoint of CH_4 Emissions	Upper Bound
Africa				
Congo	0.01	0.01	<0.01	<0.01
Egypt	0.32	0.05	0.10	0.15
Gambia	0.01	0.00	0.00	0.00
Ghana	0.05	0.01	0.02	0.02
Kenya	0.09	0.01	0.03	0.04
Liberia	0.01	0.00	0.00	0.01
Morocco	0.12	0.02	0.04	0.06
Nigeria	0.48	0.07	0.15	0.22
South Africa (Customs Union)	0.43	0.07	0.13	0.20
Sudan	0.09	0.01	0.03	0.04
Tanzania, United Republic of	0.09	0.01	0.03	0.04
Uganda	0.06	0.01	0.02	0.03
Zimbabwe	0.08	0.01	0.02	0.03
Other Africa	1.27	0.19	0.39	0.58
TOTAL AFRICA	3.11	0.46	0.96	1.42
Asia				
Bangladesh	0.42	0.06	0.13	0.19
China (Mainland, NMP)	3.87	0.59	1.18	1.77
India	0.80	0.12	0.24	0.37
Iran, Islamic Republic of	0.31	0.05	0.09	0.14
Iraq	0.12	0.02	0.04	0.05
Israel	0.05	0.01	0.01	0.02
Japan	1.04	0.28	0.50	0.72
Korea, People's Demo. Rep.	0.14	0.02	0.04	0.06
Korea, Republic of	0.64	0.10	0.20	0.29
Kuwait	0.02	0.00	0.00	0.01
Malaysia	0.08	0.01	0.02	0.04
Mongolia	0.01	0.00	0.00	0.00
Myanmar	0.16	0.02	0.05	0.07
Pakistan	0.75	0.11	0.23	0.34
Philippines	0.42	0.06	0.13	0.19
Saudi Arabia	0.10	0.01	0.03	0.04
Sri Lanka	0.10	0.01	0.03	0.04
Thailand	0.28	0.04	0.09	0.13
Turkey	0.20	0.07	0.12	0.18
United Arab Emirates	0.01	0.00	0.00	0.01
Vietnam	0.22	0.03	0.07	0.10
Other Asia	2.04	0.32	0.62	0.94
TOTAL ASIA	11.8	1.93	3.82	5.70

Table 2. Methane emission estimates from buried refuse using the OECD and regression models

Country	OECD Method	Regression Model Method		
		Lower Bound	Midpoint of CH_4 Emissions	Upper Bound
Latin America				
Argentina	0.28	0.04	0.09	0.13
Brazil	2.23	0.34	0.68	1.02
Colombia	0.44	0.07	0.13	0.20
Costa Rica	0.02	0.00	0.01	0.01
Mexico	0.65	0.10	0.20	0.30
Venezuela	0.05	0.01	0.02	0.02
Other South America	0.45	0.07	0.13	0.21
TOTAL LATIN AMERICA	4.12	0.63	1.26	1.89
North America				
Canada	2.02	0.47	0.84	1.21
United States of America	21.04	3.86	8.00	12.14
Other North America	0.36	0.06	0.11	0.17
TOTAL NORTH AMERICA	23.4	4.39	8.95	13.5
Europe				
Albania	0.01	0.00	0.00	0.00
Austria	0.17	0.04	0.08	0.12
Belgium	0.14	0.04	0.07	0.10
Bulgaria	0.06	0.01	0.02	0.02
Czechoslovakia	0.15	0.02	0.04	0.06
Denmark	0.07	0.02	0.04	0.05
Finland	0.24	0.06	0.12	0.17
France	0.88	0.23	0.42	0.61
German Democratic Republic	0.17	0.02	0.04	0.06
Germany, Federal Republic of	1.80	0.48	0.87	1.25
Greece	0.48	0.13	0.23	0.33
Hungary	0.24	0.03	0.06	0.09
Ireland	0.11	0.03	0.05	0.08
Italy	1.45	0.39	0.70	1.01
Netherlands	0.44	0.12	0.21	0.31
Norway	0.11	0.03	0.05	0.07
Poland	0.37	0.05	0.10	0.14
Romania	0.13	0.02	0.03	0.05
Spain	0.85	0.23	0.41	0.59
Sweden	0.08	0.02	0.04	0.06
Switzerland & Liechtenstein	0.12	0.03	0.06	0.08
U. S. S. R.	2.49	0.32	0.65	0.97
United Kingdom	2.85	0.76	1.37	1.98
Yugoslavia	0.17	0.02	0.04	0.07
Other Europe	0.08	0.02	0.04	0.05

Table 2. Methane emission estimates from buried refuse using the OECD and regression models

Country	OECD Method	Regression Model Method		
		Lower Bound	Midpoint of CH_4 Emissions	Upper Bound
TOTAL EUROPE	13.7	3.12	5.74	8.32
Oceania				
Australia	1.16	0.27	0.48	0.70
New Zealand*	0.15	0.04	0.07	0.11
Other Oceania	0.03	0.00	0.01	0.01
TOTAL OCEANIA	1.34	0.31	0.56	0.82
GLOBAL TOTAL	57	11	21	32

*This source has been estimated to emit 0.2 Tg/yr by Lassey et al. (1992).

A layer of sludge accumulates on the bottom of septic tanks where anaerobic processes occur. CH_4 may be released from septic systems through manhole openings, cracked lids, and vent pipes. Because these tanks are underground and typically receive sewage directly from a source in close proximity, it is not likely that the system will be upset by external factors. For the anaerobic process to produce significant amounts of CH_4, however, temperatures inside the tank would need to be above 15°C. Below 15°C, the anaerobic process would slow to a point where the septic tank would merely act as a sludge storage area.

5. Livestock waste

A review of CH_4 emissions from livestock waste is provided by Johnson et al. (this volume). They estimate that less than 5% of animal CH_4 is accounted for by the anaerobic treatment of animal excreta. The fermentation of sewage sludge may be extensive, such as in anaerobic digesters designed for CH_4 recovery and utilization (Thorneloe, 1992). CH_4 from the fermentation of excreta from free-range livestock is thought to be inconsequential (Johnson et al., this volume). The majority of the world's livestock are free ranging. Recent research by Lodman et al., (1993) indicates that CH_4 emissions from livestock waste represent less than 1% of the potential CH_4 resulting from anaerobic lagoons.

Another recent study (Williams, 1993) found that CH_4 emissions from dairy cow patties contribute less than 1% of the potential. In the U.S., livestock accounts for about 6 Tg/yr of CH_4, most of which originates in the rumen. CH_4 emissions from livestock manure are estimated to be 0.6 Tg/yr in the U.S. (Johnson et al., this volume).

In another study, Safley et al., (1992) estimated that CH_4 emissions from livestock manure contribute 21 to 35 Tg/yr, with an average of 28 Tg/yr. The estimate for the U.S. is 4 Tg/yr, as compared to Johnson et al.'s estimate of 0.6 Tg/yr. The major difference between these estimates is the use of an emission factor representing free-range livestock waste. The factor used in the Safley study is at least 10 times higher than that derived by Lodman et al., (1993) and Williams (1993). Using Safley et al.'s methodology and reducing the emission factors for pasture/range, drylot, and daily spread animal waste disposal by a factor of 10, leads to a global estimate of approximately 15 Tg/yr from animal waste.

The principal factors controlling potential CH_4 production from manure include the quantity and characteristics of the animal waste, the type of waste management system, and the temperature and moisture content of the waste (Safley et al., 1992). Cattle in the U.S. produce larger quantities of organic waste than any other type of livestock. The average head of cattle produces 24 kg of wet feces per day (Overcash et al., 1983), including 2.8 kg of organic matter. A portion of this organic matter can be decomposed by methanogenic bacteria. Other animals such as sheep, goats, horses, and fowl, have a larger fraction of volatile solids in their feces, but because cattle produce a larger quantity of manure per individual and are more populous than other livestock, they contribute the largest portion of CH_4 from livestock manure. Volatile solids are that portion of organic matter that can be decomposed by microorganisms (Safley et al., 1992). Safley et al. estimate that global cattle populations contribute 53% of the CH_4 emissions from livestock waste.

In addition to the quantity and characteristics of livestock manure, animal waste management systems largely determine the potential for CH_4 generation. Livestock are managed in conditions ranging from open pasture and range to complete confinement. Concurrently, manure may simply be left on the pasture or range where it is deposited, or it may be prepared using dry storage methods or liquid treatments. The CH_4 production potential of all waste management systems is highly dependent on temperature and moisture (Safley and Westerman, 1987; Johnson et al., this volume). Generally, warm temperatures and high moisture content provide for maximum CH_4 production.

Deep pit stacking and daily spreading of solid and semi-solid manure have the least CH_4-producing potential of all livestock waste management systems (Safley et al., 1992). Under these two management systems, manure generally has a very low moisture content. By contrast, anaerobic lagoons have the greatest CH_4-producing potential of all livestock waste management systems. Due to the high moisture content of the waste, almost all of the CH_4-producing potential of waste can be realized with proper design and operation of the anaerobic lagoons. Safley et al., (1992) estimate that the CH_4-producing potential of anaerobic lagoons is 70% higher than any other form of livestock waste management system. In warm and tropical latitudes, anaerobic lagoons have their greatest CH_4-producing potential. Although liquid/slurry management systems have 70% less CH_4-producing potential than anaerobic lagoons (Safley et al., 1992), they rank second among livestock waste management systems for CH_4-producing potential, because the moisture content of the waste is high. Unfavorable temperatures and short residence time of the waste in storage are probably the factors that limit the CH_4 production potential of liquid/slurry systems. By virtue of their relatively widespread use in western and eastern Europe, Asia, and North America, liquid/slurry management systems are estimated to contribute 26% of the CH_4 from livestock waste management systems. Together, liquid/slurry systems and anaerobic lagoons have been estimated to account for 10 Tg/year, or almost 36% of the total CH_4 emissions from livestock waste management (Safley et al., 1992).

Because more livestock waste is deposited in pasture and range systems than in any other management system globally, assumptions regarding the CH_4-producing potential of this waste have significant effects on the total emission estimate. Research being conducted at Colorado State University indicates that manure CH_4 production varies under feedlot conditions and under simulated grazing conditions. Temperature, moisture, and animal diet were the variables that had the greatest influence on CH_4 production. More accurate estimates of how manure-handling systems affect fermentation would help obtain a better estimate of CH_4 production.

Uncertainty in the estimates of global CH_4 emissions from livestock waste results primarily from limitations in available data. Data are particularly limited for developing countries and for free-range livestock waste management. Even in developed countries uncertainty is associated with animal population estimates, animal sizes and diet, and the types and numbers of animal waste management systems. In addition, refinements to the current estimates using field test data on the CH_4-production potential of livestock waste both under free range conditions and in livestock management facilities would be of value. Currently, the only published global estimate for this source is by Safley et al. Initial revised estimates by EPA/AEERL suggest that this source contributes 2 to 5 Tg/yr of CH_4 globally. Data from field and laboratory studies could help reduce the current uncertainty of this estimate.

6. Summary

Global and U.S. estimates of CH_4 emissions from landfills, wastewater treatment, and livestock waste are presented in Table 3. The recent estimate of 11-33 Tg/yr by EPA/AEERL for landfills is thought to more accurately reflect CH_4 emissions from landfilled waste than previous estimates. The global estimates for wastewater treatment and livestock waste are very uncertain due to lack of data, particularly on the emission potential of wastewater treatment lagoons and free-range livestock waste. There is also a lack of country-specific data for this source category. Uncertainties in these estimates result from optimistic assumptions regarding the extent of anaerobic decomposition and limitations in available data characterizing (1) waste quantities and composition, and (2) treatment or disposal practices.

Using the ranges presented in Table 3, two alternative estimates of the contribution of waste sources to global CH_4 emissions can be derived by calculating the joint probability distributions of the estimates as described by Khalil (1992). Assuming that any value within the range shown for landfills, wastewater and sewage treatment, and livestock waste are equally probable, a Monte Carlo model was used to generate random values for each source. Repeating this process for 200 iterations gives a distribution of estimates. Using (1) Bingemer and Crutzen's (1987) estimate for landfills, (2) Orlich's (1990) estimate for wastewater treatment, and (3) Safley et al.'s (1992) estimate for livestock waste, a global estimate of 103 Tg/yr with a 95% confidence interval of

75 to 103 Tg/yr results. Using the EPA/AEERL estimate for landfills (i.e., 11 to 33 Tg/yr) and Orlich's and Safley's estimates, results in a global estimate for waste management of 72 Tg/yr with a 95% confidence interval of 54 to 95 Tg/yr. Using Lodman et al.'s (1993) and Williams' (1993) emission factors with Safley et al.'s methodology, plus EPA/AEERL's landfill estimate and Orlich's wastewater estimate, gives a global estimate of 60 Tg/yr.

Table 3. Global and U.S. estimates of CH_4 emissions from waste management

	Avg.	Range	
	Global (Tg/yr)		
Landfills	22 [a,c]	11-13	Thorneloe, this work
Wastewater Treatment	25 [a]	12-38 [d]	Orlich, 1990
Livestock Waste	25	20-35	Safley et al., 1992
	15 [e]		
	U.S. (Tg/yr)		
Landfills	6 [a]	3-8 [b]	Augenstein & Pacey, 1990
	23 [c,a]	--	Bingemer & Crutzen, 1987 [c]
	9 [a]	4-14	Thorneloe, this work
Livestock Waste	4	--	Safley et al., 1992
	0.7	--	Lodman et al., 1993

a   Potential emissions, not corrected for amount that is flared or utilized. Approximately 1.2 million tonnes of CH_4 is being recovered from U.S. landfills (Thorneloe, 1992).
b   Uses estimated annual placement rates from 1950 to 1990.
c   Previous global estimate are 50 Tg/yr (30-70 Tg/yr range) by Bingemer and Crutzen (1987), which comes to 60 Tg/yr with country-specific data on MSW generation. The estimate of Richards (1989) is 15 Tg/yr (10-20 Tg/yr range).
d   Assumes 50% uncertainty of Orlich's (1990) estimates of 25 Tg/yr.
e   Uses Safley et al.'s (1992) methodology and Lodman et al.'s (1993) pasture/range emission factor.

References

Ahmed, M.F. 1986. Recycling of solid wastes in Dhaka. In: *Waste Management in Developing Countries, 1* (K.J. Thome-Kozmiensky, ed.), EF-Verlag fur Energie und Umvelttechnik GmbH, Berlin. pp. 169-173.

Alter, H. 1991. The future course of solid waste management in the U.S. *Waste Mgmt. & Res.,* 9:3-20.

Augenstein, D., J. Pacey. 1990. Modeling landfill CH_4 generation. In: *International Conference on Landfill Gas: Energy and Environment,* 10/17/90, Bournemouth, England.

Barlaz, M.A. 1985. Factors affecting refuse decomposition in sanitary landfills. M.S. Thesis, Dept. of Civil and Environmental Engineering, Univ. of Wisconsin - Madison.

Barlaz, M.A. 1988. Microbiological and chemical dynamics during refuse decomposition in a simulated sanitary landfill. Ph.D. Dissertation, Department of Civil and Environmental Engineering, University of Wisconsin - Madison.

Barlaz, M.A., M.W. Milke, R.K. Ham. 1987. Gas production parameters in sanitary landfill simulators. *Waste Mgmt. & Res.,* 5:27.

Barlaz, M.A., D.M. Schaefer, R.K. Ham. 1989a. Bacterial population development and chemical characteristics of refuse decomposition in a simulated sanitary landfill. *Appl. Env. Microbiol.,* 55 (1):55-65.

Barlaz, M.A., R.K. Ham, D.M. Schaefer. 1989b. Mass balance analysis of decomposed refuse in laboratory scale lysimeters. *ASCE J. of Environ. Engineering,* 115 (6):1,088-1,102.

Barlaz, M.A, R.K. Ham, D.M. Schaefer. 1990. CH_4 production from municipal refuse: A review of enhancement techniques and microbial dynamics. *CRC Critical Reviews in Environmental Control, 19,* Issue 6.

Bartone, C.R. 1990a. Economic and Policy Issues in Resource Recovery from Municipal Solid Wastes. *Resour. Conserva. Recycl,* 4:7-23.

Bartone, C.R. 1990b. Investing in Environmental Improvements Through Municipal Solid Waste Management. Paper presented at the WHO/PEPAS Regional Workshop on National Solid Waste Action Planning, Kuala Lumpur, 02/26/90-03/02/90.

Bartone, C.R. 1990c. Urban wastewater disposal and pollution control: Emerging issues for sub-Saharan Africa. In: *Proceedings of the African Infrastructure Symposium,* The World Bank, Baltimore, Maryland, 01/08-09/90, p. 6.

Bartone, C.R., C. Haley. 1990. The Bled Symposium: Introduction. *Resour. Conserva. Recycl.,* 4:1-6.

Bartone, C.R., L. Leite, T. Triche, R. Schertenleib. 1991. Private sector participation in municipal solid waste service: Experiences in Latin America. *Waste Mgmt. & Res.,* 9:495-509.

Bateman, C.S. 1988. Landfill gas development in Australia. In: *Proceedings of the International Conference on Landfill Gas and Anaerobic Digestion of Solid Waste* (Y.R. Alston and G.E. Richards, eds.), October 4-7, Harwell Laboratory, Oxfordshire, UK. pp. 156-161.

Beker, D. 1990. Sanitary landfilling in the Netherlands. In: *International Perspectives on Municipal Solid Wastes and Sanitary Landfilling* (J.S. Carra and R. Cossu, eds.), Academic Press, New York, NY. pp. 139-155.

Betts, M.P. 1984. Trend in solid waste management in developing countries. In: *Managing Solid Wastes in Developing Countries* (J.R. Holmes, ed.), John Wiley & Sons, Ltd. Chichester, England, pp. 291-302.

Bhide, A.D., B.B. Sundaresan. 1983. Solid waste management in developing countries. In: *Indian National Scientific Documentation Centre*, New Delhi, India.

Bhide, A.D., S.A. Gaikwad, B.Z. Alone. 1990. CH_4 from land disposal sites in India. In: *Proceedings of the International Workshop on CH_4 Emissions from Natural Gas Systems, Coal Mining and Waste Management Systems*. Environment Agency of Japan, the U.S. Agency for International Development, and the U.S. Environmental Protection Agency, Washington, D.C., 04/09-13/90.

Bingemer, H.G., P.J. Crutzen. 1987. The production of CH_4 from solid wastes. *J. Geophys. Res., 92* (D2):2,181-2,187.

Bogner, J.E. 1990. Controlled study of landfill biodegradation rates using modified BMP assays. *Waste Mgmt. & Research*, 8:329-352.

Bonomo, L., A.E. Higginson. 1988. *International Overview on Solid Waste Management*. Harcourt Brace Jovanovich, New York, NY, p. 268.

Boyner, J., M. Vogt, R. Piorkowski, C. Rose, M. Hou. 1988. U.S. Landfill Gas Research. In: *Proceedings of the International Conference on Landfill Gas and Anaerobic Digestion of Solid Waste* (Y.R. Alston, and G.E. Richards, eds.), 10/04-07/88, Harwell Laboratory, Oxfordshire, UK, pp. 313-338.

Buivid, M.G., et al. 1981. Fuel gas enhancement by controlled landfilling of municipal solid waste, *Resource Recovery and Conservation*, 6:3.

Campbell, D., D. Epperson, L. Davis, R. Peer, W. Gray. 1991. *Analysis of Factors Affecting Methane Gas Recovery from Six Landfills*. Prepared for Air and Energy Engineering Research Laboratory, U.S. Environmental Protection Agency, Research Triangle Park, NC. EPA-600/2-91-055 (NTIS PB92-101351).

Carra, J.S., R. Cossu (editors). 1990. *International Perspectives on Municipal Solid Wastes and Sanitary Landfilling*, Academic Press, New York, NY, pp. 1-14.

Cointreau, S.J. 1982. *Environmental Management of Urban Solid Wastes in Developing Countries*. A Project Guide, Urban Development Technical Paper Number 5, World Bank, Washington, D.C., pp. 1-17.

Cointreau, S.J. 1984. Solid waste collection practice and planning in developing countries. In: *Managing Solid Wastes in Developing Countries* (J.R. Holmes,

ed.), John Wiley & Sons, Ltd. Chichester, England, pp. 151-182.
Cointreau, S.J. 1987. *Solid Waste Management Study for the Greater Banjul Area, The Gambia.* Ministry of Economic Planning and Industrial Development, Banjul, The Gambia. 09/87.
Conner, M.A. 1978. Modern Technology for Recovering Energy and Materials from Urban Wastes - Its Applicability in Developing Countries. *Resour. Conserv. Recycl.,* 2:85-92.
Cossu, R. 1990a. Sanitary landfilling in Japan. In: *International Perspectives on Municipal Solid Wastes and Sanitary Landfilling* (J.S. Carra and R. Cossu, eds.), Academic Press, New York, NY, pp. 110-138.
Cossu, R. 1990b. Sanitary landfilling in the United Kingdom. In: *International Perspectives on Municipal Solid Wastes and Sanitary Landfilling* (J.S. Carra and R. Cossu, eds.), Academic Press, New York, NY, pp. 199-220.
Cossu, R., G. Urbini. 1990. Sanitary landfilling in Italy. In: *International Perspectives on Municipal Solid Wastes and Sanitary Landfilling* (J.S. Carra and R. Cossu, eds.), Academic Press, New York, NY, pp. 94-109.
Curi, K. 1988. Comparison of solid waste management in tourisitic areas of developed and developing areas. In: *Proceedings of the 5th International Solid Waste Conference, Internal Solid Waste and Public Cleansing Association,* Copenhagen, Denmark, 09/88.
Diaz, L.F., C.G. Golueke. 1987. Solid waste management in developing countries. *BioCycle,* 28:50-55.
El-Halwagi, M.M., S.R. Tewfik, M.H. Sorour, A.G. Abulnour. 1986. Municipal solid waste management in Egypt: Practices and trends. In: *Waste Management in Developing Countries, 1* (K.J. Thome-Kozmiensky, ed.), EF-Verlag fur Energie und Emvelttechnik GmbH, Berlin, pp. 283-288.
El-Halwagi, M.M., S.R. Tewfik, M.H. Sorour, A.G. Abulnour. 1988. Municipal solid waste management in Egypt. In: *Proceedings of the 5th International Solid Waste Conference, International Solid Waste and Public Cleansing Association,* Copenhagen, Denmark, September, pp. 415-424.
El Rayes, H., W.C. Edwards (B.H. Levelton & Associates, Ltd.) . 1991. Inventory of CH_4 Emissions from Landfills in Canada. Prepared for Environment Canada, Hull, Quebec, pp. 25-69.
Emberton, J.R. 1986. The biological and chemical characterization of landfills. In: *Proceedings of Energy from Landfill Gas,* Solihull, West Midlands, UK, 10/30-31/86.
EMCON Associates. 1982. CH_4 *Generation and Recovery from Landfills,* Ann Arbor Science Publishers, Inc., Ann Arbor, MI.
Ernst, A. 1990. A review of solid waste management by composting in Europe. *Resour. Conserva. Recycl.,* 4:135-149.
Ettala, M.O. 1990. Sanitary landfilling in Finland. In: *International Perspectives on Municipal Solid Wastes and Sanitary Landfilling* (J.S. Carra and R. Cossu, eds.), Academic Press, New York, NY, pp. 67-77.

Federal Register, Vol. 56, No. 104. May 30, 1991, pp. 24,468-24,528.
Frantzis, I. 1988. Recycling in Greece. *BioCycle*:30-31.
Gadi, M.T. 1986. In: *Waste Management in Developing Countries, 1* (K.J. Thome-Kozmiensky, ed.), EF-Verlag fur Energie and Umvelttechnik GmbH, Berlin. pp. 188-194.
Gandolla, M. 1990. Sanitary landfilling in Switzerland. In: *International Perspectives on Municipal Solid Wastes and Sanitary Landfilling* (J.S. Carra and R. Cossu, eds.), Academic Press, New York, NY, pp. 190-198.
Gloyna, E.F. 1971. Waste stabilization ponds. World Health Organization, Geneva, p. 76.
Halvadakis, C.P., et al. 1983. Landfill Methanogenesis: Literature Review and Critique, Technical Report No. 271, Department of Civil Engineering, Stanford University.
Hayakawa, T. 1990. The status report on waste management in Japan - special focus on methane emission prevention. In: *Proceedings of the International Workshop on Methane Emissions from Natural Gas Systems, Coal Mining, and Waste Management Systems*, Washington, D.C., April 9-13, pp. 509-523.
Holmes, J.R. 1984. Solid waste management decisions in developing countries. In: *Managing Solid Wastes in Developing Countries* (J.R. Holmes, ed.), John Wiley & Sons, Ltd. Chichester, England, pp. 1-17.
IPCC. 1992. Climate Change 1992. The Supplementary Report to the IPCC Scientific Assessment. Published for The Intergovernmental Panel on Climate Change (IPCC), World Meteorological Organization/United Nations Environment Programme. Cambridge University Press. Edited by J.T. Houghton, G.J. Jenkins, and J.J. Ephraums.
International Energy Agency. 1992. Landfill Gas Enhancement Test Cell Data Exchange--Final Report of the Landfill Gas Expert Working Group. Editor: Pat Lawson, AEA-EE-0286.
Jenkins, R.L., J.A. Pettus. 1985. In: *Biotechnological Advances in Processing Municipal Wastes for Fuels and Chemicals* (A.A. Antonopoulos, ed.), Argonne Natl. Lab. Report ANL/CNSV - TM - 167, p. 419.
Kaldjian, P. 1990. *Characterization of Municipal Solid Waste in the United States: 1990 Update*, EPA-530/SW-90-042, PB90-215112. Office of Solid Waste, Washington, DC.
Kaltwasser, B.J. 1986. Solid waste management in medium sized towns in the Sahel area. In: *Waste Management in Developing Countries, 1* (K.J. Thome-Kozmiensky, ed.), EF-Verlag für Energie und Umvelttechnik GmbH, Berlin, pp. 299-307.
Kessler, T. 1990. Brazilian trends in landfill gas exploitation. *ETATEC Consultores s/c Ltda*, Sào Paulo, Brazil.
Khalil, M.A.K. 1992. A statistical method for estimating uncertainties in the total global budgets of atmospheric trace gases. *J. Environ. Sci. Health, A27*:777-770.

Khalil, M.A.K., R.A. Rasmussen, M.-X. Wang. 1990. Emissions of trace gases from Chinese rice fields and biogas generators: CH_4, N_2O, CO, CO_2, chlorocarbons, and hydrocarbons. *Chemosphere*, 20:207-226.

Kinman, R.N., et al. 1987. *Waste Management and Research,* 5:13.

Lassey, K.R., D.C. Lowe, M.R. Manning. 1992. A Source Inventory for Atmospheric CH_4 in New Zealand and Its Global Perspective. *J. Geophys. Res.*, 97:3,751-3,765.

Lawson, P. 1992. Landfill gas expert working group summary report, 1989-1991. *International Energy Agency*, AEA-EE-0305.

Lodman, D.W., M.E. Branine, B.R. Carmean, P. Zimmerman, G.M. Ward, D.E. Johnson. 1993. Estimates of CH_4 Emissions from Manure of U.S. Cattle. *Chemosphere*, 26 (1-4):189-200.

Lohani, B.N., N.C. Thanh. 1980. Problems and practices of solid waste management in Asia. *J. Envion. Sci.*, 06/80, pp. 29-33.

Maniatis, K., S. Vanhille, A. Hartawijaya, A. Buekens, W. Verstraete. 1987. Solid waste management in Indonesia: Status and Potential. *Resour. Conserva. Recycl.*, 15 (87):277-290.

Mei-Chan, L. 1986. Waste management in the Taiwan area. In: *Waste Management in Developing Countries, 1* (K.J. Thome-Kozmiensky, ed.), EF-Verlag fur Energie und Umvelttechnik GmbH, Berlin, pp. 247-250.

Mnatsaknian, R.A. 1991. Legislation and public control of waste sites in the U.S.S.R. In: *Proceedings of the Third International Landfill Symposium*, Sardinia, Italy, 10/91, pp. 1,747-1,753.

Monney, J.G. 1986. Municipal solid waste management--Ghana's experience. In: *Waste Management in Developing Countries, 1* (K.J. Thome-Kozmiensky, ed.), EF-Verlag fur Energie und Umvelttechnik GmbH, Berlin, pp. 32-1327.

Mwiraria, M., J. Broome, R. Semb, W.P. Meyer. 1991. Municipal Solid Waste Management in Uganda and Zimbabwe. Draft report of the United Nations Development Program and The World Bank. 05/18/91.

Nozhevnikova, A.N., A.B. Lifshitz, V.S. Lebedev, G.A. Zavarzin. 1993. Emission of methane into the atmosphere from landfills in the former USSR. *Chemosphere*, 26 (1-4):401-418.

Oluwande, P.A. 1984. Assessment of solid waste management problems in China and Africa. In: *Managing Solid Wastes in Developing Countries* (J.R. Holmes, ed.), John Wiley & Sons, Ltd., Chichester, England, pp. 71-89.

OECD. 1991. Estimation of Greenhouse Gas Emissions and Sinks. Final Report from the OECD Experts Meeting, 02/18-21/91. Prepared for the IPCC. Revised 08/91.

Orlich, J. 1990. CH_4 Emissions from Landfill Sites and Waste Water Lagoons. Proceedings of the International Workshop on CH_4 Emissions from Natural Gas Systems, Coal Mining, and Waste Management Systems, 04/09-13/90. Environment Agency of Japan, U.S. Agency for International

Development, and the U.S. Environmental Protection Agency, Washington, DC.

Overcash, M.R., F.J. Humenik, J.R. Miner. 1983. *Livestock Waste Management*. Vol. II. CRC Press, Boca Raton, FL.

Pacey, J. 1989. Enhancement of degradation: Large-scale experiments. In: *Sanitary Landfilling: Process Technology and Environmental Impact* (T. Christensen, R. Cossu, and R. Stegmann, eds.), Academic Press, London, pp. 103-119.

Pairoj-Boriboon, S. 1986. State-of-the-art of waste management in Thailand. In: *Waste Management in Developing Countries, 1* (K.J. Thome-Kozmiensky, ed.), EF-Verlag für Energie und Umvelttechnik GmbH, Berlin, pp. 208-219.

Papachristou, E. 1988. Solid wastes management in Rhodos. In: *Proceedings of the 5th International Solid Wastes Conference, International Solid Waste and Public Cleansing Association*, Copenhagen, Denmark, September.

Parkin, G.F., W.F. Owen. 1986. Fundamentals of anaerobic digestion of wastewater sludges. *J. Environmental Engineering Division, ASCE, 112* (5), p. 867.

Peer, R.L., D.L. Epperson, D.L. Campbell, P. Von Brook. 1992. Development of an empirical model of methane emissions from landfills, EPA-600/R-92-037, PB92-152875. Prepared for the U.S. Environmental Protection Agency, Air and Energy Engineering Research Laboratory, Research Triangle Park, NC.

Peer, R.L., S.A. Thorneloe, D.L. Epperson. 1993. A comparison of methods for estimating global methane emissions from landfills. *Chemosphere, 26* (1–4):387-400.

Peterson, C., A. Perlmutter. 1989. Composting in the Soviet Union. *BioCycle*, July, 74-75.

Pohland, F.G. 1975. Sanitary landfill stabilization with leachate recycle and residual treatment. EPA-600/2-75-043, PB248524.

Pohland, F.G., S.R. Harper. 1987. Critical review and summary of leachate and gas production from landfills. EPA/600/2-86/073, PB86-240181.

Population Reference Bureau, Inc. 1989. World Population Data Sheet of the Population Reference Bureau, Inc. Demographic data and estimates for the countries and regions of the world. Washington, D.C.

Rathje, W.L. 1991. Once and future landfills. *National Geographic, May*, pp. 117-134.

Richards, K.M. 1988. Landfill gas - A global review. In: *Biodeterioration 7* (D.R. Houghton, R.N. Smith, and H.O.W. Eggins, eds.), Elsevier Applied Science, London, pp. 774-790.

Richards, K.M. 1989. Landfill gas: Working with Gaia. *Biodeterioration Abstracts, 3*:317-331.

Safley, L.M., P.W. Westerman. 1987. Biogas production from anaerobic lagoons.

Biological Wastes, 23:181.

Safley, L.M., M.E. Casada, J.W. Woodbury, K.F. Roos. 1992. Global CH_4 emissions for livestock and poultry manure. EPA/400/1-91/048, U.S. Environmental Protection Agency, Air and Radiation, Washington, D.C, 02/92.

Scheepera, M.J.J. 1990. Landfill gas in the Dutch perspective. *International Conference on Landfill Gas: Energy and Environment '90*, Session 3.2.

Shelton, D.R., J.M. Tiedje. 1984. General method for determining anaerobic biodegradation potential. *Appl. Env. Microbiol.*, 47:850-857.

Stegmann, R. 1990. Sanitary landfilling in the Federal Republic of Germany. In: *International Perspectives on Municipal Solid Wastes and Sanitary Landfilling* (J.S. Carra and R. Cossu, eds.), Academic Press, New York, NY, pp. 51-66.

Swartz, A. 1989. Overview of International Solid Waste Management Methods. State Government Technical Brief 98-89-MI-2. The American Society of Mechanical Engineers, Washington, D.C.

Thorneloe, S. June 1992. Landfill Gas Recovery/Utilization - Options and Economics, Published in Proceedings for the 16th Annual Conference by the Institute of Gas Technology on Energy from Biomass and Waste, 03/92, Orlando, FL.

Thorneloe, S.A. August 1992. Emissions and Mitigation at Landfills and Other Waste Management Facilities. Presented at the EPA Symposium on Greenhouse Gas Emissions and Mitigation Research, Washington, DC. To be published in conference proceedings.

Toprak, H. 1993. CH_4 emissions originating from the anaerobic waste stabilization ponds. *Chemosphere*, 26 (1-4):633-640.

United Nations. 1989. City Profiles. Prepared by the United Nations Centre for Regional Development and the Kitakyushu City Government. 64 pp.

United Nations. 1990. 1988 Demographic Yearbook. Fortieth Issue. Department of International Economic and Social Affairs, Statistical Office, New York, NY.

United Nations Development Programme (UNDP), The World Bank, and the Canadian International Development Agency. 1987. Master Plan for Resource Recovery and Waste Disposal, City of Abidjan. Final Report. Prepared by Roche Ltd. Consulting Group, Sainte-Foy, Quebec, Canada. February.

United Republic of Tanzania. 1989. Masterplan of Solid Waste Management for Dar es Salaam. Volume II: Annexes. Ministry of Water, Department of Sewerage and Sanitation. Prepared by HASKONING, Royal Dutch Consulting Engineers and Architects, Nijmegen, The Netherlands, and M-Konsult Ltd., Consulting Engineers, Dar es Salaam, Tanzania. pp. 1-39, 47-49, 74-94, 109-124, 142-145.

U.S. Environmental Protection Agency. 1987. *Report to Congress, Municipal*

Wastewater Lagoon Study, Office of Municipal Pollution Control.
U.S. Environmental Protection Agency. 1988. *Report to Congress, Solid Waste Disposal in the United States*, Volume 1, EPA/530-SW-88-011, PB89-110381, Office of Solid Waste, Washington, DC.
U.S. Environmental Protection Agency. 1991. *Air Emissions from Municipal Solid Waste Landfills - Background Information for Proposed Standards and Guidelines*, EPA-450/3-90-011a, PB91-197061. Office of Air Quality Planning and Standards, Research Triangle Park, NC.
Verrier, S.J. 1990. Urban Waste Generation, Composition and Disposal in South Africa. In: *International Perspectives on Municipal Solid Wastes and Sanitary Landfilling* J.S. Carra and R. Roccu, eds.), Academic Press, Harcourt Brace Jovanovich, London, England, pp. 161-176.
Vogler, J.A. 1984. Waste recycling in developing countries: A review of the social, technological, and market forces. In: *Managing Solid Wastes in Developing Countries* (J.R. Holmes, ed.), John Wiley and Sons, Ltd., Chichester, England, pp. 241-266.
Whalen, S.C., W.S. Reeburgh, K.S. Sanbeck. 1990. Rapid methane oxidation in a landfill cover. *Applied and Environmental Microbiology*, 56:3,405-3,411.
Williams, D.J. 1993. Methane emissions from the manure of free range dairy cows. *Chemosphere*, 26 (1-4):179-188.
Willumsen, H. 1990. Landfill gas. *Resour. Conserva. Recycl.*, 4:121-133.
Wolfe, R.S. 1979. Methanogenesis. In: *Microbial Biochemistry, International Review of Biochemistry*, Vol. 21 (J.R. Quayle, ed.), University Park Press, Baltimore, MD.
Wolin, M.J. and T.L. Miller. 1985. In: *Biology of Industrial Microorganisms* (A.L. Dernsin and N.A. Solomon, eds.), Benjamin/Cumings Publishing Company, Inc., Menlo Park, CA, pp. 189-221.
World Bank, The. 1985. Metropolitan Area of Douala. Study of Waste Management and Resource Recovery. Part A. Phase I. Prepared by Motor Columbus, Consulting Engineers, Inc., CH-5401 Baden, Switzerland.
World Resources Institute. 1990. *World Resources 1990-91*. Oxford University Press, New York, NY.
Wright, F., C. Bartone, S. Arlosoroff. 1988. Integrated resource recovery: optimizing waste management. In: *Proceedings of the 5th International Solid Wastes Conference, International Solid Wastes and Public Cleansing Association*, Copenhagen, Denmark, pp. 619-624, 09/88.
Xianwen, C., Z. Yanhua. 1991. Landfill gas utilization in China. In: *Proceedings of the Third International Landfill Symposium*, Sardinia, Italy, October, pp. 1,747-1,753.
Yepes, G., T. Campbell. 1990 (draft). Assessment of Municipal Solid Waste Services in Latin America. Report in progress prepared for The World Bank, Technical Department, Infrastructure and Energy Division, Urban Water Unit, Latin America and the Caribbean Region. pp. 1-6, 20-26.

Young, L.Y., A.C. Frazer. 1987. *Geomicrobiology J., 5*:261.

Zehnder, A.J.B., et al. 1982. Microbiology of CH_4 bacteria. In: *Anaerobic Digestion* (D.E. Hughes, ed.), Elsevier Biomedical Press B.V., Amsterdam, p. 45.

Zsuzsa, K.P. 1990. Possibilities for utilization of the energy content of the solid wastes of settlements. *Resour. Conserva. Recycl., 4*:173-180.

[Editor's Note on the correlation between Average Recovery Rate and Amount of Refuse: In Figure 1, the correlation is greatly affected by a single point at the upper right hand corner. Without it the correlation for the remaining 20 landfills is only 0.15. This matter was discussed with Dr. Susan Thorneloe who informed us that she has put together new data from more than 100 landfills and the correlation is similar to that reported here (for all points) and more robust. These results are still being analyzed and could not be completed for this chapter but will be published elsewhere.]

Chapter 17

Industrial Sources

L. LEE BECK,[1] STEPHEN D. PICCOT,[2] AND DAVID A. KIRCHGESSNER[1]

[1]*United States Environmental Protection Agency, Air and Energy Engineering Research Laboratory, Research Triangle Park, North Carolina 27711*

[2] *Southern Research Institute, Research Triangle Park, North Carolina 27709*

1. Introduction

This chapter identifies and describes major industrial sources of methane (CH_4) emissions. For each source type examined, CH_4 release points are identified and a detailed discussion of the factors affecting emissions is provided. A summary and discussion of available global and country-specific CH_4 emissions estimates are also presented.

The major emission sources examined include coal mining operations and natural gas production and distribution systems. However, a variety of minor industrial sources are also examined because their collective contributions to the global CH_4 budget may be significant. Although the treatment of these minor sources may not be comprehensive, the limited available data are presented for several different sources. Among the minor industrial sources examined here are: coke production facilities, chemical manufacturing operations, peat mining operations, light water nuclear reactors, fossil fuel combustion equipment (boilers and automobiles), geothermal electricity generation facilities, salt mining operations, residential refuse burning, and shale oil mining operations.

2. Coal mining operations

Historically, coal use in the United States suffered a nearly catastrophic decline after World War II. It was not until the mid 1970s that bituminous coal production once again equalled the levels seen in the mid 1940s. Production in 1970 was about 545 million tonnes and had grown more or less consistently to 830 million tonnes by 1987. This growth will undoubtedly be mirrored in countries that have significant coal reserves and are undergoing industrialization. The trend is perhaps epitomized in China where coal production rose from 620 to 1050 million tonnes between 1980 and 1989.

The global environmental effects of coal production and use will increase as coal production increases. The underground mining of coal is accompanied by the emission of CH_4 and the production of large quantities of waste water and coal wastes. The use of longwall mining often results in subsidence at the surface, which can produce significant property damage and can be costly to prevent. Surface mining can produce significant scarring of the land, but, if proper land reclamation techniques are applied, surface mine sites can be restored to near original conditions. Because of the high cost of these techniques, this restoration may not be done in all countries. Combustion of coal results in the release of large quantities of "conventional" pollutants including sulfur dioxide, nitrogen oxides, and particulate matter. It is also one of the largest sources of anthropogenic carbon dioxide emissions. Although several of these pollutants can be controlled with today's pollution control technologies, the large capital investments required may not always be available in developing countries.

2.1 Sources of methane emissions in the coal mining industry. Three categories of mines within the coal mining industry are known to emit CH_4 to the atmosphere: underground mines, surface mines, and abandoned or inactive mines. Much more is known about the emissions from underground mines than from any other mining category. Based on currently available data, underground mines are generally believed to be the most significant source of CH_4 within the coal mining industry.

In general, CH_4 can be released into a coal mine from all of the seams disturbed during the mining process. Seams other than the one being mined may be disturbed by the mechanical or blasting operations which occur at underground and surface coal mines. In an underground longwall mine, some studies suggest that this zone of disturbance may extend up to 200 meters (m) into the roof rock and 100 m below the worked seam (Creedy, 1983). For safety reasons, underground mines use various methods to remove CH_4 from the mine; however, CH_4 is not a safety concern in surface mines. In some countries, a portion of the CH_4 removed from underground mine workings is burned for energy, but in most cases the CH_4 is released to the atmosphere.

Methane from underground mines can be released from three sources: ventilation shafts; gas drainage systems; and coal crushing and handling operations (Boyer et al., 1990; Piccot et al., 1990). Figure 1 illustrates these sources. Ventilation air, although generally containing 1 percent or less CH_4, is known to contribute the majority of underground mine emissions because of the enormous volume of air used to ventilate mines. Gas drainage wells are drilled into the area immediately above the seam being mined. They provide conduits for venting CH_4, which accumulates in the rubble-filled areas formed when the mine roof subsides following longwall mining. Other types of drainage systems are used that extract CH_4 from coal seams well in advance of mining operations. The purpose of all gas drainage systems is to remove CH_4 that would otherwise have to be removed by larger and more costly shaft ventilation systems. Currently, few published data exist for the release of CH_4 from gas drainage systems. However, unpublished data obtained from industry representatives indicate that drainage well CH_4 emissions may account for a significant fraction of the total emissions associated with longwall mines (Boyer et al., 1990; Kirchgessner et al., 1993a). In some parts of the world, such as western and eastern Europe, CH_4 from gas drainage systems is utilized and is not released to the atmosphere.

Very few measurements have been taken at surface mines. Data from six surface mines in the United States suggest that the primary sources of emissions are exposed coal surfaces, in particular the areas fractured by coal blasting (Piccot et al., 1991; Kirchgessner et al., 1993b). In general, the strata overlying the coal do not appear to be a significant source of emissions, but, as in underground mines, emissions may be contributed by underlying seams or faults.

Emissions from abandoned mines may come from unsealed mine shafts or from vents installed to prevent the buildup of CH_4 in the mines. There has been little research to characterize this potential source. However, in a continuing study by the U.S. Environmental Protection Agency (USEPA), CH_4 emissions have been measured as they escape from vents installed at abandoned underground mines. Although only a small number of mines have been examined, preliminary results indicate that emissions vary significantly and that the rate of emissions from an abandoned mine may be strongly influenced by barometric pressure changes. For several mines, emission rates were negligible while at one mine emissions were almost half of those produced when the mine was active.

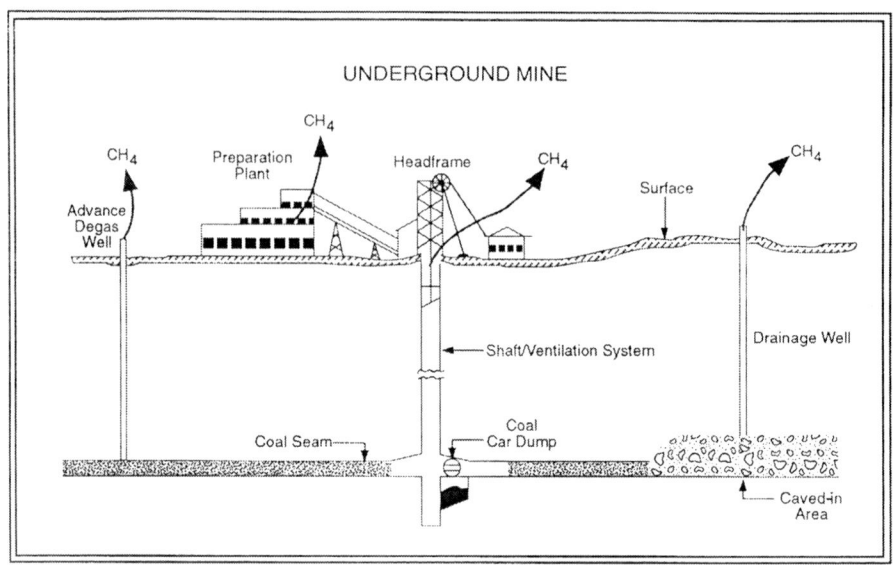
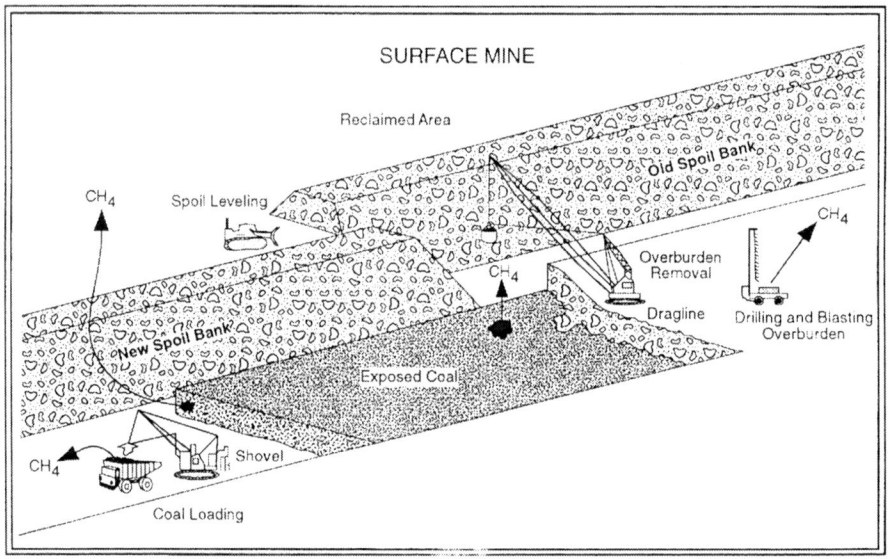

Figure 1. Sources of methane emissions at underground and surface coal mines.

The rate of release of CH_4 from mined coal varies depending on coal type, local geology, geologic history, and other factors. Methane can be released quickly, in a matter of hours, or slowly, over a period of several months. Since mined coal is typically removed within a day, there is a potential for emissions to occur in the post-mining operations. Post-mining operations can include coal breaking, crushing, drying, storage, and transportation. Although measurement data are very limited, Boyer et al. (1990) estimate that on average 25 percent of the CH_4 contained in the mined coal could be emitted after the coal has left the underground mine. Researchers at British Coal suggest that coal handling and transport operations in Great Britain may produce about 2 m^3/tonne of coal when a coal with a CH_4 content of 5 m^3/tonne is mined (i.e., 40 percent of the CH_4 contained in the mined coal is released after it leaves the mine; Watt Committee on Energy, 1991; Creedy, 1993).

3. Factors affecting methane emissions from coal mines.

Numerous studies have examined the physical factors which control the production and release of CH_4 by coal. These studies have been conducted to evaluate the potential of coalbed CH_4 resources, enhance the safety of underground mines, or estimate global CH_4 emissions. Generally, the studies address one of two topics: controlling the CH_4 content of coals, or controlling the concentration of CH_4 in the mine atmosphere and mine ventilation air. Most of these studies have focused on underground mining operations.

Studies of coalbed CH_4 contents have identified pressure, coal rank, and moisture content as important determinants of coalbed CH_4 content. Kim (1977) related gas content to coal temperature and pressure, and in turn to coal depth. After including coal analysis data to represent rank, Kim produced a diagram relating gas content to coal depth and rank. Although the validity of the rank relationship has been questioned, it generally appears to have been accepted by recent authors (Schwarzer and Byrer, 1983; Lambert et al., 1980; Murray, 1980; Ameri et al., 1981). Independently of Kim's work, Basic and Vukic (1989) established the relationship of CH_4 content with depth in brown coals and lignite.

Several studies have recognized the decrease in CH_4 adsorption on coal as moisture content increases in the lowest moisture regimes (Anderson and Hofer, 1965; Jolly et al., 1968; Joubert et al., 1974). Moisture content appears to reach a critical value above which further increases produce no significant change in

CH_4 content. Coals studied by Joubert et al. (1974) showed critical values ranging from 1 to 3 percent.

In a recent study by Kirchgessner et al. (1993a), a set procedure was developed which integrates the influences of several factors known to affect CH_4 content in coalbeds. Two equations were produced for estimating coalbed CH_4 contents in cubic meters per tonne at the basin or seam level: one for coals with heating values less than 34,860 joules/gram (J/g) (Eqn. (1) below), and one for coals with heating values equal to or greater than 34,860 J/g (Eqn. (2) below). The equations for estimating coal bed CH_4 content (IS) were developed by performing multivariate regression analyses on a database of 137 U.S. coal samples. The R^2 for Eqn. (1) is 0.56, and the R^2 for Eqn. (2) is 0.71. Coal properties in the equations have sound geological bases for inclusion and the parameters included follow patterns predicted in the literature. Coal depth (D) in meters appears in both equations. Moisture content (M) in percent and parameters closely tied to coal rank such as heating value (HV) in joules/gram, and the ratio of carbon content to volatile matter (FR) are also included. As a final step in the procedure, the IS values obtained from the equations are multiplied by a factor to yield an estimate of total CH_4 content (i.e., anthracite 1.11; low volatile bituminous 1.10; medium volatile bituminous 1.20; high volatile B and C bituminous 1.12).

$$IS = 0.0159D + 2.781\left(\frac{1}{M^2}\right) - 2.228 \qquad (1)$$

$$IS = 0.0136D + 0.0015HV + 2.6809FR - 56.490 \qquad (2)$$

Although more difficult to quantify, the amount of CH_4 contained in a given quantity of coal may be influenced by the burial and erosional history of a coal seam (Watt Committee on Energy, 1991). In some geological configurations, the CH_4 content of coal increases with coal rank but decreases on approach to the Permo-carboniferous erosion surface.

Early investigations in the United States which attempt to identify correlates of CH_4 emissions from coal mine ventilation air include those by Irani et al. (1972) and Kissel et al. (1973). Irani et al. developed a linear relationship between CH_4 emissions and coal seam depth for mines located in five seams.

Kissel et al. demonstrated a linear relationship between CH_4 emissions and coalbed CH_4 content for six mines. Although both studies suffer from a paucity of mines and seams in their analyses, Kissel et al. made the important observation that mine emissions greatly exceed the amount of CH_4 associated with the mined coal seam alone. Emissions are produced not only by the mined coal, but also by the coal left behind, overlying and underlying seams, and nearby gas deposits. For the six mines studied, emissions per tonne mined exceeded coalbed CH_4 per tonne mined by factors of six to nine.

In studies conducted by Boyer et al. (1990) and Kirchgessner, et al. (1993a), regression equations were developed relating coal production rate and coalbed CH_4 content to the emissions from underground mines. These equations were developed to estimate global emissions from underground mines, using similar data and techniques. To develop these equations, multivariate regression analyses were performed using mine-specific data for the United States. In the analysis performed by Kirchgessner et al., a database of 269 mine-specific emission measurements was used to produce an equation with an R^2 value of 0.59. This means that about 60 percent of the variation in CH_4 emissions from the mines in the database can be explained by the independent variables included in the equation. In the analyses performed by Boyer et al., a database of about 60 mine-specific observations was used to produce an equation with an R^2 value of 0.35. Although not fully understood, one likely reason for the difference in R^2 values between the two studies is that the equation developed by Kirchgessner et al. was estimated using a database which contained over 4 times more individual mine measurements than the Boyer equation.

3.1 Summary of global emissions estimates. Over the past 30 years there have been several attempts to estimate the global emissions of CH_4 from coal mining operations. Although we do not address them all, the estimates presented represent the approximate range of published emission rates. This section summarizes these estimates, describes and compares the basic assumptions used in their development, and identifies key relationships that exist among them.

Table 1 presents a summary of global CH_4 emissions estimates developed by various researchers for coal mining operations. Estimates range from 7.9 to 64 teragrams (Tg) per year. The lower estimate of 7.9 Tg/yr is unrealistic. Although the specific assumptions used in developing this estimate are not clear, it appears to be based on the implicit assumption that emissions from coal mines

Table 1. Summary of estimates of global methane emissions from coal mining operations.

Source	Year for Estimate	CH$_4$ Emissions (Tg/yr)	Comments
Koyama (1963, 1964)	1960	20	Includes hard coal only (lignite not included).
Hitchcock and Wechsler (1972)	1967	7.9 to 27.7	Includes the emissions from hard coals (6.3 to 22 Tg/year) and lignite (1.6 to 5.7 Tg/year). Based on Bates and Witherspoon, and Koyama.
Ehhalt and Schmidt (1978)	1967	7.9 to 27.7	Estimates taken from Hitchcock and Wechsler.
Seiler (1984)	1975	30	Results based on an extrapolation of Koyama's results using updated coal production. Analysis by Boyer et al., suggests this study includes only hard coal.
Tolkachev and Zimakov (1983)[a]	1979	20 to 25	Basis unknown.
Crutzen (1987)	not specified (assume middle 1980s)	34	Does not include emissions from the mining of lignite coals. Based on work by Crutzen, and Ehhalt and Schmidt.
Cicerone and Oremland (1988)	not specified (assume middle 1980s)	25 to 45 (average 35)	Results based on analyses conducted by Koyama, Ehhalt, and Seiler. Analysis by Boyer et al., suggest this study includes only hard coal.
Boyer et al. (1990)	1987	33 to 64 (average 47.4)	Includes all coal types and emissions from mining and post-mining operations.
Fung et al. (1991)	not specified (assume middle 1980s)	35	Estimate developed by comparing model concentrations derived from published budgets with atmospheric methane measurements.
Kirchgessner et al. (1993a)	1989	45.6	Includes all coal types & emissions from mining & post-mining.

a. Results are as cited in Okken and Kram (1989).

are equal to the amount of CH_4 trapped in the coal removed from the mine (Hitchcock and Wechsler, 1972; Bates and Witherspoon, 1952). Although this trapped CH_4 is released when coal is fractured and removed from the mine, this assumption fails to account for other CH_4 release mechanisms that occur. These release mechanisms, which were described earlier, can significantly contribute to the total emissions from mining operations. A second low estimate of 13.1 Tg/year reported by Darmstadter et al. (1984) for 1980 appears to be based on an unrealistically low emission factor.

A review of the global estimates presented in Table 1 reveals that many are closely related; that is, the basis of several estimates can be traced back to key assumptions made by some of the earliest researchers. Figure 2 illustrates the relationships which exist between various global estimates of CH_4 emissions from coal mines. As the figure shows, many published estimates have been based primarily on methodologies developed by Koyama (1963, 1964) and Bates and Witherspoon (1952). In general, estimates developed by Seiler (1984), Hitchcock and Wechsler (1972), Ehhalt (1974), Ehhalt and Schmidt (1978), Crutzen (1987), and Cicerone and Oremland (1988) were developed based in large part on CH_4 emission factors developed by Bates and Witherspoon (1952) and Koyama (1963, 1964). Koyama's emission factors have been used by Hitchcock and Wechsler, and Ehhalt and Schmidt to estimate an upper end of the range of coal mine emissions.

Estimates by Boyer et al. (1990) and Kirchgessner et al. (1993a) are on a country-specific basis and were not developed based on other researchers' emission factors. Instead, new emissions relationships were developed based on measurement data contained in databases on coal properties and mine emissions rates.

The coal mine estimate of Fung et al. (1991) presented in Table 1 was developed differently from the other estimates discussed here. Fung et al. used a combination of global CH_4 mass balances and atmospheric modeling techniques to infer a budget for all CH_4 sources including coal mines. In general, several CH_4 budget scenarios were constructed and tested to determine which was best able to reproduce the meridional gradient and seasonal variations of CH_4 concentrations observed in the atmosphere. One budget scenario for all CH_4 sources, including coal mines, was selected by Fung et al. because it was judged to reproduce the atmospheric record best. The coal mine estimate associated

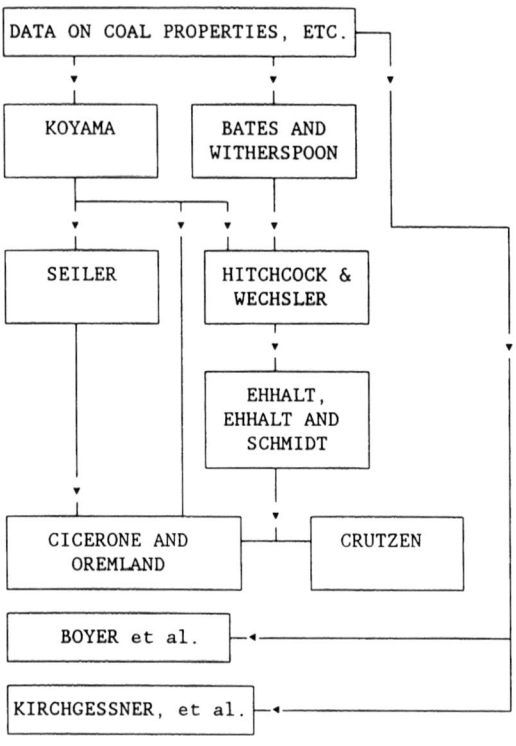

Figure 2. Relationships among various global estimates of methane emissions from coal mines.

with this budget scenario is shown in Table 1. There are methodological and other differences among the estimates that contribute to the variations observed in the results in Table 1. First, the estimates are developed for different years: estimates from 1960 through 1989 are presented. Since coal production increased significantly during this period, an increase in emissions is expected and a direct comparison of these estimates cannot be made. Second, Table 1 shows that several estimates fail to account for all the coal produced globally. Estimates developed by Koyama (1963, 1964), Crutzen (1987), and Cicerone and Oremland (1988) are known to account for the emissions associated with hard coal production only and do not include the emissions associated with brown or lignite coals. Although these types of coals typically do not contain much CH_4, their contribution to global emissions cannot be neglected. Estimates developed by Hitchcock and Wechsler (1972), Ehhalt and Schmidt (1978), Boyer et al. (1990), and Kirchgessner et al. (1993a) include emissions associated with all coal types. Another difference among these estimates is that many do not appear to include the global emissions associated with post-mining operations (i.e., crushing, grinding, handling, and transport). The estimates developed by Boyer et al. and Kirchgessner et al. are the only ones that specifically include an estimate of the emissions from post-mining operations. None of the estimates includes the emissions associated with abandoned mines.

A simple evaluation of the global estimates presented in Table 1 could lead to the potentially erroneous conclusion that the two-fold increase in CH_4 emissions since 1960 can be explained primarily by the two-fold increase in global coal production that occurred during the same period. However, the actual change in global coal mine emissions cannot be determined from the estimates in Table 1 because significant differences exist in the emissions rates used to develop those estimates, and many estimates do not include the emissions from lignite coals. Table 2 compares CH_4 emission rates associated with a variety of studies. These data indicate that the emission rates used to estimate emissions in the most recent studies tend to be lower than those used in early studies.

Table 2. Summary of methane emission rates associated with various global emissions estimates.

Source	Year	Emission rate (m^3 methane/tonne coal mined)
Koyama (1963, 1964)	1960	19.5
Hitchcock and Wechsler (1972)	1967	5.0 to 17.5
Seiler (1984)	1975	19.5
Crutzen (1987	not specified; assume middle 1980s	18 to 19[a]
Boyer et al. (1990)	1987	14.2
Kirchgessner et al. (1993a)	1989	13.8

[a] As cited in Boyer et al. (1990)

In two recent studies, emissions from coal mines were estimated on a country-specific basis (Boyer et al., 1990; Kirchgessner et al., 1993a). Both studies represent relatively comprehensive attempts to characterize key country-specific factors which may significantly influence emissions by (1) estimating country-level or basin-level coalbed CH_4 contents, (2) estimating the emissions associated with different mining techniques (i.e., surface mining and underground mining), and (3) subtracting CH_4 recovered and used at coal mines from the global estimate. Summaries of the country-specific estimates from both studies are presented in Table 3. A comparison of the two studies shows that total emissions are relatively similar but significant differences exist in the emission estimates for key countries and regions. For China, the United States, South Africa, and India, Boyer et al. produce estimates that are about two times higher than Kirchgessner et al. Conversely, estimates for the "Rest of the World" and "Surface Mining" are higher in Kirchgessner et al. by a factor of about two. The reasons for these differences have not been determined.

Other researchers have also estimated emissions for individual countries. Although these estimates are not global in nature, they are reported here because they generally represent the results of a focused and detailed assessment of country-specific coal and mine emission characteristics. As a result, they can be used to independently examine the representativeness of the country-specific estimates presented in Table 3. The results are summarized in Table 4. The reader should be cautioned that Table 4 is not intended to be a comprehensive summary of individual country estimates. Although the table includes mainly those results obtained from participants in the NATO Advanced Research Workshop on the Global Methane Cycle (this volume and in Khalil and Shearer, 1993), other independent assessments are also included.

The country-specific estimates in Table 4 generally agree with the estimates in Table 3. The estimates of Kirchgessner et al. and Boyer et al. generally agree with estimates developed for Poland and the former Soviet Union by Pilcher et al. (1991) and Andronova and Karol (1993). However, estimates for Australia and the United Kingdom developed by Kirchgessner et al. and Boyer et al. are about two times higher than the estimates presented in Table 4. Estimates for the United Kingdom developed by Kirchgessner et al. are based on coalbed CH_4 content measurements for seams mined in the United Kingdom, and on mine emissions relationships developed from 260 U.S. coal mine emissions

Table 3. Country-specific emission estimates developed by Boyer et al. (1990) and Kirchgessner et al. (1993a).

Country	Boyer et al. (1990)		Kirchgessner et al. (1993a)	
	1987 Coal Production (10^6 tonnes)	1987 CH_4 Emissions (teragrams)	1989 Coal Production (10^6 tonnes)	1989 CH_4 Emissions (teragrams)
Underground Mining				
China	891	16.0	1,053	9.3
Former Soviet Union	429	7.7	418	7.9
United States	337	6.1	356	3.5
Poland	193	3.3	181	3.6
South Africa	111	2.0	115	0.7
India	85	1.5	95	0.7
United Kingdom	86	1.5	71	1.3
West Germany	79	1.4	73	1.1
Australia	47	0.9	59	1.1
Czechoslovakia	26	0.4	-	-
Rest of the World	Not Reported	2.9	567	6.8
Subtotal	-	43.7[a]	2,988	36.0
Surface Mining				
Subtotal	Not Reported	3.7[b]	2,154	6.9
Post-Mining Emissions	-	Included above	-	2.7
Total Mining	4,630[c]	47.4	5,142	45.6

[a] A range of emissions from 30.3 to 59.1 million tonnes was established in this study for underground mines.
[b] A range of emissions from 2.6 to 5.0 million tonnes was established in this study for surface mines.
[c] This value cannot be obtained by summing the production rates listed above because values for Surface Mining and the Rest of the World were not reported by Boyer et al.; however, total production was reported.

measurements. The Creedy estimate is based on relationships developed from United Kingdom measurements data supplied by British Coal for both coalbed CH_4 content and mine emissions rates. The discrepancy between the two estimates lies in the assumptions that relate coalbed CH_4 content by depth to mine production to calculate total emissions from underground mines.

Table 4. Summary of other country-specific mine emission estimates.

Country	Emissions (tg/yr)	Source	Comments
Australia	0.45	D.J. Williams, CSIRO, Minerals Research Laboratories, Australia	Preliminary 1989 estimate for underground mines only.
Former Soviet Union	3.5 to 11.2 (average 7.4)	Andronova and Karol (1993)	Estimate is for 1988. The maximum potential emissions are 17.3 Tg.
Poland	3.3	Pilcher et al. (1991)	Estimate is for 1988.
Turkey	0.22	Personal communication with H. Köse and T. Onargan[a]	Estimate is for 1990.
United Kingdom	0.75 ± 0.1	Creedy (1993)	This measurements-based estimate includes emissions from all coal types and coal handing and transport losses (1990/1991).
United States	3.0	Piccot and Saeger (1990)	Estimate includes emissions for under- ground coal mines only (estimate for 1985)

[a] Both are from the University of The 9 Eylul, Izmir, Turkey.

4. Natural gas production and distribution

Natural gas has long been recognized as the environmentally preferred fossil fuel. It produces virtually no sulfur dioxide or particulate emissions, and far fewer nitrogen oxide and carbon monoxide emissions than other fossil fuels. For this reason, and because it is widely available, relatively easily recovered, and readily usable, the global consumption of natural gas has approximately doubled since 1970. Among the industrialized nations, the United States is an exception to this pattern in that it consumes about 10 percent less natural gas today than in 1970. This has variously been attributed to an excessively restrictive regulatory structure (DOE, 1991) and to a misconception of the future natural gas price structure stemming from an underestimate of available U.S. reserves in the 1970s (Hay et al., 1988). The U.S. National Energy Strategy produced in 1991 has recommended removing or revising excessive regulation inhibiting natural gas transactions (DOE, 1991), and the Gas Research Institute has stated that domestic natural gas reserves are sufficient for the next several decades (Hay et al., 1988). Both of these factors should accelerate the slow rate of increase in U.S. domestic natural gas utilization which is already occurring.

Natural gas emits about half as much carbon dioxide per unit of energy output as coal, and about two-thirds as much as oil. Recognizing this, the Intergovernmental Panel on Climate Change has formalized the recommendation to switch to natural gas as fuel where possible to achieve short term mitigation of the global climate change problem (Environment Agency of Japan, 1990). It must first be demonstrated, however, that CH_4 leakage from the increased production and utilization of natural gas would not nullify the benefit of decreased carbon dioxide production.

4.1 Sources of methane emissions in the natural gas industry. The natural gas industry can be broadly divided into the production, transmission, and distribution sectors diagramed in Figure 3. Each of these sectors can contribute steady or fugitive CH_4 emissions and intermittent emissions. Fugitive emissions result from normal operations and result primarily from leaking components such as valves, flanges, and seals. Intermittent emissions result from routine maintenance procedures, system upsets, and occasional large scale accidents.

Methane emissions from the production sector usually include those from well drilling, gas extraction, and field separation facilities. In this discussion gas processing plants are also included. Emissions from well drilling result primarily from occasional venting and flaring employed to prevent blowouts. During extraction, CH_4 may be emitted by natural-gas-fired engines used for power generation, various wellhead components collectively referred to as the "Christmas tree," and occasional venting and flaring when gas volumes do not warrant recovery. Field separation may involve gas heating, gas or liquid separation, and gas dehydration. Principal sources of emissions are fugitive leaks,

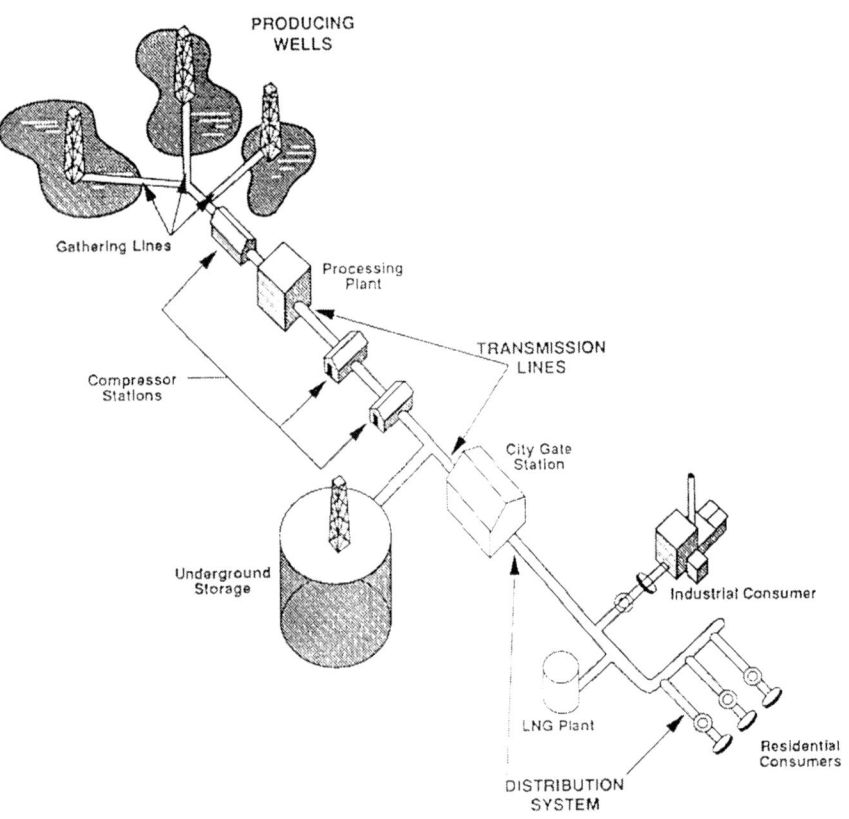

Figure 3. Natural gas pipeline system.

venting and flaring, natural-gas-powered pneumatic devices, and combustion losses from heaters and dehydrators. Gas processing plants are usually located close to the production area and may be regarded as part of the production process. Gas plants are used to separate natural gas liquids from the gas stream and to fractionate the liquids into their components. The processes which are currently most commonly used in these plants are cryogenic expansion, refrigeration, and refrigerated absorption. Primary emissions sources from gas processing plants are fugitive losses, compressor exhaust, and venting and flaring.

Methane emissions associated with the transmission sector are produced by the pipelines, compressor stations, and metering and pressure regulation stations. Leaks from the pipelines are caused by corrosion, material and construction defects, miscellaneous leaks at valves, flanges, and fittings, and earth movement which can cause strains and cracks. Venting can occur at points in the pipeline where residual liquids collect and must be drained. Pneumatic devices powered by natural gas are found throughout the transmission sector and are typically vented to the atmosphere. Maintenance procedures such as pipe scraping result in emissions during launching and retrieving of the scraper. Dehydrators must receive periodic blowdowns and purges which are vented, and pipelines must occasionally be purged during installations, abandonments, replacements, repairs, and emergency shutdowns. Compressor stations produce fugitive emissions from the usual sources (e.g., flanges, seals), occasional unflared venting from system overpressure, and gas turbine start-up and operating emissions.

The primary sources of emissions from the distribution system, which delivers natural gas to the end users, are pipeline leaks. These leaks result, in varying degrees, from all of the same causes as leaks in transmission pipelines. Gas is intentionally vented after isolating segments of lines for repair, and is used to purge air from the pipeline after repair. Blow and purge operations on meters and regulators are typically vented to the atmosphere. The distribution system, because of its size, is generally regarded as the most significant source of CH_4 emissions in the natural gas network.

Injection facilities can be located at various points in the system, depending upon the facilities' function. Gas is frequently reinjected at the production site to maintain oil or gas reservoir pressure. Gas is also injected into underground reservoirs for storage. Normal operations at these facilities produce the usual fugitive emissions, releases during routine maintenance, and venting for overpressure protection of compressors, scrubber vessels, and wellhead injection stations.

The final category of emission sources (not discussed under the three-part industry breakdown above) is liquefied natural gas (LNG) facilities. Functions performed at an LNG facility include receiving, storage, and regasification. Equipment consists of unloading piping, pumps, insulated storage tanks for LNG, and heaters and compressors for regasification. During normal operation, fugitive releases occur but, because of the nature of these facilities, maintenance can be scheduled well in advance and the necessary controlled venting can be directed to the flare system. Pressure relief system releases are typically flared as well.

4.2 Summary of emission estimates. Numerous estimates of emissions for the natural gas industry are available. Global emissions estimates from as long as 20 years ago have been produced, primarily for the purpose of determining global balances of atmospheric trace gases, but more recently for assessing global climate change issues. In the past few years, as a result of the awareness of methane's role in global climate change, country-specific and sector-specific estimates of emissions have also become available. Estimates of global emissions from the natural gas industry are summarized in Table 5. Estimates produced during the 1980s typically range from 25 to 50 Tg/yr and assume leakage rates of from 1 to 4 percent. Ehhalt and Schmidt's (1978) estimate of 7 to 21 Tg/yr is a notable exception but is explained by their acceptance of Hitchcock and Wechsler's (1972) estimate. This estimate would correspond with the others if expanded to 1985 gas production values. Seiler's (1984) estimate of 19 to 29 Tg/yr is also low but is explained by his use of 1975 gas production data. Estimates at the higher end of the typical range by Sheppard et al. (1982), Blake (1984), and Cicerone and Oremland (1988) are derived by adding assumed values for vented gas to the calculated values for gas leakage. Keeling (1973) assumed a leakage rate of 6-10 percent and estimated emissions of 40-70 Tg/yr.

Sheppard et al. (1982) and subsequently Blake (1984) estimate emissions from venting and flaring at wellheads to be about 30 Tg/yr. Cicerone and Oremland (1988) provide a later, independent estimate of 14 Tg/yr. It can be inferred from the discussions that these estimates are for both gas and oil fields and it is assumed that oil fields produce the majority of emissions. The estimates are not separated by industry, however, and such matters as flaring efficiencies and venting versus flaring practices by individual countries are not discussed. There are currently no reliable data on global venting and flaring emissions from oil and gas fields.

Table 5. Estimates of global emissions for the natural gas industry.

Source	Reported Year	Estimate (Tg/yr)	Assumed Loss Rates %
Hitchcock and Wechsler (1972)	1968	7-21	1-3
Keeling (1973)	1968	40-70	6-10
Ehhalt and Schmidt (1978)	1968	7-21	1-3
Sheppard et al. (1982)	1975	50	2 (leakage) + 25% for vented and flared
Blake (1984)	1975	50-60	2-3 (leakage) + 30 Tg for vented
Seiler (1984)	1075	19-29	2-3
Bolle et al. (1986)	Not specified	35	3-4
Crutzen (1987)	Not specified	33	4
Darmstadter et al. (1984)	1980	10	1
Cicerone and Oremland (1988)	Early 1980s	25-50	2.5 (leakage) + 14 Tg for vented & flared
Fung et al. (1991)	1986	40	Unknown

While it is not always explicitly stated in the literature, the assumed leakage rates fall into the estimate that the gas industry refers to as "unaccounted for" gas (UAG). UAG is the difference between the volume of gas that a utility reports as purchased versus the volume sold, less any company use or interchange. It is a statistical figure attributable to numerous diverse components, including meter inaccuracies, gas theft, variations in temperature and pressure, billing cycle differences, and gas leakage or other actual losses. In the United States roughly 2 percent of gas marketed annually is classified as UAG (American Gas Association, 1986). Because UAG reflects nuances of gas companies' accounting systems rather than actual gas losses, it is generally recognized that emissions estimates should not be based on these statistics.

Table 6. Summary of studies showing gas losses by industry sector.

Source	Location & Year	Summary of Findings		
		Description	Emissions	
Tilkicioglu and Winters (1989)	USA-1988	Gas field and field separation facilities	1.12 Tg/yr	
		Gas transmission	0.97 Tg/yr	
		Gas distribution	0.43 Tg/yr	
		Gas process plants	0.31 Tg/yr	
		Total	2.83 Tg/yr	
Cottengham et al. (1989)	PG&E[a]	Distribution	498,000 Mcf/yr[b]	
		Transmission	42,000 Mcf/yr	
		Dig-ins	106,000 Mcf/yr	
		Total	646,000 Mcf/yr or 0.08% of adjusted operating receipts	
Chem Systems International Ltd. (1989)	USA-1988	Production	% UAG[c] 0.13	% Loss
		Transmission	0.54	0.13
		Distribution	2.2	
		Overall		0.5-1.0
	W. Germany 1989	Production		0.16
		Transmission		0.01
		Overall		<1.0
	Netherlands 1989	Overall	Negative[d]	
	UK 1989	Overall		<1.0
	Former USSR 1989	Transmission		~2.0
		Distribution		>2.0

[a] Pacific Gas and Electric (PG&E) is a gas distributor in the United States.
[b] 1 ft^3 = 28.3L (0.0283 m^3).
[c] Unaccounted for gas.
[d] An apparent gain was indicated.

Table 7. 1985 transmission/distribution system losses reported by IEA in Piccot et al. (1990).

Country	Estimated Pipeline Throughput[a] (1,000 TOE)[b]	Transmission and Distribution Losses (1,000 TOE)	% Loss or Emission Factor
Low Reported Losses			
Tunisia[c]	637	1.6	0.2
West Germany	43.4	0.16	0.4
Japan	35.6	0.21	0.6
Italy	28.1	0.22	0.8
Brazil	2,054	24	1.2
France	25.7	0.32	1.2
Moderate Reported Losses			
Austria	4.7	0.08	1.7
Czechoslovakia	7,748	136	1.8
Spain	2.2	0.05	2.3
Former Soviet Union	1,530,148	34,428	2.3
Hungary	8,327	198	2.4
Poland	9,060	236	2.6
Pakistan	6,820	176	2.6
High Reported Losses			
Chile	189	6.1	3.2
United Kingdom	48	1.7	3.6
East Germany	7,820	365	4.7
Argentina	15,355	851	5.5

a. The sum of indigenous production and imports.
b. TOE = tonnes of oil equivalents (1 TOE = 10^7 kcal).
c. Based on 1984 data since 1985 data appear to be atypical.

Table 6 summarizes those studies that break down gas losses by industry sector. Although the number of countries covered is limited, the estimates in these studies suggest that, except for the former Soviet Union, gas actually leaked to the atmosphere consists of less than 1 percent of throughput. The most detailed study is specific to the Pacific Gas and Electric Company in the United States which suggests that actual leakage represents only 0.08 percent of gas received. These estimates could vary considerably if a larger sample of countries is considered.

Table 7 is reproduced from a report by Piccot et al. (1990) and is derived from International Energy Agency energy balances for selected countries. The average loss weighted by each country's throughput is estimated to be 2.3 percent. This table reports a percentage loss for the United Kingdom of 3.6 percent which is considerably higher than the value reported in Table 6. Unfortunately, the source of this discrepancy cannot be determined from available information. It is possible that for some of countries included in Table 7 the estimates may suffer from the same shortcoming as UAG statistics in that they may not provide sufficient detail to allow differentiation of actual leakage from other accounting losses. In those cases, the estimates could be high.

Table 8. Country-specific gas emission estimates.

Source	Year	Estimates
Mitchell et al. (1990)	Not specified	United Kingdom Low: 1.9% of supply Medium: 5.3% of supply High: 10.8% of supply (preferred estimate is medium to high)
Selzer (1990)	Not specified	West Germany: 0.3 Tg/yr
Tilkicioglu (1990)	1988	U.S.A.: 3.1 Tg/yr
Dixon (1990)	1987-1988	Australia: 2% of production

Table 8 contains country-specific emission estimates from various other sources. Again, a relatively high estimate of 5.3 to 10.8 percent of supply for the United Kingdom emerges, as does another low estimate of 0.3 Tg/yr for West Germany.

Clearly, the wide range of sometimes conflicting estimates reflects the high degree of uncertainty associated with current global estimates of CH_4 emissions from natural gas extraction, processing, transmission, and distribution systems. Despite the plethora of emissions estimates available for the gas industry, it is clear that they are based on very little hard data. It is likely that the loss rates for natural gas systems will vary significantly among countries because the types of systems, system operating characteristics, and system ages likely vary significantly from country to country.

5. Minor industrial sources

Industrial sources of CH_4 that have been given the greatest attention in the literature include coal mining operations and the production and distribution of natural gas. Logically, attention has been focused here because both sources are responsible for producing significant quantities of CH_4. However, other industrial sources also release CH_4 into the atmosphere. Individually these sources emit minor quantities of CH_4 but collectively their contribution to the global budget may be significant. Very little research has been done to characterize the CH_4 emissions from these minor industrial sources, and they are rarely included in assessments of the global CH_4 cycle. Although current estimates of CH_4 emissions for minor industrial sources require more study to reduce uncertainties, estimates developed so far suggest that the combined emissions from all sources may be as significant as the more traditional sources (i.e., coal mines, natural gas production and distribution systems, and solid waste disposal landfills). These estimates are briefly discussed here.

An early attempt to estimate CH_4 emissions from industrial and other sources located in urban areas was conducted by Blake (1984). In Blake's analysis, a limited number of ambient air samples were collected in several cities, and it was observed that these samples routinely exhibited elevated levels of CH_4. Based on these measurements, an emission flux rate for the world's cities was first calculated ($0.06/m^2/day$) and then multiplied by an estimate of the land surface area covered by the world's cities. Based on these rather crude calculations, Blake estimated that non-automobile-related emissions from the world's urban areas are about 10 Tg/yr. Urban CH_4 sources can include a variety of industrial and other activities.

At a recent international workshop on methane and nitrous oxide emission sources, new emissions estimates for minor sources were reported by Piccot and Beck (1993) and by Berdowski et al. (1993). Sources examined in both studies included industrial activities such as coke production facilities, petroleum refineries, printing operations, gasoline storage and marketing facilities, fossil fuel combustion, organic chemical manufacturing operations, and others. Non-industrial sources such as the combustion of solid waste in the residential sector were also examined.

The approach used to estimate global emissions in both studies was based on the use of source-specific emission factors (i.e., CH_4 emissions per unit of source-specific activity). For example, emissions from petroleum refineries were estimated by Piccot and Beck by multiplying an individual country's refinery crude oil throughput by an emission factor that quantifies the amount of CH_4 emitted from all refinery processes per tonne of crude oil throughput. In both studies many of the CH_4 emissions factors were based on the emissions characteristics of industrial and other operations in the United States. In some cases an attempt was made to represent specific factors influencing emission rates in individual countries (e.g., Piccot and Beck estimated the emissions reductions due to the use of 3-way catalytic emission controls on light duty motor vehicles in different countries). Although differences in the types of industrial processes and pollution control equipment used among countries may affect the magnitude of the emission factors used, most studies have not attempted to rigorously characterize these differences. At this point there can only be speculation about how these differences may affect the emission estimates.

Estimates of the CH_4 emissions from selected minor sources examined by Piccot and Beck and Berdowski et al. are compared in Table 9. The base year for both sets of estimates is 1990. The most significant source categories identified include residential on-site waste burning (estimated by Piccot and Beck only), mobile sources, fossil fuel combustion, coke production and iron and steel processes, and petroleum refining. Although emission estimates for these sources differ somewhat between the two studies, it appears that their overall contribution to the global budget is close to 10 Tg/yr. Emissions from coke production facilities were also estimated by Darmstadter et al. (1984) to be 4 Tg for 1980--higher than that estimated by both Piccot and Beck and Berdowski et al.

Table 9. Emissions estimated by Piccot and Beck (1993) and Berdowski et al. (1993) for minor industrial and other sources of methane.

Source Description	Berdowski et al. (Tg/yr)	Piccot and Beck (Tg/yr)
Residential on-site waste burning	Not estimated	3.3
Mobile sources	3.0 (± 2)	1.8
Fossil fuel combustion	4.5 (± 1.5)	1.6
Petroleum refining	0.5	0.6
Selected industrial sources Organic chemical manufacturing	0.2	0.1
Coke production and iron and steel processes.	< 2.0	0.3
Other miscellaneous sources (industrial waste, forest fires and managed burning, sewage treatment plants, kraft paper manufacturing, printing and publishing, petroleum storage and distribution).	--	1.4
TOTAL		9.1

Table 10. Emissions estimated by Lacroix (1993) for various industrial sources of methane.

Source Description	Emissions (Tg/yr)
Combustion of Fossil Fuels	4.6 ± 0.4
Petroleum Refining	3.5 ± 0.5
Petrochemicals	2.0 ± 1.0
Peat Mining	2.0 ± 1.0
Geothermal Electricity Production	1.2 ± 0.7
Total Emissions	13.3 ± 3.6

Global estimates of CH_4 emissions from several minor industrial sources have also been estimated by Lacroix (1993). Table 10 summarizes the emissions estimated by Lacroix for five industrial sources of CH_4. Total emissions from the five sources are 13.3 Tg/yr. The approach used by Lacroix is based on emission factors (JEA-EPA, 1990) and commercial fuel use for 1987 in compilations of the International Energy Agency adjusted for a world average energy growth rate of 2 percent per year to the present.

The estimates of Lacroix include several sources addressed by Piccot and Beck and by Berdowski et al. For most sources examined, agreement in the emissions estimates among the three studies is variable. For example, the combined average emissions from both petrochemical plants and refining operations estimated by Lacroix range from 4.0 to 7.0 Tg/yr. Both Piccot and Beck and Berdowski et al. estimate that emissions from these two sources are less than 1 Tg/yr. Agreement on the emissions from fossil fuel combustion-related activities is only moderate among all of the studies. Piccot and Beck estimate that emissions from all fossil fuel combustion-related activities (i.e., stationary and mobile sources) are 3.5 Tg/yr while Berdowski et al. estimate the value to be much higher -- 7.5 Tg/yr. Lacroix estimates combustion-related emissions to be 4.6 Tg/yr, closer to the value reported by Piccot and Beck. However, there is significant disagreement in the distribution of combustion-related emissions between stationary and mobile sources emissions of 3.0 to 4.3 Tg/yr, much higher than estimates developed by Hitchcock and Wechsler (1972) (0.5 Tg/yr) and by Piccot and Beck (1.8 Tg/yr).

Lacroix includes estimates for other industrial sources not addressed in most studies, including peat mining and geothermal electricity production. The total emissions from these two sources are 3.2 Tg/yr.

There is some evidence that mining oil shale and salt may release CH_4 into the atmosphere. Although we are not aware of attempts to estimate the global emissions from these mining operations, measurements collected by the U.S. Bureau of Mines (USBOM) suggest that some salt and oil shale deposits contain CH_4 that can be released during mining. Data provided by the USBOM show that normal production-grade salt adjacent to anomalous salt zones (i.e., zones with brine seeps, gas seeps, and other factors that differ from normal salt) can contain CH_4 in quantities of 0.1 m^3 of CH_4 per tonne of salt mined. Salt within the anomalous zones can contain between 0.4 and 1.8 m^3 of CH_4 per tonne mined. With oil shale, USBOM data show that oil shale samples at two mines in the United States have CH_4 contents of from 0.195 to 1.3 m^3 per tonne mined. In general, these CH_4 contents are much lower than the CH_4 contents typically encountered in coal produced at underground mines.

6. Radiocarbon emissions estimates for industrial sources.

In a recent, Cicerone and Oremland (1988) recognized that emissions from sources of radiocarbon-free CH_4 may be understated. Radiocarbon-free CH_4 sources include CH_4 hydrate deposits and many of the industrial sources included here (i.e., coal mines, natural gas processing and transmission systems, and a few of the minor industrial sources). Cicerone and Oremland cite studies suggesting that sources of radiocarbon-free CH_4 may release twice the emissions than earlier studies by Koyama (1963), Ehhalt (1974), Sheppard et al. (1982), Seiler (1984), Crutzen (1987) and others have estimated. It was suggested that collectively these sources contribute up to 50 Tg/yr more than has previously been estimated (i.e., total emissions from fossil fuel sources, CH_4 hydrates, and others are 135 Tg/yr).

Several other studies also suggest that the CH_4 budgets for sources of radiocarbon-free CH_4 may have been underestimated in the early studies cited above. In a recent study conducted by the National Aeronautics and Space Administration, model calculations based on carbon-14 data, atmospheric CH_4 concentrations, and other information indicate that 123 Tg/yr of atmospheric CH_4 was emitted from fossil carbon sources during 1987 (Whalen et al., 1989). This is about 50 percent higher than Cicerone and Oremland's estimate for coal mines and natural gas drilling, venting, and transmission systems. If the more recent coal mine estimates by Boyer et al. (1990) are considered, and if the emissions from minor industrial sources are taken into account, the gap begins to close between Whalen's estimate of 123 Tg/yr and the "bottom-up" estimates for sources of radiocarbon-free CH_4.

7. Summary

Over the past 30 years there have been several attempts to estimate the global emissions of CH_4 from coal mining operations. The estimates presented in this section are representative of the range of emission estimates published. Emissions estimates range from 7.9 to 47.2 Tg/yr for various years between 1960 and 1989. The estimates vary because of differences in assumptions made for emission factors (i.e., emissions/tonne of coal mined) and coal production rates. Although it is not possible to identify the most "correct" estimate, it is likely that for the late 1980s emission estimates which are less than 25 to 30 Tg/yr are unrealistically low. We believe a more realistic range for this period would be

35 to 48 Tg/yr. Several recent studies have attempted to characterize country-specific emissions and have included mine emission sources not previously addressed (i.e., coal handling operations). Emission estimates from these studies, which are at the upper end of the range cited above, may be the most representative available in spite of the significant uncertainties which still remain. None of the estimates presented here include emissions from abandoned underground coal mines, so the range of 35 to 48 Tg/yr may still be low.

Numerous estimates of emissions for the natural gas industry are available. Global emissions estimates from as long as 20 years ago have been produced, primarily for the purpose of determining emission inventories and global balances of atmospheric trace gases. Methane emissions estimates produced during the 1980s typically range from 25 to 50 Tg/yr and assume leakage rates of from 1 to 4 percent. Clearly, the wide range of sometimes conflicting estimates reflects the high degree of uncertainty associated with current global estimates of CH_4 emissions from natural gas extraction, processing, transmission, and distribution systems. Despite the plethora of emissions estimates available for the gas industry, it is clear that they are based on very little hard data. More detailed country-specific assessments will be needed before a "best estimate" or a narrow range of emissions can be established. It is likely that the loss rates for natural gas systems will vary significantly among countries because the types of systems, system operating characteristics, and system ages likely vary significantly from country to country.

Several minor industrial sources also release CH_4 into the atmosphere. Individually these sources emit small quantities of CH_4 but collectively their contribution to the global budget may be significant. Although current estimates of CH_4 emissions for minor industrial sources are limited and highly uncertain, estimates suggest that the combined emissions from all minor industrial sources identified so far may be between 10 and 15 Tg/yr, nearly as significant as some of the more traditional CH_4 sources. It is likely that emissions from these sources account for some of the difference observed between the total fossil fuel contribution estimated by carbon-14 analysis, and the bottom-up estimates of major industrial sources. Based on the literature cited here the most significant minor sources include residential on-site waste burning, peat mining and geothermal power production, mobile sources, stationary source fossil fuel combustion, coke production and iron and steel processes, and petroleum refining. Estimates of emissions from specific minor sources vary widely between different studies indicating that further work is needed to develop more representative estimates.

References

Ameri, S., F.T. Al-Sandoon, C.W. Byrer. 1981. Coalbed methane resource estimate of the Piceance Basin (*Report No. DOE/METC/TPR/82-6*). U.S. Department of Energy, Morgantown, West Virginia, 44 p.
American Gas Association. 1986. Lost and unaccounted for gas. In: *Issue Brief 1986-28*. American Gas Association, Arlington, Virginia, 4 p.
Anderson, R.B., L.J.E. Hofer. 1965. Activation energy of diffusion of gases into porous solids. *Fuel, 44*:303.
Andronova, G.A., I.L. Karol. 1993. The contribution of USSR sources to global methane emission. *Chemosphere, 26* (1-4):111-126.
Basic, A., M. Vukic. 1989. Dependence of methane contents in brown coal and lignite seams on depth of occurrence and natural conditions. In: *Proceedings of the 23rd International Conference of Safety in Mines Research Institutes*. U.S. Department of the Interior, Bureau of Mines, Washington, D.C.
Bates, D.R., A.E. Witherspoon. 1952. The photo-chemistry of some minor constituents of the Earth's atmosphere (CO_2, CO, CH_4, N_2O). *Roy. Astronom. Soc. Monthly Not.*, 112.
Berdowski, J.J.M., J.G.J. Olivier, C. Veldt. 1993. Methane emissions from fuel combustion and industrial processes. In: *Proceedings of the International Workshop on Methane and Nitrous Oxide: Methods in National Inventories and Options for Control*, February 1993, Amersfoort, The Netherlands.
Blake, D.R. 1984. Increasing Concentrations of Atmospheric Methane, 1979-1983 (Ph.D. Dissertation). University of California, Irvine, California.
Bolle, H.J., W. Seiler, B. Bolin. 1986. Other greenhouse gases and aerosols. In: *The Greenhouse Effect, Climatic Change, and Ecosystems* (B. Bolin, ed.), John Wiley and Sons, New York, 157-203.
Boyer, C.M., J.R. Kelafant, V.A. Kuuskraa, K.C. Manger, D. Kruger. 1990. Methane emissions from coal mining: issues and opportunities for reduction (EPA *Report No. EPA-400/9-90/008*). U.S. Environmental Protection Agency, Office of Air and Radiation, Washington, D.C.
Chem Systems International, Ltd. 1989. *Methane Losses from Natural Gas Utilization*, prepared by Chem Systems International, Ltd., for the National Energy Admin., Sweden, 42 p.
Cicerone, R.J., R. Oremland. 1988. Biogeochemical aspects of atmospheric methane. *Global Biogeochem. Cycles, 2*:299-327.

Cottengham, T.L., R.M. Cowgill, J.B. Godkin, J.R. Grinstead, D.J. Luttrell, F.A. Nelson, R.H. Noistering, J.D. Peterson, R. Quintanilla, J.L. Robertson, E.R. Walden, R.L. Waller, R.E. Wlasenko, R.M. Wong. 1989. *Unaccounted-for Gas Project*, Pacific Gas and Electric Company, San Francisco, California.

Creedy, D.P. 1983. Seam gas-content data-base aids in firedamp prediction. *The Mining Engineer*, August 1983:79-82.

Creedy, D.P. 1993. Methane emissions from coal-related sources in Britain: Development of a methodology. *Chemosphere*, 26 (1-4):419-440.

Crutzen, P.J. 1987. Role of the tropics in atmospheric chemistry. In: *Geophysiology of Amazonia* (R. Dickinson, ed.), John Wiley and Sons, New York.

Darmstadter, J., L. Ayres, R.U. Ayres, W.C. Clark, P. Crosson, P. Crutzen, T.E. Graedel, R. McGill, R.F. Richards, J.A. Tarr. 1984. *Impacts of world development on selected characteristics of the atmosphere: an integrative approach, volume 2*-appendices (Report No.ORNL/Sub/86-22033/1/V2). Oak Ridge National Laboratory, Oak Ridge, Tennessee.

Department of Energy. 1991. National energy strategy (*Report No. DOE/S-0082P*). U.S. Department of Energy, Washington, D.C., 217 p.

Dixon, B.F. 1990. Methane losses from theAustralian natural gas industry. In: *International Workshop on Methane Emissions from Natural Gas Systems, Coal Mining, and Waste Management Systems.* Environment Agency of Japan, U.S. Agency for International Development, and U.S. Environmental Protection Agency, Washington, D.C., 709 p.

Ehhalt, D.H. 1974. The atmospheric cycle of methane. *Tellus, 26*:58-70.

Ehhalt, D.H., U. Schmidt. 1978. Sources and sinks of atmospheric methane. *Pageoph, 116*:452-463.

Environment Agency of Japan. 1990. *International workshop on methane emissions from natural gas systems, coal mining, and waste management systems.* Environment Agency of Japan, U.S. Agency for International Development, and U.S. Environmental Protection Agency, Washington, D.C., 709 p.

Fung, I., J. John, J. Lerner, E. Matthews, M. Prather, L.P. Steele, P.J. Fraser. 1991. Three-dimensional model synthesis of the global methane cycle. *J. Geophys. Res., 96*:13,033-13,065.

Hay, N.E., P.L. Wilkinson, W.M. James. 1988. *Global climate change and emerging energy technologies for electric utilities: the role of natural gas.* American Gas Association, Arlington, Virginia, 30 p.

Hitchcock, D.R., A.E. Wechsler. 1972. Biological cycling of atmospheric trace gases (*Report No. NASW-2128*). National Aeronautic and Space Administration, Washington, D.C., 415 p.

Irani, M.C., E.D. Thimons, T.G. Bobick. 1972. Methane emission from U.S. coal mines, a survey (*Report No. IC 8558*). U.S. Department of the Interior, Bureau of Mines, Pittsburgh, Pennsylvania.
JEA-EPA. 1990. *Methane emissions and opportunities for control: workshop results of intergovernmental panel on climate change, response strategies working group*. Japan Environment Agency/United States Environmental Protection Agency, Washington, D.C.
Jolly, D.C., L.H. Morris, F.B. Hinsely. 1968. An investigation into the relationship between the methane sorption capacity of coal and gas pressure. *The Mining Engineer 127*:539.
Joubert, J.I., C.T. Grein, B. Bienstock. 1974. Effect of moisture on the methane capacity of American coals. *Fuel, 53*:186.
Keeling, C.D. 1973. Industrial production of carbon dioxide from fossil fuels and limestone. *Tellus, 25*:174-198.
Khalil, M.A.K., M.J. Shearer, Eds. 1993. *Chemosphere, 26* (1-4).
Kim, A.G. 1977. Estimating methane content of bituminous coalbeds from adsorption data (*Report No. RI 8245*). U.S. Department of the Interior, Bureau of Mines, Pittsburgh, Pennsylvania.
Kirchgessner, D.A., S.D. Piccot, J.D. Winkler. 1993a. Estimate of global methane emissions from coal mines. *Chemosphere, 26* (1-4):453-472.
Kirchgessner, D.A., S.D. Piccot, A. Chadha, T. Minnich. 1993b. Estimation of methane emissions from a surface coal mine using open-path FTIR spectroscopy and modeling techniques. *Chemosphere, 26* (1-4):23-44.
Kissel, F.N., C.M. McCulloch, C.H. Elder. 1973. The direct method of determining methane content of coalbeds for ventilation design (*Report No. RI 7767*). U.S. Department of the Interior, Bureau of Mines, Pittsburgh, Pennsylvania.
Koyama, T. 1963. Gaseous metabolism in lake sediments and paddy soils and the production of atmospheric methane and hydrogen. *J. Geophys. Res., 68*:3,971-3,973.
Koyama, T. 1964. Biogeochemical studies on lake sediments and paddy soils and the production of hydrogen and methane. *Recent Researches in the Fields of Hydrosphere, Atmosphere, and Geochemistry*, 143-177.
Lacroix, A.V. 1993. Unaccounted-for sources of fossil fuel and isotopically-enriched methane and their contribution to the emissions inventory: A review and synthesis. *Chemosphere, 26* (1-4):507-558.
Lambert, S.W., M.A. Trevits, P.F. Steidl. 1980. Vertical borehole design and completion practices to remove methane gas from minable coalbeds (*Report No. DOE/CMTC/TR-80/2*). U.S. Department of Energy, Washington, D.C., 163 p.
Mitchell, C., J. Sweet, T. Jackson. 1990. A study of leakage from the UK natural gas distribution system. *Energy Policy*, November 1990:809-818.

Murray, D.D. 1980. *Methane from coalbeds - A significant undeveloped source of natural gas*. Colorado School of Mines Research Institute, Golden, Colorado, 37 p.

Okken, P.A., T. Kram. 1989. CH_4/CO_2-emission from fossil fuels global warming potential. In: *Proceedings of the International Energy Agency Workshop on Greenhouse Gases*, June 1989, Paris.

Piccot, S.D., M. Saeger. 1990. National- and state-level emissions estimates of radiatively important trace gases (RITG) from anthropogenic sources (*Report No. EPA-600/8-90-073*; NTIS PB91-103572). U.S. Environmental Protection Agency, Research Triangle Park, North Carolina.

Piccot, S.D., L. Beck. 1993. Estimation of methane emissions from minor anthropogenic sources. In: *Proceedings of the International Workshop on Methane and Nitrous Oxide: Methods in National Inventories and Options for Control*, February 1993, Amersfoort, The Netherlands.

Piccot, S.D., A. Chadha, J. DeWaters, T. Lynch, P. Marsosudiro, W. Tax, S. Walata, J. D. Winkler. 1990. Evaluation of significant anthropogenic sources of radiatively important trace gases (*Report No. EPA-600/8-90-079*; NTIS PB91-127753). U.S. Environmental Protection Agency, Research Triangle Park, North Carolina.

Piccot, S.D., A. Chadha, D.A. Kirchgessner, R. Kagenn, M.J. Czerniawski, T. Minnich. 1991. Measurement of methane emissions in the plume of a large surface coal mine using open-path FTIR spectroscopy. In: *Proceedings of 84th Annual Meeting of the Air and Waste Management Association*, June 1991, Vancouver, British Columbia, Canada.

Pilcher, R.C., C. Bibler, R. Glickert, L. Machesky, J. Williams. 1991. Assessment of the potential for economic development and utilization of coalbed methane in Poland (*Report No. EPA-400/1-91-032*). U.S. Environmental Protection Agency, Office of Air and Radiation, Washington, D.C.

Schwarzer, R.R., C.W. Byrer. 1983. Variation in the quantity of methane adsorbed by selected coals as a function of coal petrology and coal chemistry (*Report No. DE-AC21-80MC14219*). U.S. Department of Energy, Morgantown, West Virginia.

Seiler, W. 1984. Contribution of biological processes to the global budget of CH_4 in the atmosphere. In: *Current Perspectives in Microbial Ecology* (M.J. Klug and C.A. Reddy, eds.), American Society for Microbiology, Washington, D.C.

Selzer, H. 1990. Anthropogen methane emission. In: *International Workshop on Methane Emissions from Natural Gas Systems, Coal Mining, and Waste Management Systems*, Environment Agency of Japan, U.S. Agency for International Development, and U.S. Environmental Protection Agency, Washington, D.C., 709 p.

Sheppard, J.C., H. Westberg, J.F. Hopper, K. Ganesan. 1982. Inventory of global methane sources and their production rates. *J. Geophys. Res.*, 87:1305-1312.

Tilkicioglu, B.H., and D.R. Winters. 1989. *Annual Methane Emission Estimate of the Natural Gas and Petroleum Systems in the United States*, draft report prepared for the U.S. Environmental Protection Agency Office of Air and Radiation by Pipeline Systems Incorporated, Walnut Creek, CA, 101 pages.

Watt Committee on Energy. 1991. Quantification of methane emissions from British coal mine sources (Draft Report).

Whalen, M., N. Tanaka, R. Henry, B. Deck, J. Zeglen, J.S. Vogel, J. Southon, A. Shemesh, R. Fairbanks, W. Broecker. 1989. Carbon-14 in methane sources and in atmospheric methane: the contribution from fossil carbon. *Science*, 245:286-290.

Chapter 18

Minor Sources of Methane

A.G. JUDD,[1] R.H. CHARLIER,[2] A. LACROIX,[3] G. LAMBERT,[4] AND C. ROULAND[5]

[1] *School of the Environment, University of Sunderland, U.K.;* [2] *University of Brussels (V.U.B.), Belgium;* [3] *Intevep, Caracas, Venezuela;* [4] *Centre des Faibles Radioactivités, Gif-sur-Yvette, France;* [5] *Université Paris XII - Val de Marne, France*

1. Introduction

This chapter deals with those sources that are commonly regarded as making a minor contribution to the total atmospheric concentration of methane. The definition of "minor" is somewhat arbitrary; that is, they are sources not covered elsewhere in this volume or identified in the other chapters as being of only "minor significance." In isolation the majority of these sources may be regarded as inconsequential; however, in combination they provide a significant proportion of atmospheric methane. In certain cases it is possible that the true significance of their contribution is yet to be realized, and some may have been overlooked hitherto. More detailed discussions of these sources are provided by Hovland et al. (1993), Lacroix (1993), Lambert and Schmidt (1993), Rouland et al. (1993), Smith et al. (1993), and Khalil et al. (1993).

Both natural and anthropogenic minor sources are discussed. The natural sources are considered under the headings "geological" and "biological." Further discussion is also presented about pathways by which this methane may enter the atmosphere (for example, via seepages or mud volcanoes, or after sequestration by gas hydrates, ice, or permafrost) and about the contribution to atmospheric methane made by the oceans. Minor anthropogenic sources are principally produced by geological resource extraction and by industry. The latter, plus other anthropogenic sources (including the combustion of fossil fuels by non-industrial end users), are reviewed elsewhere in this volume (see Beck et al., this volume).

2. Natural sources

 2.1 Geological sources. The principal geological sources of methane are identified in Table 1.

Table 1. Geological origins of natural gas.

Microbial degradation of organic matter in sediments
Thermal degradation of organic matter in sediments
Thermal alteration of oil
Maturation of coal
Volcanic, igneous, and hydrothermal activity
Primordial gas derived from the mantle

Sources: MacDonald, 1983; Schoell, 1988

2.2 Hydrocarbon-bearing sediments. "Bacterial" (or "biogenic") methane is derived by the microbial degradation of organic matter in shallow sediments. "Thermogenic" methane is generated principally from organic matter buried to great depths in sedimentary rocks by the "cracking" of organically-derived materials to form petroleum hydrocarbons. It is also derived during the process of coal formation and during the thermal alteration of oil (MacDonald, 1983; Schoell, 1988). Once generated, the methane rises towards the surface (or the seabed), either by the bulk flow of bubbles (driven by buoyancy) or by diffusion within pore waters, to escape as either macro or micro seepage, unless it becomes entrapped en route. The processes of thermogenic methane generation and migration have been in progress over a time-scale measured in tens of millions of years. Migration may be continuous or periodic. Individual events may be triggered, for example, by earthquake activity.

Hunt (1979) estimated that there is 1.4×10^{18} to 2.8×10^{19} g of carbon present as dispersed gas in sedimentary basins. Although much of this gas will be contained within petroleum reservoirs, significant quantities will by-pass or spill out of any trap structures. It may be anticipated that a surface (or seabed) gas flux will occur in most areas underlain by petroliferous sediments. A discussion of such seepages is presented below.

2.3 Volcanic and related emissions. Various gases are known to emanate from centers of igneous activity and, in many circumstances, these gases include methane. Because of the infrequency of volcanic events it is unrealistic to imply a steady input of methane to the atmosphere. According to Lacroix (1993), estimates of contributions made by individual eruptions vary from 0.003 Tg CH_4 (Mount Etna in 1974) to 1.94 Tg CH_4 (Santorini in 1470 BC). An average of 0.13 Tg CH_4 year^{-1} (conservative, 0.02; liberal 2.33) for the years 1800 to 1969 was presented by Leavitt (1982), whilst Lacroix (1993) estimated the annual input as 3.5 ± 2.7 Tg CH_4. However, these estimates considered only volcanic events on land.

The ocean crust is essentially of igneous origin, although sediments have accumulated above the igneous rocks away from active spreading centers. These spreading centers extend over a distance of some 55,000 km. Emission of methane from the ocean spreading centers was estimated by Welhan and Craig (1983) as $1.6 \times 10^8 m^3$ (about 0.1 Tg CH_4 year^{-1}); however, this may be a conservative estimate (Malahoff, 1985). Seepages of fluids, including methane, of igneous origin have also been reported from, for example, subduction zones (Suess and Massoth, 1984; Kulm et al., 1986) and back-arc spreading centers (Horibe et al., 1986). Submarine volcanoes, like their counterparts on land, probably release substantial quantities of gas, including methane, during eruptions, and steadily release small quantities of gas for extended periods after (or between) eruptions. The number of submarine volcanoes (and seamounts) is very large: in the South Pacific Ocean there are an estimated 8 per 10,000 km^2 (Hekinian, 1984), so there may be about 130,000 in the Pacific Ocean alone.

Although it is likely that large volumes of methane are produced by these deep ocean sources, it is generally considered that almost all is consumed in the water.

2.4 Hydrothermal sources. Hydrothermal emanations occur wherever water from the surface or porewaters are heated or superheated in areas with a high geothermal gradient. They commonly contain a variety of low molecular weight hydrocarbons, including methane (Des Marais et al., 1988). Hydrothermal activity occurs both at active plate margins (e.g., the ocean spreading centers) and other areas of volcanic or igneous activity on land (e.g., Iceland, Italy, Yellowstone Park, Wyoming, U.S.A.) and offshore (e.g., the White Island geothermal field, New Zealand; Guaymas Basin, Gulf of California). Glasby (1971) reported that the principal constituents of gas samples collected from geothermal sources in the Bay of Plenty, New Zealand were hydrogen sulphide and methane. Des Marais et al. (1988) reported that gases from the Cerro Prieto geothermal field (Baja California Norte, Mexico) included methane in the concentration range 37,000 to 48,000 ppmv. According to Welhan (1988), the principal sources of methane in hydrothermal fluids are:

(1) thermogenic;
(2) biogenic;
(3) outgassing from the mantle (see below); and
(4) inorganic synthesis at high temperatures.

Of these, (1) and (4) are the most important in high temperature systems. Lacroix (1993) estimated that 2.3 ± 1.4 Tg CH_4 yr^{-1} is derived from these sources. The fate of hydrothermal methane entering the ocean environment is discussed below.

2.5 Crystalline basement sources. Sherwood et al. (1988) reported that methane is "ubiquitous and discharging freely" from boreholes drilled for metalliferous ore exploration on the Canadian Shield where it represents up to 80% of the free gas phase. They also referred to similar occurrences from the Baltic Shield and the Kola Peninsula, U.S.S.R. Söderberg and Flóden (1990) described submarine gas (mainly methane) seeps associated with tectonic lineaments in crystalline rocks in the Stockholm Archipelago. The origin of these gases is unclear; however, they may have been derived from organic precursors, by migration in gaseous form, or from migrating methane-bearing connate waters (Parnell, 1988; Sherwood et al., 1988; Parnell and Swainbank, 1990). It is probable that flux rates are not high; however, rocks of crystalline basement cover extensive areas of the world's surface; no estimate of the global emission of methane from this source is offered. This source is discussed in greater detail by Lacroix (1993).

2.6 "Deep earth gas." Gold and Soter (1980, 1982) considered that large volumes of methane, remnants of the primitive volatile constituents of the Earth, may be emanating from the mantle, particularly along deep-seated faults. Gas derived from a deep crustal or mantle source has been reported by several authors (for example, Lawrence and Taviani, 1988; see also Lacroix, 1993), although alternative derivations may be offered. For example, gas that seeps from the Zambales ophiolites in the Philippines may be of mantle origin, although Abrajano et al. (1988) concluded that it may equally be derived by the reduction of water and carbon during the low temperature serpentinization of the ophiolites. Seepage gases from this site contain up to 55% methane.

Some authors (Cicerone and Oremland, 1988, for example) vigorously discount the validity of the abiogenic genesis of "primordial" methane by the outgassing of the mantle. There has, as yet, been no definite evidence to irrefutably support the claims of Gold and his associates, nor has their claim been totally disproved.

3. Biological sources

The principal biological methane sources (ruminants, rice agriculture, landfills, and wetlands) receive attention in separate chapters of this volume. The following are examples of relatively minor biological sources of methane.

3.1 Termites. Termites have a wide geographical distribution; however, they are particularly abundant in tropical areas where the biomass has been estimated by several authors (Wood et al., 1977; Abe, 1979; Collins, 1981; Wood et al., 1982). The termite population of tropical areas is relatively stable from one biotope to another within the range 18.9 g m^{-2} for agrosystems to 5.2 g m^{-2} for dry savannas.

In tropical areas the termites, because of their numbers and their nutrition mode, have a great effect on the turn-over of organic matter and on humus formation. Their ability to degrade the various polysaccharidic plant components like cellulose or hemicellulose is, to a great extent, due to symbiotic relationships with anaerobic bacteria located in their proctodeum (Breznak, 1975).

Termites are plant-eating insects, but different groups have different modes of nutrition:

    Xylophageous termites eat wood at different stages of decomposition.

    Soil-feeding termites are humivorous or sometimes geophagus; they dig galleries in the soil and, like earthworms, feed upon decomposing organic matter.

    Fungus-growing termites have a symbiotic relationship with the fungus *Termitomyces sp.*.

The digestive metabolism of termites is directly correlated to their nutrition mode (Rouland et al., 1986; Brauman et al., 1987; Rouland et al., 1989; Brauman, 1989). In particular, it appears that the production of methane is clearly different from one alimentary group to another as shown in Table 2 (Brauman et al., 1992):

    wood-eating termites produce little methane, 0.01 to 0.16 µM g^{-1} termite per hour;

    fungus-growing termites produce 0.43 to 0.67 µM g^{-1} termite per hour;

soil-feeding termites, the most active, produce 0.69 to 1.09 µM g^{-1} termite per hour.

When estimating the global production of methane by termites it is very important to take into account the considerable variation in biomass representing each of the nutrition modes in the biotopes studied, as shown in Table 2. According to these observations, the global potential methane emission by termites is estimated to be 27 Tg year^{-1} (using the method of calculation of Fraser et al., 1986). However, the uncertainties of this estimate are very substantial, in particular because of the difficulty in estimating the real populations of termites in each of the alimentary groups. The possible range is 15 to 35 Tg year^{-1} (Rouland et al., 1993).

Table 2. Methane production and biomass consumed by different feeding groups of termites in different tropical biotopes.

	Humivorous	Fungus-Growing	Xylophageous
	µM/g termite/hour		
Biomass consumed[a]	0.69-1.09	0.43-0.67	0.01-0.67
	gm^{-2}		
Rainforests[b]			
Congo	4.3	4	0.23
Ivory Coast	5.64	1.22	0.65
Malaysia	2.52	6.12	0.64
S. Guinea Savannas[c]			
Ivory Coast	8.43	3.98	0.98
Nigeria	0.66	6.39	3.58
Agro-Ecosystem			
SUCO[d]	0.01	8.5	0
SOGESCA[e]	0.2	10.8	0

[a] Bauman et al., 1991; [b] Matsumoto, 1976; [c] Wood et al., 1977; [d] Mora et al., 1990; [e] Renoux et al., 1991.

This value is slightly larger than those indicated by other authors (Seiler et al., 1984; Fraser et al., 1986; Khalil et al., 1990) because the methane production by fungus-growing termites, which represent the most important biomass in several biotopes, may be underestimated by those authors.

4. *Release pathways for methane*

4.1 Natural coal seam fires. Bustin and Matthews (1985) reported that methane composes 0.3 to 4.0% of the total gaseous emissions from natural coal seam fires. The global contribution made by these fires is probably less than 1 Tg CH_4 yr^{-1}.

4.2 Seepages. Seepages are known to be widespread in both land and marine environments (Link, 1952; Landes, 1973; Wilson et al., 1974; Kvenvolden and Harbaugh, 1983; Clarke and Cleverly, 1991). Indeed petroleum seepages gave the first indications of the presence of petroleum in most of the world's petroleum provinces (Link, 1952) and are still used in petroleum exploration (Philp, 1987). Seepages may occur in any environment in which methane is generated. The methane may be of any origin, but seepages are most commonly associated with hydrocarbon bearing sediments in which thermogenic or bacterial methane is generated.

On land oil seepages are easier to observe than gas seepages; furthermore, onshore and offshore oil seeps attract more attention as they are sources of pollution. However, gas seepages on land have been reported (for example, by Simoneit et al., 1979), and it is probable that atmospheric concentrations of methane are higher over areas characterized by methane-bearing rocks. Data presented by Stadnik et al. (1986) show that the measured atmospheric methane levels in eleven regions of the U.S.S.R. were consistently higher in petroliferous areas (1.97 to 6.6 ppm; mean 3.47 ppm) compared to the regional background (1.15 to 2.9 ppm; mean 1.85). During hydrocarbon prospecting programs seasonal surveys of the methane concentration of snow accumulations in the U.S.S.R. have shown that thermogenic methane passes through the underlying permafrost, particularly in fissured and faulted areas (Glotov et al., 1985; Bordkov et al., 1988; and Vyshemirskiy et al., 1989). Andronova and Karol (1993) estimated this flux at 0.2 to 0.9 Tg CH_4 yr^{-1} from the permafrost regions of the U.S.S.R.

Of the numerous sub-seabed sources of methane, seepages of bacterially- and thermogenically-derived methane on the continental shelves are the most likely to contribute significant quantities to the seawater and subsequently the atmosphere. A significant proportion of the oceans' continental shelves consist of sedimentary basins. Estimates of the areal extent of marine petroliferous sedimentary basins vary, for example, 27×10^6 km^2 (Earney, 1980) and 35×10^6 km^2 (Trotsyuk and Avilov, 1988). Bacterial methane may be formed in marine areas with organic-rich sediments (Floodgate and Judd, 1992). The necessary conditions are commonly found in embayments and indentations, fjords, and basins (enclosed or with narrow communication to the open sea); the resulting methane accumulations are widespread and large (Rice and Claypool, 1981).

The presence of shallow gas, mainly methane (either bacterial or thermogenic), is not uncommon beneath the seabed. Similarly, there is ample evidence of the escape of gas through the seabed. This takes the form of topographic features (e.g., mud diapirs and pockmarks), bacterial mats, and associated carbonate precipitates. In addition, gas seepages, identified on echo sounder, sonar, and seismic records, have been reported from many parts of the world (Hovland and Judd, 1988). However, a relatively small proportion of the world's continental shelves has been surveyed in detail and much of the relevant data are held in confidence by the petroleum industry. Where information is available, it is evident that seepage areas may produce considerable volumes of methane at the seabed.

Measurements of methane flux rates from seepages are not numerous. Clarke and Cleverly (1991), who noted that seepage rates could be measured most easily offshore, reported measurements of 28 gassy seeps that were found to produce between 10 m^3 and 3×10^6 m^3 (6.5×10^3 to 1.95×10^9 g) CH_4 per year. The most prolific area of marine petroleum seepage in the world is probably the Santa Barbara Channel, California, U.S.A., where tar, crude oil, and gas emanate from petroleum-bearing rocks overlain by only a thin veneer of surficial sediments. Gas seepage rates off Coal Oil Point may be as high as 400 g CH_4 m^{-2} year^{-1} (Fischer and Stevenson, 1973). Seepages occur in the Gulf of Mexico in association with organic-rich sediments derived from the Mississippi, with salt emplacement structures on the continental shelf and with destabilizing gas hydrates beyond the shelf. Estimates of the numbers of seeps in two areas reach 19,000 and 8,600 respectively (Watkins and Worzel, 1978; Addy and Worzel, 1979). Flow rates of 1 ml min^{-1} to 50 l min^{-1} were reported, so flux rates may be in the order of 0.01 to 56g CH_4 m^{-2} year^{-1}.

Hovland et al. (1993) present a crude estimate of the total flux on the world's continental shelves based on the assumptions that gas seepage rates are lognormally distributed and that measured rates represent seepages that have a low probability of occurrence. They suggest that seabed seepages may produce between 8 and 65 Tg CH_4 year^{-1}. Estimates of seepage methane entering the atmosphere have been undertaken by various authors. Ehhalt and Schmidt (1978) used the Wilson et al. (1974) study of the world-wide distribution of oil seepages for their calculation of the oceanic flux of methane. Trotsyuk and Avilov (1988) measured the disseminated flux of methane in the Black Sea and extrapolated world wide. Lacroix (1993) calculated hydrocarbon reservoir depletion and migration rates and, by estimating the rate of removal by oxidation, etc., derived an estimate of the rate of emission to the atmosphere. The resultant figures are presented in Table 3.

Table 3. Estimates of methane emissions from natural seepages.

Author(s)	Estimate Tg CH_4 yr^{-1}	Note
Trotsyuk and Avilov (1988)	1.9	continental shelves
Hovland et al. (1993)	8 - 65	continental shelves (at seabed)
Ehhalt and Schmidt (1978)	1.3 - 16.6	world's oceans
Lacroix (1993)	17 ± 14	global

4.3 Mud volcanoes. Mud volcanoes, which vary in height from less than one meter to 100 meters or more, are formed by the diapiric rise of mud or rock, often in association with gas. They are found in various environments; for example: areas of rapid sedimentation (e.g., the Mississippi delta), areas of rapid sediment compaction, which may be induced by tectonic activity (e.g., Raukumara Peninsula, New Zealand), and areas of shallow hydrocarbon entrapment (e.g., the Baku region of Azerbaijan, U.S.S.R.). They occur in many parts of the world, both on land and on the continental shelf. Reviews have been presented by Gansser (1960), Ridd (1970), Higgins and Saunders (1974), and Hedberg (1980).

Gas probably plays a critical role in mobilizing sediments, and considerable volumes of gas are thought to escape during periods of activity. This is suggested not only by reports of the noise associated with mud volcanism but also by the considerable volumes of ejecta, which may be dispersed over surrounding areas.

The gases evolved may be carbon dioxide-dominated; however, in many areas they are methane-dominated (Reitsema, 1979). It enters the atmosphere by slow seepage in some instances and by explosive activity in others. Despite the significance of this source, no realistic estimate of global methane production is possible at present. However, the following two figures suggest that mud volcanoes represent a sizeable source: Sokolov et al. (1969) reported that during quiet activity in the Baku region, groups of mud volcanoes emit 1-3,000 m^3 or more in a day and that eruptive phases may produce "some hundred millions of m^3 in 1-2 days" (although the erupted gas usually catches fire). They estimated that a total of 10^5 Tg CH$_4$ has been emitted from this area alone during the Quaternary Period (a time interval of approximately 1.64 × 10^6 years).

4.4 Gas hydrates. Gas hydrates, crystalline ice-like compounds composed of water and natural gas (principally methane), occur in a zone that is strictly controlled by pressure and temperature regimes. They are extensive in or near the permafrost regions of the world's land areas and under vast areas of the world's deep oceans (generally in water depths of > 500 m where the sediment temperature is low but the pressure is high (Matthews, 1986)). These occurrences represent an enormous reservoir of methane: 2 × 10^{18} to 4 × 10^{21} g (best estimate: 1 × 10^{19} g) carbon as methane (Kvenvolden, 1988). In many cases the hydrates also represent an impermeable layer under which gas may be trapped (Dillon and Paull, 1983).

The gas sequestered by or trapped beneath these hydrates has evidently been derived elsewhere in the sediments, and may be of thermogenic or bacterial origin.

Giant gas mounds and pockmarked areas in the Gulf of Mexico occur in probable association with destabilizing gas hydrates. The former are diapiric structures from which seabed gas hydrates have been trawled and from which gas is or has been seeping (Brooks et al., 1986); an example of the latter was described by Prior et al. (1989). Evidence such as this indicates that gas from hydrates may escape to the seabed and possibly to the atmosphere. Similar escape may also occur on land. Kvenvolden (1988) estimated that 3 Tg CH$_4$ yr^{-1} from gas hydrates enters the atmosphere.

5. The oceans

The contribution to atmospheric methane made by the ocean is not in proportion to the ratio of sea to land. Nevertheless, it is by no means negligible, and the oceans occupy the majority of the earth's surface (73%). Oceanic sources of methane are both natural (physical and biological) and anthropogenic; methane is also generated beneath the seabed by both bacterial and geological processes and may be introduced into the oceans by rivers (Scranton and

McShane, 1991). Several of the sources that are active in (or beneath) the ocean environment have been discussed above (hydrocarbon-bearing sediments, volcanic and related emissions, hydrothermal sources, marine algae, seepages, mud volcanoes, and gas hydrates). In cases where intermittent events cause large volumes of methane to enter the water, it is probable that the majority passes to the atmosphere, but in most cases a significant proportion goes into solution in the seawater. Thus, each of the listed sources makes a contribution to the methane concentration of ocean waters.

5.1 Geological sources. Geological sources contribute significant volumes of methane at the floor of the deep oceans and on the continental shelves.

It has been suggested (above) that significant volumes of methane may enter the seawater from natural gas seepages. Depending upon the water depth and bubble sizes, a large proportion of this gas may reach the sea surface and then escape to the atmosphere because the turnover time of methane in shallow waters (approximately 5 to 100 years) is long relative to the sea-air exchange time (as shown by Jones, 1991).

Methane, along with ^{3}He, has been found to be a volatile component of hydrothermal fluids escaping into the marine environment (Lilley et al., 1982; Vidal et al., 1982; Welhan and Craig, 1983; Horibe et al., 1986; Jean-Baptiste et al., 1990). As methane is easily measurable at sea, it has even been suggested that it could be used as a tool for mapping hydrothermal fields (Charlou et al., 1987, 1988). In fact, very high concentrations, up to 1.2×10^{6} nl STP per liter of water, have been reported in hydrothermal plumes very near to source (Merlivat et al., 1987). Charlou et al. (1991) reported that more common concentrations are 100 to 1,000 nl STP per liter of water, compared to about 50 nl STP per liter in normal seawater. Despite these huge methane concentrations, the proportion able to reach the sea surface is undoubtedly small. This is because of its relatively short life time (5 to 100 years according to Jones, 1991) compared to the transit time toward the sea surface. For this reason most methane profiles sharply decrease from the bottom to depths of about 100 to 200 m. Several authors (Winn et al., 1986, for example) have reported that, in the vicinity of hydrothermal sources, important colonies of methanotrophic bacteria are present. Their influence is demonstrated by the decay in the $CH_{4}:^{3}He$ ratio along the plumes (as reported by Gamo et al., 1987, and Charlou et al., 1991). However, it is noteworthy that, in the same areas, active bacteria able to live in hot waters, with temperatures in excess of 100°C, produce methane, along with CO and H_2 (Baross et al., 1982).

5.2 Sea surface methane. It has been known for some time that methane is present in sea surface waters. Dunlap et al. (1960) observed methane in seawater in areas of oil or gas seepages. Methane was observed in Lake Nitinat, an anoxic fjord, by Richards et al. (1965), suggesting possible formation in the marine environment by anaerobic fermentation. The distribution of methane dissolved in sulphide-bearing waters of the Black Sea was reported by Atkinson and Richards (1967). However, the very first measurements of methane present in seawater, by Lamontagne et al. (1973), made possible a first evaluation of the flux of methane released by the ocean, at approximately 10 Tg per year Ehhalt (1974). This figure has also been used by Khalil and Rasmussen (1983), Cicerone and Oremland (1988), Fung et al. (1991), and others.

The number of measurements is steadily increasing, allowing better estimates to be made at present. According to Lambert and Schmidt (1993), most of the experimental data relating to the ocean surface layers can be grouped into two categories. The first group deals essentially with the open ocean and indicates about 60 nl of methane STP per liter of water, with a supersaturation coefficient that remains constant at roughly 1.3 (Lamontagne et al., 1973, 1974; Brooks et al., 1974; Sackett and Brooks, 1975; Scranton and Brewer, 1977; Tranganza et al., 1979; Lilley et al., 1982; Burke et al., 1983; Gamo et al., 1987, 1988; Charlou et al., 1988, 1991; Jean-Baptiste et al., 1990; Jones, 1991). The second group deals with specific geographical areas, such as the Gulf of Mexico, the Black Sea, and generally the so-called "coastal ocean." It is typified by very variable concentrations, ranging between 40 and 13,000 nl/l (Scranton and Farrington, 1977; Brooks, 1979; Brooks et al., 1981; Belviso et al., 1987; Reeburgh et al., 1991). This latter group could well represent a source with a magnitude of 6 Tg CH_4 per year, but this figure should be taken with great reservations. The supersaturation found in these zones is so high that this source is not susceptible to significant change as there is a gradual increase in the mixing ratio of the methane in the atmosphere. In contrast, where the first group is concerned, the relatively uniform supersaturation permits the same authors to make a reasonably precise evaluation by using the mean exchange velocity of CO_2 between the sea and the atmosphere. This approaches 13 cm per hr (Etcheto et al., 1991). It indicates that the open oceans are the source of 3.5 Tg CH_4 per yr.

Table 4a. Contributions to atmospheric methane levels from "minor" sources.
Natural Sources

Source	Tg CH_4 year^{-1}	Notes
Geological		
Hydrocarbon-bearing sediments	n/a	see note a
Volcanic and related emissions	3.5 ± 2.7	land events only
Hydrothermal sources	2.3 ± 1.4	land events only
Crystalline basement sources	--	not available
"Deep Earth Gas"	--	not available
Biological		
Termites	20(15 - 35)	
Pathways		
Natural coal seam fires	< 1	
Seepages	± 1	
Mud volcanoes	--	not available
Gas hydrates	3	
The Oceans		
The open ocean	3.5	see note b
Total natural sources	34	

Note a: Methane from hydrocarbon-bearing sediments may escape via several pathways (seepages, mud volcanoes, etc.); consequently, it is not appropriate to estimate the emission from this source.

Note b: The figure of 3.5 Tg CH_4 yr^{-1} refers only to the open ocean; sources in shallow (continental shelf) waters and "point" sources (seepages, hydrothermal plumes, etc.) may make significant additional contributions.

It should be understood that, inasmuch as the superficial methane concentration of the ocean would not be substantially modified in the future, as an increase occurs in atmospheric methane, the flux direction would be inverted for a mixing ratio of about 2.25 ppmv. This oceanic sink would become important for higher values of the atmospheric mixing ratio, reaching the present rate for a mixing ratio of 8 ppmv. Taking into account the low methane mixing ratio of 0.35 ppmv at the time of the peak of the last Quaternary glaciation (Chappellaz et al., 1990), the same hypothesis of a constant methane concentration at the ocean's surface leads to an estimation of the open ocean sources to be in the order of 20 Tg per year, this being approximately ten percent of the global source at the time of the last glaciation (Lambert and Schmidt, 1993). Sea-air exchanges have been discussed by several authors in the volume edited by Buat-Ménard (1986), and more recently the results have been published of SEAREX, a ten-year study of air-sea interchanges, particularly in the Pacific Ocean (Duce, 1989). Various papers discuss contributions of several ecosystems including rivers and the open oceans. The rivers' input affects primarily the coastal ocean area, but sea-air exchange has an immediate and important effect on open ocean regions. SEAREX looked at natural as well as anthropogenic contributions, attempted to identify sources of substances found in the "marine" atmosphere, and one chapter (by D.L. Savoir) in particular addressed itself, besides non-sea salt sulphates and nitrates, to methanosulphate generated by marine biological sources.

As has been shown, methane is emitted by various sources in the oceans and, although serious uncertainties remain over the ability of methane at the seabed to escape, the oceans do contribute to the methane content of the atmosphere. The magnitude of the emissions is, at best, roughly estimated. Nevertheless, the amount released may be significant, and the oceans should not be ignored in the atmospheric budget.

6. Anthropogenic sources

The various "minor" anthropogenic sources of methane that have been identified relate to man's industrial activities. Methane emissions from "minor industrial sources" have been discussed by various authors. Figures derived from Piccot et al. (1990) and Lacroix (1993) are included in Table 4b; however, a review of these sources is presented elsewhere in this volume by Beck et al. Two recent assessments (Smith et al., 1993; Khalil et al., 1993) have been made of the possible methane emissions for inefficient fuel combustion in cooking stoves. Wood fuel is usually considered to be part of the biomass burning sources (see Levine et al., this volume). However, for fuels including charcoal, kerosene, and liquefied petroleum gas, Smith et al. estimated less than 1 Tg/yr emissions. Khalil et al. estimated a global source of 16 (7-30) Tg/yr from low temperature coal burning for cooking and home heating.

7. The extractive industries

The mining of coal and the exploitation of petroleum account for the most significant emissions of methane (see Beck et al., this volume); however, the extraction of certain other natural resources produce minor amounts of methane.

Table 4b. Contributions to atmospheric methane levels from "minor" sources.
Anthropogenic Sources

Source	Tg CH_4 year^{-1}	Notes
Industrial sources		see note c
Coke production	0.3	
Residential waste burning	3.3	not landfill
Mobile sources	1.8	
Fossil fuel burning	1.6	
Refining, storage, and distribution		
of petroleum products	0.6	
Petrochemical/organic chemical manufacture	0.1	
Other industrial and utility activities	1.4	
Total Industrial Sources	9.1	
Combustion of fossil fuels	4.6 ± 0.4	see note d
Petroleum refining	3.5 ± 0.5	see note d
Petrochemicals	2.0 ± 1.0	see note d
The extractive industries		
Groundwater abstraction	1.1 ± 0.8	
Peat mining	2.0 ± 1.0	
Mining of other geological resources		not available
Total Extractive Industries	3	
Geothermal energy production	1.2 ± 0.7	see note d
Home cooking and heating	17 (7 - 30)	see note e
Total anthropogenic sources	30 (36)	see note f
Total "Minor" Sources	64	

Note c: Figures from Beck et al. (this volume).

Note d: Figures from Lacroix (1993). Lacroix's category: "Combustion of fossil fuels" is approximately equivalent to Beck et al.'s "Fossil fuel burning" + "Mobile sources"; similarly with Lacroix's "Petroleum refining" and "Petrochemicals" categories and Beck et al.'s "Refining, storage ..." and "Petrochemical/organic ..." categories. Calculations were done separately.

Note e: Figures from Smith et al. (1993) for all fuels except wood and Khalil et al. (1993) for low temperature coal cooking and heating.

Note f: Total uses Beck et al.'s industrial sources; in parentheses is the total calculated using Lacroix's estimates.

7.1 Groundwater abstraction. Zor'kin et al. (1985) estimated that there is 10^{16} to 10^{17} m^3 methane dissolved in the Earth's waters (the "hydrosphere"). Groundwater may be a source of atmospheric methane (Aravena et al., 1989; Geyh and Softner, 1989; Kimmelmann et al., 1989), and Coleman et al. (1988) showed that methane may be released during groundwater abstraction. Lacroix (1993) estimated that 1.1 ± 0.8 Tg CH$_4$ yr^{-1} is derived from groundwater on the basis of average methane contents of groundwater and global rates of groundwater usage.

7.2 Peat mining. Using the methane content of peat presented by Glotov et al. (1985) and an estimated global production rate of 23 × 10^{6}t yr^{-1}, Lacroix (1993) suggested that peat mining may be responsible for 2 ± 1 Tg CH$_4$ yr^{-1}. However, it is acknowledged that this estimate is only a crude approximation and that further work is required.

7.3 Mining of other geological resources. Methane is a common gaseous component of igneous, sedimentary, and metamorphic rocks, and emissions of methane from uranium, salt, and metal sulphide mines have been reported (Sokolov et al., 1972; Hyman, 1987; Lacroix, 1993). At present it is not possible to provide a realistic estimate of the methane emissions from these mining activities; however, one example indicates that mining may represent a not-insignificant source. According to Sokolov et al. (1972) 0.3 Tg CH$_4$ yr^{-1} is emitted from the Witwatersrand gold mines of South Africa.

8. Summary

The contributions of the sources described in this chapter, and those "minor" sources identified elsewhere in this volume, are summarized in Table 4. In many cases inadequate data have prevented reliable estimation, and in several cases the estimates presented are crude. Further work is therefore required. Individually the sources are small in comparison to those which form the subject of other chapters in this volume, yet in total these "minor" sources account for a contribution to the atmospheric methane budget, which should not be ignored. For example, the total emission from the "minor" anthropogenic sources, 30-35 Tg CH$_4$ yr^{-1}, is comparable to some "major" sources.

The significance of these sources lies not only in the volumes of methane emitted but also in the isotopic composition and the variability and unpredictability of the flux rates.

Geological methane sources, both on land and in the oceans, can be expected to be active over long periods of geological time. However, emission rates may vary considerably on a human time scale, for example, with cycles of volcanic activity, the triggering of gas escapes from petroleum reservoirs by seismic activity, or the destabilization of gas hydrates. Consequently it is difficult to assess the true contribution they make to atmospheric levels.

Recent models combining ^{13}C and ^{14}C contents of atmospheric methane and its source imply that fossil or ^{14}C-depleted sources constitute about 20% (Whalen et al., 1989) to 30% (Lowe et al., 1988) of the total atmospheric budget. The budget of Cicerone and Oremland (1988) proposes a total methane source strength of 540 Tg CH_4 year^{-1}, only 80 Tg CH_4 year, or 15%, of which are identified as arising from fossil sources. Thus, about 30 to 80 Tg CH_4 year^{-1} of ^{14}C-depleted methane cannot be accounted for. Evidence alluded to in this paper suggests that the so-called "minor" sources may account for a significant proportion of the missing fossil methane, as well as making a contribution to the recently-derived (biological) concentrations.

References

Abe, T. 1979. Studies on the distribution and ecological role of termites in a lowland rain forest of west Malaysia. *J. Ecol. Japan, 292*:121-136.

Abrajano, T.A., N.C. Sturchi, J.K. Bohlke, G.L. Lyon, R.J. Poreda, C.M. Stevens. 1988. Methane-hydrogen gas seeps, Zambales Ophiolites, Philippines: deep or shallow origin? In: *Origins of Methane in the Earth* (M. Schoell, ed.), *Chem. Geol., 72*:211-222.

Addy, S.K., J.L. Worzel. 1979. Gas seeps and sub-surface structure off Panama City, Florida. *Am. Assoc. Petrol. Geol. (Bull.), 63*:668-675.

Andronova N.G., I.L. Karol. 1993. The contribution of USSR sources to global methane emission. *Chemosphere, 26* (1-4):111-126.

Aravena, R., L.I. Wassenaar, J.F. Baricek. 1989. Investigating carbon sources for methane and dissolved organic carbon in a regional confined aquifer using ^{14}C, radiocarbon. *Proc. 14th Internat. Radiocarbon Conf., 31*:170-171.

Atkinson, L.P., A.F. Richards. 1967. The occurrence of methane in the marine environment. *Deep Sea Res., 14*:673-684.

Baross, J.A., M.D. Lilley, L.I. Gordson. 1982. Is the CH_4, H_2 and CO venting from submarine hydrothermal systems produced by thermophilic bacteria? *Nature, 298*:366-368.

Belviso, S., P. Jean-Baptiste, B.C. Nguyen, L. Merlivat, L. Labeyrie. 1987. Deep methane maxima and ^{3}He anomalies across the Pacific entrance to the Celebes Basin. *Geochem. Cosmochim. Acta, 51*:2,373-2,680.

Bordkov, Yu K., V.I. Yefimov, J.B. Timkia. 1988. Result of a gas- biochemical survey of snow cover for direct exploration for hydrocarbon deposits in the Venisey-Khatanga Downwarp (U.S.S.R.). *Petrol. Geol., 22*:203-205.

Brauman, A. 1989. Etude du métabolisme bactérien de termites supérieurs à régimes alimentaires différenciés. Thèse d'Université, Aix-Marseille II, 168 p.

Brauman, A., M. Labat, P. Methener, C. Rouland, J.L. Garcia. 1987. Etude de la microfflore hétérotrophe de termite supérieur en fonction de leur régime alimentaire. In: Diversité microbienne. *A Coll. Soc. Fr. Microbiol.* 18-21.

Brauman, A., M.D. Kane, M. Labat, J.A. Breznak. 1992. Genesis of acetate and methane by gut bacteria of nutritionally diverse termites. *Science, 257*:1,384-1,387.

Breznak, J.A. 1975. Symbiotic relationship between termites and their intestinal microbiota. In: *Symbiosis* (D.H. Jennin and D.L. Lee, eds.), Cambridge University Press.

Brooks, J.M. 1979. Deep methane maxima in the northwest Caribbean sea: possible seepage along the Jamaica Ridge. *Science, 206*:1,069-1,072.

Brooks, J.M., J.R. Gormly, W.M. Sackett. 1974. Molecular and isotopic composition of two seep gases from the Gulf of Mexico. *Geophys. Res. Let., 1*:213-217.

Brooks, J.M., D.F. Reid, B.B. Bernard. 1981. Methane in the upper water column of the northwestern Gulf of Mexico. *J. Geophys. Res., 86*:11,029-11,040.

Brooks, J.M., H.B. Cox, W.B. Bryant, M.C.I.I. Kennicutt, R.G. Mann, T.J. McDonald. 1986. Association of gas hydrates and oil seepage in the Gulf of Mexico. In: *Advances in Organic Geochemistry 1985, Part I Petroleum Geochemistry* (D. Leythauser and J. Rullkotter, eds.), 221-234.

Buat-Ménard, P. 1986. *The Role of Air-Sea Exchange in Geochemical Cycling. NATO ASI Ser. C, 185*, Reidel, Dordrecht.

Burke, R.I. Jr., D.F. Reld, J.M. Brooks, D.M. Lavole. 1983. Upper water column methane geochemistry in the eastern tropical North Pacific. *Limnol. Oceanogr., 28*:19-32.

Bustin, R.M., W.H. Matthews. 1985. In situ gassification of coal, a natural example: additional data on the Aldridge Creek coal fire, south-eastern British Columbia. *Can. J. Earth Sci., 22*:1,858-1,864.

Chappellaz, J., J.M. Barnola, D. Raynaud, Y.S. Korotkevich, C. Lorius. 1990. Ice-core record of atmospheric methane over the past 160,000 years. *Nature, 345*:127-131.

Charlou, J.L., P. Rona, H. Bougault. 1987. Methane anomalies over TAG hydrothermal field on Mid Atlantic Ridge. *J. Marine Res., 45*:461-472.
Charlou, J.L., L. Dmitriev, H. Bougault, H.D. Needham. 1988. Hydrothermal CH_4 between 12°N and 15°N over the Mid-Atlantic Ridge. *Deep Sea Res., 35*:121-131.
Charlou, J.L., H. Bougault, P. Appriou, P. Jean-Baptiste, J. Etoubleau, A. Birolleau. 1991. Water column anomalies associated with hydrothermal activity between 11°40'N and 13°N on the east Pacific Rise: discrepancies between tracers. *Deep Sea Res., 38*:569-596.
Cicerone, R.J., R.S. Oremland. 1988. Biogeochemical aspects of atmospheric methane. *Global Biogeochem. Cycles, 2*:299-327.
Clarke, R.H., R.W. Cleverly. 1991. Petroleum seepage and post-accumulation migration. In: Petroleum Migration, *Geol. Soc. Sp. Publ. No. 59* (W.A. England and A.J. Fleet, eds.), Geological Society of London.
Coleman, D.D., C. Liu, K.M. Riley. 1988. Microbial methane in the shallow paleozoic sediments and glacial deposits of Illinois, USA. *Chem. Geol., 71*:23-40.
Collins, N.M. 1981. The role of termites in the decomposition of wood and leaf litter in the southern guinea savanna of Nigeria. *Oecologia, 51*:389-399.
Des Marais, D.J., M.L. Stallard, N.L. Nehring, A.H. Truesdell. 1988. Carbon isotope geochemistry of hydrocarbons in the Cerro Prieto geothermal field, Baja California Norte, Mexico. In: *Origins of Methane in the Earth* (M. Schoell, ed.), *Chem. Geol. 71*:159-167.
Dillon, W.P., C.K. Paull. 1983. Marine gas hydrates - II: geophysical evidence. In: *Natural Gas Hydrates: Properties, Occurrence and Recovery* (J.L. Cox, ed.), Butterworth, Boston, p 73-90.
Duce, R.A. (ed.). 1989. SEAREX. The Air/Sea Exchange Program. *Chem. Oceanogr. 10*:404.
Dunlap, H.F., J.S. Bradley, T.F. Moore. 1960. Marine seep detection - a new reconnaissance exploration method. *Geophysics, 25*:275-282.
Earney, F.C.F. 1980. *Petroleum and Hard Minerals from the Sea.* V.H. Winston and Sons, New York, 291p.
Ehhalt, D.H. 1974. The atmospheric cycle of methane. *Tellus, XXVI*(1-2):58-70.
Ehhalt, D.H., U. Schmidt. 1978. Sources and sinks of atmospheric methane. *Pure Appl. Geophys., 116*:452-464.
Etcheto, J., J. Boutin, L. Merlivat. 1991. Seasonal variation of the CO_2 exchange coefficient over the global ocean using satellite wind speed measurements. *Tellus, 43B*:247-255.
Fischer, P.J., A.J. Stevenson. 1973. Natural hydrocarbon seeps along the northern shelf of the Santa Barbara Channel, California. *Paper 1728, Offshore Technol. Conf.*, Houston, Texas.
Floodgate, G.D., A.G. Judd. 1992. The origins of shallow gas. *Continental Shelf Res., 12*:1,145-1,156.

Fraser, P.J., R.A. Rasmussen, J.W. Creefield, J.R.J. French, M.A.K. Khalil. 1986. Termites and global methane: another assessment. *J. Atmos. Chem.*, *4*:295-310.
Fung, I., J. John, J. Lerner, E. Matthews, M. Prather, L.P. Steele, P.J. Fraser. 1991. Three-dimensional model synthesis of the global methane cycle. *J. Geophys. Res.*, *96*:13,033-13,065.
Gamo, T., J.-I. Ishibashi, H. Sakai, B. Tilbrook. 1987. Methane anomalies in seawater above the Loihi submarine summit area, Hawaii. *Geochem. Cosmochim. Acta*, *51*:2,857-2,864.
Gamo, T., J.-I. Ishibashi, K. Shitashima, M. Kinoshita, M. Watanabee, E. Nakayama, Y. Sohrin, E.-S. Kim, T. Mazuzawa, K. Fujioka. 1988. Anomalies of bottom CH_4 and trace metal concentrations associated with high heat flow at the *Calyptogena* community off Hatsushima Island, Sagami Bay, Japan: A preliminary report of Tansei Maru KT-88-1 cruise Leg 1. *Geochem. J.*, *22*:215-230.
Gansser, A. 1960. Über Schlammvulkane und Saldome (Mud volcanoes and salt domes). *Naturf Gesell Zürich Vierteljahrssch*, *105*:1-46.
Geyh, M.A., B. Softner. 1989. Groundwater analysis of environmental carbon and other isotopes from the Jakarta Basin Aquifer, Indonesia. *Radiocarbon*, *31*:919-925.
Glasby, G.P. 1971. Direct observations of columnar scattering associated with geothermal gas bubbling in the Bay of Plenty, New Zealand. *N.Z. J. Mar. Freshwater Res.*, *5*:483-496.
Glotov, V., V.V. Ivanov, N.A. Shilo. 1985. Migration of hydrocarbons through permafrost rock. Trans. (Doklady) *U.S.S.R. Acad. Sci., Earth Sci. Sect.*, *285*:192-194.
Gold, T., S. Soter. 1980. The deep earth gas hypothesis. *Sci. Am.*, *242*:154-161.
Gold, T., S. Soter. 1982. Abiogenic methane and the origin of petroleum. *Energy Exploration and Exploitation*, *1*:89-104.
Hedberg, H.D. 1980. Methane generation and petroleum migration. In: *Problems of Petroleum Migration. Am. Assoc. Petrol. Geol., Studies in Geology No. 10* (W.H. Roberts, III, and R.J. Cordell, eds.), pp 179-206.
Hekinian, R. 1984. Undersea volcanoes. *Sci. Am.*, *251*:46-55.
Higgins, G.E., J.B. Saunders. 1974. Mud volcanoes - their nature and origin. *Verhandl Naturforschung Gellschaft, Basel*, *84*:101-154.
Horibe, Y., K. Kim, H. Craig. 1986. Hydrothermal methane plumes in the Mariana back-arc spreading centre. *Nature*, *324*:131-133.
Hovland, M., A.G. Judd. 1988. *Pockmarks and Seabed Seepages: Impact on Geology, Biology and the Marine Environment*. Graham & Trotman, London.
Hovland, M., A.G. Judd, R.A. Burke. 1993. The global flux of methane from shallow submarine sediments. *Chemosphere*, *26* (1-4):559-578.

Hunt, J.M. 1979. *Petroleum Geochemistry and Geology*, W.H. Freeman, San Francisco.
Jean-Baptiste, P., S. Belviso, G. Alaux, B.C. Nguyen, N. Mihalopoulos. 1990. ^{3}He and methane in the Gulf of Aden. *Geochim. Cosmochim. Acta*, *54*:111-116.
Jones, R.D. 1991. Carbon monoxide and methane distribution and consumption in the photic zone of the Sargasso Sea. *Deep Sea Res.*, *38*:625-635.
Khalil, M.A.K., R.A. Rasmussen. 1983. Sources, sinks and seasonal cycles of atmospheric methane. *J. Geophys. Res.*, *88*:5,131-5,144.
Khalil, M.A.K., R.A. Rasmussen, J.R.J. French, J.A. Holt. 1990. The influence of termites on atmospheric trace gases: CH_4, CO_2, $CHCl_3$, N_2O, H_2 and light hydrocarbons. *J. Geophys. Res.*, *95*:3,619-3,634.
Khalil, M.A.K., R.A. Rasmussen, M.J. Shearer, S. Ge, J.A. Rau. 1993. Methane from coal burning. *Chemosphere, 26* (1-4):473-478.
Kimmelmann, A.A., Aldo de Cnha, S. Reboucas, M.M. Freitas Santiago. 1989. 14-C analysis of groundwater from the Botucatu Aquifer system in Brazil. *Radiocarbon, 31*:926-933.
Kulm, L.D., E. Suess, J.C. Moore, V. Carson, B.T. Lewis, S.D. Ritger, D.C. Kadko, T.M. Thornburg, R.W. Embley, W.D. Rugh, G.J. Massoth, M.G. Langseth, G.R. Cochrane, R.L. Scamman. 1986. Oregon subduction zone: venting fauna and carbonates. *Science, 231*:561-566.
Kvenvolden, K.A. 1988. Methane hydrate: a major reserve of carbon in the shallow geosphere. In: *Origins of Methane in the Earth* (M. Schoell, ed.), *Chem. Geol., 71*:41-51.
Kvenvolden, K.A., J.W. Harbaugh. 1983. Reassessment of the rates at which oil from natural sources enters the marine environment. *Marine Env. Res., 10*:223-243.
Lacroix, A.V. 1993. Unaccounted-for sources of fossil and isotopically-enriched methane and their contribution to the emissions inventory: A review and synthesis. *Chemosphere, 26* (1-4):507-558.
Lambert, G., S. Schmidt. 1993. Re-evaluation of the oceanic flux of methane: uncertainties and long term variations. *Chemosphere, 26* (1-4):579-590.
Lamontagne, R.A., J.W. Swinnerton, V.J. Linnenbon, W.D. Smith. 1973. Methane concentrations in various marine environments. *J. Geophys. Res., 78*:5,317-5,324.
Lamontagne, R.A., J.W. Swinnerton, V.J. Linnenbon. 1974. C_1-C_4 hydrocarbons in the north and south Pacific. *Tellus, XXVI*:(1-2)71-77.
Landes, K.K. 1973. Mother nature as oil polluter. *Am. Assoc. Petrol. Geol. (Bull), 53*:2,431-2,479.
Lawrence, J.R., M. Taviani. 1988. Extreme hydrogen, oxygen and carbon isotope anomalies in the pore water and carbonates of the sediments and basalts from the Norwegian Sea: methane and hydrogen from the mantle. *Geochim. et Cosmochim. Acta, 52*:2,077-2,083.

Leavitt, S.W. 1982. Annual volcanic carbon dioxide emission: an estimate from eruption chronologies. *Environ. Geol., 4*:15-21.
Lilley, M.D., J.A. Baross, L.I. Gordon. 1982. Dissolved hydrogen and methane in Saanich Inlet, British Columbia. *Deep Sea Res., 28*:1,471-1,484.
Link, W.K. 1952. Significance of oil and gas seeps in world oil exploration. *Am. Assoc. Petrol. Geol. (Bull.), 36*:1,505-1,540.
Lowe, D.C., C.A.M. Brenninkmeijer, M.R. Manning, R. Sparks, G. Wallace. 1988. Radiocarbon determination of atmospheric methane at Baring Head, New Zealand. *Nature, 332*:522-525.
MacDonald, G.J. 1983. The many origins of natural gas. *J. Petrol. Geol., 5*:341-362.
Malahoff, A. 1985. Hydrothermal vents and polymetallic sulfides of the Galapagos and Gorda/Juan de Fuca Ridge systems and of submarine volcanoes. In: Hydrothermal Vents of the Eastern Pacific: An Overview (M.L. Jones, ed.), *Bull. Biol. Soc., Washington, 6*:19-41.
Matsumoto, T. 1976. The role of termites in an equatorial rain forest ecosystem of west Malaysia. *Oecologia, 22*:153-178.
Matthews, M. 1986. Logging characteristics of methane hydrate. *The Log Analyst* May-June 1986:26-63.
Merlivat, L., F. Pineau, M. Javoy. 1987. Hydrothermal vent waters at 13°N on the East Pacific Rise: isotopic composition and gas concentration. *Earth & Planet. Sci. Let., 84*:100-108.
Mora, P., C. Rouland, V. Dibangou, J. Renoux. 1990. Damages caused by the recent infestations of the sugar cane fields by the fungus growing termite *Pseudacanthotermes spiniger*. *Act. 11th Int. Cong. I.U.S.S.I.*, Bengalore, India, p 78.
Parnell, J. 1988. Migration of biogenic hydrocarbons into granites - a review of hydrocarbons in British plutons. *Mar. & Petrol. Geol., 5*:385-395.
Parnell, J., I. Swainbank. 1990. Pb-Pb dating of hydrocarbon migration into a bitumen-bearing ore deposit, North Wales (United Kingdom). *Geology, 18*:1,028-1,030.
Philp, R.P. 1987. Surface prospecting methods for hydrocarbon accumulations. In: *Advances in Petroleum Geochemistry, Vol. II* (J. Brooks and D. Welte, eds.), Academic Press, London, p 209-253.
Piccot, S., A. Chadha, J. Dewaters, T. Lynch, P. Marsosudiro, W. Tax, S. Walata, J.D. Winkler. 1990. Evaluation of significant anthropogenic sources of radiatively important trace gases. *EPA-600/8-90-079*, U.S. Environmental Protection Agency, Research Triangle, NC.
Prior, D.B., E.H. Doyle, M.J. Kaluza. 1989. Evidence for sediment eruption on deep sea floor, Gulf of Mexico. *Science, 243*:517-519.
Reeburgh, W.S., B.B. Ward, S.C. Whalen, K.A. Sandbeck, K.A. Kilpatrick, L.J. Kerkhof. 1991. Black Sea methane geochemistry. *Deep Sea Res. 38 Suppl. Issue 2A*:S1,198-S1,210.

Reitsema, R.H. 1979. Gases of mud volcanoes in the Copper River Basin, Alaska. *Geochim. et Cosmochim. Acta, 43*:183-187.
Renoux, J., C. Rouland, P. Mora, N. Hassen. 1991. Dégats causés par les termites champignonnistes dans les cultures de canne à sucre en Afrique Intertropicale. *Coll. Inter. Canne à Sucre*, Montpellier, France
Rice, D.R., G.C. Claypool. 1981. Generation accumulation and resource potential of biogenic gas. *Am. Assoc. Petrol. Geol. (Bull.), 65*:5-25.
Richards, A.F., J.D. Cline, W.W. Broenkow, L.P. Atkinson. 1965. Some consequences of the decomposition of organic matter in Lake Litinat, an anoxic fjord. *Limnol. Oceanogr., 10 (suppl.)*:R185-R201.
Ridd, M.F. 1970. Mud volcanoes in New Zealand. *Am. Assoc. Petrol. Geol. (Bull.) 54*:601-616.
Rouland, C., C. Chararas, J. Renoux. 1986. Etude comparée des osidases de trois espèces de termites à régimes alimentaire différent. *C.R. Acad. Sc. Paris, 302*:341-345.
Rouland, C., A. Brauman, S. Keleke, M. Labat, P. Mora, J. Renoux. 1989. Endosymbiosis and exosymbiosis in the fungus growing termites. In: *Microbiol. Poecil.* (R. Lesel, ed.), Elsevier Science, Amsterdam.
Rouland, C., A. Braumann, M. Labat, M. Lepage. 1993. The production of methane by termites in tropical area. *Chemosphere, 26* (1-4):617-622.
Sackett, W.M., J.M. Brooks. 1975. Origin and distribution of low molecular weight hydrocarbons in Gulf of Mexico coastal waters. In: *Marine Chemistry in the Coastal Environment*, Chuch Ed, Washington.
Schoell, M. 1988. Multiple origins of methane in the earth. In: *Origins of Methane in the Earth* (M. Schoell, ed.), *Chem. Geol., 71*:1-10.
Scranton, M.I., P.G. Brewer. 1977. Occurrence of methane in the near-surface of the western subtropical North Atlantic. *Deep Sea Res., 24*:127-138.
Scranton, M.I., J.W. Farrington. 1977. Methane production in the waters off Walvis Bay. *J. Geophys. Res., 82*:4,947-4,953.
Scranton, M.I., K. McShane. 1991. Methane fluxes in the southern North Sea: the role of European rivers. *Cont. Shelf Res., 11*:37-52.
Seiler, W., R. Conrad, D. Scharffe. 1984. Field studies of methane emission from termite nests into the atmosphere and measurements of methane uptake by tropical soils. *J. Atmos. Chem., 1*:171-186.
Sherwood, B., P. Fritz, S.K. Frape, J.A. Macko, S.M. Weise, J.A. Welhan. 1988. Methane occurrences in the Canadian Shield. In: *Origins of Methane in the Earth* (M. Schoell, ed.), *Chem. Geol., 71*:223-23.
Simoneit, B.R.T., P.T. Crisp, B.G. Rohrback, B.M. Didyk. 1979. Chilean paraffin dirt - II. Natural gas seepage at an active site and its geochemical consequences. In: *Physics and Chemistry of the Earth, Vol. 12*: Advances in Geochemistry 1979 (A.G. Douglas and J.R. Maxwell, eds.), (Proc. 9th Internat. Meeting on Organic Geochem., Newcastle, Sept. 1979) p 171-176.

Smith, K.R., M.A.K. Khalil, R.A. Rasmussen, S.A. Thorneloe, F. Manegdeg, M. Apte. 1993. Greenhouse gases from biomass and fossil fuel stoves in developing countries: a Manila pilot study. *Chemosphere,* 26 (1-4):479-506.
Söderberg, P., T. Flóden. 1990. Gas seepages, gas eruptions and pockmarks in the seabed along the Stromma tectonic lineament at Stavsnas in the crystalline Stockholm Archipelago, east Sweden. *Methane in Marine Sediments Conf.*, Shallow Gas Group, Edinburgh, Sept 1990.
Sokolov, V.A., Z.A. Buniat-Zade, A.A. Geodekian, F.G. Dadashev. 1969. The origin of gases of mud volcanoes and the regularities of the powerful eruptions. In: *Advances in Organic Chemistry - 1969* (P. Schenk and I. Havemar, eds.), Pergamon Press, Oxford, p 473-484.
Sokolov, V.T., V. Tichomolova, O.A. Cheremisinov. 1972. The composition and distribution of gaseous hydrocarbons and dependence on depths, as a consequence of their generation and migration. In: *Advances in Geochemistry - 1971* (H.R. Gaertner and H. Wehner, eds.), Pergamon Press, Oxford, p 479-486.
Stadnik, Ye.V., I. Ya. Sklyarenko, I.S. Guliyev, A.A. Feyzullayev. 1986. Methane distribution in the atmosphere above tectonically different regions. Trans. (Doklady) *U.S.S.R. Acad. Sci., Earth Sci. Sect.,* 289:190-192.
Suess, E., G.J. Massoth. 1984. Evidence for venting of pore waters from subducted sediments of the Oregon continental margin. (Abs.) *EOS,* 65:1,089.
Trotsyuk, V.Y., V.I. Avilov. 1988. Disseminated flux of hydrocarbon gases from the sea bottom and a method of measuring it. Trans. (Doklady) *U.S.S.R. Acad. Sci., Earth Sci. Sect.,* 291:218-220.
Vidal, F.V., J.A. Welhan, V.N.V. Vidal. 1982. Stable isotopes of helium, nitrogen and carbon in a coastal submarine hydrothermal system. *J. Volcano Geother. Res.,* 12:101-110.
Vyshemirskiy, V.S., R.S. Khakimzyanova, V.F. Shugurov. 1989. A gas survey of snow cover in the Kuznetsk Basin. Trans. (Doklady) *U.S.S.R. Acad. Sci., Earth Sci. Sect.,* 309:172-174.
Watkins, J.S., J.L. Worzel. 1978. Serendipity gas seep area, South Texas offshore. *Am. Assoc. Petrol. Geol. (Bull.),* 62:1,067-1,074.
Welhan, J.A. 1988. Origins of methane in hydrothermal systems. In: *Origins of Methane in the Earth* (M. Schoell, ed.), *Chem. Geol.,* 71:183-198.
Welhan, J.A., H. Craig. 1983. Methane hydrogen and helium in hydrothermal fluids at 21 N on the East Pacific Rise. In: *Hydrothermal Processes at Seafloor Spreading Centers* (Rana et al., eds.), Plenum Press, New York, p 391-409.
Whalen, M., M. Tanaka, B. Henry, B. Deck, J. Zeglen, J.S. Vogel, J. Southon, A. Shemesh, R. Fairbanks, W. Broecker. 1989. Carbon-14 in methane sources and in atmospheric methane: the contribution from fossil carbon. *Science,* 245:286-290.

Wilson, R.D., P.H. Monaghan, A. Osanik, L.C. Price, M.A. Rogers. 1974. Natural marine oil seepage. *Science, 184*:857-865.

Winn, C.D., D.M. Karl, G.J. Massoth. 1986. Microorganisms in deep-sea hydrothermal plumes. *Nature, 320*:744-746.

Wood, T.G., R.A. Johnson, C.E. Ohiagu. 1977. Populations of termites in natural and agricultural ecosystems in southern guinea savanna near Mokwa, Nigeria. *Geol. Ecol. Trop., 1*:139-148.

Wood, T.G., R.A. Johnson, S. Bacchus, M.O. Shittu, J.M. Anderson. 1982. Abundance and distribution of termites in riparian forest near Rabba in the Southern Guinea savanna vegetation zone of Nigeria. *Biol. Trop., 14*:25-39.

Zor'kin, L.M., F.G. Dadashev, A.A. Dadashev, Krylova. 1985. Peculiarities of the isotopic concentration of methane from petrogas-condensate and gas condensate deposits of Azerbaijan. *Dan SSR, 280*:1,225-1,228.

Chapter 19

Group Report

Sources and Sinks of Methane

CHAIRPERSONS: D. BACHELET[1] AND H.U. NEUE[2]

[1]*ManTech Environmental Technology, Inc., US EPA Environmental Research Laboratory
200 SW 35th Street, Corvallis, Oregon 97333, USA.*

[2]*International Rice Research Institute, P.O. Box 933, 1099 Manila, Philippines.*

Group Members:
D. Augenstein, L. Beck, J. Bogner, B. Bonsang, B. Callander, J. Chappellaz, R. Charlier, Z-L. Chen, R. Delmas, R. Eisma, W-M. Hao, D. Johnson, A. Judd, D. Klass, H. Köse, A. Lacroix, G. Lambert, C-I. Lin, E. Matthews, M. Miah, N. Miller, C. Mitchell, A. Mosier, D. Olszyk, S. Piccot, C. Rouland, P. Sage, A. Smith, K. Smith, P. Soot, S. Thorneloe, H. Toprak, M. Torn, H. Tyrrell, D. Williams.

1. Introduction

Methane (CH_4) concentrations in the atmosphere have increased from about 0.75 to 1.7 ppmv since preindustrial times (Steele et al., 1987; Khalil and Rasmussen, 1990). The current annual rate of increase of about 0.8% year^{-1} is attributed to increases in industrial and agricultural emissions since some key natural sources (e.g., wetlands and marshes) have been reduced due to development pressure, decreasing their area in various parts of the world (see, for example, Lelieveld et al., 1993). We have tried, in this chapter, to concisely summarize the discussions that took place at the NATO-ARW and to quantify the size of the global "Methane Sources and Sinks" that may contribute to the atmospheric increase. Several "specialty" groups emerged during the workshop, and it is their conclusions that are presented here. Each paragraph is also the focus of an individual chapter and usually of several papers that have appeared in *Chemosphere Vol.* 26(1-4). We have tried to cite these documents in the relevant sections, and we refer the reader to these sources for detailed explanations of each source and sink.

2. Ruminants

The principal uncertainties of methane emission from enteric fermentation concern the number of animal species recorded by country, the amount and type of their daily diet, and the methane yield per diet energy, which is defined as the amount of methane produced as a percentage of the gross food energy intake of the animal (Gibbs et al., 1989). A methane yield of 9%, as has been applied to cattle in developing countries, was believed to be too high for global extrapolation. Six percent may be more realistic, but reliable measurements for various species and diets are needed, especially for buffaloes because of their global importance. Back calculations could be used to estimate methane emissions from required diets for measured animal production yields.

Lodman et al. (1993) estimated emissions from cattle manure in the U.S.A. at 0.6 Tg or approximately 10% of direct animal emission estimates. Minor amounts appear to arise from feedlot or grazing animal manure, major amounts from anaerobic lagoon disposal of manure. The greatest uncertainty is the lack of information on what fraction of manure from cattle and other species is disposed of by this means. Williams (1993) reported results from a study on Australian cattle where estimates of methane emissions from dairy cows need to be increased by about 5% to allow for partial disposal by anaerobic lagoon (cesspit) emissions. Again, manure deposited on pasture by grazing animals produced little methane.

More research is needed to elucidate antagonistic effects of acetogens and methanogens in their competition for available hydrogen in gastrointestinal tracts of insects and animals.

3. Rice cultivation

Rice-related global databases for water regimes and especially cultural practices are lacking. The effects of these factors on methane emission have to be reliably estimated and verified. Preliminary results using soil characteristics estimate global emissions from paddies to be as low as 25-60 Tg year^{-1} (Neue et al., 1990). Since yield or biomass data are more easily accessible, the possibility of using these data to estimate methane emission rates seems promising and should be explored further. It may then be possible to link rice growth models with models of methane fluxes. More baseline research is needed to explore the effects of root growth, root exudates, and the development and structure of the vascular system on methane fluxes. Because of the lack of baseline data, no process-oriented model has yet been developed to explain differences in emission rates.

It is essential to develop mitigation options that are in accord with the needed increase of rice production. Based on initial results from various researchers, a 20-30% reduction on methane emission from rice fields through changes in cultural practices and breeding of rice cultivars with a lower methane emission potential seems to be feasible concurrent with an increase in rice yield (Neue and Roger, this volume).

4. Biomass burning

Methane emission from biomass burning comes from clearing and agricultural practices in tropical ecosystems, forest fires in temperate areas, agricultural waste burning, and from domestic fires (fuelwoods, stoves, charcoal). The amount of methane emitted is generally calculated by multiplying the amount of biomass burnt by the efficiency of combustion (ratio between the amount of carbon dioxide released over the initial total carbon content, or CO_2/total carbon) and the emission factor of methane (ratio of the amount of methane emitted over the amount of carbon dioxide emitted, or CH_4/CO_2). Emission factors of methane have been measured for different ecosystems and fuel material. For savanna grass, a realistic emission factor is 0.3%, which is much lower than the commonly used factor of 1% (Bonsang et al., 1991). The emission factor for fuel wood can be as high as 2%. These factors are highly dependent on the kind of fuel and the efficiency of combustion (flaming and smoldering). The total amount of biomass burnt is estimated using biomass density, the fraction of above ground biomass burnt, and total area burnt each year. It is particularly difficult to determine the area burnt in savannas because the burning period lasts several months and fires are scattered.

To reduce the uncertainty associated with the global estimate of methane emissions due to biomass burning, each fuel source (e.g., grass or wood) and its respective emission factor was considered. Global estimates amount to 25 Tg year^{-1} with 6.1 Tg year^{-1} coming from tropical forest fires, 6.8 Tg year^{-1} from tropical savanna fires, 5 Tg year^{-1} from temperate and boreal forest fires, 5 Tg year^{-1} from charcoal burning, 1.8 Tg year^{-1} to 0.3 Tg year^{-1} from agricultural wastes burning. Estimates for fuelwood and charcoal emissions are still very uncertain due to the lack of reliable data from tropical regions.

5. Landfills

Little data exist to directly quantify methane emission rates from landfills and the factors that control them, such as moisture content. A better understanding of the rates of biochemical reactions and physical transport processes is needed. Since published landfill methane generation and emission rates vary over several orders of magnitude, more field measurements and supporting studies are required to improve emission estimates and to evaluate the effects of various landfill management practices (Bogner and Spokas, 1993; Nozhevnikova et al., 1993). In particular, additional studies addressing controls on rates of subsurface microbial methane oxidation are needed to evaluate net methane emissions.

Published estimates of worldwide landfill methane emissions vary between 9 and 70 Tg year^{-1} (Bingemer and Crutzen, 1987; Richards, 1989; Orlich, 1990; Thorneloe et al., this volume). These estimates were developed from calculations for worldwide refuse quantities, the fraction landfilled, and assumed methane generation/emission rates. Refuse quantities were based directly on refuse statistics, or, alternatively, they were derived indirectly from empirical relationships to general economic indicators (gross domestic product), economic production/consumption models, or from refuse statistics for countries with similar economies (per capita basis).

Current thinking favors an annual worldwide emission rate in the middle of the range cited above. Peer et al. (1993) discussed some of the key variable assumptions used to develop worldwide estimates. Augenstein and Pacey (1991) presented calculations for U.S. emissions, which they estimated to total 3-10 Tg year^{-1}. Calculations for the republics of the former USSR suggest a total of 1-2 Tg year^{-1} (Nozhevnikova et al., 1993). The major mitigative strategy for decreasing landfill methane emissions is increased methane recovery (for commercial use, if possible).

6. Mining and energy

The proportion of atmospheric methane produced from fossil carbon sources has been estimated by isotope studies. About 20% of the total annual emissions are from fossil carbon sources (coal mining operations as well as abandoned mines, natural gas production and use, oil exploring and production operation). Using country-specific data on coal characteristics and coal depth, Kirchgessner et al. (1993) developed regression equations to predict CH_4 emissions from underground coal mining. He estimated that 36 Tg year^{-1} were produced globally from underground mines in 1989. The reliability of this estimate is directly related to the quality of country-specific data and the issue is discussed at length by the authors. Kirchgessner et al. (1993) also presented a gross estimate of surface mine emissions, based on minimal data, of about 7 Tg year^{-1} for 1989. Data on emissions from coal handling operations and abandoned mines were judged too scarce to use as a basis to estimate their global contribution.

Mitchell (1993) presented estimates of coal bed methane emissions for the United Kingdom as reported by the US EPA (1.7 Tg, Boyer et al., 1990) and by British Coal (0.77 Tg, Creedy, 1993). She also reported her calculations of natural gas leakage rates in the United Kingdom. She found that rates were greater than 1.9% in contrast with the 1% rate reported by British Gas, and considered them in the range of 5-11% (Mitchell et al., 1990).

Beck (1993) presented an initial estimate of methane emissions from natural gas system leakage in the United States of 4.4 Tg year^{-1}. The data are the interim results of a research program being conducted by the US EPA and the Gas Research Institute. Methane emissions from natural gas systems are broken down according to emissions from production, transmission, and distribution losses, with further disaggregation by process.

7. Wetlands

Methane emissions from natural wetlands are estimated to be one of the largest source terms in the methane budget (~ 100 Tg year^{-1}). While temperature and water status are recognized as major controllers of methane flux in wetland habitats, the expanding suite of measurements confirms the importance of local ecology (topography, water table, organic matter content, vegetation), as well as the differential relationship between controls and flux among various wetland ecosystems. In addition, episodic events such as ebullition, degassing, and changes in hydrostatic pressure can account for the majority of seasonal methane emissions in both tropical and boreal wetlands. There is general agreement on the global area and distribution of wetlands (Matthews and Fung, 1987; Aselmann and Crutzen, 1989). Most major wetland environments have been measured; full season and multi-year measurements in a variety of habitats have confirmed and characterized some of the large spatial, seasonal, and inter-annual variability of fluxes. Isotopic, atmospheric, and emission measurements, as well as models, confirm the importance of low latitude wetlands in global emissions.

There are as yet no published measurements of methane flux from the Siberian lowlands (one of the largest wetland regions on earth) nor from the bog forests of Southeast Asia; measurements are sparse for African wetlands and absent for South American wetlands not associated with the Amazon River Basin. Uncertainties remain in characterizing seasonal cycles of methane-producing conditions in wetlands, in the annual magnitude of fluxes from extensive wetland environments not yet covered by full-season or inter-annual measurements, and in the response of wetlands to future climate change. Improvements in quantifying the extent and seasonal variability of some wetland environments can be achieved via remote sensing. For wetlands with poor measurement coverage, field campaigns during wet and dry seasons (tropical and subtropical), and throughout a growing season (temperate and boreal) will reduce the uncertainties in emission estimates. Expanded measurements in representative wetlands can improve the generalizations and extrapolations, made from available databases on wetland areas, distributions, and environmental characteristics, that form the basis for global emissions estimates. A recent compilation of flux measurements may be found in Bartlett and Harriss (1993).

8. Other sources

A detailed account of the potential importance of so-called minor sources can be found in Lacroix (1993). Judd et al. (this volume) discuss the role of hydrocarbon-bearing sediments, mud volcanoes, volcanic and related emissions, hydrothermal sources, crystalline basement, and deep sea earth gas. Biological sources such as termites and marine algae are examined. The importance of the oceans as a source of methane from seabed seepages, gas hydrates, and igneous methane sources is detailed. Minor anthropogenic methane sources are described in detail in Lacroix (1993) and Beck et al. (this volume).

8.1 Oceans. Previous estimates of the oceans' contribution to the global methane budget, based on a few supersaturation measurements by Lamontagne et al. (1973), have suggested that oceans play only a minor role in global methane emissions, contributing about 10 Tg CH_4 year^{-1}. These estimates referred to the role of ocean waters or of deep ocean geological sources and algal blooms. As new data have become available, these estimates can be somewhat refined. Open ocean measurements lead to estimates of 3.5 Tg year^{-1} while measurements from specific geographical areas (Gulf of Mexico, Black Sea, etc.) indicate that oceans emit about 6 Tg year^{-1}. However, the low supersaturation of the open ocean surface (about 30%) makes emissions very sensitive to changes in the atmosphere mixing ratio, and the open ocean could become a significant sink of methane in the future.

Methane is in fact emitted by various sources in the oceans, and the magnitude of these emissions is, at best, roughly estimated. For example, data are virtually missing to estimate emissions from marine macroalgae and algal blooms. Suitable conditions for the existence of large reservoirs of gas hydrates exist over vast areas of the ocean floor. Vast but unquantified volumes of methane are also thought to be generated by volcanic and hydrothermal activity in the deep oceans, but they are generally believed to become dissolved within the seawater and are usually oxidized before reaching the surface.

Seepage of biogenically- and thermogenically-derived methane on the continental shelves most likely contribute significant quantities to the atmosphere. On the world's continental shelves, methane is generated by bacterial activity near the seabed in organic-rich marine sediments and at great depth in petroleum-bearing sedimentary basins. Although methane seepages are known to be widespread and occur in large numbers in certain areas (e.g., Santa Barbara channel, California, and the Gulf of Mexico), a relatively small proportion of the world's continental shelves has been surveyed in detail, and much of the relevant data are now held in confidence by the petroleum industry. Seepage flow-rate measurements made in the North Sea have been used as the basis for a crude estimation of the global flux from the seabed of continental shelves, which amounts to 73 Tg CH_4 year^{-1}. The amount reaching the atmosphere is unknown. A more thorough discussion of this issue can be found in Judd et al. (this volume).

8.2 Termites. Methane is produced in the hindgut of termites, which contains an abundant and wide variety of anaerobic symbiotic microflora, particularly methanogenic bacteria. Methane production by termites is directly related to their abundance and to their mode of nutrition: one gram of wood-eating termites produces 0.01 to 0.016 μM CH_4 hour^{-1}, one gram of fungus-growing termites produces 0.4 to 0.7 μM hour^{-1}, and one gram of soil-feeding termites produces 0.7 to 1.1 μM CH_4 hour^{-1}. Accounting for the biomass of all the various dietary groups of termites in a given biotope, a potential methane production of 5.7 μM m^{-2} h^{-1} can be estimated for tropical forests, 7.7 μM m^{-2} h^{-1} for wet savannas, 4.2 μM m^{-2} h^{-1} for dry savannas, and 0.2 μM m^{-2} h^{-1} for temperate lands. These estimates were derived from in vitro measurements or from direct flux measurements in or over epigeous termite nests. There are no data available to estimate fluxes from hypogeous nests of soil-eating termites or to estimate the potential oxidation by the soil of methane escaping from the nests through several soil layers. Using Zimmerman et al.'s (1982) ecological regions, the global annual production of methane by termites can then be estimated to 27 Tg CH_4 per year. If all of the methane produced in hypogeous nests of soil-eating termites were oxidized during its diffusion through soil, emissions would decrease to 21 Tg per year. Because of the uncertainties attached to the estimation of population sizes of each dietary group, the possible range of emissions was determined to be between 10 and 35 Tg year^{-1}. The transformation of natural savanna or forest into agricultural land is currently greatly modifying termite populations (number and species) increasing the uncertainty of this estimate.

Table 1. Global methane emissions estimates from various sources (Crutzen, 1991; Cicerone and Oremland, 1988; Khalil and Rasmussen, 1983) and revised workshop estimates.

Sources/Sinks	Crutzen 1991	Cicerone and Oremland 1988	Khalil and Rasmussen 1983	Uncertainties	Revisions
Sinks					
OH[a]	420 ± 80				
Soils	30 ± 15				
Stratospheric Cl[b] and O[c]	10 ± 5				
Atmospheric loading	45 ± 5				
TOTAL	505 ± 105	535	553		
Biogenic Sources					
Ruminants	80	80	120	Buffaloes	15-25
Landfills	50 ± 20	40		Lack of data	75
Oceans and insects	30	10	13	Increased seepages; termites	15-35
Biomass burning	30 ± 15	55	25	Area burnt	25
Rice and wetlands	215 ± 50	110	95	Soil effect	25-60
		115	150	Siberia	
Other		45	50		
TOTAL	405 ± 85	455	451		
Anthropogenic Sources					
Coal mines	25 ± 5	35		Industry release	43
Hydrates	5				
Leaks in natural gas distribution systems, natural gas and oil wells	70 ± 15	45	40	Industry release	
Other			60		
TOTAL	100 ± 20	80	100		

[a] OH = hydroxyl; [b] Cl = chlorine; [c] O = oxygen; [d] increased seepages

9. Sinks

The main sink of methane is its reaction with OH radicals in the troposphere. The current annual removal of methane by OH is estimated to be 420 ± 80 Tg CH_4 (Crutzen 1991). The annual uptake by soils is estimated to be about 30 Tg. Photochemical removal in the stratosphere is about 10 Tg per year. This subject is further discussed in another chapter of this book.

10. Summary: revised estimates of the global methane budget

To conclude this chapter we have put together Table 1 including the revised methane emissions values for each of the known sources of methane.

Acknowledgements. This document has been prepared at the EPA Environmental Research Laboratory in Corvallis, Oregon, through contract #68-C8-0006 to ManTech Environmental Technology, Inc. It has been subjected to the Agency's peer and administrative review and approved for publication. Mention of trade names or commercial products does not constitute endorsement or recommendation for use.

References

Augenstein, D., J. Pacey. 1991. US landfill methane emissions estimates. Poster presentation, NATO-ARW, Timberline Lodge, OR, October 5-11, 1991.
Aselmann, I., P.J. Crutzen. 1989. Global distribution of natural freshwater wetlands and rice paddies, their net primary productivity, seasonality and possible methane emissions. *J. Atmos. Chem., 8*:307-358.
Bartlett, K.B., R.C. Harriss. 1993. Review and assessment of methane emissions from wetlands. *Chemosphere, 26* (1-4):261-320.
Beck, L. 1993. A global methane emissions program for landfills, coal mines and natural gas systems. *Chemosphere, 26* (1-4):447-452.
Bingemer, H.G., P.J. Crutzen. 1987. The production of methane from solid wastes. *J. Geophys. Res., 92*:2,181-2,187.
Bogner, J., K. Spokas. 1993. Landfill CH_4: Rates, fates, and role in global carbon cycle. *Chemosphere, 26* (1-4):369-386.
Bonsang, B., C. Boissard, U. Bassler, G. Lambert. 1991. Methane production by savanna fires. Poster presentation, NATO-ARW, Timberline Lodge, OR, October 5-11, 1991.

Boyer, C.M., J.R. Kelafant, V.A. Kuuskraa, K.C. Manger, D. Kruger. 1990. Methane emissions from coal mining: Issues and opportunities for reduction. U.S. EPA, Office of Air and Radiation, Washington D.C., *EPA 400/9-90/008*, 3pp.

Cicerone, R.J., R.S. Oremland. 1988. Biogeochemical aspects of atmospheric methane. *Global Biogeochem. Cycl.,* 2:299-327.

Crutzen, P.J. 1991. Methane's sinks and sources. *Nature, 350*:380-381.

Gibbs, M.J., L. Lewis, J.S. Hoffman. 1989. Reducing methane emissions from livestock: Opportunities and issues. U.S. EPA, Office of Air and Radiation, Washington DC, *EPA 400/1-89/002*, 50pp.

Khalil, M.A.K., R.A. Rasmussen. 1983. Sources, sinks, and seasonal cycles of atmospheric methane. *J. Geophys. Res., 83*:5,131-5,144.

Khalil, M.A.K., R.A. Rasmussen. 1990. Atmospheric methane: Recent global trends. *Environ. Sci. Tech.,* 24:549-553.

Kirchgessner, D.A., S.D. Piccot, J.D. Winkler. 1993. Estimate of global methane emissions from coal mines. *Chemosphere,* 26 (1-4):453-472.

Lacroix, A.V. 1993. Unaccounted-for sources of fossil and isotopically-enriched methane and their contribution to the emissions inventory: A review and synthesis. *Chemosphere,* 26 (1-4):507-558.

Lamontagne, R.A., J.W. Swinnerton, V.J. Linnenbon, W.D. Smith. 1973. Methane concentrations in various marine environments. *J. Geophys. Res.,* 78:5,317-5,324.

Lelieveld, J., P.J. Crutzen, C. Brühl. 1993. Climate effects of atmospheric methane. *Chemosphere,* 26 (1-4):739-768.

Lodman, D.W., M.E. Branine, B.R. Carmean, P. Zimmerman, G.M. Ward, D.E. Johnson. 1993. Estimates of methane emissions from manure of U.S. cattle. *Chemosphere,* 26 (1-4):189-200.

Matthews, E., I. Fung. 1987. Methane emission from natural wetlands: Global distribution, area, and environmental characteristics of sources. *Global Biogeochem. Cycl.,* 1:61-86.

Mitchell, C. 1993. Methane emissions from the coal and natural gas industries in the UK. *Chemosphere,* 26 (1-4):441-446.

Mitchell, C., J. Sweet, T. Jackson. 1990. A study of leakage from the UK natural gas distribution system. *Energy Policy* (November 1990):809-818.

Neue, H.U., P. Becker-Heidmann, H.W. Scharpenseel. 1990. Organic matter dynamics, soil properties and cultural practices in rice lands and their relationship to methane production. In: *Soils and the Greenhouse Effect* (A.F. Bouwman, ed.), John Wiley, New York, pp 457-466.

Nozhevnikova, A.N., V.S. Lebedev, A.B. Lifshitz, G.A. Zavarzin. 1993. Emission of methane from landfills into atmosphere in the USSR. *Chemosphere, 26* (1-4):401-418.

Orlich, J. 1990. Methane emissions from landfill sites and wastewater lagoons. In: *Proceedings International Workshop on Methane Emissions from Natural Gas Systems, Coal Mining, and Waste Management Systems,* April 9-13, 1990. U.S. EPA, Washington DC, pp 465-472.

Peer, R.L., S.A. Thorneloe, D.L. Epperson. 1993. A comparison of methods for estimating global methane emissions from landfills. *Chemosphere, 26* (1-4):387-400.

Richards, K. 1989. Landfill gas: Working with Gaia. *Biodeterioration Abstracts 3*:525-539.

Steele, L.P., P.J. Fraser, R.A. Rasmussen, M.A.K. Khalil, T.D. Conway, A.J. Crawford, R.H. Gammon, K.A. Masarie, K.W. Thoning. 1987. The global distribution of methane in the troposphere. *J. Atmos. Chem., 5*:125-171.

Williams, D.J. 1993. Methane emissions from manure of free-range dairy cows. *Chemosphere, 26* (1-4):179-188.

Zimmerman, P.R., J.P. Greenberg, S.O. Wandiga, P.J. Crutzen. 1982. Termites: a potentially large source of atmospheric methane, carbon dioxide and molecular hydrogen. *Science, 218*:563-565.

Chapter 20

The Role of Methane in the Global Environment

DONALD J. WUEBBLES AND JOHN S. TAMARESIS

Lawrence Livermore National Laboratory
7000 East Avenue / L-262, Livermore, CA 94550

1. Introduction

The increasing concentration of methane (CH_4) in the atmosphere is of concern because of the potential effects that it can have on global atmospheric chemistry and climate. Due to its relatively long atmospheric lifetime, methane emissions do not have an appreciable effect on local or regional air pollution. However, methane chemistry does have an important influence on the global atmosphere, affecting the amount of ozone (O_3) in both the troposphere and stratosphere, the amount of hydroxyl (OH) in the troposphere, and the amount of water vapor (H_2O) in the stratosphere. The oxidation of methane is an important source of atmospheric carbon monoxide (CO) and formaldehyde (CH_2O). Methane is the most abundant reactive trace gas in the troposphere. In addition, methane is a greenhouse gas, and its increasing concentrations are of interest to concerns about climate change.

This chapter discusses the effects of methane on atmospheric chemistry and climate. Discussion focuses on three major topics: effects on tropospheric chemistry, effects on stratospheric and upper atmospheric chemistry, and effects on climate.

2. Effects on tropospheric chemistry

2.1 Oxidizing capacity of the atmosphere. Photochemistry in the troposphere generates oxidants that are important to destroying many important gases emitted into the atmosphere. This self-cleansing feature of the atmosphere is called the oxidizing capacity. Several chemical species determine the oxidizing capacity of the troposphere. In descending order of reactivity, the most important oxidants

NATO ASI Series, Vol. I 13
Atmospheric Methane: Sources, Sinks, and Role in Global Change
Edited by M. A. K. Khalil

are hydroxyl, ozone, hydroperoxyl radical (HO_2), and organic peroxy radicals (RO_2). The most reactive oxidant is hydroxyl; it is responsible for the oxidation of the majority of the gases emitted into the atmosphere and is the primary scavenger for methane, most of the higher hydrocarbons (referred to as nonmethane hydrocarbons or NMHCs), carbon monoxide, methylchloroform (CH_3CCl_3), methyl chloride (CH_3Cl), methyl bromide (CH_3Br), hydrogen sulfide (H_2S), and sulfur dioxide (SO_2). Therefore, the atmospheric hydroxyl concentration determines the lifetime and hence the atmospheric abundance of these compounds. The reactions of these gases with hydroxyl generally result in the formation of compounds that can be removed by wet deposition. As will be discussed in detail later, ozone is the primary driver of photochemical processes that recycle gases because its photolysis controls hydroxyl formation. Hydroperoxyl radicals oxidize nitric oxide (NO) to nitrogen dioxide (NO_2), which allows ozone formation to occur via nitrogen dioxide photolysis. Nitric oxide is often the limiting species in this process.

The main factors that regulate the oxidizing capacity of the atmosphere are:

1. Species such as methane, higher hydrocarbons, carbon monoxide, and nitrogen oxides ($NO_x = NO + NO_2$) undergo chemical reactions that produce and destroy important oxidizing species.
2. Stratospheric ozone controls the penetration of ultraviolet radiation, which determines photochemical activity in the troposphere.
3. Climate characteristics such as atmospheric temperatures and humidity also affect photochemical activity in the troposphere.

The underlying phenomenon that links all of these processes is photochemistry. If the oxidizing capacity of the troposphere is perturbed, then this would produce a change in the levels of trace species. As an example, a decrease in oxidizing capacity implies that trace species would have longer residence times, permitting transport of pollutants over long distances and resulting in perturbations over remote regions.

As the most abundant organic species in the atmosphere, methane plays an influential role in determining the tropospheric oxidizing capacity. An important series of reactions is initiated via the reaction of methane with hydroxyl. This consumption of methane is so effective that 80–90% of the methane destruction occurs in the troposphere (Cicerone and Oremland, 1988). A rise in the background level of methane — due to growing emissions — can reduce hydroxyl, which would result in a further increase in the methane concentration. Therefore, a positive feedback exists in this chemical cycle, which could lead to an overall decrease in the oxidizing capacity of the troposphere. The relationship between methane and hydroxyl will be discussed further in the next section.

Is the oxidizing capacity of the atmosphere changing? Isaksen (1988) concludes that although changes have been observed in key atmospheric species, no direct evidence exists from the trace gas budgets that the oxidizing capacity of the atmosphere has changed on a global scale. Basically, sufficient measurements of the tropospheric concentrations of hydroxyl, ozone, and the other oxidants do not exist to determine whether the oxidizing capacity is changing. Model results produced by Lu and Khalil (1991) and Pinto and Khalil (1991) suggest that mean hydroxyl concentrations may have changed only slightly even though climatic conditions and trace gas (e.g., methane) levels vary tremendously between ice ages, interglacial epochs, and the present time. In both studies, the authors found that increases in hydroxyl destruction due to rising methane and carbon monoxide levels are offset by increases in the production processes. As a result of the increasing atmospheric emissions and concentrations of methane, carbon monoxide, and several other gases, it is likely that changes in the oxidizing capacity of the atmosphere have occurred but the magnitude of the perturbation will be regionally dependent and remains poorly understood.

Generally, changes in oxidizing capacity would be expected to vary in accordance with a location's proximity to major pollution sources. Also, Thompson et al. (1989) and Crutzen and Zimmermann (1991) note that, due to high concentrations of hydroxyl and ozone in the tropics, this region may be very important to future changes in global oxidant levels.

2.2 Methane, carbon monoxide, and hydroxyl. On a global scale, methane oxidation is one of the major reaction pathways affecting atmospheric concentrations of hydroxyl. Depending on nitric oxide levels, methane oxidation can be either a production or destruction process for odd hydrogen (OH +

HO_2). Thus, different chemically coherent (Thompson et al., 1989) regions can be distinguished on the basis of concentrations of nitrogen oxides. Polluted (high NO_x) environments where odd hydrogen is produced include the temperate zone of the Northern Hemisphere and planetary boundary layer of the tropics during the dry season. Unpolluted (low NO_x) environments where odd hydrogen is destroyed include marine areas, the free troposphere over the tropics, and most of the Southern Hemisphere (e.g., see measurements of McFarland et al., 1979, and Ridley et al., 1987). For odd hydrogen production to occur, nitrogen oxide concentrations must exceed 5-20 pptv (Fishman et al., 1979; Crutzen, 1988; Cicerone and Oremland, 1988; WMO, 1991). Under warm, humid conditions, nitrogen oxide levels must be substantially higher.

2.3 High NO_x areas. To illustrate this qualitative description, we will consider the details of the methane oxidation cycle. In areas characterized by high concentrations of nitrogen oxides, methane oxidation primarily occurs via the following mechanism:

$$CH_4 + OH = CH_3 + H_2O \quad (1)$$
$$CH_3 + O_2 + M = CH_3O_2 + M \quad (2)$$
$$CH_3O_2 + NO = CH_3O + NO_2 \quad (3)$$
$$CH_3O + O_2 = CH_2O + HO_2 \quad (4)$$
$$HO_2 + NO = OH + NO_2 \quad (5)$$
$$(2x) \quad NO_2 + h\nu = NO + O \ (\lambda \leq 400 \text{ nm}) \quad (6)$$
$$(2x) \quad O + O_2 + M = O_3 + M \quad (7)$$

$$\text{Net:} \quad CH_4 + 4 O_2 = CH_2O + H_2O + 2 O_3 \quad (8)$$

Initially, methane is attacked by hydroxyl to form a highly reactive methyl (CH_3) radical. Note that nitric oxide is oxidized to nitrogen dioxide twice in the mechanism via the reduction of a peroxy-radical to an oxy-radical. One step provides a pathway for methylperoxyl radical (CH_3O_2) to form methoxy radical (CH_3O), which in turn produces formaldehyde (CH_2O). The other step reduces hydroperoxyl to hydroxyl. These reactions generate nitrogen dioxide, which then photolyzes to provide the oxygen atom needed for ozone production. Summing the reactions algebraically, we find that the products of the net reaction are ozone and formaldehyde (CH_2O).

There are three reaction pathways for the oxidation of formaldehyde to carbon monoxide. The first is the direct photolysis of formaldehyde to carbon monoxide and molecular hydrogen:

$$CH_2O + h\nu = CO + H_2 \; (\lambda \leq 350 \text{ nm}) \qquad (9)$$

This reaction is independent of nitric oxide levels and occurs at wavelengths shorter than 350 nanometers (nm). The second also begins with photolysis but it has a different product channel:

$$CH_2O + h\nu = CHO + H \; (\lambda \leq 350 \text{ nm}) \qquad (10)$$

$$CHO + O_2 = CO + HO_2 \qquad (11)$$

$$H + O_2 + M = HO_2 + M \qquad (12)$$

$$(2x) \quad HO_2 + NO = OH + NO_2 \qquad (13)$$

$$(2x) \quad NO_2 + h\nu = NO + O \; (\lambda \leq 400 \text{ nm}) \qquad (14)$$

$$(2x) \quad O + O_2 + M = O_3 + M \qquad (15)$$

$$\text{Net:} \quad CH_2O + 4\,O_2 = CO + 2\,OH + 2\,O_3 \qquad (16)$$

The third pathway is initiated by hydroxyl attack:

$$CH_2O + OH = CHO + H_2O \qquad (17)$$

$$CHO + O_2 = CO + HO_2 \qquad (18)$$

$$HO_2 + NO = OH + NO_2 \qquad (19)$$

$$NO_2 + h\nu = NO + O \; (\lambda \leq 400 \text{ nm}) \qquad (20)$$

$$O + O_2 + M = O_3 + M \qquad (21)$$

$$\text{Net:} \quad CH_2O + 2\,O_2 = CO + H_2O + O_3 \qquad (22)$$

In the troposphere, Crutzen (1988) calculates that the averaged relative fractions of these three formaldehyde oxidation pathways are roughly 60–50 %, 20–25 %, and 20–30 %, respectively. Provided that sufficient formaldehyde survives atmospheric removal processes such as wet deposition to undergo further reaction, odd hydrogen and carbon monoxide are formed as end products of methane oxidation. Figure 1 provides a graphical summary of the methane oxidation pathway in high NO_x regions. The odd hydrogen species are circled to emphasize the point in the mechanism where they are either produced or consumed. Also, note the variety of loss processes that impact formaldehyde: photolysis, chemical reaction with hydroxyl, and heterogeneous removal.

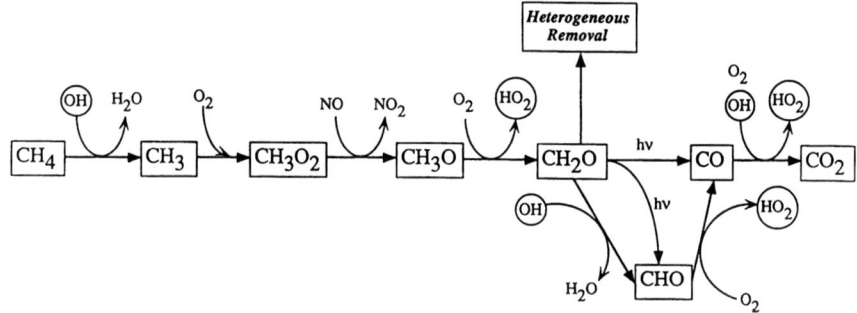

Figure 1. Methane oxidation pathway in the presence of sufficient nitrogen oxides. Odd hydrogen species are circled in order to indicate steps involving production or destruction.

2.4 Low NO_x areas. The photochemistry of methane in the unpolluted troposphere can be divided similarly into several reaction pathways. In these areas characterized by low levels of nitrogen oxides, methane oxidation occurs primarily by the following reaction sequence:

$$CH_4 + OH = CH_3 + H_2O \tag{23}$$

$$CH_3 + O_2 + M = CH_3O_2 + M \tag{24}$$

$$CH_3O_2 + HO_2 = CH_3O_2H + O_2 \tag{25}$$

$$CH_3O_2H + OH = CH_2O + OH + H_2O \tag{26}$$

$$\text{Net:} \quad CH_4 + OH + HO_2 = CH_2O + 2\,H_2O \tag{27}$$

Here we observe that hydroperoxyl reacts with methylperoxyl to form methyl peroxide (CH_3O_2H). In turn, methyl peroxide reacts with hydroxyl to produce formaldehyde. The methyl peroxide formed in this sequence also participates in a catalytic subcycle that consumes odd hydrogen:

$$CH_3O_2 + HO_2 = CH_3O_2H + O_2 \qquad (25)$$

$$CH_3O_2H + OH = CH_3O_2 + H_2O \qquad (28)$$

$$\text{Net:} \quad OH + HO_2 = H_2O + O_2 \qquad (29)$$

From this subcycle, we see that methyl peroxide can produce methylperoxyl radical. According to DeMore et al. (1990), the recommended branching ratios for the $CH_3O_2H + OH$ reaction are 70% ($CH_3O_2 + H_2O$) and 30% ($CH_2O + OH + H_2O$).

In addition to reaction (9), there are two more reaction pathways. One involves formaldehyde photolysis with a different product channel:

$$CH_2O + h\nu = CHO + H \ (\lambda \leq 350 \text{ nm}) \qquad (30)$$

$$CHO + O_2 = CO + HO_2 \qquad (31)$$

$$H + O_2 + M = HO_2 + M \qquad (32)$$

$$(2x) \quad HO_2 + O_3 = OH + 2\,O_2 \qquad (33)$$

$$\text{Net:} \quad CH_2O + 2\,O_3 = CO + 2\,OH + 2\,O_2 \qquad (34)$$

The other pathway is initiated by hydroxyl attack on formaldehyde:

$$CH_2O + OH = CHO + H_2O \qquad (35)$$

$$CHO + O_2 = CO + HO_2 \qquad (36)$$

$$HO_2 + O_3 = OH + 2\,O_2 \qquad (37)$$

$$\text{Net:} \quad CH_2O + O_3 = CO + H_2O + O_2 \qquad (38)$$

The branching ratios for these three pathways vary with altitude. In the troposphere (surface to 10 km), we calculate with the LLNL one-dimensional model that the range of relative fractions for these three formaldehyde oxidation paths are approximately 39%-60%, 22%-30%, and 39%-10%, respectively.

As a result of these mechanisms there is a net destruction of odd hydrogen by the oxidation of methane in low NO_x regions. Figure 2 provides a graphical summary of the methane oxidation pathway in unpolluted environments. The odd hydrogen species are circled to emphasize the point in the mechanism where they are either produced or consumed. Another interesting feature is the subcycle — reactions (25) and (28) — involving methylperoxyl radical and methyl peroxide. Two odd hydrogen species are destroyed each time the cycle turns over. Also, note that methyl peroxide and formaldehyde are affected by several loss processes. In particular, formaldehyde can be destroyed by photolysis, chemical reaction with hydroxyl, and heterogeneous removal.

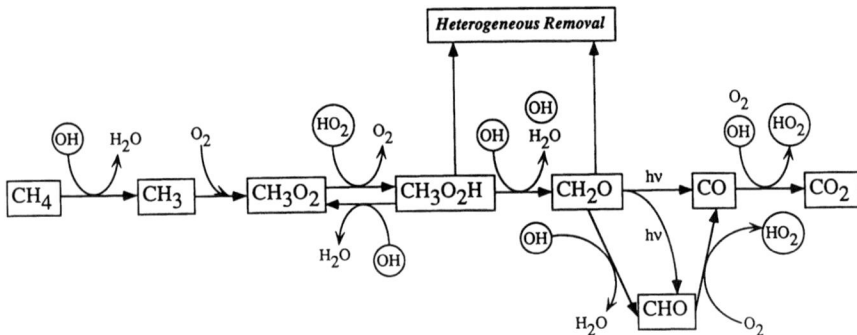

Figure 2. Methane oxidation pathway in the case of insufficient levels of nitrogen oxides. Odd hydrogen species are circled in order to indicate steps involving production or destruction.

The methane oxidation cycle is an important source of carbon monoxide, accounting for roughly a quarter of the carbon monoxide in the troposphere. Carbon monoxide concentrations are rather variable due to its relatively short atmospheric lifetime (approximately 1-3 months) and because of the variety of natural and anthropogenic sources that contribute to its budget. These sources include fossil fuel combustion, biomass burning in the tropics, and the oxidation of natural hydrocarbons (those emitted by vegetation such as isoprene). As in the case of methane, the carbon monoxide oxidation cycle also depends on the levels of nitric oxide present in the atmosphere. For polluted regions (high NO_x):

$$CO + OH = CO_2 + H \quad (39)$$
$$H + O_2 + M = HO_2 + M \quad (40)$$
$$HO_2 + NO = OH + NO_2 \quad (41)$$
$$NO_2 + h\nu = NO + O \;(\lambda \leq 400 \text{ nm}) \quad (42)$$
$$O + O_2 + M = O_3 + M \quad (43)$$

$$\text{Net:} \quad CO + 2\,O_2 = CO_2 + O_3 \quad (44)$$

As before, we note that hydroperoxyl oxidizes nitric oxide to form nitrogen dioxide, which in turn is photolyzed to produce the oxygen atom required for ozone formation. When nitrogen oxide levels are low, there is no net consumption of odd hydrogen:

$$CO + OH = CO_2 + H \quad (45)$$
$$H + O_2 + M = HO_2 + M \quad (46)$$
$$HO_2 + O_3 = OH + 2\,O_2 \quad (47)$$

$$\text{Net:} \quad CO + O_3 = CO_2 + O_2 \quad (48)$$

Regardless of the nitric oxide levels, one important result of carbon monoxide oxidation is the production of carbon dioxide (CO_2), which is a very important greenhouse gas.

Based on the reaction sequences for methane, formaldehyde, and carbon monoxide in polluted as well as unpolluted environments, the effect of the complete oxidation of one mole of methane on odd hydrogen is summarized in Table 1. According to our calculations, the catalytic subcycle involving reactions (25) and (28) has a sizeable impact on the amount of odd hydrogen loss in the low NO_x case. Crutzen included this cycle in his results whereas Cicerone and Oremland did not calculate its effect. Our values are substantially in agreement with Crutzen's for both cases; the slight discrepancy in the low NO_x case can be attributed to the use of different branching ratios. Cicerone and Oremland (1988) estimate that one to two moles of odd hydrogen are consumed under low

NO_x conditions; our results for the case without reaction (29) fall in the same range. They note that the resultant destruction of odd hydrogen will depend on the hydroxyl concentration, the methyl peroxide chemical reaction pathways, and the heterogeneous removal rates of important intermediate species.

Table 1. Estimated changes (in moles) of odd hydrogen (HO_x) and ozone resulting from the complete oxidation of one mole of methane. In this work, Crutzen's (1988) relative fractions for formaldehyde oxidation are used in the high NO_x case and the LLNL one-dimensional model is used to calculate these branching ratios for the low NO_x case. Results for this work are quoted as a range to emphasize that the relative fractions vary with altitude. Reaction (29) only affects odd hydrogen loss in the low NO_x case.

Reference	High NO_x		Low NO_x	
	ΔHO_x	ΔO_3	ΔHO_x	ΔO_3
Crutzen (1988)	+0.5	+3.7	-3.5	-1.7
Cicerone & Oremland (1988)	--	--	-1 to -2	--
This work w/o reaction (29)	+0.4 to +0.5	+3.6 to +3.8	-1.4 to -1.6	-1.7 to -1.8
This work w/reaction (29)	+0.4 to +0.5	+3.6 to +3.8	-3.7 to -3.9	-1.7 to -1.8

These reaction mechanisms are extremely important for the photochemistry of the unpolluted troposphere because methane and carbon monoxide are the main reaction partners of hydroxyl. Observational data from ground-based stations show that the global level of methane is increasing (primary references include Rasmussen and Khalil, 1981; Khalil and Rasmussen, 1983; Steele et al., 1987; Blake and Rowland, 1988; scientific reviews can be found in WMO, 1989; IPCC, 1990). The rate of increase for methane during the last decade is approximately 15-17 ppbv (~1%) per year (Khalil et al., 1989; Khalil and Rasmussen, 1990; Wallace and Livingston, 1990). Polar ice cores provide a record of the atmospheric concentration of methane over much longer time scales (i.e., present time to 100,000 years ago). Based on measurements from these cores, the mixing ratio of methane started to increase about 200 years ago (Craig and Chou, 1982; Khalil and Rasmussen, 1982, 1989). Determining the trend for carbon monoxide is difficult because it has a short tropospheric lifetime and the ground-based observational network is inadequate (Khalil and Rasmussen, 1984; WMO, 1991). Due to the likelihood of local contamination, the data are noisy and the trends may not be globally representative (Zander et al., 1989). For example, the observed trend for carbon monoxide in the Southern Hemisphere varies from year to year (WMO, 1991). An analysis by Khalil and Rasmussen (1988) indicates that hemispheric and global average concentrations of carbon monoxide show small increasing trends of about 1 ppbv per year. Their investigation also supports Zander et al.'s statement that the rate of change of carbon monoxide varies between different locations. Khalil and Rasmussen conclude that their data evaluation may indicate an increase in the global trend of carbon monoxide, but accurate determinations of the rate of change will require systematic data from a much longer period of time.

Another point to consider is the nonlinear nature of the coupling between methane, carbon monoxide, hydroxyl, ozone, and nitrogen oxides. A greatly simplified form (Thompson et al., 1989) of the photochemical steady-state equation for hydroxyl concentration is:

$$[OH] = \frac{2k[O(^1D)][H_2O]}{k[CO] + k[CH_4] + k[NMHC]}$$

where k = chemical reaction rate constant; [X] = species concentration.

In the unpolluted troposphere, an increase in carbon monoxide, methane, and non-methane hydrocarbons decreases hydroxyl. At higher concentrations of nitrogen oxides, ozone formation increases with methane and carbon monoxide and feeds back to form hydroxyl via the reaction between excited oxygen atoms ($O(^1D)$) and water vapor; the numerator and denominator both increase in this case. As the nitrogen oxide levels get even higher, hydroxyl again decreases as carbon monoxide is added. At any given time a perturbation in emissions of one trace gas produces a change that depends on the ambient levels of other species. This implies that a particular species may not be perturbed in the future the same way that it has been in the past. Therefore, we see that predictions of ozone and hydroxyl become very difficult when emission rates change.

Interest in the chemical cycles of trace species has elicited much research into the effects of methane perturbations. Several studies (e.g., Khalil and Rasmussen, 1985; Levine et al., 1985; Thompson and Cicerone, 1986a, b; Thompson and Kavanaugh, 1986; Hough, 1991) have concluded that the increasing mixing ratio of methane may be caused by both its increasing emission rate and the downward perturbation of the tropospheric hydroxyl concentration. Khalil and Rasmussen (1985) conclude that some depletion of hydroxyl — on the order of 20% — is likely to have occurred over the past 100-200 years. They suggest that most of the increase in methane (about 70%) is due to the increase in emissions from anthropogenic sources. Recent one-dimensional model calculations by Pinto and Khalil (1991) and Lu and Khalil (1991) also suggest that changes in OH have played only a minor role in explaining the increases in methane over the last few centuries.

Anthropogenic emissions of methane, nitrogen oxides, carbon monoxide, and non-methane hydrocarbons in the Northern Hemisphere appear to have increased substantially over the last several decades, with corresponding increases in their concentrations (Isaksen, 1988; Penkett, 1988; WMO, 1989, 1991; IPCC, 1990). Liu et al. (1988) argues that there should have been a decrease of hydroxyl levels in the Southern Hemisphere due to increases in methane and carbon monoxide. They attribute this to low nitrogen oxide emissions. It is not possible to imply such trends of hydroxyl in the Northern Hemisphere. Unfortunately, only a few measurements of hydroxyl exist and they are tainted by uncertainty. Existing hydroxyl measurements only provide an indication of local hydroxyl levels and have not yet provided useful information about global distributions and their trends.

Model studies of the relationship of hydroxyl to other species provide the best alternative to understanding its atmospheric behavior provided that the chemistry is well understood. Isaksen (1988) determined that the hydroxyl distribution calculated in atmospheric models agrees reasonably well with indirect determinations obtained from measurements of trace species oxidized by hydroxyl. Thompson et al. (1989), using a one-dimensional model of global tropospheric chemistry, show that increasing carbon monoxide and methane emissions, while holding nitrogen oxide levels constant, will decrease hydroxyl and increase ozone in all remote regions they have analyzed. Isaksen and Hov (1987) use their two-dimensional model to show that hydroxyl levels exhibit a strong seasonal and latitudinal variation. Based on their calculated hydroxyl distribution, the oxidation rate also changes dramatically with season and latitude. The largest changes are evident at latitudes greater than 50° in the winter hemisphere. Over the year, hydroxyl levels change by an order of magnitude and even methane varies noticeably. Isaksen and Hov estimate that global hydroxyl levels may have changed only slightly in recent decades. Lu and Khalil (1991) and Pinto and Khalil (1991) reached similar conclusions. However, it is difficult to judge the extent of the perturbation until the distributions and trends of other relevant constituents (e.g., nitrogen oxides) are better understood. Ehhalt et al. (1991) generally find a consistency between their limited hydroxyl measurement database and regional model calculations.

Estimates of future trends in hydroxyl concentrations have been obtained from one-dimensional and two-dimensional atmospheric models. Thompson et al. (1989) studied future atmospheres characterized by higher methane levels and increased anthropogenic emissions of carbon monoxide and nitric oxide with a one-dimensional model. Specifically, they found that most remote regions will lose hydroxyl as a result of increasing carbon monoxide and methane if nitrogen oxide levels remain about the same. They state that this should be the case in the mid- and higher-latitude Southern Hemisphere, which is mostly ocean, and perhaps over large regions of the tropics where 72% of the 60°N-60°S total of hydroxyl is found. In addition, near areas where biomass burning takes place, emissions of carbon monoxide, nitric oxide, methane, and non-methane hydrocarbons may suppress hydroxyl.

In a similar effort, Thompson et al. (1990) used a one-dimensional model to calculate future changes in tropospheric ozone and hydroxyl due to carbon monoxide, methane, and nitrogen oxide emissions for chemically coherent regions during the years from 1985 to 2035. One of the scenarios studied was an increase in methane mixing ratio of 0.8% per year and carbon monoxide mixing ratio of 0.5% per year. They found that initially there was an increase in hydroxyl for a

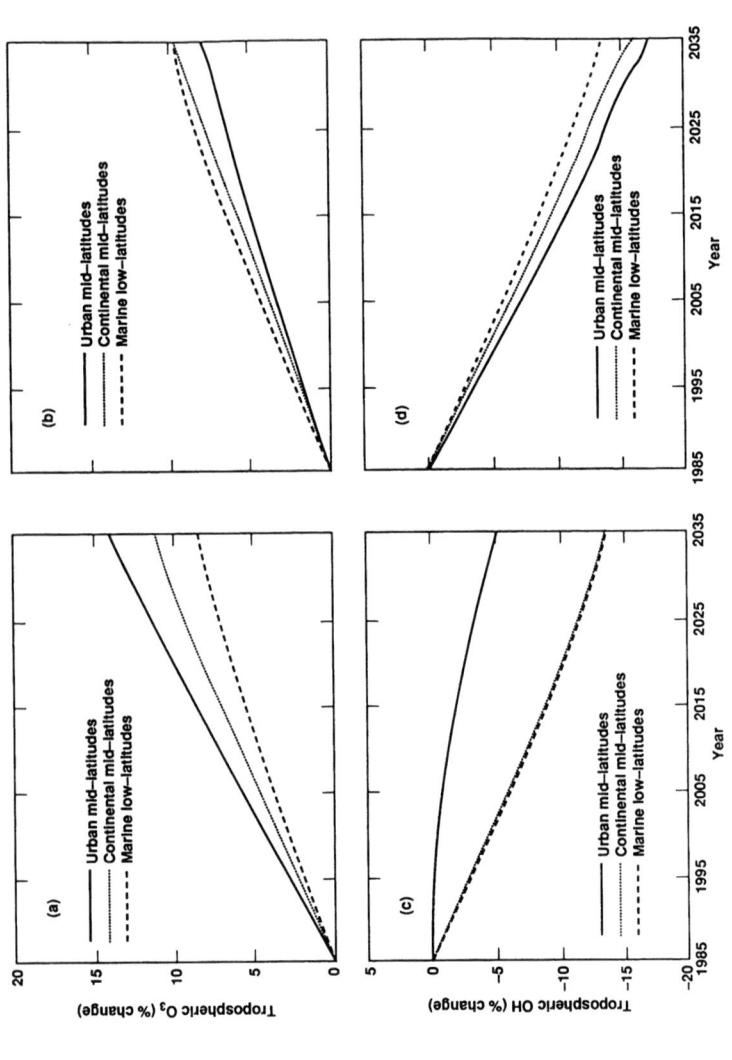

Figure 3. Calculated percentage change in tropospheric ozone (a and b) and hydroxyl (c and d) during the years 1985 to 2035 for an increase in the methane mixing ratio of 0.8% per year. Parts (a) and (c) depict urban as well as continental mid-latitudes and marine low-latitude regions. Parts (b) and (d) are for marine as well as southern hemisphere mid-latitudes and continental low-latitudes (Thompson et al., 1990).

few years in the urban mid-latitude region followed by a decrease, resulting in only a few percent change by the end of the simulation (see Figure 3). In all the lower nitrogen oxide regions, hydroxyl losses were dominant and regions with the smallest ozone increases had the greatest loss of hydroxyl. They found that urban and nonurban regions may respond quite differently to chemical and climatic perturbations which is due to differences in nitrogen oxide levels. Also, the effects of large-scale stratospheric ozone change or other chemical changes in the troposphere may counteract the influences of increased methane and carbon monoxide emissions. Isaksen and Hov (1987) used their two-dimensional tropospheric model, which has a domain from the surface up to 14 km, to examine a scenario for a 1.5% per year increase in the emission of methane for the years 1950–2010. They found that this gave rise to an average increase in ozone of 0.45% per year while the hydroxyl radical concentration dropped by about 0.4% annually.

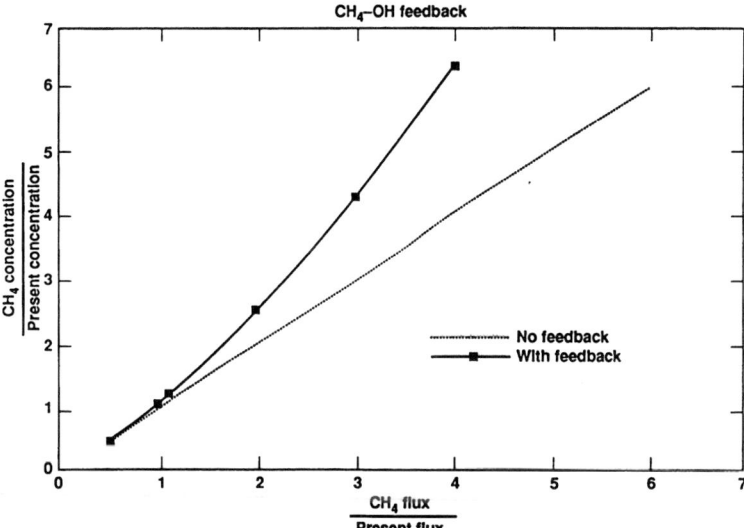

Figure 4. Calculated changes in average global mean concentration of CH_4 as a function of changes in fluxes. The dotted line represents no feedback between OH and CH_4 (from WMO, 1991).

As an indication of the possible importance of the relationship between methane and hydroxyl, Figure 4 shows calculated changes in global mean concentration of methane as a function of changes in the methane flux (relative to the present emissions). In these model calculations, only the methane flux was allowed to change. The importance of hydroxyl feedback is demonstrated by the non-linear response in the calculated methane concentration for a change in the surface flux. As described above, the actual response would depend on the changes occurring in emissions and concentrations of other atmospheric constituents as well.

3. Background on tropospheric ozone

Approximately 10% of the ozone in the atmosphere is located in the troposphere. The downward transport of ozone from the stratosphere traditionally was thought to be the major source of tropospheric ozone (Crutzen, 1988, and references therein). It is now generally regarded that the net tropospheric photochemical production of ozone is of similar magnitude to the downward transport source (e.g., Fishman et al., 1979; Fishman, 1985; WMO, 1985; Isaksen, 1988; Penkett, 1988; Hough and Derwent, 1990). Further support is lent to this notion from model calculations cited by Isaksen (1988) and Penkett (1988) that show the column-integrated photochemical production of ozone is of the same order of magnitude as the stratospheric ozone flux.

Nitrogen dioxide photolysis is the only known photochemical mechanism for producing ozone in the troposphere. This implies that the generation rate is roughly proportional to the concentration of nitric oxide. High concentrations of nitrogen oxides over the continental boundary layer signify that this region is likely a net source of ozone. Increases in nitric oxide emissions may lead to further ozone increases, especially in the tropics. However, the magnitude of odd oxygen $(O + O(^1D) + O_3)$ production is ultimately limited by the supply of carbon monoxide, methane, and nonmethane hydrocarbons. The oxidation of one mole of carbon monoxide molecules can form one mole of ozone. In contrast, the complete oxidation of a mole of methane can produce 3 to 4 moles of ozone. Because there are insufficient nitrogen oxides present in the background troposphere, only about 10% of the potential tropospheric ozone production is being realized.

Ozone can be removed from the troposphere by dry deposition and photochemical mechanisms. Dry deposition occurs when an atmospheric constituent is transferred directly to a surface such as leaves or soils. Photochemical ozone destruction in the troposphere occurs through a variety of processes. The two primary mechanisms are photolysis and reaction with hydroperoxyl. According to the studies performed by Fishman et al. (1979) and Hough and Derwent (1990), these two reactions account for most (i.e., 70-90%) of the destruction of ozone in the troposphere. Ozone also can react with hydroxyl, but this chemical sink plays a minor role based on the calculations in Fishman's paper. In addition, the rate of loss of ozone is almost independent of nitrogen oxide levels for concentrations below 200 parts per trillion by volume, pptv (Fishman et al., 1979; WMO, 1985). Oceans or regions of the world characterized by low nitric oxide concentrations are probably a net photochemical sink of odd oxygen (Liu et al., 1983; WMO, 1985).

Ozone is the primary driver of the photochemistry that recycles gases emitted from natural and anthropogenic sources because its photolysis regulates hydroxyl formation. Initially, ozone is photolyzed by ultraviolet light:

$$O_3 + h\nu = O(^1D) + O_2 \ (\nu \leq 310nm) \qquad (49)$$

Most of the excited oxygen atoms are collisionally deactivated by nitrogen and oxygen (collectively represented as M) back to the ground state:

$$O(^1D) + M = O + M \qquad (50)$$

The remainder react with water vapor to form hydroxyl:

$$O(^1D) + H_2O = 2 \ OH \qquad (51)$$

This reaction constitutes the primary tropospheric source of hydroxyl radicals. Due to this interconnection, the present composition of the earth's atmosphere would be totally different if ozone were not present because hydroxyl removes many trace gases from the troposphere.

4. Methane and tropospheric ozone

The methane oxidation pathways regulate the tropospheric ozone budget. For this analysis, we will sum the net reactions and take the formaldehyde branching ratios into account to get a stoichiometric relation involving methane and carbon dioxide. In regions characterized by high nitric oxide concentrations, the net methane oxidation reaction is:

$$CH_4 + 7.2\text{-}7.5\ O_2 = CO_2 + 0.4\text{-}0.5\ OH + 3.6\text{-}3.8\ O_3$$
$$+ 1.2\text{-}1.4\ H_2O + 0.5\text{-}0.6\ H_2 \qquad (52)$$

We see that complete methane oxidation results in net production of 3.6–3.8 moles of ozone in the polluted troposphere. For areas with low nitric oxide levels, the net methane oxidation reaction is:

$$CH_4 + 1.5\text{-}1.7\ OH + 2.2\ HO_2 + 1.7\text{-}1.8\ O_3 =$$
$$CO_2 + 3.3\text{-}3.6\ H_2O + 3\ O_2 + 0.4\text{-}0.6\ H_2 \qquad (53)$$

In this case, complete methane oxidation results in the net destruction of approximately 1.7–1.8 moles of ozone. The effect of methane on tropospheric ozone is summarized in Table 1.

The formation of tropospheric ozone occurs primarily through smog formation-type mechanisms that involve peroxy radicals, including hydroperoxyl, methylperoxyl, or any complex organic peroxy radical. Recall that formation of tropospheric ozone occurs because nitric oxide reacts with peroxy radicals (illustrated with hydroperoxyl) to form nitrogen dioxide:

$$HO_2 + NO = OH + NO_2 \qquad (41)$$

Nitrogen dioxide can photolyze to produce an oxygen atom and nitric oxide. The oxygen atom then combines with molecular oxygen to form ozone:

$$NO_2 + h\nu = NO + O\ (\lambda \leq 400\ nm) \qquad (42)$$
$$O + O_2 + M = O_3 + M \qquad (43)$$

The nitric oxide molecule can react again with peroxy radicals to repeat this process. Thus, nitrogen oxides act as a catalyst to produce ozone in the troposphere. At the low nitric oxide levels characteristic of the remote troposphere, peroxy radicals react directly with ozone to destroy it:

$$HO_2 + O_3 = OH + 2\ O_2 \qquad (47)$$

Current data show that the rate constant for reaction (41) is about 4000 times greater than that for reaction (47). Therefore, the formation cycle dominates the destruction cycle if the concentration ratio of nitric oxide to ozone exceeds 1:4000. For example, model results cited by Fishman et al. (1979), Crutzen (1988), and Cicerone and Oremland (1988) indicate that the nitric oxide concentration must be greater than 5-10 pptv in the lower troposphere and above 20 pptv in the upper troposphere for this to happen.

Over the last several years many researchers have focused their attention on global changes in tropospheric ozone. The studies performed to date show that the effects of nitrogen oxide levels are crucial, particularly where their concentrations are low: hydroxyl concentrations tend to decrease, hydroperoxyl and hydrogen peroxide increase, and ozone appears to be insensitive to methane changes. Isaksen (1988) and Penkett (1988) have arrived at several conclusions based on their assessment:

1. Nitrogen oxides, carbon monoxide, methane, and nonmethane hydrocarbons are the major ozone precursors. Nitrogen oxides are usually the limiting species due to their short lifetime and widely scattered sources.
2. Ozone produced from anthropogenic nitrogen oxide emissions is a significant contributor to the budget in the Northern Hemisphere and is the dominant term in the budget of the planetary boundary layer over industrialized areas.
3. Ozone concentrations measured in the boundary layer over industrialized areas during the last forty years exhibit increasing trends that range from 20-100%. Northern Hemispheric observations made at remote stations and in the free troposphere also show an increase of 0.5-2% per year during the past several decades. An increase greater than a factor of two is obtained by comparing observations made during the end of the nineteenth century at a station near Paris to present-day observations made at stations with similar ambient conditions (Bojkov, 1986; Volz and Kley, 1988).
4. The observed seasonal variations of surface ozone and peroxyacetyl nitrate tend to indicate that the spring ozone maximum in the Northern Hemisphere could be influenced significantly by photochemical production during the winter and spring.

Thompson et al. (1989) concluded from their modelling study that nitric oxide, carbon monoxide, and methane emission increases will suppress hydroxyl and increase ozone. These trends may be opposed by stratospheric ozone depletion and climate change. Stratospheric ozone depletion would tend to decrease ozone (except where nitrogen oxide levels are high) and increase hydroxyl through enhanced ultraviolet photolysis. Increased levels of water vapor (one possible result of climate change) also would decrease ozone and increase hydroxyl. Liu et al. (1988) have come to the conclusion that tropospheric ozone has increased in the Northern Hemisphere. Crutzen (1988) also suspects that ozone and hydroxyl concentrations are decreasing in clean atmospheric environments and increasing at midlatitudes in the Northern Hemisphere. He cites a previous study that indicates average surface ozone concentrations at remote locations are changing in the expected directions.

Long-term ozonesonde balloon measurement programs are in operation at a small number of ground stations (Tiao et al., 1986) located at northern midlatitudes (there is also one in the southern hemisphere). Ozone soundings at these sites mostly began in the early 1970s, although some started in the late 1960s (Tiao et al., 1986). By analyzing the ozone time series from the beginning of the data sets and averaging over all the stations, a statistically significant increasing trend in the tropospheric ozone concentrations at northern midlatitudes is observed (Logan, 1985; WMO, 1985, 1989, 1991; Tiao et al., 1986; Lacis et al., 1990; Miller et al., 1992). The ozone increase from these measurements is largest in the lower troposphere, with ozone amounts increasing about 8% per decade in the lowest kilometer of the atmosphere. However, the limitations (e.g., sparsity of the stations, sign of the tropospheric trend differs for each station) are such that these data sets do not necessarily provide conclusive evidence for a global tropospheric ozone increase.

5. Connections with other cycles

Regardless of the nitric oxide levels, methane oxidation is responsible for much of the formaldehyde in the atmosphere. Based on the estimate of Lowe and Schmidt (1983), the formaldehyde production rate is on the order of 100 Tg per year. The other major photochemical precursors of formaldehyde consist of nonmethane hydrocarbons and higher aldehydes. The nonmethane hydrocarbons are emitted from natural and anthropogenic sources while the higher aldehydes are produced by in situ photochemistry. There are appreciable anthropogenic sources of formaldehyde as well (e.g., automobile exhaust).

In order to analyze the photochemical mechanisms that produce formaldehyde, we must distinguish between different regions on the basis of nitrogen oxide levels. As we can see from the methane oxidation mechanism under low nitrogen oxide conditions, methyl peroxide reacts with hydroxyl to produce formaldehyde:

$$CH_3O_2H + OH = CH_2O + OH + H_2O \qquad (26)$$

Since methyl peroxide has a long lifetime against photolysis, heterogeneous processes (which occur on aerosols) can serve as a loss mechanism for this species in unpolluted environments. This implies that formaldehyde production is controlled by aerosol concentrations in remote regions. However, in polluted environments formaldehyde formation does not involve methyl peroxide. In this case, it would be regulated by nitric oxide concentrations because methylperoxyl is reduced to methoxyl. Methoxyl then reacts with oxygen to produce formaldehyde:

$$CH_3O_2 + NO = CH_3O + NO_2 \qquad (54)$$

$$CH_3O + O_2 = CH_2O + HO_2 \qquad (55)$$

Once it is generated, formaldehyde is destroyed rather quickly under direct insolation. During the daytime, photolysis is the dominant loss process for formaldehyde in the remote troposphere while in polluted atmospheres reaction with hydroxyl and hydroperoxyl will be comparable to photolytic degradation. At night, the only appreciable formaldehyde consumption is by reaction with nitrate radical (NO_3), but this is slow in comparison to the daytime reactions with hydroxyl and hydroperoxyl. In summary, formaldehyde is an important intermediate in the removal processes of hydrocarbons as well as the general chemical reactivity of the troposphere.

6. *Effects on upper atmospheric chemistry*

Although about 85% of the total emissions of methane is consumed by reaction with tropospheric hydroxyl (Cicerone and Oremland, 1988), the remaining methane flux, on average about 60 teragrams of methane per year (Tg CH_4/year), enters the stratosphere. In the stratosphere and above, the reaction with OH continues to be the dominant sink, but reactions with chlorine atoms and excited oxygen atoms are also important.

6.1 Background on stratospheric ozone. In order to put the role of methane in stratospheric chemistry into context, it is useful to first discuss the importance of ozone and the changes occurring in its distribution. Changes in the distribution and amount of ozone in the global troposphere and stratosphere have received much attention. Much of the concern about ozone has centered on the importance of ozone as an absorber of ultraviolet radiation; its concentrations determine the amount of ultraviolet radiation reaching the Earth's surface. Absorption of solar radiation by ozone also explains the increase in temperature with altitude found in the stratosphere. Finally, ozone is also a greenhouse gas and can influence climate.

Approximately 90% of the ozone in the atmosphere is contained in the stratosphere. In the stratosphere, the production of ozone begins with the photodissociation of oxygen (O_2) at ultraviolet wavelengths less than 242 nm. This reaction produces two ground-state oxygen atoms that can react with oxygen to produce ozone. Since an oxygen atom is essentially the same as having an ozone, it is common to refer to the sum of the concentrations of ozone and oxygen atoms (both ground state and excited state) as odd oxygen. The primary destruction of odd oxygen in the stratosphere comes from catalytic mechanisms involving various free radical species. Nitrogen oxides, chlorine oxides, and hydrogen oxides participate in catalytic reactions that destroy odd oxygen. The odd nitrogen (or nitrogen oxides) cycle above is believed to be responsible for about 70% of the total odd oxygen destruction (but depends on uncertainties in the importance of heterogeneous chemistry on stratospheric particles and aerosols). While most of the nitrogen oxides in the stratosphere are naturally occurring, the increasing concentration of nitrous oxide is leading to increased amounts of nitrogen oxides and increased effectiveness of the catalytic reactions that destroy stratospheric ozone.

The chlorine catalytic mechanism is particularly efficient. Because of the growing levels of reactive chlorine in the stratosphere resulting from emissions of trichlorofluoromethane ($CFCl_3$), difluorodichloromethane (CF_2Cl_2), and other halocarbons, this mechanism has been the subject of much study due to the potential effects on concentrations of stratospheric ozone. The chlorine catalytic cycle can turn over thousands of times before the catalyst is converted to a less reactive form. Because of this cycling, relatively small concentrations of reactive chlorine can have a significant impact on the amount and distribution of ozone in the stratosphere. The total amount of chlorine in the current stratosphere is about 3 parts per billion by volume (ppbv), much of which is in the form of less reactive compounds like hydrochloric acid (HCl).

Methane plays an important role in the chlorine chemistry of the stratosphere, serving both as a source and a sink in key reactions affecting reactive chlorine. The direct reaction of methane with a chlorine atom is the primary source of hydrochloric acid, the primary chlorine reservoir species. However, hydroxyl produced through the oxidation of methane in the stratosphere can react with the hydrochloric acid to return the chlorine atom, thus reinitiating the chlorine catalytic mechanism.

In addition to being involved in the reaction taking reactive chlorine to the less reactive hydrochloric acid, methane has several other effects on stratospheric ozone. Hydrogen oxides produced from the dissociation of methane can react catalytically with ozone, particularly in the upper stratosphere. In the lower stratosphere, the primary effect of these hydrogen oxides is to react with nitrogen oxides and reactive chlorine, reducing the effectiveness of the ozone destruction catalytic cycles involving nitrogen oxides and chlorine oxides.

Measurements of ozone from ground-based stations and from satellites indicate that concentrations of ozone in the stratosphere are decreasing. Ozone at 3 millibars (mbars), about 40 km altitude, is decreasing globally by 3-4% per decade, in good agreement with the model calculations of the expected effects from chlorofluorocarbons (CFCs) and other trace gas emissions (WMO, 1989; DeLuisi et al., 1989; Wuebbles et al., 1991a). Satellite and surface-based measurements of the total ozone column indicate that stratospheric ozone is decreasing throughout much of the world, with the effects increasing with latitude in both hemispheres (WMO, 1989, 1991; Stolarski et al., 1991). The ozonesonde and the SAGE satellite data sets (WMO, 1991; Miller et al., 1992) suggest that a significant fraction of this ozone change is occurring in the lower stratosphere and also indicate that ozone in the lower stratosphere is decreasing at a faster rate than can be explained by current theory. Part of this lower stratospheric ozone decrease can be explained by the dilution of the Antarctic ozone hole after its late springtime breakup, but the rest of the lower stratospheric ozone decrease is still not understood. Recent studies (WMO, 1991) suggest that heterogeneous chemistry on stratospheric sulfate aerosols may at least partially explain this discrepancy. The role of increasing atmospheric methane in the changes occurring in the ozone distribution is poorly understood.

Beginning in the late 1970s, a special phenomenon began to occur in the springtime over Antarctica, referred to as the Antarctic ozone "hole" (Solomon, 1988). A large decrease in total ozone is occurring over Antarctica beginning in early spring. Decreases in total ozone column of more than 50% as compared to historical values have been observed by both ground-based and satellite techniques. Measurements made in 1987 indicated that more than 95% of the ozone over Antarctica at altitudes from 15 to 20 km had disappeared during September and October (WMO, 1989). The Antarctic ozone hole was smaller in 1988 than in 1987. In general, the odd years appear to have larger ozone decreases than the alternate years; dynamical effects resulting from the quasi-biennial oscillation seem to explain these variations (Garcia and Solomon, 1987). The Antarctic ozone holes in 1989 and in 1990 were similar in magnitude to that in 1987.

Measurements also indicate that the unique meteorology during the winter and spring over Antarctica sets up special conditions producing a relatively isolated air mass (the polar vortex). Polar stratospheric clouds form if the temperatures are cold enough in the lower stratosphere, a situation that often occurs within the vortex over Antarctica. Heterogeneous reactions can occur between atmospheric gases and the particles composing these clouds. Measurements indicate that reactions of hydrochloric acid and chlorine nitrate ($ClONO_2$) on these particles can release reactive chlorine once the sun appears in early spring. Thus, the reactions on the cloud particles allow chlorine to be in a very reactive state with respect to ozone. The ozone hole ends in late spring with the breakup of the vortex. The weight of scientific evidence strongly indicates that man-made chlorinated (produced from CFCs) and brominated chemicals are primarily responsible for the substantial decreases of stratospheric ozone over Antarctica in springtime (WMO, 1989, 1991). A similar phenomenon may happen, but probably to a lesser degree, in the arctic region of the Northern Hemisphere.

7. Calculated effects of methane on ozone

As indicated above, the increasing atmospheric concentrations of methane, carbon dioxide, carbon monoxide, nitrous oxide, and various chlorinated and brominated compounds are all thought to be affecting the distribution of ozone in the troposphere and stratosphere. In recent years, there have been a number of research studies using numerical models to examine the combined effects on ozone from the increases occurring in concentrations of methane and the other gases listed above (Wuebbles et al., 1983, 1991b; WMO, 1985, 1988, 1989; Stordal and Isaksen, 1987). When combined with the effects of the other trace gas emissions, it is difficult to evaluate the role of methane in the observed and projected ozone trends. For this reason, it is useful to examine studies that have only considered the effects of increasing methane on ozone.

Numerical models of atmospheric chemical and physical processes generally calculate that increasing methane concentrations result in a net ozone production in the troposphere and lower stratosphere and net ozone destruction in the upper stratosphere (Owens et al., 1982, 1985; WMO, 1985, 1991; Isaksen and Stordal, 1986). The net effect from these calculations has been that methane by itself causes a net increase in ozone. For a doubling of the methane concentration (early papers went from 1.6 to 3.2 ppmv, while recent analyses assume 1.7 to 3.4 ppmv), published effects on the calculated change in total ozone range from +0.3% (Prather, in WMO, 1985) to +4.3% (Owens et al., 1985). With radiative feedback effects included, the published model results tend to be in the upper end of this range (Owens et al., 1985; WMO, 1985; Isaksen and Stordal, 1986).

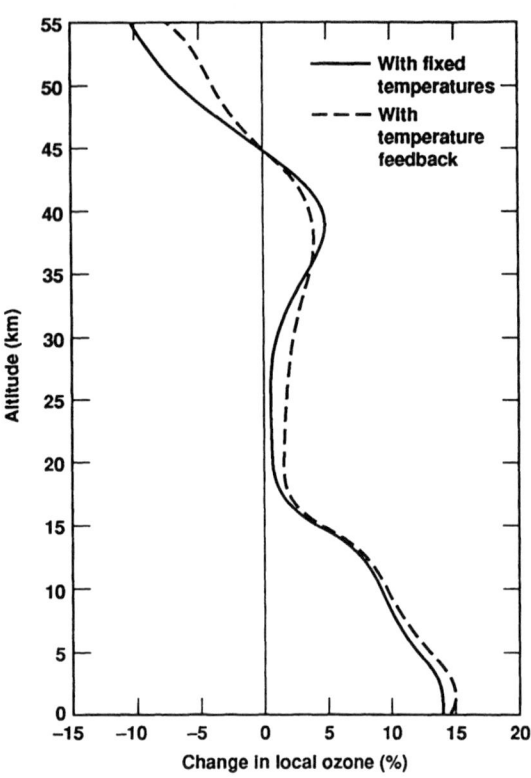

Figure 5. Calculated percentage change in local ozone for a doubling in the concentration of atmospheric methane from 1.6 to 3.2 ppmv. Profiles obtained with the 1985 version of the Lawrence Livermore National Laboratory one-dimensional model (WMO, 1985).

As an example, the derived change in ozone from the Lawrence Livermore National Laboratory (LLNL) one-dimensional model as reported in WMO (1985) is shown in Figure 5. The calculated change in total ozone for a doubling of the methane concentration from this model (with temperature feedback) was +2.9%. Calculated changes in ozone with altitude from other published studies (e.g., Owens et al., 1985; Isaksen and Stordal, 1986) are very similar. Recent calculations using the LLNL zonally-averaged two-dimensional chemical-radiative-transport model to evaluate the doubling of methane concentrations give a similar picture, as shown in Figures 6 and 7. Figure 6 gives the changes in total ozone as a function of latitude and season, while Figure 7 shows the change in ozone as a function of latitude and altitude for July. This 2-D model calculation gives a 3.6% increase in globally-averaged total ozone for a doubling of methane.

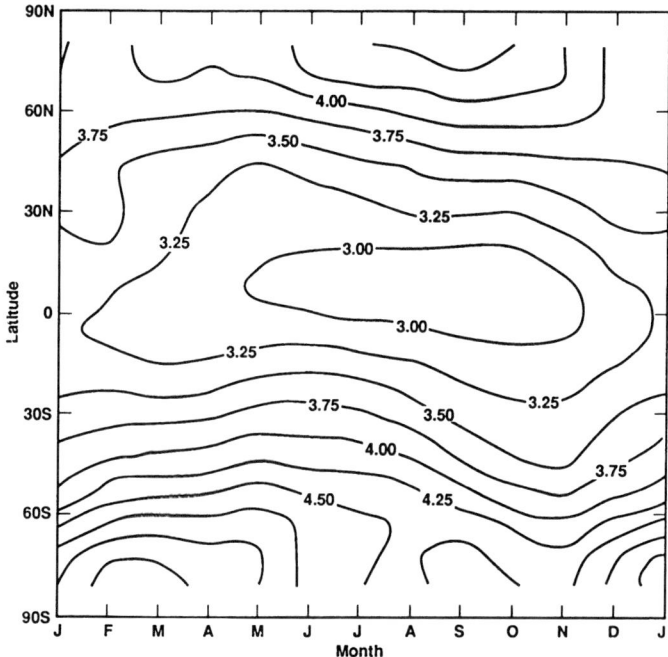

Figure 6. Calculated percentage change of total ozone for a doubling in the concentration of atmospheric methane from 1.7 to 3.4 ppmv. Based on the 1991 version of the Lawrence Livermore National Laboratory two-dimensional model (Wuebbles et al., 1991).

The calculations of increasing methane give a small percentage increase in lower stratospheric ozone, a larger increase near 40 km altitude, and a decrease in ozone above 45 km. As mentioned earlier, the hydrogen oxides produced by methane oxidation affect the efficiency of the nitrogen oxide and chlorine oxide catalytic ozone destruction mechanisms. However, the effect of methane in the lower stratosphere will depend on the efficiency of the nitrogen oxide catalytic cycle; if the amount of reactive odd nitrogen is reduced, then the additional hydrogen oxides from methane could destroy ozone in this region. The recently suggested importance of heterogeneous chemistry processes in the lower stratosphere could be important to determining the effect of increasing methane concentrations on lower stratospheric ozone. In the upper stratosphere, the additional hydrogen oxides reacts catalytically with ozone, leading to the decrease in ozone determined at these altitudes in the model calculations.

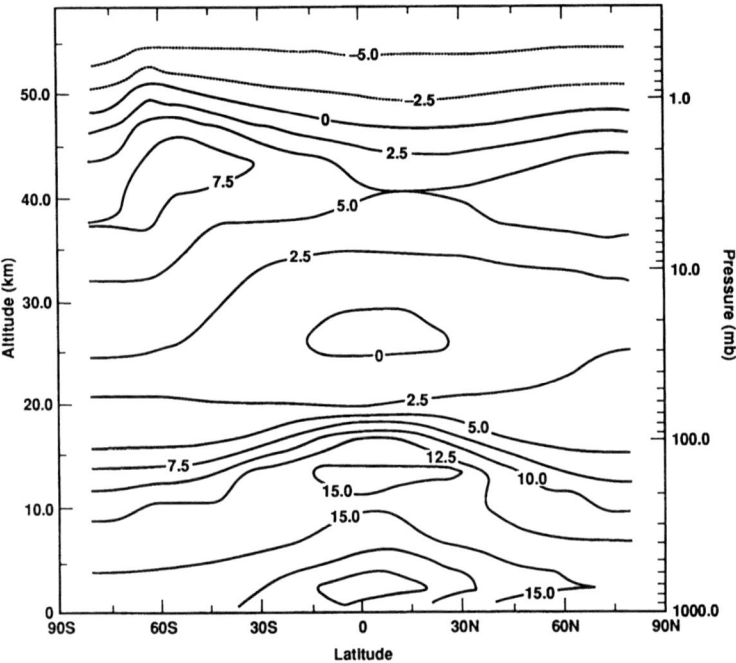

Figure 7. Calculated percentage change of local ozone in July for a doubling in the concentration of atmospheric methane from 1.7 to 3.4 ppmv. Based on the 1991 version of the Lawrence Livermore National Laboratory two-dimensional model (Wuebbles et al., 1991).

Figure 8, which is adapted from WMO (1991), compares the calculated globally and annually averaged increase in tropospheric and lower stratospheric ozone from four different models for a doubling of the methane surface flux. While several of the models agree reasonably well with each other, there is more than a factor of two difference in the calculated response of ozone between some of the models. In these model calculations, the globally-averaged increase in tropospheric ozone for the methane increase ranges from 10.4 to 15.2 % (WMO, 1991). Even larger differences are found in the calculated changes in tropospheric hydroxyl from these models, with globally averaged hydroxyl decreases ranging from -10.2 to -17.7 % for a doubling of methane (WMO, 1991). The differences in these results are indicative of the large remaining uncertainties in modeling of tropospheric chemical and physical processes.

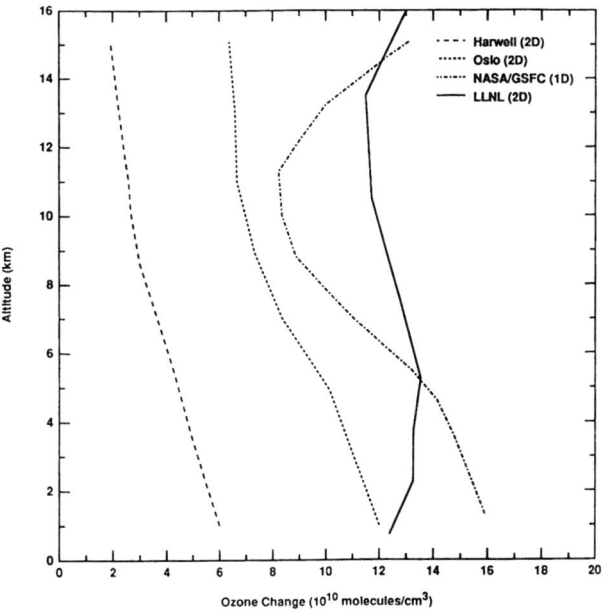

Figure 8. Global annual average height profiles of O_3 changes for doubled CH_4 surface flux based on model calculations. Symbols in parentheses after model names indicate whether it is one-dimensional (1D) or two-dimensional (2D). (Adapted from WMO, 1991. LLNL curve was prepared for this chapter by the authors.)

Figure 9, from WMO (1991), shows the change in tropospheric ozone at two latitudes (40° N and S) and two seasons (summer and winter) calculated with the University of Oslo model for a doubling of the methane flux. These results, along with those from the LLNL model in Figure 6, indicate that significant differences may be expected as a function of season and latitude in the effect on ozone from increasing methane.

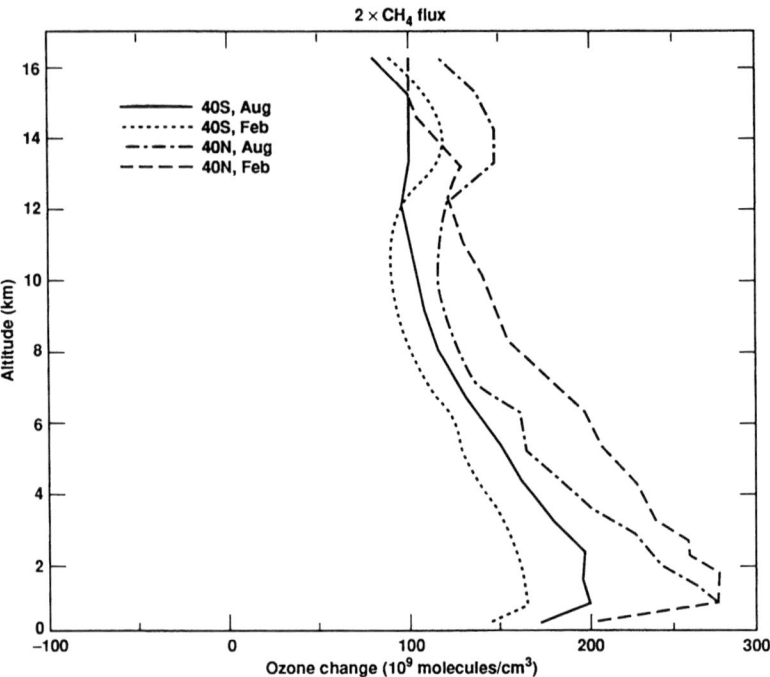

Figure 9. Vertical profiles of ozone increases at 40°N latitude and 40°S latitude during February and August for doubled CH_4 emission (based on WMO, 1991).

8. Stratospheric water vapor

Concentrations of water vapor in the atmosphere vary from as much as 15,000 ppmv near the surface in the tropics to as low as 3 ppmv in the lower stratosphere (Ellsaesser et al., 1980; WMO, 1985; Parameswaran and Krishna Murthy, 1990; Schwab et al., 1990). The spatial distribution of water vapor in the troposphere is primarily determined by evaporation, condensation, and transport processes. Human activities are currently thought to have little effect on tropospheric water vapor concentrations. Increased water vapor concentrations as a result of global warming is a well recognized climatic feedback process; increasing temperatures allow more water vapor to remain in the atmosphere, but, since water vapor is one of the most important greenhouse gases, the added water vapor further enhances the greenhouse radiative forcing.

Very little of the tropospheric water vapor penetrates into the stratosphere. The mechanism limiting the transport of tropospheric water vapor into the stratosphere is still not well understood. As a consequence, it is not known how water vapor concentrations in the lower stratosphere will respond to climate change effects on tropospheric water vapor concentrations. Concentrations of water vapor increase with altitude in the stratosphere, from 3 ppmv in the lower stratosphere to about 6 ppmv in the upper stratosphere. This increase in concentration with altitude occurs as a result of the oxidation of methane.

The methane oxidation reactions roughly produce two moles of water vapor for each mole of methane that is destroyed. Stratospheric water vapor concentrations should increase as concentrations of methane increase. Since methane concentrations have increased from about 0.7 ppmv in the pre-industrial atmosphere to the current concentration of 1.7 ppmv, this implies that upper stratospheric water vapor concentrations have increased by roughly 2 ppmv over this time period. Actually the increase in water vapor should be somewhat less than this due to methane reactivity with chlorine and oxygen atoms. Both modeling and data analysis studies (e.g., Le Texier et al., 1988; Hansen and Robinson, 1989) are in agreement with this conclusion, indicating that the overall stratospheric water vapor yield from methane is somewhat less than two.

Figure 10 shows the increase in stratospheric water vapor calculated with the LLNL one-dimensional chemical-radiative-transport model for a 30 % and 100 % increase in the surface mixing ratio of methane (WMO, 1991). The calculated concentration of stratospheric water vapor at 50 km increases by approximately 3 ppmv for a doubling of the methane surface mixing ratio from 1.7 to 3.4 ppmv.

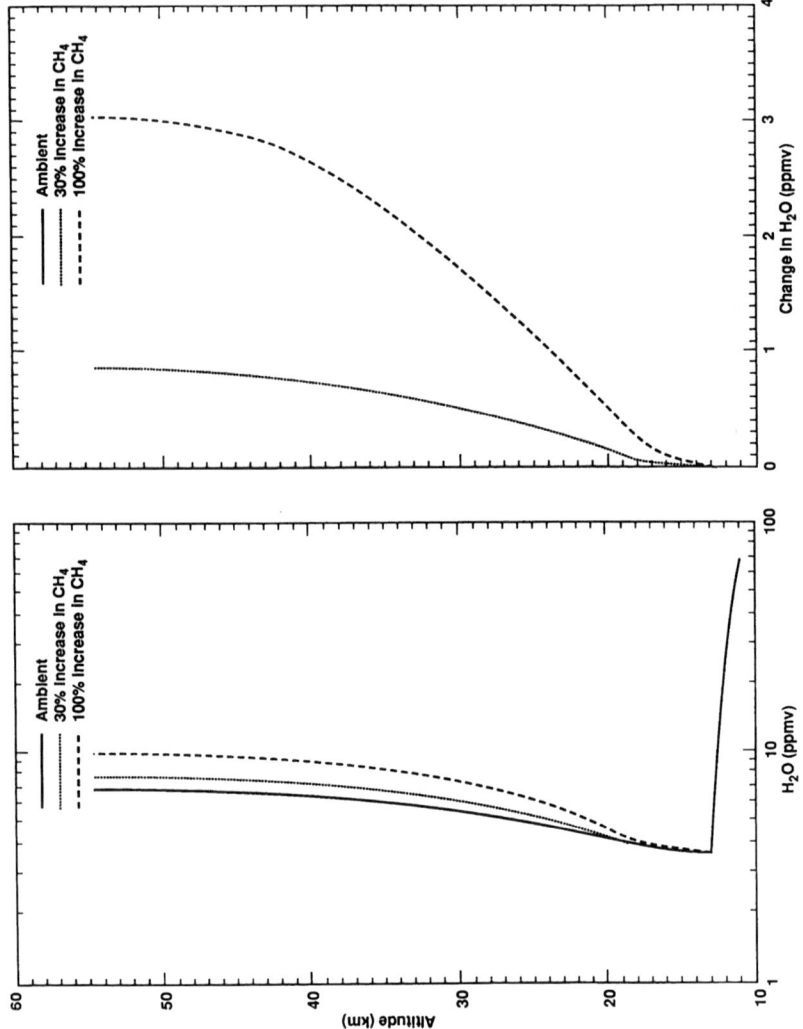

Figure 10. Calculated change in stratospheric water vapor due to a 30% and 100% increase in methane (Wuebbles and Grant, 1991; WMO, 1991).

9. Polar stratospheric clouds

As discussed above, the occurrence of polar stratospheric clouds is thought to be an important element in the formation of the Antarctic ozone hole. Increases in stratospheric water vapor concentrations from methane oxidation could contribute to the formation of stratospheric clouds in the polar lower stratosphere, and to increased effectiveness of the heterogeneous processes determining the destruction of ozone presently occurring in the Antarctic (the springtime *ozone hole*) and, to a lesser degree, in the arctic (Blake and Rowland, 1988; Ramanathan, 1988a, b; WMO, 1989). Cooler stratospheric temperatures as a result of increasing concentrations of carbon dioxide and other greenhouse gases could also enhance the formation of polar stratospheric clouds (Ramanathan, 1988a,b; Shine, 1989; Blanchet, 1989). The implications of these processes are still highly uncertain; however, increasing concentrations of methane may be partially responsible for the apparent increase in the frequency of polar stratospheric clouds during the last decade.

10. Mesosphere and above

Increasing methane concentrations should also be leading to increasing concentrations of water vapor in the mesosphere. Increasing concentrations of methane and carbon dioxide should also lead to cooler temperatures in the stratosphere, mesosphere, and thermosphere (WMO, 1985, 1989; Brasseur and Hitchman, 1988; Roble and Dickinson, 1989). With the increase in water vapor and the cooler temperatures, there is the potential for increased occurrences of noctilucent clouds near the mesopause (Thomas et al., 1989). Roble and Dickinson (1989) also point out that other changes in mesospheric and thermospheric composition should occur due to both the cooler temperatures and direct chemical effects from increased concentrations of methane and carbon dioxide. In addition, exospheric hydrogen will increase with increasing methane (Ehhalt, 1986).

11. Effects on climate

The concern that human activities may be affecting global climate has largely centered around carbon dioxide because of its importance as a greenhouse gas and also because of the rapid rate at which its atmospheric concentration has been increasing. However, research over the last decade has shown that other greenhouse gases are contributing about half of the overall increase in the greenhouse radiative forcing on climate. Methane is a greenhouse gas. The increasing concentration of methane is a major contributor to the increase in the greenhouse effect. In addition to its direct radiative forcing effect on climate, methane can also influence climate indirectly through chemical interactions; these will be discussed below.

Like other greenhouse gases, methane absorbs infrared radiation (also called longwave or terrestrial radiation) emitted by the relatively warm planetary surface and emits radiation to space at the colder atmospheric temperatures, leading to a net trapping of infrared radiation within the atmosphere. This is called the greenhouse effect. The balance between the absorbed solar radiation and the emitted infrared radiation determines the net radiative forcing on climate.

Although its atmospheric abundance is less than 0.5 % that of carbon dioxide, methane is an important greenhouse gas. Donner and Ramanathan (1980) calculated that the presence of methane at current levels causes the globally averaged surface temperature to be about 1.3 K higher than it would be without methane. On a molar basis, an additional mole of methane in the current atmosphere is about 21 times more effective at affecting climate than an additional mole of carbon dioxide (Ramanathan et al., 1985, 1987; IPCC, 1990). Correspondingly, on a mass basis, an additional kilogram of methane is about 58 times more effective as a greenhouse gas than a kilogram of carbon dioxide. It should be noted, however, that the actual lifetime of carbon dioxide in the atmosphere is much longer than the atmospheric lifetime of methane (IPCC, 1990). But it should also be recognized that the dissociation of methane eventually produces carbon dioxide, leading to additional climatic forcing from the original emission of methane.

The strongest bands for absorption by methane in the infrared are in the short wavelength edge of the window region. The most important infrared spectral feature of methane is the 7.66 µm (1306 cm^{-1}) absorption band. Due to saturation of the line cores for methane and emission from the pressure-broadened Lorentz line wings, radiative forcing from methane increases approximately as the square root of its concentration (IPCC, 1990; Wigley, 1987). Overlap with absorption by water vapor and other species (particularly nitrous oxide) also affects the efficiency of methane absorption.

Wang et al. (1991) show that the greenhouse radiative forcing for methane has different effects on climate than carbon dioxide, and that methane needs to be accounted for explicitly when attempting to predict the climate response to increasing concentrations of greenhouse gases. However, few calculations with three-dimensional global climate models (GCMs) have yet considered the explicit effects of methane on climate. As far as we are aware, Hansen et al. (1988) and Wang et al. (1991) are the only two groups to have explicitly considered the effects of methane in GCM studies. The vast majority of the climate modeling studies that have included methane have been done with radiative-convective models and other models with simplified treatments of climatic processes.

Increasing concentrations of methane are thought to be a significant fraction of the increase in radiative forcing from greenhouse gases over the last two centuries. IPCC (1990) calculated that the direct radiative effect of the increase in methane since the mid-1700s has accounted for an increase in radiative forcing of 0.42 Wm^{-2} (climate models indicate that the 4 Wm^{-2} associated with a doubling of CO_2 from 300 to 600 ppmv would give approximately a 1.5 to 4.5 K increase in surface temperature). This is about 17% of the total change in radiative due to CO_2 and other greenhouse gases over this time period (other research studies have found similar percentages for the effect of methane over this period: e.g., Rodhe (1990) derived 15 %, while Hansen et al. (1989) and MacKay and Khalil (1991) got about 22 %). Inclusion of indirect effects on stratospheric water vapor, ozone, and carbon dioxide (see below) could increase this percentage appreciably; inclusion of the stratospheric water vapor effect in IPCC(1990) increases this percentage to 23%.

Over the last decade, the change in atmospheric methane concentration is calculated to increase the radiative forcing by about 0.06 Wm^{-2}, about 11 % of the total increase in radiative forcing from greenhouse gases over this time period. Hansen et al. (1988) determined that methane was 12.2 % of the total change in radiative forcing over this period (note that these estimates ignore effects on radiative forcing due to changes in global ozone and aerosols over this time period).

Various studies have evaluated the potential effects on radiative forcing and surface temperature from a doubling of methane concentrations using radiative-convective models. For a doubling of methane concentration from 1.6 to 3.2 ppmv, effects on surface temperature range from 0.2 K to 0.3 K (Wang et al., 1976; Donner and Ramanathan, 1980; Lacis et al., 1981; Owens et al., 1985; Ramanathan et al., 1987; MacKay and Khalil, 1991), with the differences in model results primarily relating to uncertainties in the band strengths for methane infrared absorption. For a doubling from 1.7 to 3.4 ppmv, Owens et al. (1985) calculate a direct 0.34 K increase in surface temperature, along with an additional 0.26 K due to indirect effects from methane-induced effects on carbon dioxide and ozone. For a 25 % increase in methane concentrations, Ramanathan et al. (1985) determine a 0.08 K increase in surface temperature when overlap with the radiative absorption with other greenhouse gases is included, and a 0.19 K increase in surface temperature without overlap.

Other modeling studies have included increasing methane concentrations in studies evaluating scenarios for potential future changes in radiative forcing and global temperatures (Wang and Molnar, 1985; Ramanathan et al., 1985, 1987; WMO, 1985; Dickinson and Cicerone, 1986; Wang et al., 1986; Wigley, 1987; Hansen et al., 1988, 1989; IPCC, 1990). Table 2 shows the effect on radiative forcing due to assumed methane increases for several scenarios evaluated by IPCC (1990). For the IPCC Business-as-Usual scenario, the direct radiative effect of increasing methane accounts for 15.7 % of the increase in radiative forcing from 1765 to 2025 (14 % from 1990 to 2025). Similar effects to those in Table 2 were found in the other published scenario studies cited above.

Methane releases from northern gas fields and from gas hydrates may have been a significant contributor to the warming at the end of the last major glacial period (Nisbet, 1990a, b; MacDonald, 1990). Global warming could destabilize the storage of extensive amounts of methane in methane hydrates and clathrates, and lead to increased emissions of methane into the atmosphere, adding further to the greenhouse forcing (MacDonald, 1990). Climate change could also affect other natural sources of methane (Lashof, 1989; Moraes and Khalil, 1993).

Table 2. Change in radiative forcing (Wm-2) due to increasing methane compared to total change due to all greenhouse gases as derived by IPCC (1990) for several future scenarios. The percentage effect of methane on the total change in radiative forcing is also shown (only direct methane effects are included).

	Scenarios											
	BaU			B			C			D		
Years	CH_4	All	%	CH_4	All	%	CH_4	All	%	CH_4	All	%
1765-2025	0.72	4.59	15.7	0.56	3.8	14.7	0.51	3.63	14.0	0.47	3.52	13.3
1765-2050	0.90	6.49	13.9	0.65	4.87	13.3	3.53	4.49	11.8	0.43	3.99	10.8
1990-2025	0.30	2.14	14.0	0.14	1.35	10.4	0.09	1.18	7.6	0.05	1.07	4.7
1990-050	0.48	4.04	11.9	0.23	2.42	9.5	0.11	2.04	5.4	0.01	1.54	0.6

12. Indirect effects on climate

There are several ways that methane, through its chemical interactions in the atmosphere, can indirectly influence climate. Oxidation of methane leads eventually to carbon dioxide, one of the most important greenhouse gases. About 450 Tg/yr of methane are destroyed by reaction with hydroxyl and converted to carbon dioxide, accounting for production of 340 Tg C/yr as carbon dioxide; in contrast, the production of carbon dioxide from anthropogenic fossil fuel use and cement manufacturing is about 6000 Tg C/yr. Other indirect effects on climate resulting from methane include: production of stratospheric water vapor, changes in tropospheric and stratospheric ozone, and changes in concentrations of tropospheric hydroxyl.

Even though the concentration of stratospheric water vapor is appreciably smaller than that in the troposphere, water vapor is such an important greenhouse gas that changes in the concentrations of stratospheric water vapor can influence global radiative forcing on climate. Therefore, an increase in stratospheric water vapor concentrations resulting from increasing methane concentrations further enhances the greenhouse effect, increasing the radiative forcing from the added methane. Calculations by Lacis and Wuebbles using the NASA Goddard Institute for Space Studies radiative-convective model (as cited in Wuebbles et al., 1989) determined that the increased stratospheric water vapor should increase the radiative forcing from methane increase by about an additional 30 %. This result was used in the calculations of radiative forcing for the 1990 international assessment on climate change (IPCC,1990). A recent study by Wuebbles and Grant (1991), described in WMO (1991), using the LLNL radiative transfer models determined that the added water vapor should give about 22 % additional forcing, while Lelieveld and Crutzen (1991) got 6-7 %. The differences in these results need to be further examined.

As discussed earlier, increasing water vapor from methane could be leading to an increased amount of polar stratospheric clouds. Ramanathan (1988b) notes that both water and ice clouds, when formed at cold lower stratospheric temperatures, are extremely efficient in enhancing the atmospheric greenhouse effect. He also notes that there is a distinct possibility that large increases in future methane may lead to a surface warming that increases nonlinearly with the methane concentration.

The indirect effects on climate due to changes in the distributions of ozone and hydroxyl are still poorly understood. Research studies are currently underway towards determining the potential magnitude of these effects.

13. Global Warming Potentials

Global Warming Potentials (GWPs) have been developed as an analysis tool for policymakers in their evaluations of possible policy actions related to emissions of greenhouse gases. The GWP of a greenhouse gas is defined as the time-integrated commitment to climatic forcing from the instantaneous release of a kilogram of the gas relative to the climatic forcing from the release of 1 kg of carbon dioxide. Under this measure, the GWP for methane after a 100-year integration for the direct methane effects is a value of 11 (compared to 1.0 for carbon dioxide). As shown in IPCC (1990), indirect effects could more than double the methane GWP value. Published GWPs are derived for integration periods from 20 to 500 years, but 100-year values are thought to provide a balanced representation of the various time horizons for climatic response.

14. Summary

The growing concentrations of atmospheric methane require that the potential environmental effects from methane be understood. The necessity for furthering this understanding is made all the more urgent when one considers the variety of roles that methane plays in both the chemistry of the global atmosphere and in affecting the Earth's radiation balance. Research studies over the last few decades have shown that methane influences tropospheric ozone, hydroxyl radical, formaldehyde, and carbon monoxide concentrations, stratospheric chlorine and ozone chemistry, and, through its infrared absorption, the radiative forcing on climate. This chapter established that much has been learned over the last few decades about the effects of methane on atmospheric chemistry and climate. However, there are still large uncertainties associated with determining the effects of methane on the global environment that require further study.

Acknowledgements. Work performed under the auspices of the U.S. Department of Energy by the Lawrence Livermore National Laboratory under contract No. W-7405-Eng-48 and was supported in part by the U.S. Department of Energy (DOE) Office of Environmental Analysis, the DOE Environmental Sciences Division plus the U.S. Environmental Protection Agency under Interagency Agreement DW-899-32676.

References

Blake, D. R., F. S. Rowland. 1988. Continuing worldwide increase in tropospheric methane 1978 to 1987. *Science, 239*:1,129-1,131.
Blanchet, J. P. 1989. The response of polar stratospheric clouds to increasing carbon dioxide. In: *Proceedings of the International Radiation Symposium, Lille, France,* August, 1988, A. Deepak Publishing, Hampton, Va.
Bojkov, R. J. 1986. Surface ozone during the second half of the nineteenth century. *J. Clim. Appl. Meteor., 25*:343-352.
Brasseur, G., M. H. Hitchman. 1988. Stratospheric response to trace gas perturbations: Changes in ozone and temperature distribution. *Science, 240*:634-637.
Cicerone, R. J., R. S. Oremland. 1988. Biogeochemical aspects of atmospheric methane. *Global Biogeochem. Cycles, 2*:299-327.

Craig, H., C. C. Chou. 1982. Methane: the record in polar ice cores. *Geophys. Res. Lett.,* 9:1,221-1,224.

Crutzen, P. J. 1988. Tropospheric ozone: An overview. In: *Tropospheric Ozone: Regional and Global Scale Interactions* (I.S.A. Isaksen, ed.), D. Reidel, Boston, pp. 3-11.

Crutzen, P. J., P. H. Zimmermann. 1991. The changing photochemistry of the troposphere. *Tellus,* 43:136-151.

DeLuisi, J. J., D. U. Longenecker, C. L. Mateer, D. J. Wuebbles. 1989. An analysis of northern mid-latitude Umkehr measurements corrected for stratospheric aerosols for 1979-1986. *J. Geophys. Res.,* 94:9,837-9,845.

DeMore, W. B., S. P. Sander, D. M. Golden, M. J. Molina, R. F. Hampson, M. J. Kurylo, C. J. Howard, A. R. Ravishankara. 1990. Chemical kinetics and photochemical data for use in stratospheric modeling, *JPL, 90-1.* NASA Jet Propulsion Laboratory, Pasadena California, pp. 47-48.

Dickinson, R. E., R. J. Cicerone. 1986. Future global warming from atmospheric trace gases. *Nature,* 319:109-115.

Donner, L., V. Ramanathan. 1980. Methane and nitrous oxide: Their effects on the terrestrial climate. *J. Atmos. Sci.,* 37:119-124.

Ehhalt, D. H. 1986. On the consequence of a tropospheric CH_4 increase to the exospheric density. *J. Geophys. Res.,* 91:2,843.

Ehhalt, D. H., H-P. Dorn, D. Poppe. 1991. The chemistry of the hydroxyl radical in the troposphere. *Proc. Royal Soc. Edinburgh,* 97B:17-34.

Ellsaesser, H. W., J. E. Harries, D. Kley, R. Penndorf. 1980. Stratospheric H_2O. *Planet. Space Sci.,* 28:827-835.

Fishman, J. 1985. Ozone in the troposphere. In: *Ozone in the free atmosphere* (R. C. Whitten and S. S. Prasad, eds.), Van Nostrand Reinhold, New York, pp. 161-194.

Fishman, J., S. Solomon, P. J. Crutzen. 1979. Observational and theoretical evidence in support of a significant in-situ photochemical source of tropospheric ozone. *Tellus,* 31:432-446.

Garcia, R. R., S. Solomon. 1987. Interannual variability in Antarctic ozone and the quasi-biennial oscillation. *Geophys. Res. Lett.,* 14:848-851.

Hansen, A. R., G. D. Robinson. 1989. Water vapor and methane in the upper stratosphere: an examination of some of the Nimbus 7 measurements. *J. Geophys. Res.,* 94:8,474-8,484.

Hansen, J., I. Fung, A. Lacis, D. Rind, S. Lebedeff, R. Ruedy, G. Russell, P. Stone. 1988. Global climate changes as forecast by Goddard Institute for Space Studies three-dimensional model. *J. Geophys. Res.,* 93:9,341-9,364.

Hansen, J., A. Lacis, M. Prather. 1989. Greenhouse effect of chlorofluorocarbons and other trace gases. *J. Geophys. Res.,* 94:16,417-16,421.

Hough, A. M. 1991. The development of a two-dimensional global tropospheric model: the model chemistry. *J. Geophys. Res.*, 96:7,325-7,362.

Hough, A. M., R. G. Derwent. 1990. Changes in the global concentration of tropospheric ozone due to human activities. *Nature*, 344:645-648.

Intergovernmental Panel on Climate Change, Climate Change: *The IPCC Scientific Assessment, (1990)* (J.T. Houghton, G. J. Jenkins, and J. J. Ephraums, eds.), Cambridge University Press, Cambridge.

Isaksen, I. S. A. 1988. Is the oxidizing capacity of the atmosphere changing. In: *The Changing Atmosphere* (F.S. Rowland and I.S.A. Isaksen, eds.), John Wiley & Sons, New York, pp. 141-157.

Isaksen, I. S. A., F. Stordal. 1986. Ozone perturbations by enhanced levels of CFCs, N_2O, and CH_4: A two-dimensional diabatic circulation study including uncertainty estimates. *J. Geophys. Res.*, 91:5,249-5,263.

Isaksen, I. S. A., O. Hov. 1987. Calculation of trends in the tropospheric concentration of O_3, OH, CO, CH_4, and NOx. *Tellus*, 39B:271-285.

Khalil, M. A. K., R. A. Rasmussen. 1983. Sources, sinks, and seasonal cycles of atmospheric methane. *J. Geophys. Res.*, 88:5,131-5,144.

Khalil, M. A. K., R. A. Rasmussen. 1984. Carbon monoxide in the earth's atmosphere: increasing trend. *Science*, 224:54-56.

Khalil, M. A. K., R. A. Rasmussen. 1985. Causes of increasing atmospheric methane: depletion of hydroxyl radicals and the rise of emissions. *Atmos. Environ.*, 13:397-407.

Khalil, M. A. K., R. A. Rasmussen. 1988. Carbon monoxide in the earth's atmosphere: indications of a global increase. *Nature*, 332:242-245.

Khalil, M. A. K., R. A. Rasmussen. 1989. Temporal variations of trace gases in ice cores. In: *The environmental record in glaciers and ice sheets* (H. Oeschger and C. C. Langway, Jr., eds.) John Wiley and Sons Limited, New York, p. 193-205.

Khalil, M. A. K., R. A. Rasmussen. 1990. Atmospheric methane: recent global trends. *Environ. Sci. Tech.*, 24:549-553.

Khalil, M. A. K., R. A. Rasmussen, M. J. Shearer. 1989. Trends of atmospheric methane during the 1960s and 1970s. *J. Geophys. Res.*, 94:18,279-18,288.

Lacis, A., J. Hansen, P. Lee, T. Mitchell, S. Lebedeff. 1981. Greenhouse effect of trace gases, 1970-1980. *Geophys. Res. Lett.*, 8:1,035-1,038.

Lacis, A. A., D. J. Wuebbles, J. A. Logan. 1990. Radiative forcing of climate by changes in the vertical distribution of ozone. *J. Geophys. Res.*, 95:9,971-9,981.

Lashof, D. A. 1989. The dynamic greenhouse: feedback processes that may influence future concentrations of atmospheric trace gases and climatic change. *Climatic Change*, 14:213-242.

Le Texier, L., S. Solomon, R. R. Garcia. 1988. The role of molecular hydrogen and methane oxidation in the water vapor budget of the stratosphere. *Q. J. Roy. Meteorol. Soc., 114*:281-296.

Lelieveld, J., P. Crutzen. 1991. The role of clouds in tropospheric photochemistry. *J. Atmos. Chem., 12*:229-267.

Levine, J. S., C. P. Rinsland, G. M. Tennille. 1985. The photochemistry of methane and carbon monoxide in the troposphere in 1950 and 1985. *Nature, 318*:254.

Liu, S. C., M. McFarland, D. Kley, O. Zafiriou, B. J. Huebert. 1983. Tropospheric NOx and O_3 budgets in the equatorial Pacific. *J. Geophys. Res., 88*:1,360-1,368.

Liu, S. C., R. A. Cox, P. J. Crutzen, D. H. Ehhalt, R. Guicherit, A. Hofzumahaus, D. Kley, S. A. Penkett, L. F. Phillips, D. Poppe, F. S. Rowland. 1988. Group report: oxidizing capacity of the atmosphere. In: *The Changing Atmosphere* (F. S. Rowland and I. S. A. Isaksen, eds.), John Wiley & Sons, New York, pp. 219-232.

Logan, J. A. 1985. Tropospheric ozone: Seasonal behavior, trends, and anthropogenic influence. *J. Geophys. Res., 90*:10,463-10,482.

Lowe, D. C., U. Schmidt. 1983. Formaldehyde (HCHO) measurements in the nonurban atmosphere. *J. Geophys. Res., 88*:10,844-10,858.

Lu, Y., M. A. K. Khalil. 1991. Tropospheric OH: model calculations of spatial, temporal, and secular variations. *Chemosphere, 23*:397-444.

MacDonald, G. J. 1990. Role of methane clathrates in past and future climates. *Climatic Change, 16*:247-281.

MacKay, R. M., M. A. K. Khalil. 1991. Theory and development of a one dimensional time dependent radiative convective climate model. *Chemosphere, 22*:383-417.

McFarland, M., D. Kley, J. M. Drummond, A. L. Schmeltekopf, R. H. Winkler. 1979. Nitric oxide measurements in the equatorial Pacific. *Geophys. Res. Lett., 6*:605-608.

Miller, A. J., R. M. Nagatani, G. C. Tiao, X. F. Niu, G. C. Reinsel, D. J. Wuebbles, K. Grant. 1992. Comparisons of observed ozone and temperature trends in the lower stratosphere. *Geophys. Res. Lett., 19*:929-932.

Nisbet, E. 1990a. Did the release of methane from hydrates accelerate the end of the last ice age? *Can. J. Earth Sci., 27*:148-157.

Nisbet, E. 1990b. Climate change and methane. *Nature, 347*:23.

Owens, A. J., J. M. Steed, D. L. Filkin, C. Miller, J. P. Jesson. 1982. The potential effects of increased methane on atmospheric ozone. *Geophys. Res. Lett., 9*:1,105-1,108.

Owens, A. J., C. H. Hales, D. L. Filkin, C. Miller, J. M. Steed, J. P. Jesson. 1985. A coupled one-dimensional radiative-convective, chemistry-transport model of the atmosphere: 1. Model structure and steady state perturbation calculations. *J. Geophys. Res., 90*:2,283-2,311.

Parameswaran, K., B. V. Krishna Murthy. 1990. Altitude profiles of tropospheric water vapor at low latitudes. *J. Appl. Meteor., 29*:665-679.

Penkett, S. A. 1988. Indications and causes of ozone increase in the troposphere. In: *The Changing Atmosphere* (F. S. Rowland and I. S. A. Isaksen, eds.), John Wiley & Sons, New York, pp. 91-103.

Pinto, J. P., M. A. K. Khalil. 1991. The stability of tropospheric OH during ice ages, inter-glacial epochs and modern times. *Tellus, 43B*:347-352.

Ramanathan, V. 1988a. The greenhouse theory of climate change: A test by an inadvertent global experiment. *Science, 240*:293-299.

Ramanathan, V. 1988b. The radiative and climatic consequences of the changing atmospheric composition of trace gases. In: *The Changing Atmosphere* (F. S. Rowland and I. S. A. Isaksen, eds.), John Wiley & Sons, New York, pp. 159-186.

Ramanathan, V., R. J. Cicerone, H. B. Singh, J. T. Kiehl. 1985. Trace gas trends and their potential role in climate change. *J. Geophys. Res., 90*:5,547-5,566.

Ramanathan, V., L. Callis, R. Cess, J. Hansen, I. Isaksen, W. Kuhn, A. Lacis, F. Luther, J. Mahlman, R. Reck, M. Schlesinger. 1987. Climate-chemical interactions and effects of changing atmospheric trace gases. *Rev. Geophys., 25*:1,441-1,482.

Rasmussen, R. A., M. A. K. Khalil. 1981. Atmospheric methane (CH_4): trends and seasonal cycles. *J. Geophys. Res., 86*:9,826-9,832.

Ridley, B. A., M. A. Carroll, G. L. Gregory. 1987. Measurements of nitric oxide in the boundary layer and free troposphere over the Pacific ocean. *J. Geophys. Res., 92*:2,025-2,047.

Roble, R. G., R. E. Dickinson. 1989. How will changes in carbon dioxide and methane modify the mean structure of the mesosphere and thermosphere? *Geophys. Res. Lett., 16*:1,441-1,444.

Rodhe, H. 1990. A comparison of the contribution of various gases to the greenhouse effect. *Science, 248*:1,217-1,219.

Schwab, J. J., E. M. Weinstock, J. B. Nee, J. G. Anderson. 1990. In situ measurement of water vapor in the stratosphere with a cryogenically cooled Lyman-alpha hygrometer. *J. Geophys. Res., 95*:13,781-13,796.

Shine, K. P. 1989. The greenhouse effect. In: *Ozone Depletion: Health and Environmental Consequences* (R. R. Jones and T. Wigley, eds.), John Wiley & Sons, New York, pp. 71-83.

Solomon, S. 1988. The mystery of the Antarctic ozone hole. *Rev. Geophys., 26*:131-148.

Steele, L. P., P. J. Fraser, R. A. Rasmussen, M. A. K. Khalil, T. J. Conway, A. J. Crawford, R. H. Gammon, K. A. Masarie, K. W. Thoning. 1987. The global distribution of methane in the troposphere. *J. Atmos. Chem.,* 5:125-171.

Stolarski, R. S., P. Bloomfield, R. D. McPeters, J. R. Herman. 1991. Total ozone trends deduced from Nimbus 7 TOMS data. *Geophys. Res. Lett.,* 18:1,015-1,018.

Stordal, F., I. S. A. Isaksen. 1987. Ozone perturbations due to increases in N_2O, CH_4, and chlorocarbons: two-dimensional time-dependent calculations. *Tellus, 39B*:333-353.

Thomas, G. E., J. J. Olivero, E. J. Jensen, W. Schroeder, O. B. Toon. 1989. Relation between increasing methane and the presence of ice clouds at the mesopause. *Nature, 338*:490-492.

Thompson, A. M., R. J. Cicerone. 1986a. Possible perturbations to atmospheric CO, CH_4, and OH. *J. Geophys. Res., 91*:10,853-10,864.

Thompson, A. M., R. J. Cicerone. 1986b. Atmospheric CH_4, CO, and OH from 1860 to 1985. *Nature, 321*:148-150.

Thompson, A. M., M. Kavanaugh. 1986. Tropospheric CH_4/CO/NOx: The next fifty years. In: *Effects of Changes in Stratospheric Ozone and Global Climate, vol. II.* United Nations Environmental Program Report.

Thompson, A. M., R. W. Stewart, M. A. Owens, J. A. Herwehe. 1989. Sensitivity of tropospheric oxidants to global chemical and climate change. *Atmos. Environ., 23*:519-532.

Thompson, A. M., M. A. Huntley, R. W. Stewart. 1990. Perturbations to tropospheric oxidants, 1985-2035: 1. Calculations of ozone and OH in chemically coherent regions. *J. Geophys. Res., 95*:9,829-9,844.

Tiao, G. C., G. C. Reinsel, J. H. Pedrick, G. M. Allenby, C. L. Mateer, A. J. Miller, J. J. DeLuisi. 1986. A statistical trend analysis of ozonesonde data. *J. Geophys. Res., 91*:13,121-13,136.

Volz, A., D. Kley. 1988. Evaluation of the Monteouris series of ozone measurements made in the nineteenth century. *Nature, 332*:240-242.

Wallace, L., W. Livingston. 1990. Spectroscopic observations of atmospheric trace gases over Kitt Peak: 1. Carbon dioxide and methane from 1979 to 1985. *J. Geophys. Res., 85*:9,823-9,827.

Wang, W. C., G. Molnar. 1985. A model study of the greenhouse effects due to increasing atmospheric CH_4, N_2O, CF_2Cl_2, and $CFCl_3$. *J. Geophys. Res., 90*:12,971-12,980.

Wang, W. C., Y. L. Yung, A. A. Lacis, T. Mo, J. E. Hansen. 1976. Greenhouse effects due to man-made perturbations of trace gases. *Science, 194*:685-690.

Wang, W. C., D. J. Wuebbles, W. M. Washington, R. G. Isaacs, G. Molnar. 1986. Trace gases and other potential perturbations to global climate. *Rev. Geophys., 24*:110-140.

Wang, W. C., M. P. Dudek, X. Z. Liang, J. T. Kiehl. 1991. Inadequacy of effective CO_2 as a proxy in simulating the greenhouse effect of other radiatively active gases. *Nature, 350*:573-577.

Wigley, T. M. L. 1987. Relative contributions of different trace gases to the greenhouse effect. *Climate Monitor, 16*:14-28.

World Meteorological Organization, Atmospheric Ozone. 1985. WMO *Global Ozone Research and Monit. Proj. Report, 16*, Geneva: WMO.

World Meteorological Organization, Report of the International Ozone Trends Panel. 1988. *Global Ozone Research and Monit. Proj. Report, 18*, Geneva: WMO.

World Meteorological Organization, Scientific Assessment of Stratospheric Ozone. 1989. *Global Ozone Research and Monit. Proj. Report, 20*, Geneva: WMO.

World Meteorological Organization, Scientific Assessment of Ozone Depletion. 1991. *Global Ozone Research and Monit. Proj. Report, 25*, Geneva: WMO.

Wuebbles, D. J., K. E. Grant. 1991. Indirect effects on climatic forcing from stratospheric water vapor resulting from increased concentrations of CH_4 and H_2. Lawrence Livermore National Laboratory, as reported in IPCC, 1992.

Wuebbles, D. J., F. M. Luther, J. E. Penner. 1983. Effect of coupled anthropogenic perturbations on stratospheric ozone. *J. Geophys. Res., 88*:1,444-1,456.

Wuebbles, D. J., K. E. Grant, P. S. Connell, J. E. Penner. 1989. The role of atmospheric chemistry in climate change. *JAPCA, 39*:22-28.

Wuebbles, D. J., D. E. Kinnison, K. E. Grant, J. Lean. 1991a. The effect of solar flux variations and trace gas emissions on recent trends in stratospheric ozone and temperature. *J. Geomagnetism and Geoelectricity, 43*:709-718.

Wuebbles, D. J., J. S. Tamaresis, D. E. Kinnison. 1991b. Effects of increasing methane on tropospheric and stratospheric chemistry. Poster presentation, NATO Advanced Research Workshop on the Atmospheric Methane Cycle: Sources, Sinks, Distributions, and Role in Global Change, Portland, Oregon.

Zander, R., P. H. Demoulin, D. H. Ehhalt, U. Schmidt, C. P. Rinsland. 1989. Secular increase of the total vertical column abundance of carbon monoxide above central Europe since 1950. *J. Geophys. Res., 94*:11,021-11,028.

Chapter 21

Working Group Report

The Current and Future Environmental Role of Atmospheric Methane: Model Studies and Uncertainties

JOSEPH P. PINTO[1], CHRISTOPH H. BRÜHL[2], AND ANNE M. THOMPSON[3]

[1] *US EPA, MD-84, Research Triangle Park, NC 27711, USA*
[2] *Max Planck Institute for Chemistry, P.O. Box 3060, D-6500, Mainz, Germany*
[3] *NASA/Goddard Space Flight Center, Code 916, Greenbelt, MD 20771, USA*

Group Members:
D. Ahuja, N. Andronova, V. Aneja, C. Atherton, C. Brühl, P. Guthrie, R. Harriss, M. Kanakidou, M. Kandlikar, K. Law, C-I. Lin, N. Miller, C. Mitchell, L. Morrissey, D. Murdiyarso, R. Peer, R. Pelet, J. Pinto, J. Robinson, J. Rotmans, S. Sadler, A. Thompson, X. Tie, D. Wuebbles

1. Introduction

Concern over increasing levels of methane in the atmosphere centers on its radiative and chemical properties. Methane absorbs terrestrial infrared radiation and contributes to the greenhouse effect. Effects on other greenhouse absorbers (e.g., O_3, H_2O, and CO_2) as the result of its oxidation must also be considered. These indirect effects have made the quantification of the total climatic effects of chemically active gases, such as CH_4, much more difficult than if direct radiative effects are considered alone. The oxidation of methane also exerts a controlling influence on atmospheric OH levels and is a major source of carbon monoxide. The variations in OH induced by changing CH_4 levels feed back onto the lifetime of methane and the abundance of CO (Sze, 1977; Chameides et al., 1977). However, there is a shortage of intercompared model results documenting the effects of CH_4 and nonmethane hydrocarbon (NMHC) additions on tropospheric OH levels. Most analyses to date have relied on

NATO ASI Series, Vol. I 13
Atmospheric Methane: Sources, Sinks, and Role in Global Change
Edited by M. A. K. Khalil

analyses of gas phase reaction sequences for methane oxidation (e.g., Crutzen, 1987, 1988), without considering the numerous feedbacks on atmospheric chemistry. More complete modeling studies are needed because OH levels also depend on the emissions of CO, NMHCs, and NO_y (NO_x + NO_3 + $2N_2O_5$ + $CH_3CO_3NO_2$ (PAN) + HNO_3 + HNO_4 + $ClNO_3$ + NO_3^-), where NO_x is NO + NO_2 and NO_y and NX are interchangeable terms. Furthermore, analyses which simulate the role of climate in controlling CH_4 emissions from various natural sources (e.g., wetlands) are critical for attempting to predict the response of atmospheric methane levels to future climate change.

2. Calculation of tropospheric OH levels

The major sink for methane involves reaction with hydroxyl (OH) radicals in the troposphere ($\sim$ 90%), followed by reactions with stratospheric OH ($\sim$ 7%), chlorine atoms and singlet atomic oxygen (e.g., Lelieveld et al., 1993). Accurate calculations of OH fields are necessary, because the calculations of the total methane sink are used to constrain estimates of the total source strength of methane (Khalil and Rasmussen, 1990; Fung et al., 1991; Tie and Mroz, 1993). The accurate calculation of globally averaged OH values for use in budget studies requires knowledge of the spatial distribution of a number of species such as O_3, H_2O, CO, CH_4, and NO_x. A number of calculations of OH radical distributions in the atmosphere were presented at the NATO-Advanced Research Workshop. Calculations were based on a variety of one-, two-, and three- dimensional models. The one-dimensional models consider only vertical variations of horizontally averaged quantities. Such an approach is valid provided constituents are well-mixed horizontally or if horizontal transport can be neglected compared to vertical transport. This is not the case for many short-lived species in the atmosphere, especially for nitrogen oxides whose sources are highly heterogeneous. Some improvement can be achieved using hemispherical or semi-hemispherical domains, which makes allowance for spatial variations in short-lived species (e.g., NO_x and CO). An alternative approach is to consider the chemistry of chemically coherent regions separately in a one-dimensional framework (Thompson et al., 1990). Two-dimensional models consider latitudinal variations while averaging emissions zonally (around a latitude belt). However, the short atmospheric lifetime of NO_x (about a day or two) insures that there will be large variations around a latitude belt, since the zonal mixing time is of the order of two weeks. Finally, three-dimensional models consider the full three-dimensional nature of atmospheric motions and distribution of sources, but there are still questions about the adequacy of the grid resolution necessary to describe the variability of NO_x. Results presented at the NATO Workshop allow intercomparisons of the predictions of globally averaged OH levels produced by these different classes of models. These differences can be interpreted in terms of model dimensionality and grid spacing, amount and distribution of emissions, and uncertainties in the chemical kinetics data used. Values for tropospheric mean OH levels determined by models presented at the workshop are shown in Table 1. These values are simple volume averages over the troposphere and also represent annual averages. The "hemispheric" one-dimensional models are formulated for both the northern and southern hemispheres and include interhemispheric transport. The model based on chemically coherent regions is as mentioned earlier. In general, the model results all agree to better than a factor of two. Within specific classes of models, differences are more directly related to the chemical kinetics included, methods for diurnal averaging and for radiative transfer, and the choice of boundary conditions, particularly for nitrogen oxides. Further work is needed to quantify and fully understand the roles of all these factors.

All modeling studies report the strong dependence of OH levels on assumed NO_x and CO levels. The relation with NO_x levels is complex, because OH levels are suppressed at very high NO_x levels (1 ppb or greater), characteristic of source regions, whereas in regions characterized by lower NO_x levels the reverse is true. At high NO_x levels the termolecular recombination reaction of NO_2 with OH to form HNO_3 is a major sink for OH and the subsequent loss of HNO_3 from the atmosphere represents a net loss for HO_x (OH + HO_2) radicals. Hence, OH levels tend to decrease with increasing NO_x. At lower NO_x levels this reaction decreases in importance for controlling the abundance of OH because other reactions dominate. In this regime, the recycling of HO_2 to OH by reaction with NO results in an increase in calculated OH with increasing NO_x (Hameed et al., 1979; Logan et al., 1981). At levels of NO $>\sim$

10 ppt, or $NO_x > \sim$ 30 ppt, the oxidation of CO, CH_4, and NMHCs tends to increase OH, whereas at low NO_x levels, their oxidation tends to consume OH (e.g., Logan et al., 1981; Crutzen, 1988; Thompson et al., 1989; Wuebbles and Tamaresis, this volume). The large degree of heterogeneity of NO_x sources, coupled with its relatively short lifetime, results in calculated NO_x levels, which are much higher in source regions than outside. Major source regions for NO_x are located in northern mid-latitudes, where the dominant source is fossil fuel combustion and in the tropics where the major source of NO_x is biomass burning (e.g., Dignon and Penner, 1991). On the other hand, large areas of the world's oceans in the tropics and southern hemisphere are characterized by very low levels of NO_x.

Table 1. Model calculated OH values ($\times 10^{-6}$ molecules/cm^3)

	Global Average	Surface Average
One-Dimensional Models		
Hemispheric		
MPI (Brühl, personal communication, 1991)	0.85	2.1
AREAL (Pinto and Khalil, 1991)	0.59	1.2
OGI (Lu and Khalil, 1991)	0.80	
Chemically Coherent Regions		
GSFC (Thompson et al., 1993)	0.60	
Two-Dimensional Models		
LLNL (Wuebbles, personal communication)	0.82	
CAM (Law and Pyle, 1991)	0.95	1.6
Three-Dimensional Models		
MPI (Kanakidou, personal communication)	0.80	
	0.91 (zonally av. NO_x)	

Kanakidou and Crutzen (1993) conducted a study of the effects of zonally averaging emissions of NO_x and NMHCs on globally averaged OH levels in a three-dimensional model. OH levels computed using a longitudinally uniform distribution of emissions in the model were substantially higher near the surface in latitude belts where there are major NO sources, such as in northern mid-latitudes and in the tropics and 14% higher in the global average than using a ($10° \times 10°$) longitudinally varying distribution of emissions. The overall result was a 20% overestimate in the net destruction rate for the case of the zonally uniform emissions. Zonal averaging overestimates NO_x and NO_y (NX) levels over background areas (the oceans and remote continental areas) whereas NO_x levels over less widespread source regions (industrialized areas of the Northern Hemisphere and tropical areas of active biomass burning) tend to be underestimated. It seems clear that these differences in the calculated OH levels have implications for the applicability of two-dimensional models to global problems and should be borne in mind when making predictions of future change in tropospheric composition. It may be that appropriate averaging techniques, such as prescribing correction factors to zonally averaged reaction rates and species abundances or incorporating modules to transform NO_x to other forms (such as PAN) at source latitudes in two-dimensional models, could be developed to provide a better comparison with three-dimensional results. Indeed, there may be more similarities in the behavior of one- and two-dimensional models, in this regard, than with three-dimensional models because the characteristic chemistry of NO_x and OH in source regions has already been averaged out in two-dimensional models. It should be noted that problems related to the dimensionality of models and grid spacing are not really independent, since the same effects should be seen within three-dimensional models as the horizontal resolution is varied.

The discussion given above refers only to differences between models based on their formulation not to differences based on uncertainties in their chemical kinetics data bases. There are also inherent uncertainties in calculations of OH due to imprecisions in the rate coefficients, UV absorption cross sections, and quantum yields used in models (NASA/JPL, 1990). For example, where NO_x levels are relatively high, the computed uncertainty in a one-dimensional model calculation of boundary layer OH due to kinetics imprecisions as specified in the standard kinetics data base (NASA/JPL, 1990) is as high as 70% (Thompson and Stewart, 1991). If fairly accurate measurements of O_3, CO and NO_2 are available to constrain the model calculations of OH levels, the uncertainty in OH can be cut to 20%. In regions characterized by lower levels of NO_x, kinetics uncertainties introduce ~20% into the calculation of boundary layer OH. Averaging over regions that differ in NO_x emissions suggests that a typical estimate of global OH has an uncertainty of about 25% due to imprecisions in the rate coefficients for reactions used in most photochemical models (Thompson and Stewart, 1991). The largest uncertainty comes from ozone photodissociation to $O(^1D)$, the initial precursor for OH formation in the troposphere. The treatment of clouds in model radiation routines is a major source of uncertainty in calculating the photolysis of ozone and other photochemically active species in a given model and of differences among model calculations of OH radical concentrations.

A further, perhaps important, uncertainty in understanding the atmospheric effects of methane oxidation arises from heterogeneous reactions of the radicals involved in the CH_4 oxidation chain, especially for CH_3O_2, in cloud droplets, (Lelieveld and Crutzen, 1990). A critical question concerns the fractional yield of formic acid which can be rained out, thereby reducing the yield of H_2CO, H_2 and CO from methane oxidation. The neglect of OH and HO_2 scavenging (also O_3) may introduce uncertainties into calculation of gas-phase concentrations. The possible heterogeneous conversion of labile NO_x reservoirs, such as N_2O_5, into soluble species such as HNO_3 tends to shift the partitioning of NO_y away from NO_x, thereby affecting the overall level of OH radicals and ozone formation.

Apart from direct calculation based on kinetic data and data for species abundances, OH values appropriate for estimating lifetimes of CH_4 and other long-lived species have also been derived from the simulation of gases such as methyl chloroform (Khalil and Rasmussen, 1984; Prinn et al., 1987). Current simulations of the distribution of CH_3CCl_3 yield globally averaged OH concentrations of $8.1 \pm 0.9 \times 10^5/cm^3$ (where the $\pm$ value is 1 standard deviation) and an increasing trend for OH of $1 \pm 0.8\%$ per year (Prinn et al., 1992). Although this method is highly useful, it is not free of uncertainties, since its success depends on accurate values for rate coefficients for reaction with OH and on accurately knowing source strengths and additional loss processes.

Recently it has been established that there may be an oceanic sink for CH_3CCl_3 (Butler et al., 1991), which may require downward revisions of previous values of OH of from 5 to 11%.

There are a number of sources of uncertainty involved in assessing the effects of future increases of methane in the atmosphere. Predictions are highly sensitive to the total levels of NO_x and CO chosen and their future trends. Furthermore, there are also uncertainties in the prediction of penetrating UV-B radiation due to changes in stratospheric ozone (Liu and Trainer, 1988; Thompson et al., 1989; Thompson, 1991; Madronich, 1992) and in feedback processes from global warming, e.g., changes in cloudiness and tropospheric H_2O. In particular, models that have high globally averaged NO_x levels will tend to systematically overestimate globally averaged OH as methane levels increase. This in turn may lead to a systematic underestimate of the atmospheric lifetime and subsequently to an underestimate in the rate of increase of CH_4 concentrations for a given scenario of future methane emissions (Guthrie and Yarwood, 1991). Because of the complex set of chemical interactions involving CH_4, NO_x, O_3 and OH, it is inappropriate to prescribe the evolution of OH levels independently from the NO_x growth scenario used in calculations of the impact of anthropogenic activities on future atmospheric composition.

3. Assessment of climatic effects of methane

The concept of Greenhouse Warming Potentials (GWPs) has been introduced to assess the role of methane relative to CO_2 in the global warming problem (IPCC, 1990). As mentioned earlier, the total climatic effect from increasing methane also includes a number of indirect effects arising from its oxidation products, such as CO_2 and stratospheric H_2O. The effects on tropospheric ozone are more complex with possible production in some regions and destruction in others, depending on the local NO_x level. These effects should also include feedbacks on OH levels and hence on the lifetime of methane itself. Estimates of Greenhouse Warming Potentials (GWPs), e.g., by Rodhe (1990), IPCC (1990), Brühl (1993) have attempted to include these indirect effects.

$$\int Q_x(t)R_x(t)dt = \int Q_{xi} e^{-t/\tau} \, dt$$

$$\int Q_{CO2}(t)R_{CO2}(t)dt = \int Q_{CO2} e^{-t/\tau} \, dt$$

where T is the time following the unit pulse input at T = 0, ΔQ the change in radiative forcing per unit increase in methane concentration; the summation is given over the direct and indirect effects (Wuebbles and Tamaresis, this volume). However, while better than considering the radiative forcing from various greenhouse gases alone, there are still a number of difficulties to be overcome in developing a suitable definition for Greenhouse Warming Potentials. The decay of a CO_2 perturbation is too complex to be described by a single lifetime (Maier-Reimer and Hasselmann, 1987; Rodhe, 1990). The above equation is often applied by using a constant lifetime for methane. However, the lifetime of CH_4 may change with the scenario considered for future growth of CH_4, CO, and NO_x emissions because of feedbacks involving changing CH_4, CO, and NO_x concentrations on atmospheric OH levels. The radiative forcing terms can also change with time because of spectral overlap between absorption bands of greenhouse gases (e.g., CH_4, H_2O, and N_2O).

The generally used approximation for the decay of the pulse, $e^{-t/\tau}$ with a fixed lifetime, τ, is therefore not appropriate in many cases. To estimate indirect effects more accurately, a coupled chemistry-climate model could be used. A better way than assuming a constant decay of direct and indirect radiative forcing may be to superimpose an emission pulse to different scenarios and calculate the decay of the radiative forcing perturbation from a coupled chemistry-climate model (Brühl, 1993). Obviously, because a calculated GWP for methane will be scenario-dependent, a single number for the GWP of CH_4 is not enough, apart from any questions about model uncertainties. Instead, a range must be used which brackets extreme cases. Two cases, which demonstrate this point, are considered in Brühl (1993); IPCC scenario B and a case with constant background CH_4 concentration and CO and NO_x fluxes. For an integration time of 50 years, the GWP in the IPCC scenario was about 30% larger than in the second case and, of course, as the integration time in the calculation is increased, the uncertainty in the resulting GWP also increases.

4. Climatic feedbacks on methane sources

Future climate warming resulting from the addition of greenhouse gases to the atmosphere could also feed back onto natural sources of methane, quite apart from any human mitigation efforts. The primary factors controlling biogenic methane production in terrestrial environments are soil temperature, moisture, the availability of carbon substrates, oxygen (methane is produced only in the absence of oxygen) and microbial populations. The interaction of atmospheric variables with soil climate and methane production dynamics is illustrated with a simplified schematic model in Figure 1. The number of potential positive and negative methane responses to climate variables is sufficiently large that it is beyond the capability of current state-of-the-art models to accurately calculate methane source feedbacks to climate variability at regional or global scales. Most studies to date have only considered the influence of temperature, arising from global warming on methane emissions (e.g., Hameed and Cess, 1983; Harriss and Frolking, 1991). Guthrie (1986) proposed that increased photosynthesis from enhanced atmospheric carbon dioxide levels would be coupled with increased inputs of organic detritus to methanogenic environments and subsequent rates of methane emissions to the global atmosphere. Lashof (1989) has presented a qualitative summary of some of the feedback processes that may influence future concentrations of methane and other trace gases.

Advances in understanding the response of methane emissions to global warming will require a combination of field, laboratory and modeling studies. The field and laboratory studies will have to define the response functions of methanogenesis to each individual soil parameter influenced by climate variability. For example, Livingston and Morrissey (1992) have examined the interannual variability of methane emissions from Alaska arctic tundra and have pointed out the potential importance of this sort of study for assessing the role of CH_4 emissions in climate change. Modeling studies will have to provide the framework for determining the net effect of the numerous atmospheric-biospheric interactions which interact to determine methane production and emissions (Figure 1).

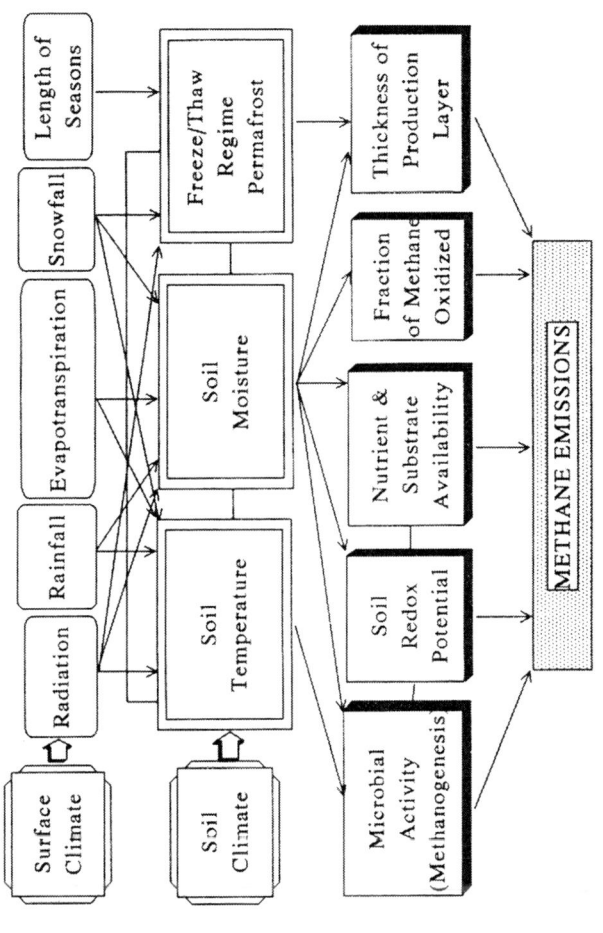

Figure 1. The interaction of atmospheric variables with soil climate and methane production dynamics.

Further insight into the response of methane fluxes to climate change can be obtained by examination of the ice core record of atmospheric methane variations from the Last Glacial Maximum (LGM), when globally averaged surface temperatures were about 5°C lower than today. Although methane levels during the LGM were approximately 350 ppbv (e.g., Raynaud and Chappellaz, this volume), this does not necessarily mean that the sources of methane were a factor of five lower than today. Variations in the atmospheric sink through reaction with OH radicals must also be considered (Khalil and Rasmussen, 1989; Valentin, 1990; Pinto and Khalil, 1991; Thompson et al., 1993) and reconciled with estimates of source strengths during the LGM (Chappellaz et al., 1993). The model studies, mentioned above, calculated OH levels during the LGM about 30-50% higher than today. However, when the effects of lower temperatures on the rate coefficient for reaction of CH_4 with OH during the LGM are considered, it can be seen that an even greater fraction of the methane variations seen in the ice core record are source driven.

Chappellaz et al. (1993) propose that most of the reduction in methane flux was from mid latitude wetlands, arising from temperature and moisture mediated changes in emission rates. Pinto and Khalil (1991) pointed out that changes in emissions rates of methane might well apply to emissions of other species as well (e.g., NO_x and CO), resulting in a stabilizing mechanism for OH levels during climate changes.

Predicting the response of the flux of methane from natural sources, such as wetlands, to future global warming is highly problematical. A good deal of attention has focused on the role of high latitude wetlands, because all general circulation models applied to the problem predict the largest temperature rises to occur at high latitudes. However, there are still substantial differences between model predictions regarding even the sign of changes in ground wetness in high latitudes. Although temperature plays an important role in determining the methane flux through the surface, ground wetness determines the direction of the flux. Observations in a high latitude wetland area around Hudson Bay (Roulet et al., 1992) show a high degree of spatial variability in both magnitude and direction in the methane flux through the surface, with wetter soils acting as sources and drier soils acting as sinks for CH_4. Uncertainties in future climate predictions, therefore, have consequences not only for the magnitude of any feedbacks in methane fluxes from wetlands but also for their direction.

5. Methane control strategies

Human activities that produce CH_4 as a byproduct are rice cultivation and cattle raising, natural gas production and distribution, and waste management. Hogan et al. (1991) have estimated that a reduction in agricultural and industrial emissions of CH_4 of only about 5 to 15% may be sufficient to stabilize its atmospheric level. Brühl (1993) has considered different scenarios involving controls on emissions of NO_x and CO in his analysis. He estimated that stabilization of atmospheric CH_4 levels can be achieved with only a 5% reduction in its emissions, provided NO_x emissions are frozen at current levels along with a 20% reduction in CO emissions. For the case of NO_x and CO emissions, held constant at present values, a reduction in CH_4 emissions of about 9% is required. If anthropogenic CO emissions are held constant and anthropogenic NO_x sources are cut by 80% (corresponding to a reduction in total NO_x emissions of 40%), a reduction in CH_4 emissions of at least 15% is needed for stabilization. These results demonstrate the importance of considering interactions between chemical species in the atmosphere when developing control strategies. It should also be borne in mind, when reading these figures, that they are also subject to the same uncertainties discussed earlier with regard to model formulation.

Controls on emissions of greenhouse gases should be viewed in the larger context of the tradeoffs involving the contributions of other greenhouse gases to global warming. The replacement of coal by natural gas results in a tradeoff of CO_2 emissions for CH_4 emissions. The energy yield of natural gas per kg CO_2 emitted is 1.64 times that for coal (Rodhe, 1990; Lelieveld et al., 1993). However, the favorable impact of a switch from coal to natural gas on total radiative forcing is offset by leakages during gas drilling and distribution. The break even point is at a gas leakage rate of about 5%. This value was estimated by using GWPs with short integration times, as the integration time increases the critical leakage rate also increases (Lelieveld et al., 1993). The average gas leakage rate is 2% or less in western industrialized countries with relatively new distribution systems and can range up to about 5% in areas with older distribution systems such as in the United Kingdom and in cities such as Paris and Brussels (Mitchell et al., 1990; Beck et al., this volume). A leakage rate of about 5% is also estimated for the former Soviet Union (Andronova and Karol, 1993). Therefore, a switch from coal to gas in certain western areas may decrease radiative forcing associated with energy consumption, whereas a switch from coal to natural gas may actually increase radiative forcing in other areas. The consequences of a switch from coal to natural gas on atmospheric methane concentrations are discussed in Mitchell (1993) and Tie and Mroz (1993).

Tradeoffs involving issues other than global warming should also be included in discussions of CH_4 control strategies. A major question concerns the importance of CH_4 in the processes that affect stratospheric ozone depletion (Wuebbles and Tamaresis, this volume). The photochemical oxidation of methane is a major source of stratospheric water vapor. The condensation of water vapor in winter polar stratospheres is a key step in the formation of polar stratospheric clouds and hence in ozone hole formation. However, the reaction of methane with atomic chlorine is the major mechanism for recycling reactive chlorine into more stable reservoirs, such as HCl. It appears that the latter effect dominates, at least on a global scale, but more research is needed to further quantify this statement.

6. Research needs

Several areas for further research were identified at the workshop. These include the use of satellites for mapping sources, techniques for deriving regional and global source inventories from measurements made over relatively small areas and the design of future monitoring networks. A rigorous intercomparison of tropospheric chemistry models is required to explain differences in OH predictions between models (e.g.) shown in Table 1 and in general to better understand the response of these models to various perturbations. The development of coupled climate-chemistry models should also be encouraged for the calculation of Greenhouse Warming Potentials. Remote sensing, especially by satellites, has not been sufficiently utilized for its potential for gathering information about sources of methane. Instruments aboard satellites like SPOT and LANDSAT could be especially useful for mapping wetlands. LANDSAT imaging has already proven useful for mapping the areal extent of biomass burning. UARS instruments (HALOE, ISAMS) are now measuring zonal variations in upper tropospheric methane. Data from these satellites may provide much needed information about zonal variations in major methane sources. This capability is currently lacking in existing surface based monitoring networks. It should also be noted that global scale chemical tracer models (e.g., Fung et al., 1991; Tie and Mroz, 1993) will be needed to help interpret the satellite data in terms of surface fluxes.

7. Summary

In recent years, many attempts have been made to model the tropospheric chemistry of reactive gases including methane, and the possible effects of increasing methane. The models considered here are estimates of tropospheric hydroxyl radicals (OH), the calculation of Global Warming Potentials (GWPs), the magnitude of climatic feedbacks of global warming on the sources of methane, and the effects of control of methane emissions.

Estimates of the global average OH concentration from the eight modeling studies discussed here range from 0.59×10^{-6} to 0.91×10^{-6} molecules/cm^3. The models varied in their transport, input parameters and chemistry. Intercomparison of models is necessary to explain their different OH predictions, and to understand their response to perturbations.

Calculation of the GWP of methane should include both direct (e.g., radiative forcing) and indirect (e.g., oxidation to CO_2) effects. The GWP will be dependent on the scenario of methane increase considered. Coupled climate-chemistry models may be necessary to calculate GWPs and consider all indirect effects correctly.

Future climate warming may affect factors controlling natural methane sources, particularly soil temperature, moisture, oxygen, carbon content, and microbial populations. The number of potential negative and positive methane source feedbacks cannot be adequately modeled at this time. Field and laboratory studies are necessary to determine the response of the soil parameters influenced by changing climate in order to model the processes and reliably predict the outcome.

Recent studies estimate that a 5% to 15% reduction in methane source emissions is necessary to stabilize the atmospheric concentrations. The range is particularly dependent on changes in anthropogenic CO and NO_x emissions. Controls on emissions of greenhouse gases should consider the tradeoffs involving other greenhouse gases, for example, replacement of coal by natural gas trades lower CO_2 emissions for higher CH_4 emissions. Trade offs with other global environmental problems should also be considered. For instance, CH_4 is believed to be the major means of removing reactive chlorine from the atmosphere, thus limiting stratospheric ozone destruction.

Acknowledgments. We gratefully acknowledge all the contributions from working group members to this Report.

References

Andronova, N.G., I.L. Karol. 1993. The contribution of USSR sources to the global methane emission. *Chemosphere, 26* (1-4):111-126.
Brühl, Ch. 1993. The impact of future scenarios for methane and other chemically active gases on the GWP of methane. *Chemosphere, 26* (1-4):731-738.
Butler, J.H., J.W. Elkins, T.M. Thompson, B.D. Hall, T.H. Swanson, and V. Koropalov. 1991. Oceanic consumption of CH_3CCl_3: Implications for tropospheric OH. *J. Geophys. Res., 96*:22,347-22,356.
Chameides, W.L., S.C. Liu, R.J. Cicerone. 1977. Possible variations in atmospheric methane. *J. Geophys. Res., 82*:1,795-1,798.
Chappellaz, J.A., I.Y. Fung, A.M. Thompson. 1993. Atmospheric methane increase since the Last Glacial Maximum. *Tellus, 45B* (in press).
Crutzen, P.J. 1987. Role of the tropics in atmospheric chemistry. In: *Geophysiology of Amazonia* (R. Dickinson, ed.), John Wiley and Sons, New York, New York.
Crutzen, P.J. 1988. Tropospheric ozone: an overview. In: *Tropospheric Ozone: Regional and Global Scale Interactions* (I.S.A. Isaksen, ed.), D. Reidel, Boston, MA.
Dignon, J., J.E. Penner. 1991. Biomass burning: a source of nitrogen oxides in the atmosphere. In: *Proceedings of the Chapman Conference on Biomass Burning* (J.S. Levine, ed.), MIT Press, Cambridge, MA.

Fung, I., J. John, J. Lerner, E. Matthews, M. Prather, L.P. Steele, P.J. Fraser. 1991. Three-dimensional model synthesis of the global methane cycle. *J. Geophys. Res., 96*:13,033-13,065.
Guthrie, P.D. 1986. Biological methagonesis and the CO_2 greenhouse effect. *J. Geophys. Res., 91*:10,847-10,851.
Guthrie, P.D., G. Yarwood. 1991. Analysis of the Intergovernmental Panel on Climate Change (IPCC) future methane simulations, Rep. SYSAPP-91/114, Systems Applications International, San Rafael, CA.
Hameed, S., R.D. Cess. 1983. Impact of a global warming on biospheric sources of methane and its climatic consequences. *Tellus, 35B*:1-7.
Hameed, S., J.P. Pinto, R.W. Stewart. 1979. Sensitivity of the predicted CO-OH-CH_4 perturbation to tropospheric NO_x concentrations. *J. Geophys. Res., 84*:763-767.
Harriss, R.C., S.E. Frolking. 1991. The sensitivity of methane emissions from northern freshwater wetlands to global warming. In: *Global Climate Change and Freshwater Ecosystems* (P. Firth and S.F. Fisher, eds.), Springer-Verlag, NY, p. 48-67.
Hogan, K.B., J.S. Hoffman, A.M. Thompson. 1991. Methane on the Greenhouse agenda. *Nature, 354*:181-182.
IPCC. 1990. *Climate Change: The IPCC Assessment* (J.T. Houghton, G.J. Jenkins, and J.J. Ephraums, eds.), Cambridge Univ. Press, Cambridge, UK.
Kanakidou, M., H.B. Singh, K.M. Valentin, P.J. Crutzen. 1991. A two-dimensional study of ethane and propane oxidation in the troposphere. *J. Geophys. Res., 96,* 15,395-15,413.
Kanakidou, M., P.J. Crutzen. 1993. Scale problems in global tropospheric chemistry modeling: Comparison of results obtained with a three-dimensional model, adopting longitudinally uniform and varying emissions of NO_x and NMHC. *Chemosphere, 26* (1-4):787-802.
Khalil, M.A.K., R.A. Rasmussen. 1984. The atmospheric lifetime of methylchloroform (CH_3CCl_3). *Tellus, 36B*:317-322.
Khalil, M.A.K., R.A. Rasmussen. 1990. Constraints on the global sources of methane and an analysis of recent budgets. *Tellus, 42B*:229-236.
Lashof, D. 1989. The dynamic greenhouse: Feedback processes that may influence future concentrations of atmospheric trace gases and climate change. *Climatic Change, 14*:213-242.
Law, K.S., J.A. Pyle. 1991. Modeling the response of tropospheric trace species to changing source gas concentrations. *Atmos. Environ., 25A*:1,863-1,871.
Lelieveld, J., P.J. Crutzen. 1990. Influences of cloud photochemical processes on tropospheric ozone. *Nature, 343*:227-233.
Lelieveld, J., P.J. Crutzen, Ch. Brühl. 1993. Climate effects of atmospheric methane. *Chemosphere, 26* (1-4):739-768.

Livingston, G.P., L.A. Morrissey. 1992. Methane emissions from Alaska arctic tundra in response to climatic warming. *Proceedings of the International Conference on the Role of the Polar Regions in Global Change* (University of Alaska Press, Fairbanks) (in press).

Liu, S.C., M. Trainer. 1988. Responses of the tropospheric ozone and odd hydrogen radicals to column ozone change. *J. Atmos. Chem.*, 6:221-234.

Logan, J.A., M.J. Prather, S.C. Wofsy, M.B. McElroy. 1981. Tropospheric chemistry: A global perspective. *J. Geophys. Res.*, 86:7,210-7,254.

Lu, Y., M.A.K. Khalil. 1991. Tropospheric OH: Model calculations of spatial, temporal, and secular variations. *Chemosphere*, 23:397-444.

Maier-Reimer, E., K. Hasselmann. 1987. Transport and storage of CO_2 in the ocean - an inorganic ocean-circulation carbon cycle model. *Climate Dynamics*, 2:63-90.

Madronich, S. 1992. Implications of recent total atmospheric ozone measurements for biologically active ultraviolet radiation reaching the Earth's surface. *Geophys. Res. Lett.*, 19:37-40.

Mitchell, C. 1993. Methane emissions from fossil fuels - The need for independent verification. *Chemosphere*, 26 (1-4):441-446.

NASA/Jet Propulsion Laboratory. 1990. Chemical Kinetics and Photochemical data for use in stratospheric modeling, NASA Kinetics Panel Evaluation No. 9, *JPL Publ. 90-1*.

Pinto, J.P., M.A.K. Khalil. 1991. Stability of tropospheric OH levels during ice ages, interglacial epochs and modern times. *Tellus*, 43B:347-352.

Prinn, R., D. Cunnold, R. Rasmussen, P. Simonds, F. Alyea, A. Crawford, P. Fraser, R. Rosen. 1987. Atmospheric trends in methyl chloroform and the global average for the hydroxyl radical. *Science*, 238:946-950.

Prinn, R., D. Cunnold, P. Simmonds, F. Alyea, R. Boldi, A. Crawford, P. Fraser, D. Gutzler, D. Hartley, R. Rosen, R. Rasmussen. 1992. Global average concentration and trend for hydroxyl radicals deduced from ALE/GAGE trichloroethane (methyl chloroform) data for 1978-1990. *J. Geophys. Res.*, 97:2,445-2,462.

Rodhe, H. 1990. A comparison of the contribution of various gases to the greenhouse effect. *Science*, 248:1,217-1,219.

Roulet, N.T., A. Jano, C.A. Kelly, L. Klinger, T.R. Moore, R. Protz, J.A. Ritter, and W.R. Rouse. 1993. The role of the Hudson Bay Lowland as a source of atmospheric methane. *J. Geophys. Res.* (in press).

Sze, N.D. 1977. Anthropogenic CO emission: implications for atmospheric CO-OH-CH_4 cycle. *Science*, 195:673-675.

Thompson, A.M. 1991. New ozone hole phenomenon. *Nature*, 352:282-283.

Thompson, A.M., R.W. Stewart. 1991. Effect of chemical kinetics uncertainties on calculated constituents in a tropospheric photochemical model. *J. Geophys. Res., 96*:13,089-13,108.

Thompson, A.M., R.W. Stewart, M.A. Owens, J.A. Herwehe. 1989. Sensitivity of tropospheric oxidants to global chemical and climate change. *Atmos. Environ., 23*:519-532.

Thompson, A.M., M.A. Huntley, R.W. Stewart. 1990. Perturbations to tropospheric oxidants, 1985-2035. 1. Calculations of ozone and OH in chemically coherent regions. *J. Geophys. Res., 95*:9,829-9,844.

Thompson, A.M., J.A. Chappellaz, I.Y. Fung, T.L. Kucsera. 1993. Atmospheric Methane increase since the Last Glacial Maximum: 2. Effect on oxidants. *Tellus, 45B* (in press).

Tie, X.X., E.J. Mroz. 1992. The potential changes of methane due to an assumed increased use of natural gas: A global three-dimensional model study. *Chemosphere, 26* (1-4):769-776.

Valentin, K.M. 1990. Numerical modeling of the climatological and anthropogenic influences on the chemical composition of the troposphere since the Last Glacial Maximum, Ph.D. thesis, Johannes-Gutenburg-Univ. Mainz, Germany.

Appendix 1 -- List of Participants

Dr. Dilip R. Ahuja, The Bruce Company, 1100 Sixth St., SW, Suite 515, Washington, D.C. 20024, USA

Dr. Natalia Andronova, Dept. of Atmospheric Sciences, University of Illinois at Urbana-Champaign, 105 S. Gregory Ave., Urbana, IL 61801, USA

Dr. Viney Pal Aneja, Research Associate Professor, Department of Marine, Earth and Atmospheric Sciences, College of Physical and Mathematical Sciences, North Carolina State University, Box 8208, Raleigh, NC 27695-8208, USA

Dr. Cynthia S. Atherton, Lawrence Livermore National Laboratory, Atmospheric and Geophysical Sciences, Division, L-262, Livermore, CA 94550, USA

Dr. Don Augenstein, EMCON, 1921 Ringwood Ave., San Jose, CA 95131, USA

Dr. Dominique Bachelet, Research Scientist, Management Technology, 200 S.W. 35th Street, Corvallis, OR 97333, USA

Dr. Lee Beck, MD-63, Air & Energy Engineering Research Laboratory (AEERL), U.S. Environmental Protection Agency, Research Triangle Park, NC 27711, USA

Dr. Max Beran, Institute of Hydrology, Maclean Building, Crowmarsh Gifford, Wallingford, Oxfordshire OX10 8BB, UNITED KINGDOM

Dr. Peter Bergamaschi, Max Planck Institute for Chemistry, Air Chemistry Department, P.O. Box 3060, W-6500 Mainz, GERMANY

Dr. Jean Bogner, Argonne National Laboratory, ES/362, 9700 S. Cass Avenue, Argonne, IL 60439, USA

Dr. Bernard Bonsang, Centre des Faibles Radioactivites, Laboratoire Mixte CNRS-CEA, Avenue de la Terrasse, 91198 Gif-sur-Yvette Cedex, FRANCE

Dr. David Boone, Oregon Graduate Institute, 20,000 N.W. Walker Road, P.O. Box 91,000, Portland, OR 97291-1000, USA

Dr. Christoph Brühl, Abtlg. Chemie der Atmosphäre, Max Planck Institute for Chemistry, P.O. Box 3060, W-6500 Mainz, GERMANY

Dr. Bruce A. Callander, Meteorological Office, Hadley Centre, London Road, Bracknell, RG12 2SY, UNITED KINGDOM

Dr. Jerome Chappellaz, CNRS, Laboratoire de Glaciologie et Géophysique de l'Environnement, BP 96 38402, St. Martin d'Hères, Cedex, FRANCE

Dr. Roger H. Charlier, Professor, Free University of Brussels, 2 Avenue du Congo (Box 23), B-1050 Brussels, BELGIUM

Dr. Chen Zongliang, Director and Professor, Institute of Atmospheric Environment, Chinese Research Academy of Environmental Sciences, Beiyuan, Beijing 100012, P.R. CHINA

Dr. Ralf Conrad, Abteilung Biogeochemie, Max-Planck-Institut für Terrestrische Mikrobiologie, Karl-von-Frisch-Strasse, D-3550 Marburg/Lahn, GERMANY

Dr. Bruce Deck, Research Specialist, Geological Research Division, A-020, Scripps Institution of Oceanography, University of California, San Diego, La Jolla, CA 92093-0220, USA

Dr. Robert A. Delmas, Université Paul Sabatier, Laboratoire d'Aerologie, Unité Associée au CNRS, 118, Route de Narbonne, 31062 Toulouse Cedex, FRANCE

Ms. Roos Eisma, R.J. van de Graaff Laboratory, Rijks Universiteit Utrecht, Postfux 80.000, 3805 TA Utrecht, THE NETHERLANDS

Dr. Su Ge, Oregon Graduate Institute, 20,000 N.W. Walker Road, P.O. Box 91,000, Portland, OR 97291-1000, USA

Dr. Paul H. Glaser, Research Associate, Limnological Research, University of Minnesota, 220 Pillsbury Hall, 310 Pillsbury Drive, S.E., Minneapolis, MN 55455-0219, USA

Dr. Paul Guthrie, Systems Applications International, 101 Lucas Valley Road, San Rafael CA 94903, USA

Dr. Charles Hakkarinen, Environment Division, Electric Power Research Institute, 3412 Hillview Avenue, P.O. Box 10412, Palo Alto, CA 94303, USA

Dr. Wei Min Hao, U.S.D.A. Forest Service, Intermountain Fire Sciences Laboratory, P.O. Box 8089, Missoula, Montana 59807, USA

Dr. Robert Harriss, Professor, Complex Systems Research Center, University of New Hampshire, Durham, NH 03824, USA

Dr. Donald E. Johnson, 209 Animal Science Building, Colorado State University, Fort Collins, CO 80523, USA

Dr. A.G. Judd, School of Environmental Technology, Benedict Building, St. George's Way, Sunderland SR2 7BW, UNITED KINGDOM

Dr. Maria Kanakidou, Max Planck Institute for Chemistry, Atmospheric Chemistry Division, P.O. Box 3060, D-6500 Mainz, GERMANY

Dr. Milind Kandlikar, Department of Engineering & Public Policy, Carnegie-Mellon University, Pittsburgh PA 15213, USA

Dr. Michael Keller, Postdoctoral Fellow, NCAR, P.O. Box 3000, Boulder, CO 80307, USA

Dr. M.A.K. Khalil, Professor, Oregon Graduate Institute, 20,000 N.W. Walker Road, P.O. Box 91,000, Portland, OR 97291-1000, USA

Dr. Donald L. Klass, IGT, 3424 South State Street, Chicago, IL 60616, USA

Prof. Dr. Halil Köse, Dokuz Eylül University, Faculty of Architecture and Engineering, Department of Mining Engineering, Bornova/Izmir 35100, TURKEY

Dr. R.V. Krishnamurthy, Institute for Water Sciences, 1024 Trimpe, Western Michigan University, Kalamazoo, Michigan 49008, USA

Dr. Andrew Lacroix, Senior Analyst, Strategic Planning, Planning Coordination, Petroleos de Venezuela, S.A., Piso 11, Ofc. 11-40, Av. Libertador, la Campina, Apdo. 169, Caracas 1010-A, VENEZUELA

Dr. Gerard Lambert, Professor, Centre des Faibles Radioactivites, Domaine du CNRS, Gif-sur-Yvette 91198, FRANCE

Dr. Keith Lassey, Nuclear Sciences Group, DSIR Physical Sciences, P.O. Box 31-312, Lower Hutt, NEW ZEALAND

Dr. Katharine Law, Department of Physical Chemistry, University of Cambridge, Lensfield Road, Cambridge CB2 1EW UNITED KINGDOM

Dr. Ingeborg Levin, Institut für Umweltphysik, Universität Heidelberg, Im Neuenheimer Feld 366, Heidelberg, Baden-Würtemberg 6900, GERMANY

Dr. Chin-I Lin, Program Manager, Research & Development, PG&E, 3400 Crow Canyon Road, San Ramon, CA 94583, USA

Dr. Yu Lu, Oregon Graduate Institute, 20,000 N.W. Walker Road, P.O. Box 91,000, Portland, OR 97291-1000, USA

Mr. Robert MacKay, Oregon Graduate Institute, 20,000 N.W. Walker Road, P.O. Box 91,000, Portland, OR 97291-1000, USA

Dr. Elaine Matthews, NASA-GISS, 2880 Broadway, New York, NY 10025, USA

Dr. M.A. Miah, Space and Environment Studies Laboratory, University of Arkansas at Pine Bluff, Pine Bluff, AR 71601, USA

Dr. Norman L. Miller, Environmental Research Division, Argonne National Laboratory, 9700 S. Cass Avenue, Argonne, Illinois 60439, USA

Dr. Catherine Mitchell, Science Policy Research Unit, Mantell Building, University of Sussex, Brighton, East Sussex BN1 9RF UNITED KINGDOM

Mr. Francis Moraes, Oregon Graduate Institute, 20,000 N.W. Walker Road, P.O. Box 91,000, Portland, OR 97291-1000, USA

Dr. Leslie A. Morrissey, NASA Ames Research Center, MS 242-4, Moffett Field, CA 94035, USA

Dr. Arvin Mosier, USDA/ARS, P.O. Box E, Fort Collins, CO 80522, USA

Dr. Gene Mroz, Los Alamos National Laboratory, Isotope Geochemistry Group, INC-7, MS-J514, Los Alamos, NM 87545, USA

Dr. D. Murdiyarso, Environmental Research Centre, Institut Pertanian Bogor, P.O. Box 145, Bogor 16143, West Java, INDONESIA

Dr. J.C. Murrell, Department of Biological Sciences, University of Warwick, Coventry, West Midlands CV4 7AL, UNITED KINGDOM

Dr. Heinz-Ulrich Neue, Head, Soil & Water Sciences Division, International Rice Research Institute, P.O. Box 933, Manila, PHILIPPINES

Dr. David Olszyk, Research Ecologist, U.S. EPA/ERL-Corvallis, 200 S.W. 35th St., Corvallis, OR 97333, USA

Dr. Rebecca L. Peer, Radian Corporation, P.O. Box 13000, Research Triangle Park, NC 27709, USA

Dr. Régis Pelet, Institut Français du Pétrole, BP311, 92506 Rueil Cedex, FRANCE

Dr. Steve Piccot, Southern Research Institute, P.O. Box 13825, Research Triangle Park, NC 27709-3825 USA

Dr. Joe Pinto, U.S.EPA, OAQP, MD 14, Research Triangle Park, NC 27711, USA

Dr. R.A. Rasmussen, Professor, Oregon Graduate Institute, 20,000 N.W. Walker Road, P.O. Box 91,000, Portland, OR 97291-1000, USA

Dr. Dominique Raynaud, Director, Laboratoire de Glaciologie et Géophysique de l'Environnement, BP 96 38402, St. Martin d'Hères Cedex, FRANCE

Dr. William S. Reeburgh, Department of Geosciences, Physical Sciences Building, University of California, Irvine, California 92717 USA

Dr. John Ritter, Atmospheric Sciences Division, NASA Langley Research Center, Mail Stop 483, Hampton, VA 23665-5225, USA

Dr. Jeanne M. Robinson, Chemical & Laser Sciences Div., Los Alamos National Lab., P.O. Box 1663, MS J567, Los Alamos, NM 87545, USA

Dr. Jan Rotmans, National Institute of Public Health & Environmental Protection, P.O. Box 1, 3720 BA, Bilthoven, THE NETHERLANDS

Dr. Corinne Rouland, Université Paris Val de Marne, U.F.R. de Sciences, Laboratoire de Zoologie et de, Biologie des Populations, Avenue du Général de Gaulle, 94010 Creteil Cedex, FRANCE

Dr. Nigel T. Roulet, Assoc. Professor, Department of Geography, York University, 4700 Keele Street, North York, Ontario, CANADA M3J 1P3

Dr. Sam Sadler, Oregon Department of Energy, 625 Marion Street, N.E., Salem, Oregon 97310, USA

Dr. Peter W. Sage, Head of Environmental Science, British Coal Corporation, Coal Research Establishment, Stoke Orchard, Cheltenham, Glos. GL52 4RZ, UNITED KINGDOM

Dr. Robert Sepanski, Research Associate, CDIAC, Oak Ridge National Laboratory, P.O. Box 2008, Oak Ridge, Tennessee 37831-6335, USA

Ms. Martha Shearer, Oregon Graduate Institute, 20,000 N.W. Walker Road, P.O. Box 91,000, Portland, OR 97291-1000, USA

Dr. Joe Sickles, AREAL, Environmental Protection Agency, Research Triangle Park, NC 27709, USA

Ms. Amy Smith, Center for Climatic Research, Department of Geography, University of Delaware, Newark, Delaware 19716, USA

Dr. Kirk Smith, East-West Center, Environmental Policy Institute, 1777 East-West Road, Honolulu, Hawaii 96848, USA

Dr. Peet M. Soot, President, Northwest Fuel Development, Inc., P.O. Box 25562, Portland, OR 97225, USA

Dr. Bernhard Stauffer, Professor, Physikalisches Institut, Universität Bern, Sidlerstrasse 5, CH-3012 Bern, SWITZERLAND

Dr. Paul Steudler, Ecosystems Center, Marine Biological Laboratory, Woods Hole, Woods Hole, Massachusetts 02543, USA

Dr. Charles M. Stevens, Chemistry Division, Argonne National Laboratory, 9700 South Cass Avenue, Argonne, Illinois 60439, USA

Dr. Robert Striegl, U.S.G.S., Branch of Regional Research, Box 25046, MS 413, Denver, CO 80225, USA

Dr. Anne Thompson, Atmospheric Chemistry & Dynamics Branch, Goddard Space Flight Center, NASA/Code 916, Greenbelt, MD 20771, USA

Dr. Susan Thorneloe, Program Manager, EPA/Air & Energy Engineering Research Laboratory, MD-63, Research Triangle Park, NC 27711, USA

Dr. Xuexi Tie, Research Scientist, IGPP, K305, Los Alamos National Laboratory, Los Alamos, NM 87545, USA

Dr. Hikmet Toprak, Associate Professor, Dokuz Eylül University, Faculty of Architecture and Engineering, Department of Environmental Engineering, Bornova/Izmir 35100, TURKEY

Ms. Margaret Torn, Energy & Resources Group, University of California, Berkeley, CA 94720, USA

Dr. Henry F. Tyrrell, Plant and Animal Science Staff, Cooperative State Research Service, USDA, Room 330 Aerospace Building, 901 D Street, S.W., Washington, D.C. 20250-2200, USA

Dr. David Valentine, National Resource Ecology Lab., Colorado State University, Fort Collins, CO 80523, USA

Dr. Martin Wahlen, Professor, Scripps Institution of Oceanography, University of California, San Diego, La Jolla, CA 92093-0220, USA

Dr. Reiner Wassmann, Fraunhofer Institute, Kreuzeckbahnstr. 19, Garmisch-Partenkirchen 8100, GERMANY

Dr. Peter Westermann, Assoc. Professor, Department of General Microbiology, University of Copenhagen, Sølvgade 83H, Copenhagen 1307K, DENMARK

Dr. Michael J. Whiticar, School of Earth and Ocean Sciences, P.O. Box 1700, University of Victoria, Victoria, B.C. V8R 2Y2, CANADA

Dr. D.J. Williams, CSIRO Division of Coal Energy Technology, P.O. Box 136, N. Ryde, NSW 2113, AUSTRALIA

Dr. Donald J. Wuebbles, Lawrence Livermore National Laboratory, 7000 East Avenue/L-262, Livermore, CA 94550, USA

Mr. Weining Zhao, Oregon Graduate Institute, 20,000 N.W. Walker Road, P.O. Box 91,000, Portland, OR 97291-1000, USA

Working Group 1. Formation and Consumption of Methane

R. Striegl, A. Mosier, Co-Chairman W.S. Reeburgh, M.J. Whiticar, E. Matthews, P.H. Glaser, J.C. Murrell, A. Smith, P. Westermann, L.A. Morrissey, D.R. Boone, M. Torn, P. Steudler, D. Valentine, Co-Chairman N.T. Roulet

Working Group 2. Measurement and Research Methods

M. Beran, M. Keller, G. Mroz, R. Wassmann, S. Piccot, Co-Chairman R. Conrad, P. Bergamaschi, B. Deck, (Co-Chairman R.A. Rasmussen not shown), (J. Ritter not shown)

Working Group 3. Record of Methane

Co-Chairman M. Wahlen, K. Lassey, M.J. Shearer, C.M. Stevens, D. Raynaud, M.A.K. Khalil, R.V. Krishnamurthy, I. Levin, Co-Chairman B. Stauffer, R. Sepanski (F. Moraes not shown)

Working Group 4. Sources, Sinks, and Mass Balance

R. Eisma, Z.-L. Chen, H. Toprak, S. Thorneloe, D. Augenstein, A. Lacroix, D.L. Klass, S. Ge, M.A. Miah, Co-Chairwoman D. Bachelet, D.E. Johnson, H.F. Tyrrell, H. Köse, A.G. Judd, Co-Chairman H.-U. Neue, G. Lambert, J. Bogner, R.A. Delmas, B.A. Callander, W.-M. Hao, D.J. Williams, P.W. Sage, L.L. Beck, B. Bonsang, R.H. Charlier, C. Rouland, D. Olszyk (P.M. Soot not shown)

COMBINED WORKING GROUPS

Co-Chairman J.P. Pinto, X.-X. Tie, Y. Lu, C.-I. Lin, V.P. Aneja, C. Mitchell, R. Harriss, C.S. Atherton, D.J. Wuebbles, M. Kanakidou, J.M. Robinson, J. Rotmans, K. Law, P. Guthrie, N.L. Miller, D. Murdiyarso, Co-Chairman C.H. Brühl, J. Chappellaz, M. Kandlikar, Ritter, D.R. Ahuja, R.L. Peer, N.G. Andronova, W. Zhao, R. Pelet, Chairwoman A. Thompson (S. Sadler not shown)

Appendix 2

CHEMOSPHERE

Contents

MEASUREMENT AND RESEARCH TECHNIQUES

1   Murrell, J.C., V. McGowan, and D.L.N. Cardy.
    Detection of methylotrophic bacteria in natural samples by molecular probing techniques.

13 Schupp, M., P. Bergamaschi, G.W. Harris, and P.J. Crutzen.
    Development of a tunable diode laser absorption spectrometer for measurements of the $^{13}C/^{12}C$ ratio in methane.

23 Kirchgessner, D.A., S.D. Piccot, A. Chadha, and T. Minnich.
    Estimation of methane emissions from a surface coal mine using open-path FTIR spectroscopy and modeling techniques.

45 Mroz, E.J.
    Deuteromethanes: Potential fingerprints of the sources of atmospheric methane.

55 Burke, R.A., Jr.
    Possible influence of hydogen concentration on microbial methane hydrogen isotopic composition.

MASS BALANCE ISSUES

69 Kammen, D.M., and B.D. Marino.
    On the origin and magnitude of pre-industrial anthropogenic CO_2 and CH_4 emissions.

87   Ward, G.M., K.G. Doxtader, W.C. Miller and D.E. Johnson.
     Effects of intensification of agricultural practices on emission of radiatively active gases.

95   Lassey, K.R., D.C. Lowe, C.A.M. Brenninkmeijer, and A.J. Gomez.
     Atmospheric methane and its carbon isotopes in the southern hemisphere: Their time series and an instructive model.

111 Andronova, N.G. and I.L. Karol.
     The contribution of USSR sources to global methane emission.

127 Khalil, M.A.K., M.J. Shearer, and R.A. Rasmussen
     Methane sources in China: Historical and current emissions.

143 Thom, M., R. Bösinger, M. Schmidt, and I. Levin.
     The regional budget of atmospheric methane of a highly populated area.

161 Levin, I., P. Bergamaschi, H. Dörr, and D. Trapp.
     Stable isotopic signature of methane from major sources in Germany.

SOURCES: RUMINANTS AND OTHER ANIMALS

179 Williams, D.J.
     Methane emissions from manure of free-range dairy cows.

189 Lodman, D.W., M.E. Branine, B.R. Carmean, P. Zimmerman, G.M. Ward and D.E. Johnson.
     Estimates of methane emissions from manure of U.S. cattle.

SOURCES: RICE AGRICULTURE

201 Wassmann, R., H. Papen and H. Rennenberg.
     Methane emission from rice paddies and possible mitigation strategies.

219 Bachelet, D., and H.U. Neue.
Methane emission from wetland rice areas of Asia.

239 Chen Zongliang, Li Debo, Shao Kesheng, and Wang Bujun.
Features of CH_4 emission from rice paddy fields in Beijing and Nanjing.

247 Parashar, D.C., Prabhat K. Gupta, J. Rai, R.C. Sharma, and N. Singh.
Effect of soil temperature on methane emission from paddy fields.

251 Masscheleyn, P.M., R.D. DeLaune, and W.H. Patrick, Jr.
Methane and nitrous oxide emissions from laboratory measurements of rice soil suspension: effect of soil oxidation-reduction status.

SOURCES: WETLANDS

261 Bartlett, K.B., and R.C. Harriss.
Review and assessment of methane emissions from wetlands.

321 Westermann, P.
Temperature regulation of methanogenesis in wetlands.

329 Vourlitis, G.L., W.C. Oechel, S.J. Hastings, and M.A. Jenkins.
The effect of soil moisture and thaw depth on CH_4 flux from wet coastal tundra ecosystems on the north slope of Alaska.

339 Morrissey, L., D. Zobel, and G. Livingston.
Significance of stomatal control on methane release from Carex-dominated wetlands.

357 Torn, M.S., and F.S. Chapin III.
Environmental and biotic controls over methane flux from Arctic tundra.

Sources: Landfills

369 Bogner, J., and K. Spokas.
Landfill CH_4: Rates, fates and role in global carbon cycle.

387 Peer, R.L., S.A. Thorneloe, and D.L. Epperson.
A comparison of methods for estimating methane emissions from landfills.

401 Nozhevnikova, A.N., V.S. Lebedev, A.B. Lifshitz, and G.A. Zavarzin.
Emission of methane into atmosphere from landfills in the USSR.

Sources: Energy and Industrial Sources

419 Creedy, D.P.
Methane emissions from coal related sources in Britain: Development of a methodology.

441 Mitchell, C.
Methane emissions from the coal and natural gas industries in the UK.

447 Beck, L.
A global methane emissions program for landfills, coal mines, and natural gas systems.

453 Kirchgessner, D.A., S.D. Piccot, and J.D. Winkler.
Estimate of global methane emissions from coal mines.

Sources: Minor Sources

473 Khalil, M.A.K., R.A. Rasmussen, M.J. Shearer, S. Ge, and J.A. Rau.
Methane from coal burning.

479 Smith, K.R., M.A.K. Khalil, R.A. Rasmussen, S.A. Thorneloe, F. Manegdeg and M. Apte.
Greenhouse gases from biomass and fossil fuel stoves in developing countries: a Manila pilot study.

507 Lacroix, A.V.
Unaccounted-for sources of fossil and isotopically-enriched methane and their contribution to the emissions inventory: A review and synthesis.

559 Hovland, M., A.G. Judd, and R.A. Burke, Jr.
The global flux of methane from shallow submarine sediments.

579 Lambert, G., and S. Schmidt.
Reevaluation of the oceanic flux of methane: uncertainties and long term variations.

591 Rasmussen, R.A., M.A.K. Khalil, and F. Moraes.
Permafrost methane content: 1. Experimental data from sites in northern Alaska.

595 Moraes, F. and M.A.K. Khalil
Permafrost methane content: 2. Modeling theory and results.

609 Kvenvolden, K.A., and T.D. Lorenson
Methane in permafrost -- preliminary results from coring at Fairbanks, Alaska.

617 Rouland, C., A. Brauman, M. Labat, and M. Lepage.
The production of methane by termites in tropical area.

623 Martius, C., R. Wassmann, U. Thein, A. Bandeira, H. Rennenberg, W Junk, and W. Seiler.
Methane emission from wood-feeding termites in Amazonia.

633 Toprak, H.
Methane emissions originating from the anaerobic waste stabilization ponds case study: Izmir wastewater treatment system.

SINKS: REMOVAL PROCESSES

641 Lu, Y., and M.A.K. Khalil
Methane and carbon monoxide in OH chemistry: the effects of feedbacks and reservoirs generated by the reactive products.

657 Andronova, N. G., I.L. Karol, and M.E. Schlesinger.
Cause-and-effect analysis of the photochemical interactions among CH_4, CO, O_3, and OH in the global troposphere.

675 Ojima, D.S., D.W. Valentine, A.R. Mosier, W.J. Parton, and D.S. Schimel.
Effect of land use change on methane oxidation in temperate forest and grassland soils.

687 Bender, M., and R. Conrad.
Kinetics of methane oxidation in oxic soils.

697 Dörr, H., L. Katruff, and I. Levin.
Soil texture parameterization of the methane uptake in aerated soils.

715 Striegl, R.G.
Diffusional limits to the consumption of atmospheric methane by soils.

721 Yavitt, J.B., J.A. Simmons, and T.J. Fahey.
Methane fluxes in a northern hardwood forest ecosystem in relation to acid precipitation.

The Environmental Role of Methane and Future Trends

731 Brühl, C.
The impact of the future scenarios for methane and other chemically active gases on the GWP of methane.

739 Lelieveld, J., P.J. Crutzen, and C. Brühl.
Climate effects of atmospheric methane.

769 Tie, XueXi, and E.J. Mroz.
The potential changes of methane due to an assumed increased use of natural gas: a global three-dimensional model study.

777 Karol, I.L., A.A. Kiselev, and V.A. Frolkis.
The key role of methane release rate in the expected ozone content and composition changes of the greenhouse atmosphere in the next century.

787 Kanakidou, M., and P.J. Crutzen.
Scale problems in global tropospheric chemistry modeling: comparison of results obtained with a three-dimensional model, adopting longitudinally uniform and varying emissions of NO_x and NMHC.

803 Khalil, M.A.K., and R.A. Rasmussen.
Decreasing trend of methane: unpredictability of future concentrations.

(*Chemosphere* is published by Pergamon Press Ltd., Headington Hill Hall, Oxford OX3 0BW, England.)

Index

^{13}C 8-10, 12, 28, 54, 63, 64, 66, 69, 70, 72, 77, 79, 80-83, 94, 96-98, 118-121, 141, 143, 144, 146, 148, 149, 151, 152, 154, 156, 157, 159-162, 353, 448 (see also isotope)
^{13}C/^{12}C ratio 80, 118
^{14}C 8-10, 12, 13, 39, 94, 95, 121, 194, 196, 269, 271, 353, 448
^{14}C content of methane 121
^{14}C/^{12}C 121
^{222}Radon tracer technique 16
^{222}Radon 16, 17, 25, 96
abandoned mines 401, 407, 460, 461 (see also inactive mines)
abiogenic sources 94, 353
ABLE (experiments) 321, 322, 338, 339, 350
acetate 102, 105, 110-115, 119-121, 130, 132, 142, 153, 204, 213, 223, 260, 261, 266, 308, 363, 364
acetic acid 103, 104, 224, 276, 363
acetotrophic methanogens 260
acidogenesis 103
actinic flux 176
aerobic 107, 116, 120, 134, 143, 154, 256, 263, 265, 271, 279, 364, 369, 373, 382
aerobic bacteria 107
aerobic decomposition 364, 373, 382
Africa 52, 240, 243, 246, 250, 257, 300, 302, 303, 308, 309, 311, 316, 321-323, 375, 376, 381, 384, 409, 447
agricultural wastes 300, 459
air bubbles 39, 40, 91, 93 (see also ice cores)
airborne eddy correlation 18-21, 311, 338
aircraft 8, 11, 15, 17-23, 29, 316, 344, 350
alluvial formations 317, 319, 339
Amazon floodplain 66, 319, 335, 337, 342, 349
ambient wind field 20
anaerobic bacteria 28, 107, 109, 233, 263, 272, 363, 436
anaerobic decomposition 108, 262, 263, 320, 362-365, 370, 373, 389
anaerobic digesters 220, 383, 386
anaerobic fermentation 221, 265, 269, 443
animal waste management systems 387, 388
anoxic environments 102, 104
Antarctic ozone hole 496, 507
anthropogenic emissions 4, 58, 66, 483, 484
anthropogenic sources 2, 4, 10, 48, 58, 62-64, 66, 80, 82, 83, 93, 139, 185, 186, 191, 196, 432, 445-447, 465, 478, 482-484, 488, 492
Archaeobacteria 109
ash 235, 301, 304, 380
asphalt 180, 195
atmospheric methane 1, 38, 39, 41, 43, 46, 48, 51-54, 62-64, 66, 68-70, 72, 76, 77, 79, 80, 82-84, 90, 94-96, 98, 186, 257, 299, 308, 427, 447, 525
bacteria 27, 28, 102-112, 115, 116, 119-121, 131, 141, 199-201, 205, 206, 208, 233, 234, 238, 259-264, 266, 270, 272, 275, 277, 279, 308, 363, 366, 369, 382, 387, 436, 442, 463
bacterial gases 141, 143, 148
biochemical methane potential (BMP) 368
biochemical pathways 128
biochemistry of methanogenesis 109
biogas 139, 180, 195, 224, 367, 370, 378, 381, 383
biogas pits 180, 195, 383
biogenic emissions 1, 2, 12, 13, 63, 65, 69, 77, 94, 95, 98, 102, 118, 121, 139, 180, 254, 259, 308, 433, 435, 465, 523
biomass burning 62-65, 69, 72, 73, 76, 77, 79, 80, 82-84, 94, 95, 97, 98, 129, 141, 157, 160-162, 180, 191, 195, 255, 299, 300, 302, 306-311, 353, 445, 459, 465, 478, 484, 517, 519, 527
biomolecular techniques 28
blending height 21-23
BMP 368
BOD 381-383
bogs 317, 336
boreal 18, 21, 23, 131, 134, 302, 303, 307, 309, 310, 319, 322, 324, 328-339, 342, 344, 346, 348, 350-353, 459, 461, 462
boundary conditions 8, 12, 15-17, 19, 21, 29, 176, 322, 339, 472, 488, 491, 516, 519
calibration 2, 9, 46, 69, 84, 91, 170, 171
CAMREX 321, 350
carbon 10, 16, 17, 48, 53, 62-64, 66, 69, 70, 72, 77, 78, 83, 84, 92, 94, 95, 97, 103-109, 112-116, 118-121, 138, 139, 141-152, 154-157, 159, 224, 234, 254, 260, 261, 263, 265, 268, 269, 272, 274-277, 283, 285, 287, 299-301, 303-308, 321, 363, 367,

370, 379, 400, 404, 413, 414, 427,
428, 433, 435, 440, 441, 459, 460,
469-474, 478, 479, 481-484, 486,
488, 489, 491, 497, 507-513, 515,
523, 528
carbon isotopes 77, 118, 147-149
carbon monoxide 48, 53, 94, 112, 300, 308, 413,
469-474, 478, 479, 481-484, 486,
488, 491, 497, 513, 515
carbon-14 63, 69, 84, 94, 427, 428 (see also ^{14}C)
catagenesis 139
cattle 2, 3, 58, 65, 73-75, 77, 80, 82, 83, 105, 141,
180, 181, 186, 188, 190-193, 195,
201, 203, 204, 206, 208-216,
218-222, 224, 225, 387, 458, 525
CD-isotope ratio 152
cellulose 105, 200, 201, 203, 204, 208, 209, 363,
365, 367, 369, 436
CH_2O 300, 469, 473-477, 493
CH_3Br 470
CH_3CCl_3 8, 184, 185, 470, 520
CH_3Cl 94, 308, 470
chambers 15, 25, 23, 28, 29, 91, 232, 233, 316,
350
chaparral 302, 304
charcoal production 303, 307, 309, 445, 459
chemical manufacturing 2, 4, 17, 20, 25, 39, 42,
53, 54, 62, 93, 132, 169, 184, 202,
262, 264, 272, 274, 286, 288, 307,
365, 367, 368, 399, 423, 424, 446,
470, 471, 475, 477, 480-482, 486,
488, 493, 497, 499, 501, 505, 507,
508, 511, 514, 516, 519, 521, 525,
527
chlorofluorocarbons 495, 497
Cl 91, 133, 173, 275, 465
clathrate 42 (see also ocean clathrates)
clathrate formation 42
climate 8, 23, 39, 49, 51-54, 83, 84, 93, 135, 138,
187, 230, 242, 256, 287, 314, 316,
345, 346, 347, 349, 351, 352,
370-372, 414, 416, 462, 469, 470,
491, 494, 504, 507-509, 511-513,
515, 522-525, 527, 528
climatic changes 38, 48-54, 93, 169, 175, 255,
280, 335, 337, 345, 471, 486, 504,
508, 509, 513, 514, 521, 522, 527
clouds 54, 175-177, 496, 507, 512, 520, 526
coal 10, 12-14, 48, 58, 64, 65, 68, 69, 72, 74, 75,
80, 94, 95, 139, 141, 144, 157,
162, 180, 195, 196, 399-408, 407,
409, 410, 412-414, 422, 426-428,
433, 438, 439, 444-446, 460, 461,
465, 525, 526, 528
coal depth 403, 404, 460
coal mines 10, 12, 13, 180, 195, 400, 402, 403,
405, 407-409, 413, 422, 427, 428,
465
coal mining 14, 64, 65, 68, 69, 80, 95, 196, 399,
400, 405, 406, 422, 427, 460, 461
coal rank 403, 404
coke production 399, 423, 424, 428, 446
combustion 9, 14, 196, 300-306, 308, 310, 373,
380, 399, 400, 415, 423-426, 428,
432, 445, 446, 459, 478, 517
combustion products 300-302, 304, 308
complete combustion 17, 29, 49, 103, 105, 107,
108, 119, 220, 261, 300, 301, 306,
322, 387, 479, 480, 488, 489, 515
complex organic matter 102, 103
compost 110, 234, 235, 239, 285, 364
concentration 1, 4, 9, 10, 12-18, 41, 46, 49, 52,
53, 55, 62, 72, 73, 76, 77, 80,
82-84, 89-94, 96, 97, 103,
143-145, 155, 162, 169, 170, 176,
184-186, 232, 254, 261, 285, 287,
302, 345, 352, 403, 432, 434, 438,
442, 444, 469-471, 480-482, 486,
487, 490, 494, 497-501, 504, 505,
508-510, 512, 522, 527
Controlled Landfill Project 369
crop calendars 240, 241
crop wastes 310
crystalline basement sources 435, 444
D/H ratio 120
dating air trapped in ice cores 39
decomposition 51, 102-104, 107, 108, 121, 262,
263, 268-270, 272, 281, 285, 320,
336, 362-365, 367, 368, 370, 373,
382, 389, 436
deep earth gas 435, 444
deepwater rice 256, 259, 283, 287
deuterium 63, 77, 98, 120, 151
diagenetic gases 141
diffusion through the floodwater 233
distribution 7, 13, 22, 23, 29, 49, 52, 53, 56, 63,
64, 94, 97, 98, 144, 147, 168-171,
173, 174, 187, 188, 219, 257, 260,
267, 268, 280, 305, 316, 318, 320,
323, 338, 340, 344, 350, 352, 365,
389, 399, 413-415, 418, 419, 422,
425, 426, 428, 436, 440, 443, 446,
461, 462, 465, 483, 494-497, 504,
515, 516, 518, 520, 525, 526

distribution of methane 29, 98, 169, 170, 187, 188, 340, 443
distribution system (energy) 415, 419
DNA probes 28, 130
DOC 370, 374, 376, 377, 379
dryland rice 239, 240, 250, 256 (see also upland rice)
ebullition 26, 131, 233, 238, 287, 335, 337, 351, 352, 461
ecosystems 10, 23-25, 97, 98, 103, 104, 115, 117, 283, 302-304, 306, 309, 311, 315-317, 319, 320, 324, 335, 338, 339, 344-346, 349-351, 445, 459, 461
eddy accumulation 18
eddy correlation technique 17, 19
Eh 238, 272-277, 285 (see also redox potential)
electron acceptors 104-109, 113-115, 119-121, 130, 261, 265, 269
emission factors 14-17, 22, 25-28, 52, 56, 62, 66, 68, 129, 132, 141, 143, 160, 180-185, 187, 191, 193, 225, 231-234, 238, 240, 241, 245, 246, 250, 254, 255, 257, 259, 262, 263, 266-268, 271, 274, 278-280, 282, 285-288, 299, 301-309, 311, 314-316, 320, 322, 323, 335-340, 344-350, 363, 365, 370, 372, 373, 375, 377-380, 382, 383, 386-390, 399-401, 405, 407, 409, 410, 412, 416, 419, 421-425, 427, 428, 434, 435, 437, 440, 444, 447, 448, 457-460, 462, 482, 486, 491, 503, 508, 509, 522, 524
Everglades 24, 66-68
extrapolating factor 181
extrapolation process 231
extrapolation techniques 21
facultative lagoons 382
fatty acids 103-105, 119, 200, 205, 206, 208, 261, 269
feedbacks on OH 521
fens 317, 321, 336, 337, 342, 348, 349, 351
fermentation 103-105, 118, 119, 130, 141, 142, 148, 149, 151, 152, 154, 156, 199-201, 204, 206-209, 213, 215, 219-221, 223, 261, 265, 269-271, 363, 386, 388, 443, 457
fertilization 131, 233, 234, 238, 251, 264, 270, 284, 285, 287, 345
fertilizer 26, 183, 237-239, 261, 262, 269, 282-285, 288

field experiments 25, 30, 134, 135, 271
fires 48, 64, 66, 68, 69, 301, 303, 309-311, 314, 425, 438, 444, 459
firewood burning 303, 309
firn 39, 41, 42, 94
flaming phase 301
flood irrigation 236, 237, 239
flooding 26, 239, 257, 263, 265, 270, 272, 275-277, 280-282, 287
floodplain 66, 317, 319, 320, 335, 337, 342, 349
floodwater temperature 266, 267
flux 10-12, 14-30, 52, 53, 62-69, 72, 73, 76, 79, 80, 82-84, 91, 94, 95, 97-99, 129, 131, 134, 138, 139, 141, 156, 157, 159-162, 176, 181-184, 230-234, 238-244, 246, 250, 262, 263, 266, 267, 271, 278, 280, 281, 283, 284, 286, 287, 300, 308, 315, 316, 319, 320, 322, 324, 335-340, 344-352, 365, 423, 433, 435, 438-440, 443, 444, 447, 458, 461, 462-464, 487, 493, 501, 502, 524
flux gradient technique 15, 17
flux measurements 10-12, 18-21, 23, 24, 26, 27, 30, 91, 99, 231-234, 238, 240, 316, 319, 320, 322, 335-337, 340, 346, 347, 350, 462, 464
forest fires 66, 301, 309, 425, 459
forests 300, 307-311, 323, 335, 462, 464
formaldehyde 469, 473
formate 102, 104, 105, 110-112, 130, 142, 200, 260, 261, 263, 308, 364
fossil fuel combustion 196, 399, 423, 424, 426, 428, 478, 517
fossil methane 13, 94, 95, 448
fractionation 9, 63, 66, 69, 72, 77, 91, 94, 96-98, 118, 120, 121, 139, 143, 147, 148, 152, 154, 155, 157, 162
fuel wood 300, 307, 459
gas chromatography 7, 17, 92, 232
gas drainage systems 401
gas extraction 414, 422, 428
gas hydrates 139, 432, 439, 441, 442, 444, 448, 462, 463, 510
gas seepages 438, 439, 442, 443
gas transport 25, 53, 238
geographical information systems 11, 29
geological sources 432, 442, 463
geothermal 104, 139, 140, 151, 399, 425, 426, 428, 434, 446
glacial 38, 41-43, 46, 49, 51-54, 314, 511, 524
glacial-interglacial change 38, 42, 43, 49, 51, 53,

global biomass burning 299, 300
global emissions 64, 181, 184, 187, 189, 192, 194, 195, 240, 243, 248, 307, 308, 310, 378, 405, 407, 416, 417, 423, 426-428, 458, 462
global environment 469, 513
global flux 52, 80, 183, 231, 463
global warming 4, 54, 201, 504, 511, 513, 521, 523-527
Global Warming Potentials 513, 527 (see also GWPs)
global wetland areas 315, 317
gradient 8, 15, 17, 29, 49, 54, 95, 168, 352, 407, 434
grasslands 309, 310, 322, 352
green manure 285, 286
greenhouse effect 267, 508, 512, 514
Greenhouse Warming Potentials (GWPs) 521
groundwater 265
groundwater abstraction 446, 447
growing season 183, 231, 233, 234, 238, 240, 241, 243, 250, 256, 257, 262, 264, 266, 287, 462
GWPs (Greenhouse Warming Potentials) 513, 521, 525, 527
H_2 (see hydrogen)
H_2 partial pressure 106
H_2O_2 93
H_2S 105, 108, 265, 272, 470
HCHO 93
HCl 495, 526
hemicellulose 200, 363, 365, 367, 436
heterogeneous processes 493, 507
high boreal/arctic wetlands 337
high latitude wetlands 352, 524
Holocene 51, 54, 93, 314
hydrocarbon-bearing sediments 433, 442, 444, 462
hydrocarbons 11, 53, 103, 117, 139-144, 146-148, 150, 151, 175-177, 269, 300, 433, 434, 470, 478, 482-484, 488, 491-493
hydrochloric acid 495, 496 (see also HCl)
hydrogen 103, 104, 107, 120, 121, 130, 138, 139, 141, 144, 147, 149-152, 154-157, 159, 200, 204-207, 223, 224, 260, 279, 300, 308, 363, 364, 434, 458, 470, 472-480, 490, 494, 495, 500, 507
hydroperoxyl radical 470
hydrothermal 116, 139, 151, 433-435, 442, 444, 462, 463
hydrothermal sources 434, 442, 444, 462
hydroxyl radical 154-156, 173, 465, 469-477, 480-493, 495, 501, 512, 513, 515, 527 (see also OH)
ice cores 4, 38-44, 46-56, 62, 66, 68, 72, 73, 76, 77, 84, 89-94, 97, 99, 185, 186, 194, 254, 314, 481, 524
ice record 39, 47, 49
ice-flow model 39
in situ measurements 96, 132
inactive mines 400 (see also abandoned mines)
incomplete combustion 300, 301
industrial sources 180, 195, 399, 422-428, 445, 446
industrial wastes 379
infrared radiation 13, 17, 38, 306, 319, 508-510, 513, 514
inhibitors of methane formation 261
instrumental time lag 20
inter-laboratory comparisons 96
interhemispheric concentration 92
interhemispheric gradient 49, 54, 95
irrigated rice 240, 256-258, 268, 287
irrigation 183, 231, 236-239, 242, 249, 251, 256, 258, 280, 281, 285
irrigation regimes 239
isolation depth 39
isotope calibration 9
isotope fractionation 139, 147, 148, 152, 154, 162
isotope measurements 8, 10, 90, 149
isotope signatures 15, 138, 139, 144, 147, 148, 150-152, 155
isotopes 7, 9, 10, 46, 77, 118, 138, 144, 147-149, 151, 154-156, 162, 186, 314, 352
isotopic abundances 1, 8, 9, 12, 29, 39, 49, 50, 54, 62-66, 68-70, 72, 73, 76, 77, 79, 80, 82-84, 78, 90, 94, 96-99, 117-121, 147-149, 152, 186, 187, 195, 315, 350, 353, 447, 462
kerogen sources 142
kerogens 143, 148
lakes 64, 66-68, 110, 111, 180, 195, 320, 335, 337, 351
land preparation 282-284
landfill gas production 141, 353, 364-374, 377-381, 389, 446, 460
landfills 11, 13, 14, 48, 58, 62, 64, 65, 72, 73, 76, 80, 83, 84, 94, 116, 117, 157, 180, 195, 362-374, 377-381, 389, 390, 398, 422, 436, 460, 465

landscape 23, 24, 29
laser absorption spectroscopy 9, 92
last climatic cycle 49-51
Last Glacial Maximum (LGM) 524
latitudinal distribution 171, 174, 188, 318, 340
Law Dome 41, 43
leachate recycling systems 364, 366-369
LGM 52, 53, 524
lifetime 13, 53, 62, 63, 72, 73, 75, 76, 81, 84, 168, 169, 174, 184, 469, 470, 478, 481, 491, 493, 508, 515-517, 521, 522
light water nuclear reactors 399
lignin 103, 104, 130, 221, 365, 367
liquefied natural gas 416
little ice age 38, 48, 49, 93
livestock manure 219, 221, 386, 387
livestock waste 362, 363, 386-390
long path spectrometry 15
low boreal regions 336
major sources 11, 62, 63, 97, 156, 180, 181, 188, 190-192, 195
manufacturing 399, 423-425, 512
manure 111, 194, 219-222, 224, 225, 237, 239, 285, 286, 366, 386-388, 458
marine algae 442, 462
marshes 66, 110, 317, 338, 457
mass balance 4, 11, 72, 79, 156-160, 162, 168, 173, 184, 185, 367, 369, 370
mass balance model 11, 185
mass balance of methylchloroform 184
mechanistic models 26, 28, 30
medieval warming 38, 48
melt extraction 92
mesosphere 507
meteorological conditions 17, 231
methane budget 7, 8, 10, 11, 13, 29, 51, 90, 128, 129, 185, 195, 196, 240, 251, 255, 316, 321, 353, 447, 461, 462, 466
methane consumption 130, 133, 134, 154, 173
methane control strategies 525
methane emission from wetlands 316, 320, 344, 346
methane emissions 1, 4, 14-16, 24, 128, 132-135, 138, 157, 181, 185-187, 189, 191, 192, 196, 199, 202, 213-216, 219-221, 223, 225, 232, 233, 241, 243, 244, 248, 249, 254, 255, 267, 287, 310, 311, 314-316, 318, 320, 324, 342, 345, 346, 349-351, 400, 402, 403, 406, 408, 414, 415, 428, 440, 445, 447, 458-461, 463, 465, 466, 469, 483, 487, 521, 523, 527

methane flux measurements 24, 26, 234, 322, 335-337, 346, 350
methane formation 102, 104, 109, 113, 117, 119, 149, 255, 261, 265, 266, 268, 269, 271, 272, 274, 276, 277, 285, 288, 365
methane isotope pairs 152
methane oxidation 27, 97, 116, 117, 120, 129, 131, 135, 147, 152, 154, 156, 176, 233, 239, 262, 263, 265, 279, 288, 345, 460, 472, 474, 475, 477, 478, 489, 492, 500, 504, 507, 515, 520
methane production 115, 130, 147, 208-210, 266
methane record 52, 91
methane source 12, 13, 16, 118, 154, 246, 448, 523, 528
methane transport 116, 233
methane trends 1, 3, 55, 58
methane-oxidizing bacteria 27, 28, 116, 131, 262, 263, 279
methane-production seasons 338, 339
Methanobacterium 110, 259, 260
Methanobrevibacter 110, 259
Methanococcoides 110, 259
Methanococcus 110, 259
methanofuran 113, 114
methanogenesis 102, 103, 105, 107-109, 113-115, 118-120, 128-131, 135, 142, 146-149, 152, 153, 156, 204-207, 259-263, 265, 266, 281, 287, 336, 523
methanogenic bacteria 102, 108, 109, 141, 205, 206, 259, 261, 266, 275, 277, 259, 308, 366, 382, 387, 463
methanogens 102-110, 112-116, 118, 119, 121, 141, 147, 200, 205, 224, 233, 238, 239, 259-262, 266, 269, 272, 275, 277, 363, 458
Methanohalobium 110, 259
Methanohalophilus 110, 260, 259
methanol 102, 107, 110-115, 120, 121, 142, 260-263, 364
Methanolacinia 111, 259
Methanolobus 111, 259
Methanomicrobium 111, 259
Methanoplanus 111, 259
Methanopyrus 111, 259
Methanosarcina 111, 114, 115, 259, 260
Methanosphaera 111, 114, 259
Methanospirillum 112, 260, 259
Methanothermus 112, 259

Methanothrix 112, 115, 259
methanotrophy 129
methyl bromide 470
methyl chloride 262, 308, 470
methyl-coenzyme M 114, 115
methylamines 102, 142, 260-262, 364
methylchloroform 184, 470 (see also CH_3CCl_3)
Methylobacter 112
Methylococcus 112
Methylocystis 112
Methylomonas 112
Methylosinus 112
methylotrophic methanogens 260, 262
methylotrophy 128
methylsulfides 102
microbial oxidation 143, 145, 152-154, 257
microbiology of methane emission 259
micrometeorological techniques 15, 23
microorganisms 25, 28, 142, 201, 363, 365, 367, 387
mineral fertilizer application 285
mineralogy 238, 274, 276
mining 14, 64, 65, 68, 69, 80, 95, 196, 399-401, 403, 405-407, 409, 422, 425-428, 445-447, 460, 461
minor industrial sources 399, 422, 425, 427, 428, 445
minor sources 180, 181, 399, 423, 424, 428, 432, 462
mitigating biomass burning 310
modelling of consumption 134
modelling of in situ emissions 133
moisture content 19, 21, 134, 135, 221, 264, 265, 281, 283, 301, 349, 365-369, 378, 387, 388, 403, 404, 460, 523, 524, 528
monsoon 52, 282
MSW 364, 365, 367, 370, 372-381, 390
mud volcanoes 432, 440-442, 444, 462
municipal solid waste 364 (see also MSW)
natural emissions 186
natural gas leakages 58, 180, 195
natural gas production and distribution 413
natural sources 62, 66, 68, 72, 73, 76, 80, 82, 89, 185, 196, 280, 432, 444, 457, 511, 515, 523, 524
natural wetlands 10, 64, 66, 83, 94, 141, 180, 316, 318, 320, 338, 339, 349-351, 353, 461 (see also wetlands)
nitrogen oxides 25, 53, 108, 131, 175, 217, 235, 240, 261, 263, 269, 275, 277, 284, 300, 308, 400, 413, 470, 472, 473, 475, 478, 479, 481-484, 486-488, 490, 491, 492-495, 500, 516
NMHC 304, 515 (see also non-methane hydrocarbons)
noctilucent clouds 507
non-methane hydrocarbons 53, 175-177, 300, 470, 482-484, 488, 491, 492
NO_x 11, 470, 472, 474, 475, 477-480, 515-522, 524, 525, 528
nuclear reactor sources 95
nutrient concentrations 199, 223, 225, 233, 269, 270, 276, 281, 284, 310, 321, 345, 365
$O(^1D)$ 168, 173-175, 177, 482, 488, 489, 520
O_3 21, 94, 174, 175, 469, 473, 474, 476, 477, 479, 480, 488-490, 502, 514, 515, 519-521 (see also ozone)
ocean clathrates 139
oceans 52, 64, 110, 120, 141, 162, 175, 180, 187, 195, 314, 432, 438, 440-445, 448, 462, 463, 465, 488, 517, 518
odd hydrogen 472, 474-480
OH 3, 13, 51-53, 58, 63, 66, 73, 77, 96-98, 154-157, 162, 163, 168-170, 173-177, 184, 185, 187, 189, 191, 192, 194, 465, 466, 469, 472-479, 482, 487, 489, 490, 493, 494, 515-522, 524, 526, 527
one-dimensional models 515, 516, 518
open dumps 364, 365, 369, 373, 374, 378-380
organic amendments 268, 269, 285, 287
organic matter 28, 51, 95, 102-105, 107, 108, 121, 142, 143, 146, 148, 152, 153, 200, 203, 204, 217, 221, 222, 233, 237, 238, 246, 247, 261, 263, 265, 268, 269, 271, 272, 274, 285, 336, 363, 378, 387, 433, 436, 461
organic peroxy radicals 470
oxidation 25, 27, 28, 51, 54, 66, 77, 93, 97, 104-107, 114-120, 129-131, 134, 135, 143, 145, 147, 148, 151-154, 156, 162, 168, 173-176, 233, 239, 257, 262, 263, 265, 272, 279-281, 285, 288, 345, 350, 365, 369-371, 373, 440, 460, 464, 469, 470, 472-475, 477-480, 483, 488, 489, 492, 495, 500, 504, 507, 511, 514, 515, 517, 520, 521, 526, 527
oxidation of methane 27, 28, 168, 175, 263, 469, 477, 495, 504, 511, 515, 526
oxidation path-length 134; 135

oxidation-reduction 272
oxidizing capacity 53, 93, 174, 270, 470-472
ozone 2, 53, 73, 308, 321, 469-473, 479-492,
    494-503, 507, 509, 510, 512, 513,
    520, 521, 526, 528
ozone hole 496, 507, 526
ozone photodissociation 520
partial methanogenic fermentations 104
particulate carbon 300, 301, 306
peat mining 399, 425, 426, 428, 446, 447
peatlands 24, 66, 68, 315, 337, 342, 348
pectin 104
per capita waste generation 364
permafrost 52, 349, 432, 438, 441
pest control 286, 287
petroleum 142, 423-425, 428, 433, 438, 439, 445,
    446, 448, 463
petroleum refineries 423
petroleum refining 424, 425, 428, 446
petroleum seepages 438 (see also seepages)
pH 26, 116, 132, 199, 200, 205, 233, 238, 240,
    260, 263-265, 269, 270, 272-277,
    282, 285, 287, 365, 366
photochemistry 470, 471, 475, 480, 488, 492
photodissociation 494, 520
photolysis 470, 473, 475-477, 487, 488, 491, 493,
    520
pipeline leaks 414, 415, 419
plant-mediated transport 233
plants 25-28, 69, 97, 109, 117, 121, 131, 232-234,
    238, 263, 275, 278-281, 286, 287,
    380, 414, 415, 418, 425, 426
polar stratospheric clouds 496, 507, 512, 526
polar vortex 496
ponds 66, 68, 222, 336
pre-industrial atmosphere 46, 47, 49, 53, 73, 75,
    76, 84, 92, 97, 314, 504
pre-industrial ice record 47
process-oriented studies 25
production sector 414
propionic acid 104
psychrophilic 116, 266
puddling 239, 265, 282
quantum yields 519
radiative effect 53, 54, 509, 510, 515
radiative forcing 504, 508-513, 521, 522, 525-527
radiative transfer 512, 516
radiative-convective models 509, 510
radioactive dating 39
radiocarbon emissions 426
rainfall 231, 238, 239, 257, 267, 280, 283, 337
rainfed rice 239, 242, 243, 255-257, 283, 287

range of emissions 10, 14, 19, 21-23, 41, 46, 73,
    91, 92, 96, 102, 146, 152, 155,
    185, 187, 201, 208, 209, 211, 219,
    230, 243, 247, 256, 260, 274, 307,
    308, 315, 335, 336, 338, 344, 351,
    370, 372, 374, 405, 407, 422, 427,
    428, 461, 464, 477
rate coefficients 519, 520
Rayleigh distillation relationships 147
reactive chlorine 93, 129, 147, 469, 470, 473,
    494-496, 500, 526-528
redox potential 26, 199, 233, 263, 265, 272, 274,
    283, 321 (see also Eh)
refuse decomposition 365
refuse mass 371, 372
regional methane budget 11, 12, 16, 19, 23, 29,
    96, 120, 134, 180, 187-189, 231,
    232, 240, 243, 246, 250, 311, 316,
    317, 319, 321-323, 338, 342, 346,
    347, 350, 352, 380, 438, 469, 484,
    523, 526
remote sensing 14, 91, 134, 319, 350, 352, 462,
    527 (see also Thematic
    Mapper/Landsat)
residential refuse burning 399
rhizosphere 27, 263, 278-280
rice 2, 3, 10-12, 14, 17, 25-28, 30, 48, 58, 63-65,
    68, 69, 72, 73, 75-77, 80, 82-84,
    93, 94, 98, 110, 117, 129, 139,
    141, 157, 158, 160-162, 180, 181,
    183, 186, 188-193, 195, 230-244,
    246-251, 254-260, 262-272,
    275-288, 352, 353, 436, 439,
    457-459, 465, 525
rice agriculture 2, 58, 93, 180, 181, 183, 186, 188,
    195, 230, 231, 239, 240, 243, 244,
    246-248, 250, 251, 254, 255, 436
rice cultivars 234, 238, 249, 251, 258, 278, 282,
    288, 459
rice environments 255
rice growth index 277
rice paddies 12, 14, 17, 64, 69, 82, 84, 110, 117,
    139, 141, 157, 161, 233, 234, 237,
    238, 240, 243, 246, 248, 249, 263,
    353
rice soils 257, 260, 262-264, 266, 268, 270-272,
    276, 277, 282, 284, 287
rice variety 231
rice yields 28, 269, 281, 284
rumen 104, 105, 110, 111, 117, 119, 200-202,
    205, 208, 215-217, 224, 259, 261,
    259, 386

rumen fermentation 104, 105, 119
ruminants 2, 12, 48, 64, 65, 84, 94, 105, 119, 129, 180, 199-201, 206, 208, 209, 215, 220, 224, 225, 261, 436, 457, 465
sampling 7-9, 11, 12, 14, 15, 17-20, 22, 24, 27, 29, 66, 148, 170, 250, 368
sanitary landfill 116, 364, 368, 373, 380, 381
sapropelic 142-144, 260
satellite 15, 22-24, 29, 305, 306, 311, 316, 319, 495, 496, 527
satellite imagery 15, 22-24, 29
savanna 64, 68, 69, 300, 301, 303, 307-311, 436, 459, 464
sea surface methane 443
season length 240
seasonal cycles 23, 26, 39, 55, 56, 96, 98, 168-170, 172, 173, 189, 233, 240-243, 250, 266, 271, 310, 315, 317, 319, 322, 324, 335-338, 344, 345, 349-352, 407, 438, 461, 462, 483, 491
seeding 234, 267, 283
seepages 95, 143, 432-435, 438-440, 442-444, 462, 463, 465
septic tanks 383, 386
sewage 180, 195, 259, 379, 381, 383, 386, 389, 425
sewage disposal 180, 195
sewage systems 381
shale oil mining 399
sinks 3, 4, 10, 52, 54, 56, 89, 90, 96, 99, 154, 158, 162, 168, 170, 173, 174, 185, 194, 345, 350, 351, 353, 457, 465, 466, 525
sintering process 39
smoldering phase 301
SO_2 470
soil 10, 12, 17, 25-28, 51, 53, 110, 115, 117, 129, 130, 132-134, 153, 154, 168, 173, 175, 183, 231-236, 238-240, 242, 245-247, 255, 256, 258, 259, 262-288, 308, 310, 315, 319, 320, 345-349, 369, 383, 436, 437, 458, 464, 465, 523, 527, 528
soil characteristics 26, 231, 275, 276, 458
soil microbiology 231, 238
soil sink 53, 168, 173, 175
soil temperature 26, 231, 238, 239, 242, 264, 266, 267, 269, 276, 283, 346-348, 523, 528
soil texture 238, 282
soil uptake 53, 154, 308

solar radiation 231, 238, 494, 508
solid phase reactions 42
soot 305, 457
sources 2-4, 7, 9-14, 16, 29, 30, 44, 48, 49, 51, 52, 54, 56, 58, 62-64, 66, 68, 69, 72, 73, 76, 77, 79-84, 89, 90, 93-99, 118, 121, 138, 139, 141, 142-144, 148, 151, 152, 156-162, 170, 175, 180, 181, 184-196, 200, 224, 230, 249, 251, 267, 280, 285, 288, 300, 302, 307, 308, 315, 317, 319, 321, 345, 350-353, 362, 377, 378, 382, 389, 399-402, 407, 414-416, 422-428, 432-436, 438, 441, 442, 444-448, 457, 460, 462, 463, 465, 466, 472, 478, 482, 488, 491, 492, 511, 515-518, 520, 522-527
spectrometry 9, 15
stable isotope ratios 94, 118, 143, 144, 163
stable isotope signatures 138
stable isotopes 9, 138, 154, 155, 162, 186
stoichiometry of methanogenesis 107, 108, 367
stratospheric chemistry 1, 494
stratospheric ozone 470, 486, 487, 491, 494-497, 500, 501, 512, 521, 526, 528
stratospheric ozone depletion 491, 526
stratospheric water vapor 53, 54, 503-507, 509, 512, 526
subarctic boreal fens 336, 337
substrates 27, 102, 106-108, 112, 114, 115, 119, 142, 147, 152, 201, 238, 265, 266, 270-272, 276, 285, 336, 363, 370, 523
subtropical wetlands 315, 335, 350
sulfate aerosols 496
sulfate-reducing bacteria 105, 107, 120, 262, 275
sulfur dioxide 400, 413, 470 (see also SO_2)
surface mining 400, 409
surface-based methane measurement 90
swamps 189, 317, 319, 336, 338, 339
temperate freshwater swamp 337
temperature regimes of rice soils 266
temporal variation 8, 26, 48, 49, 51, 63, 69, 77, 98, 187, 189, 193, 266, 315
termites 12, 52, 64, 94, 141, 157, 162, 180, 195, 224, 436, 437, 444, 462-465
Thematic Mapper/Landsat 306 (see also remote sensing)
thermogenic 10, 63, 65, 69, 118, 121, 139, 142, 144-148, 150, 151, 156, 157, 162, 433, 435, 438, 439, 441
three-dimensional models 516, 518, 519

time series of global methane emissions 248
trace gases 8, 10, 11, 16, 17, 19, 21, 29, 48, 141, 174, 176, 416, 428, 489, 523
transmission sector 415
transplanting rice 283
trends 1-4, 7, 8, 13, 23, 28, 29, 54-58, 63, 69, 73, 77, 79-84, 96, 97, 174, 180, 191, 192, 247, 249, 336, 352, 373, 379, 481, 483, 484, 491, 497, 521
trends in global emissions (CH_4) 192
trends in waste management 379
trends of lifetime 174
tritiated methane 13
tropical wetland environments 8, 52, 69, 77, 170, 188, 218, 239, 250, 266, 269, 276, 280-283, 300, 302, 303, 307-311, 315-317, 319-322, 324, 325, 335, 336, 338, 339, 342, 344, 345, 350-352, 381, 388, 436, 437, 459, 461, 462, 464, 519
tropospheric chemistry 350, 469, 470, 483, 526, 527
tropospheric OH levels 515 (see also OH, hydroxl radical)
tropospheric ozone 53, 73, 308, 321, 484, 485, 487-492, 501, 502, 513, 521
TROPOZ campaign 321, 350
tundra 18, 24, 25, 66, 68, 94, 98, 135, 180, 195, 337, 338, 343, 344, 347, 348, 523
two-dimensional models 53, 516, 518, 519
UAG ("unaccounted for" gas) 417, 418, 421
ultraviolet radiation 470, 494 (see also UV)
unaccounted for gas (see UAG)
uncontrolled dumps 364, 367
under-determined system 186
underground mines 400, 401, 403, 405, 412, 426, 461
ungulates 105, 117
upland rice 243, 256, 257, 268, 287 (see also dryland rice)
upper atmospheric chemistry 469, 493
uptake 53, 90, 104, 134, 138, 154, 173, 175, 276, 284, 308, 466
upwind-downwind measurements 11, 29
urban areas 180, 195, 422, 423
USBOM (United States Bureau of Mines) 426
UV absorption cross sections 519
vascular transport 26, 28, 131, 458
ventilation shafts 401
vertical distribution 8, 11, 14, 16-20, 29, 131, 169-171, 176, 320, 349, 503, 515, 516

volatile fatty acids 103-105, 119, 138, 149, 200, 269-271, 387, 404, 435, 442
volcanic and related emissions 39, 235, 259, 433, 434, 442, 444, 448, 462, 463
volcanism 440
Vostok ice core 41, 49-52
wastewater lagoons 363, 381
wastewater treatment 194, 362, 381, 382, 389, 390
water control 280, 283
water level 231
water vapor 2, 17, 39, 53, 54, 174-176, 300, 469, 482, 489, 491, 503-507, 509, 512, 526
wetland classification 319, 320
wetland ecosystems 315, 316, 320, 324, 338, 344, 346, 349, 351, 461
wetlands 10, 12, 25, 51, 52, 64, 66-68, 77, 82-84, 94, 98, 102, 130, 132, 138, 139, 141, 157, 160, 180, 181, 186, 188, 189, 195, 233, 239, 256, 266, 302, 308, 314, 315-322, 335-339, 344-346, 349-354, 436, 457, 461, 462, 465, 515, 524, 525, 527

Springer-Verlag and the Environment

We at Springer-Verlag firmly believe that an international science publisher has a special obligation to the environment, and our corporate policies consistently reflect this conviction.

We also expect our business partners – paper mills, printers, packaging manufacturers, etc. – to commit themselves to using environmentally friendly materials and production processes.

The paper in this book is made from low- or no-chlorine pulp and is acid free, in conformance with international standards for paper permanency.